Space Scienc

Vol

MW00736367

For further volumes:
www.springer.com/series/6592

Michael Toplis • James Bell III • Eric Chassefière •
Christophe Sotin • Tilman Spohn • Michel Blanc
Editors

Quantifying the Martian Geochemical Reservoirs

Previously published in *Space Science Reviews* Volume 174,
Issues 1–4, 2013

Editors
Michael Toplis
Institut de Recherche
en Astrophysique et Planétologie
Toulouse, France

James Bell III
Arizona State University
Tempe, AZ, USA

Eric Chassefière
Université Paris-Sud Orsay/CNRS
Paris, France

Christophe Sotin
Caltech University
Jet Propulsion Laboratory
Pasadena, CA, USA

Tilman Spohn
German Aerospace Center (DLR e.V.)
Berlin, Germany

Michel Blanc
Institut de Recherche
en Astrophysique et Planétologie
Toulouse, France

ISSN 1385-7525 Space Sciences Series of ISSI
ISBN 978-1-4899-9305-2
Springer New York Heidelberg Dordrecht London

Cover illustration: Mosaic acquired by the Mars Exploration Rover Spirit on Sol 1933 (August, 2009), using Pancam's 601 nm, 535 nm, and 482 nm filters. For scale, the two rocks near the upper right corner of this view are each about 10 cm long and 2 to 3 cm wide. The bright disturbed soil was a key source of information about the story of water at Gusev Crater, as well as the reason Spirit would ultimately never leave this spot, the rover becoming embedded in this soil.
Image Credit: NASA/JPL-Caltech/Cornell University/Arizona State University

Printed on acid-free paper

Springer is part of Springer Science+Business Media (www.springer.com)

Contents

Space Sci Rev (2013) 174:1–3
DOI 10.1007/s11214-012-9955-4

Foreword

Roger-Maurice Bonnet · Michel Blanc

Published online: 8 December 2012

"Quantifying the Martian Geochemical Reservoirs" is the fourth publication of its kind in the Space Science Series of ISSI, published as a result of the very successful collaboration between ISSI and Europlanet.

Europlanet started in 2005, the year of the successful landing of the Huygens probe on Titan's surface, as a network of over 110 European and U.S. laboratories deeply involved in the development of planetary sciences and the support to the European programme of space exploration of planets. Since 2005, Europlanet obtained support from the European Commission to strengthen the planetary science community worldwide, and to amplify the scientific output, impact and visibility of the European space programme, essentially ESA's Horizon 2000 and Cosmic Vision programmes. Its first contract with the Commission under the FP6 programme was a "Coordination action". It extended from 2005 to 2008 and included 7 networking activities, among which the set-up of disciplinary working groups covering the main areas of planetary sciences. A new contract with the Commission, this time as a full "research infrastructure network" under the FP7 programme, extended Europlanet's activities into the period 2009–2012. For the first time ISSI became a participant in this extended programme.

With the broad community connection made through its Discipline Working Groups and other activities, Europlanet offers an ideal base from which to identify new fields of research for planetary sciences and to stimulate collaborative work among its member laboratories. For Europlanet, developing collaboration with ISSI in holding workshops and producing books on these new and emerging subjects is both natural and extremely stimulating, considering the high profile, international standing and proven success of ISSI. For ISSI, collaboration with Europlanet offers a very interesting opportunity to extend its successful series

R.-M. Bonnet
International Space Science Institute, Bern, Switzerland

M. Blanc (✉)
National Centre for Scientific Research/Research Institute in Astrophysics and Planetology, Toulouse, France
e-mail: Michel.blanc@irap.omp.eu

of workshops and books within the area of planetary sciences and to reinforce its links with this community.

Under the FP7 contract, ISSI and Europlanet committed themselves to produce four workshops (one per year) and one summer school: the 2012 Alpbach summer school on "the exploration of giant planets and their systems". The selection of workshops followed the same scheme as under FP6: the five disciplinary working groups of Europlanet were each invited to propose ISSI workshop concepts. These concepts were validated by the management of Europlanet, and then presented to the Science Committee of ISSI. The first such presentation was made to the science committee on January 13–14 2009, at the beginning of the FP7 contract. Four candidate workshops were presented. After due discussion it was recommended that the first workshop to be implemented be that on "Quantifying the Martian Geochemical Reservoirs", proposed by Dr. Michael Toplis. After two years of preparation, the workshop was held in Bern in April 2011. The workshop preparation started with a complementary analysis of the timeliness of the subject and of its relationship to the past and future programme of space missions to Mars.

Indeed, this was a very timely choice, in the tradition of Europlanet whose scientific purpose with respect to the space programme is to capitalize on the outcome of past and present space missions to help optimize the preparation of future missions. The planet Mars has now been studied for decades, by space missions of increasing complexity. Most recently, the combination of orbiting missions (Mars Global Surveyor, Mars-Odyssey, Mars Express, Mars Reconnaissance Orbiter) and lander/rover missions (Sojourner, Mars Exploration Rovers, Phoenix, to which one should now add the Curiosity rover of the Mars Science Laboratory mission which was launched after the workshop) has provided a multi-scale approach to the study of Mars, in which the local in situ measurements provided by landers and rovers could be placed in the context of remote sensing cartography from orbit. These space-based data can also be put in synergy with Earth-based measurements such as data from numerical simulations, laboratory studies, and geochemical data from the SNC meteorites which are believed to be of martian origin. Altogether, these Earth-based and Mars-based data provide complementary pieces of information on the history of the formation and evolution of Mars, from accretion times to the present age, and also make it possible to study each of the major geochemical reservoirs of the red planet (core, mantle, crust and atmosphere) and make progress in our understanding of its chemical composition and its history.

The success of the workshop on "quantifying the Martian geochemical reservoirs" and the value of the corresponding book are due to the fact that it has brought together scientists from a wide variety of scientific backgrounds, space scientists, Earth science specialists, specialists of numerical or laboratory simulations, meteoriticists specializing in SNC meteorites, to review the different geochemical reservoirs of Mars, their role in the dynamics and evolution of the planet, and to address some overall "big" questions, such as the still unsolved question of the reservoir of water on Mars and its history. By working closely together, this group of almost 50 specialists was able to summarize their views of the geochemistry of Mars in a nicely designed set of chapters which covers all main geochemical reservoirs as well as some cross-disciplinary issues. At a time when the new generation rover Curiosity has started its detailed exploration of Gale crater, we are convinced that this book will remain for some significant time a key reference for whoever will be preparing his/her involvement in the future missions to Mars: MAVEN, Exomars, Insight, and their successors.

The likely success of this book will be due to all those whose hard work and dedication made this fourth joint publication of ISSI and Europlanet possible. Many thanks, first, to the leaderships of ISSI and Europlanet who designed this collaborative project, and to the leaders of the Discipline Working Groups of Europlanet, Norbert Krupp and Ari-Matti Harri. Many thanks to the Science Committee of ISSI, chaired by Prof. Len Culhane, who made the selection and provided useful suggestions on the workshop content. Our special gratitude goes to Michael Toplis, who originally proposed the workshop's topic and masterly led the whole process, from the first Conveners Meeting to book production. The team of conveners and editors did a fantastic job in defining the structure of the book and in managing the writing and the overall review process: as usual, all chapters were carefully reviewed by independent experts to whom we would also like to extend our gratitude.

Last, but not least, we would like to extend our appreciation to the staff of ISSI, Maurizio Falanga, ISSI's programme manager, Andrea Fischer, Saliba Saliba, Irmela Schweizer, Silvia Wenger, Jennifer Zaugg who managed the organisation of this workshop with dedication and efficiency, and made it possible for this important collaboration between ISSI and Europlanet to develop and flourish over the years.

Space Sci Rev (2013) 174:5–9
DOI 10.1007/s11214-012-9951-8

Quantifying the Martian Geochemical Reservoirs: An Interdisciplinary Perspective

Michael J. Toplis · James F. Bell · Eric Chassefière · Christophe Sotin · Tilman Spohn · Michel Blanc

Published online: 14 December 2012

The planet Mars has fascinated humanity since antiquity, a fascination that persists to the present-day, fuelled by the enticing possibility that the conditions necessary for life may once have existed there, and perhaps continue to exist to the present day. Indeed, comparative planetology of terrestrial bodies large enough to retain an atmosphere is a key to understanding the formation of our own planet and the physical and chemical conditions that led to the emergence and the development of life. In more general terms, the study of rocky bodies of the solar system also provides a context for planetary formation as a whole, shedding light on the geochemical similarities and differences between the principal objects of the inner solar system, as well as the cores of the outer planets and many of their satellites.

However, it was not until the advent of space-based exploration in the 1960s that quantification of the geochemical characteristics of our planetary neighbours first became possible. Indeed, our knowledge of Mars changed dramatically when Mariner 4 took the first close-up pictures in July 1965, observing a highly cratered arid landscape. In 1976 the Viking orbiters and landers provided ground-breaking insight into surface morphology and the chemistry of the soil and atmosphere on Mars, data that also led to confirmation that the Shergottite-Nakhlite-Chassignite group of meteorites (SNC) most probably have a martian origin. This

M.J. Toplis (✉) · M. Blanc
IRAP-OMP, Toulouse, France
e-mail: michael.toplis@irap.omp.eu

J.F. Bell
ASU, Tempe, AZ, USA

E. Chassefière
Univ. Paris 11, Orsay, France

C. Sotin
JPL, Pasadena, USA

T. Spohn
DLR, Berlin, Germany

link between Mars and a group of rocks that may be studied with Earth-based analytical tools opens the possibility for precious detailed quantification of certain mineralogical and geochemical features of the red planet.

Over the last fifteen years, space-based exploration of the solar system has increased dramatically, with more and more sophisticated orbiters and landers being sent to Mars: the Sojourner rover (NASA, 1997), the Mars Exploration Rovers Spirit and Opportunity (NASA, 2004–present), the Phoenix lander (2008), the Mars Global Surveyor orbiter (NASA, 1997–2006), the Mars-Odyssey orbiter (NASA, 2001–present), the Mars Express mission (ESA, 2004–present), and the Mars Reconnaissance Orbiter (NASA, 2006–present). This intense period, rich in unprecedented scientific results, has led to immense progress in our perception of Mars and of its evolution over geological time. Thanks to the wide range of scientific instruments carried by these missions, the scientific community has access to more or less global coverage of surface morphology at $\sim$1 to 10 m/pix scales, as well as information concerning the composition and mineralogy of the surface/subsurface. In addition, the scientific payloads of the MER rovers and the Phoenix lander provide highly detailed information at the selected landing sites, also providing the opportunity for geological context to be appreciated in the case of the MER rovers. In parallel to this accumulation of remote data, advances in numerical simulations and laboratory experiments also shed new light on the geochemical evolution of the planet Mars.

It was within this context that the ISSI-Europlanet Workshop entitled "Quantifying the Martian Geochemical Reservoirs" was organized. The workshop was held in Bern in April 2011 with the objective to create a diverse interdisciplinary forum composed of scientists directly involved in space-based exploration of the Martian surface, meteoriticists studying SNC meteorites, and planetary and/or Earth scientists simulating, numerically or experimentally, the physical and chemical processes occurring on or within Mars. Considerable time was dedicated to questions and discussion throughout the workshop and during this time the participants' diverse fields of expertise led to stimulating and thought-provoking exchanges. It is those cross-disciplinary discussions that form the basis of the papers that may be found here.

More than forty individuals have collaborated to provide this overview of current knowledge of the past and present martian geochemical reservoirs. In addition to the detailed description of data from Mars and the methods used to obtain them, many of the contributions also emphasize comparison with features on Earth, providing a perspective on the extent to which our knowledge of terrestrial systems influences interpretation of data from Mars. Almost all chapters also identify areas that would benefit from future work and measurements, providing a view of the short-term and long-term future of the study of Mars.

In detail, this special volume begins with a paper by *Brasser*, describing current thoughts on the formation of Mars, the reasons for its relatively small size, the composition of its building blocks and its accretion time-scale, highlighting several recent studies that conclude that Mars may in fact be a planetary embryo that never grew to be a fully-fledged planet. The second contribution by *Mezger et al.* is largely based on isotopic studies of the SNC meteorites, which can be used to constrain the extent and time-scales of planet-wide differentiation into a core, mantle, and crust. These results highlight the fact that, in contrast to the Earth, Mars appears to have preserved chemical heterogeneities produced during the first few tens of millions of years of its history, implying that vigorous whole-mantle convection has not homogenized the martian silicate interior.

The subsequent four contributions focus on specific geochemical reservoirs and how they have evolved over geological time. The paper by *Grott et al.* considers the mantle and magmatic crust, bringing together an impressively diverse range of data, assessing to what extent the morphological, mineralogical, and compositional features of the martian surface can be reconciled with numerical models of interior convection as constrained by laboratory measurements. Overall, a very satisfactory first-order picture emerges, dominated by global cooling of a stagnant-lid planet, although the authors identify a significant number of areas for future research to confirm our understanding of the internal structure of Mars and its evolution over time. In the next contribution, *Lammer et al.* treat the question of the Martian atmosphere, and its principal sinks and sources of volatile species, in particular H_2O and CO_2. This chapter highlights the difficulties in quantifying the pressure and composition of the atmosphere in the past, although a robust case is made for loss of the original proto-atmosphere due to the high EUV flux of the young Sun, and for minimal loss of volatiles to space over the last 4 billion years. On a related topic, *Lasue et al.* focus on the martian hydrosphere/cryosphere, considering the morphological evidence for the presence of water, both at the present time and in the past. Those data are combined with constraints from orbital and *in-situ* measurements to provide quantitative estimates of the amounts of water present and the global distribution and structure of icy and/or liquid water subsurface reservoirs. That contribution is complemented by a paper by *Mousis et al.* who explore the geochemical consequences of clathrate formation in the martian sub-surface, highlighting the possibility that formation/dissociation cycles of clathrates could have potentially important implications for short-term seasonal variations in the abundance of noble gases such as Ar, Kr, and Xe, as well as the intriguing possible detections of methane that have been described over the last few years.

The last three contributions focus on secondary minerals produced by interaction of the magmatic crust and the atmosphere, providing critical markers of past conditions at the martian surface. The first of these, by *Gaillard et al.*, considers sulphates and in more general terms the sulphur cycle on Mars, considering how sulphur is partitioned between the core, mantle, and crust and in what chemical forms sulphur is transferred to the atmosphere. This chapter highlights the fact that the iron-rich nature of the martian mantle has important consequences for the concentrations of sulphur that may be transported by magmatism, and that the fate of degassed sulphur is critically dependent on a combination of the oxidation state and water content of the mantle, and/or the pressure of the atmosphere, the transition to sulphate-dominated outcrops in the Hesperian being tentatively explained by a decrease in atmospheric pressure relative to earlier epochs. The contribution of *Niles et al.* focuses on carbonates, as seen in the SNC meteorites and as observed on the surface of Mars. This review quantifies the global concentration of carbonate and highlights the diversity of carbonate chemistry observed in martian rocks or outcrops, providing important constraints on the CO_2 budget of Mars, the role of Mg- and Fe-rich fluids, and the interplay between carbonate formation and acidity. For example, despite outcrops that appear limited in spatial extent and primarily in terrains of Noachian age, evidence for active carbonate precipitation and the preservation of more ancient samples argues against global acidification of the martian surface during the Hesperian. The special issue ends with a contribution by *Ehlmann et al.* focussed on the presence of clays at the surface of Mars. This chapter reviews the chemical diversity and spatial distribution of clays, demonstrating that these volatile-bearing minerals are preferentially found in Noachian terrains, indicating a distinctive set of water-rock interaction processes during that time. However, different models of clay formation are considered, such as near-surface weathering, formation in ice-dominated near-surface ground waters, and formation by subsurface hydrothermal fluids. The extent of geochemical fractionation of the crustal reservoir and possible interactions with the atmosphere are discussed for each of these environments.

While each individual chapter in this volume provides a comprehensive look at a particular aspect of the geochemical evolution of Mars, taken as a whole, they offer a truly multidisciplinary look at the evolution of Mars over geological time, providing multiple but complementary viewpoints on numerous important issues. Of these, the question of the distribution of water over time is a case in point. For example, the contribution by *Brasser* underlines our lack of knowledge of the amount of water accreted to Mars during its growth, although *Lammer et al.*, demonstrate that this uncertainty is mitigated to some extent by the fact that the primary atmosphere of Mars should have been rapidly lost to space. The contribution of *Grott et al.* highlights that the study of SNC meteorites and numerical models of convection both argue for water in the martian mantle, but that the amount could be anywhere from 2 to 200 ppm, providing few firm constraints on the impact of volcanic outgassing. Furthermore, *Lammer et al.*, argue that it is extremely difficult to quantify the water content of the secondary atmosphere, not only due to uncertainties in the amount of water released during internal degassing, but also because the effect of meteoritic or cometary bombardment could either erode or add water to the atmosphere. In this respect, secondary minerals formed through interaction of magmatic rocks with the atmosphere/hydrosphere are important markers of surface conditions. Interestingly, all three contributions considering this approach (*Gaillard et al., Niles et al., Ehlmann et al.*) argue against the need for a 'warm and wet' early Mars, the case being made that the Noachian atmosphere was possibly of relatively low density and cold (though warmer and wetter than the current Mars atmosphere). Liquid water is proposed to have been present near the surface of Mars at certain times during the Noachian, but it is suggested that this could have occurred during transient periods of warmer conditions, possibly related to volcanism. On the other hand, *Lasue et al.*, illustrate the diverse geomorphological evidence for significant amounts of water in the geological past, such as the spectacular late Hesperian outflow channels. However, they find no current evidence in the near subsurface for a corresponding large icy or liquid water reservoir, despite the fact that *Lammer et al.* conclude that if abundant water were present on Mars at some time after the Noachian, then it must still be there, as atmospheric loss processes to space are not sufficient to explain any more than a modest reduction in atmospheric pressure. These considerations thus beg the question of just how much water was present at the end the Hesperian, and where and in what form that water has been sequestered.

The critical issue of water on Mars thus appears far from being resolved, leaving ample room for future studies to provide additional constraints. In this respect, the successful arrival of the next generation rover Mars Science Laboratory (MSL, "Curiosity") in August 2012 will provide further key observations concerning the chemical and mineralogical diversity of ancient sedimentary rocks in Gale crater, hopefully providing new insight into the relations between magmatic precursors and the formation of carbonates, clays and sulphates. In the near future, the orbiter MAVEN (that will be launched in November 2013) will provide new data on the composition and dynamical state of the martian atmosphere and solar wind interactions, while critical information on the interior structure will be revealed by the seismic observations and heat-flow measurements to be made by the recently selected NASA InSight Discovery mission. Future missions like ESA's planned ExoMars rover or other future Mars orbiters, landers, and rovers will no doubt help to address and test the kinds of questions and hypotheses described in this volume.

Our understanding of Mars based on more than four decades of incredible discovery thus continues to evolve, and it is our hope that this collection of chapters will constitute a timely perspective on current knowledge and thinking concerning the geochemical evolution of Mars, providing context and a valuable reference point for even more exciting future discoveries.

References

R. Brasser, The formation of Mars: building blocks and accretion time scale. Space Sci. Rev. (2012, this issue). doi:10.1007/s11214-012-9904-2

B. Ehlmann et al., Geochemical consequences of widespread clay mineral formation in Mars' ancient crust. Space Sci. Rev. (2012, this issue). doi:10.1007/s11214-012-9930-0

F. Gaillard et al., Geochemical reservoirs and timing of sulfur cycling on Mars. Space Sci. Rev. (2012, this issue). doi:10.1007/s11214-012-9947-4

M. Grott et al., Long-term evolution of the martian crust-mantle system. Space Sci. Rev. (2012, this issue). doi:10.1007/s11214-012-9948-3

H. Lammer et al., Outgassing history and escape of the martian atmosphere and water inventory. Space Sci. Rev. (2012, this issue). doi:10.1007/s11214-012-9943-8

J. Lasue et al., Quantitative assessments of the martian hydrosphere. Space Sci. Rev. (2012, this issue). doi: 10.1007/s11214-012-9946-5

K. Mezger et al., Core formation and mantle differentiation on Mars. Space Sci. Rev. (2012, this issue). doi: 10.1007/s11214-012-9935-8

O. Mousis et al., Volatile trapping in martian clathrates. Space Sci. Rev. (2012, this issue). doi:10.1007/s11214-012-9942-9

P. Niles et al., Geochemistry of carbonates on Mars: implications for climate history and nature of aqueous environments. Space Sci. Rev. (2012, this issue). doi:10.1007/s11214-012-9940-y

Space Sci Rev (2013) 174:11–25
DOI 10.1007/s11214-012-9904-2

The Formation of Mars: Building Blocks and Accretion Time Scale

Ramon Brasser

Received: 4 December 2011 / Accepted: 28 May 2012 / Published online: 12 June 2012

Abstract In this review paper I address the current knowledge of the formation of Mars, focusing on its primary constituents, its formation time scale and its small mass compared to Earth and Venus. I argue that the small mass of Mars requires the terrestrial planets to have formed from a narrow annulus of material, rather than a disc extending to Jupiter. The truncation of the outer edge of the disc was most likely the result of giant planet migration, which kept Mars' mass small. From cosmochemical constraints it is argued that Mars formed in a couple of million years and is essentially a planetary embryo that never grew to a full-fledged planet. This is in agreement with the latest dynamical models. Most of Mars' building blocks consists of material that formed in the 2 AU to 3 AU region, and is thus more water-rich than that accreted by Earth and Venus. The putative Mars could have consisted of 0.1 % to 0.2 % by mass of water.

Keywords Mars · Formation · Origin

1 Introduction

The formation of the terrestrial planets of the Solar System has been an outstanding problem for a long time. Significant progress has recently been made with the aid of fast computers and has led to a coherent picture and sequence of events. Put simply, the terrestrial planets formed from the accumulation of many planetesimals whose sizes ranges from hundreds of meters to hundreds of kilometres. This process has been shown to take of the order of 100 million years (Myr) (e.g. Chambers 2001; Kokubo and Ida 1998), with the Moon-forming event being the last giant impact on the Earth (Kleine et al. 2009), some 60 Myr after the formation of the Solar System.

Even though many successes have recently been made, the small size of Mars (and also Mercury) has been a sticking point for a long time (e.g. Chambers 2001; Raymond et al. 2009). In this chapter I shall review some of the recent progress that has been made towards

R. Brasser (✉)
Institute for Astronomy and Astrophysics, Academia Sinica, P.O. Box 23, Taipei 10617, Taiwan
e-mail: brasser_astro@yahoo.com

understanding the formation of the terrestrial planets, in particular focusing on the formation of Mars. A more thorough review of terrestrial planet formation can be found in Morbidelli et al. (2012). This chapter has been divided into the following sections. Section 2 gives an overview of terrestrial planet formation from a dynamical point of view. In Sect. 3 I shall focus on terrestrial planet formation from a narrow annulus, which appears to be able to reproduce the small masses of Mercury and Mars. The next section deals with the so-called 'Grand Tack' scenario, which invokes the migration of the giant planets Jupiter and Saturn to dynamically truncate the outer edge of the disc from which the terrestrial planets eventually formed. Section 5 is devoted to present cosmochemical evidence in support of the fact that Mars might be a left-over planetary embryo, as is suggested from the annulus and Grand Tack formation scenarios. In Sect. 6 I shall present a summary and conclusions.

2 An Overview of Terrestrial Planet Formation

Terrestrial planet formation proceeds through three main stages, with some overlap between them. The first stage is believed to be the settling of dust in the midplane of the solar nebula, which then accretes together to form small bodies called 'planetesimals'. This phase is not well understood, and there are several competing ideas suggesting that planetesimals either form small and systematically grow through collisions (e.g. Weidenschilling and Cuzzi 1993; Wurm et al. 2001) or that they form big ($\sim$100 km) through gravitational instability (e.g. Johansen et al. 2007) or by turbulent concentration (Cuzzi et al. 2008). What is important to point out is that once the planetesimals have reached a critical size of a few kilometres, their gravitational influence on each other is strong enough that they begin to scatter each other onto crossing orbits. This is when the second phase kicks in.

Safronov and Zvjagina (1969) pointed out that available planetesimals are systematically accumulated into fewer but larger bodies (Greenberg et al. 1978). If left unchecked, and assuming that the accretion occurs locally, all the material would eventually collide together to form a planet out of a small feeding zone. The size of this feeding zone grows slowly as the mass of the planet increases and scales with the mass of the planet as $m_p^{1/3}$. This growth process is called 'runaway accretion' because effectively $\dot{m}_p \propto m_p^{\alpha}$ and exponential growth ensues. The relative velocity between the planetesimals is governed by encounters between them. The masses of the planetesimals are very low and so their relative encounter velocities, increased slowly through self-stirring, are much lower than their orbital speeds. Once one of the planetesimals has accumulated more mass than its neighbours by colliding with another planetesimal, its gravitational cross section and mass also grow. This allows it to accrete more material from its surroundings because its feeding zone has increased in size. The stage of runaway growth typically lasts less than 1 Myr at 1 AU from the Sun (Kokubo and Ida 1995). This simple picture leads one to believe that the terrestrial planets Mercury, Venus, Earth and Mars, all grew from material that was accreted locally. As I shall show later, this is most certainly not the case.

The simple picture of local runaway accretion described above breaks down when the mass of the largest body, a 'protoplanet' or 'planetary embryo', is significantly larger than the mass of nearby planetesimals. Once a protoplanet has accreted most of the planetesimals in its vicinity, the remaining planetesimals are no longer able to damp their mutual velocities and the gravitational cross section for accretion onto the protoplanet decreases. By now the relative velocity of the planetesimals is dominated by encounters with the protoplanet rather than themselves. The encounters with the protoplanet increase the relative velocity of the planetesimals, which results in a decrease in the gravitational cross section for collision with the protoplanet. At this stage the protoplanet essentially stops its runaway growth phase and its accretion rate decreases significantly. This allows neighbouring, less massive protoplanets to catch up with the most massive protoplanet since the relative velocity of the planetesimals in their vicinity is lower than that around the most massive protoplanet. The next protoplanet catches up with the first one, and this process continues until the system reaches equilibrium which consists of a large number of roughly equal-mass protoplanets that are almost evenly spaced. The combined mass of the protoplanets is similar to the remaining mass in planetesimals. The protoplanets still accrete the planetesimals but they are all forced to grow at a similar pace. This growth phase of the protoplanets is dubbed 'oligarchic growth' (Kokubo and Ida 1998). The typical size of these protoplanets is Mercury to Mars sized and the time to reach this stage is shorter than 10 Myr (Kokubo and Ida 1998).

The system consisting of a large number of oligarchs and planetesimals is stable as long as the mass in planetesimals is larger than or equal to that of the protoplanets. The reason for this is that the planetesimals damp the eccentricities and inclinations of the protoplanets through angular momentum and energy partitioning, and this process is called 'dynamical friction' (Ida and Makino 1993). The disc starts with a certain angular momentum and energy budget. The protoplanets scatter the planetesimals onto eccentric orbits, decreasing the angular momentum budget in the planetesimals and increasing their orbital energies. The conservation of angular momentum implies that the protoplanets gain the angular momentum that the planetesimals have lost. Similarly the energy that the planetesimals have gained is compensated by a loss in energy of the protoplanets. Once the planetesimal reservoir is depleted, mutual perturbations of the protoplanets tend to increase their eccentricities and inclinations and eventually their orbits begin to cross. This instability is the onset of the third and last stage of terrestrial planet formation: the final terrestrial planets grow through mutual collisions of the protoplanets and the remaining planetesimals, which decreases the number of objects. This third stage is characterised by violent, stochastic, large collisions, one of which formed the Earth's Moon (Canup 2004). Eventually all the protoplanets have collided with one another and one is typically left with 3 to 5 terrestrial planets (Chambers 2001; O'Brien et al. 2006; Kokubo et al. 2006; Raymond et al. 2006, 2009).

The reason we generally find between 2 to 6 (mostly 3 to 5) terrestrial planets is partially the result of constrained formation and initial conditions. An earlier work by Kominami and Ida (2002, 2004) studied terrestrial planet formation from protoplanets in the presence of gas drag that was left from the primordial solar nebula in the range $\sim$0.6 AU to $\sim$1.7 AU. They generally formed 7 terrestrial planets of sub-terrestrial mass. The most likely reason for this higher number of planets is that the gas drag damped the eccentricity of the protoplanets strongly enough to prevent long-term orbit crossing and thus the violent mutual collisions that characterise the third stage. However, the final number of planets in the simulations is partially a result of the initial conditions. Extending the embryos close to Jupiter ($\sim$4 AU) does not change the outcome because the secular resonance ν_6 at 2 AU and the strong perturbations from Jupiter in the asteroid belt present an effective barrier for terrestrial planet formation beyond $\sim$1.8 AU (O'Brien et al. 2006). Most studies have truncated the inner edge of the disc, typically at 0.5 AU to 0.7 AU (Kokubo and Ida 1998; Hansen 2009; Raymond et al. 2006, 2009) or decreased its density profile inwards of 0.7 AU (Chambers 2001; O'Brien et al. 2006). Hence the amount of mass on the Sunward side was artificially restricted. In addition, most studies used a surface density value $\Sigma \sim 7$ g cm^{-2}, and the number of planets does not sensitively depend on this result (Kokubo et al. 2006).

In Fig. 1 I plot a series of snapshots of a simulation of terrestrial planet formation i.e. runaway growth, oligarchic growth and the final collisions. The picture displays the semi-major axis (a) and eccentricity (e) of the protoplanets and planetesimals. The size of the

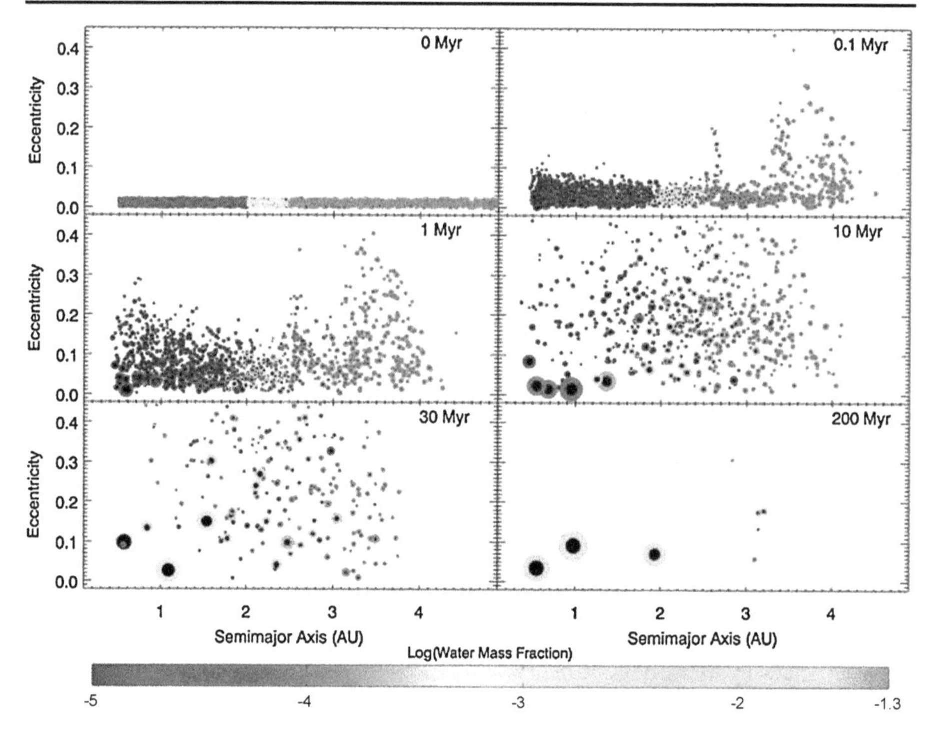

Fig. 1 Sample evolution of terrestrial planet formation. The size of each bullet scales as the mass of the object, $m^{1/3}$ and the colour is an indication of its original location. Taken from Raymond et al. (2006)

bullets representing the protoplanets in the figure, scale as their mass $m_p^{1/3}$. The colour coding in Fig. 1 denotes the initial position of each planetesimal and the final colouring of each (proto)planet is a measure of its consituents.

The simulations start with a disc which consists of a large sample of planetesimals spread out between 0.5 AU and 4.5 AU. The planetesimals perturb and collide with each other and after 1 Myr we see the formation of a few protoplanets. The protoplanets form from the inside out because the orbital time scale, and thus the collision time scale, is shorter close to the Sun than far away. After 10 Myr the system consists of a few tightly-packed red protoplanets in the inner region and a variety of smaller protoplanets farther out, all on eccentric orbits. The planetesimals are all over the place. A further 20 Myr later the number of protoplanets has decreased and the remaining ones have grown in mass. In addition, their composition has changed. Even the innermost protoplanet is no longer completely red but has become yellow, implying it has accreted material from outside its feeding zone. This radial mixing of material is an important outcome of simulations of terrestrial planet formation and could account for Earth's water content. I shall return to this topic in Sect. 4. Finally, 200 Myr after the start of the simulation, the system consists of a few remaining planetesimals and three terrestrial planets situated between 0.5 AU and 2 AU.

What is important to note is that the masses of all three terrestrial planets are comparable, as distinguished from the size of their bullets in Fig. 1. One can perform many more simulations of terrestrial planet formation with various initial conditions to determine how the end result depends on the initial conditions, but there appears little variety: there are almost always three or five planets between 0.5 AU and 2 AU with the middle two planets having masses similar to Venus and Earth (Kokubo et al. 2006; Raymond et al. 2006, 2009;

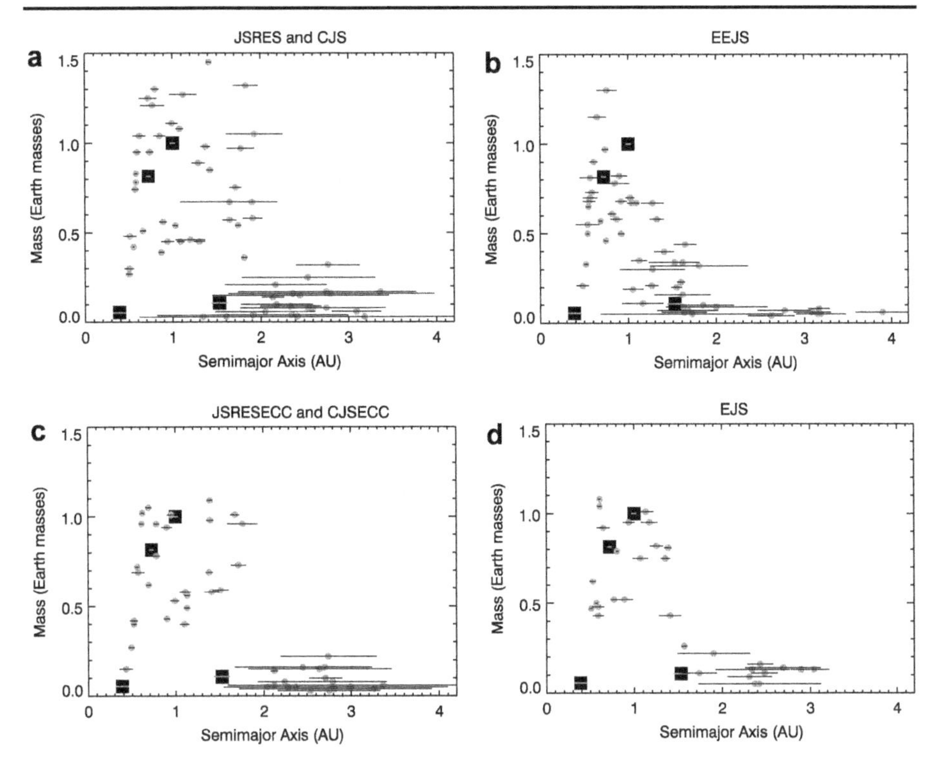

Fig. 2 Mass vs semi-major axis for a range of end states of terrestrial planet formation. Taken from Raymond et al. (2009)

O'Brien et al. 2006). However, the inner and outer planet are systematically much heavier than Mercury and Mars. Even though most of the simulated systems are compatible with the terrestrial planets in terms of their Angular Momentum Deficit (AMD)—a measure of the deviation from circular, coplanar orbits—and the spacing between the planets is also well reproduced, the concentration of mass is not.

The final masses versus semi-major axes of the end states of several terrestrial planet formation simulations from Raymond et al. (2009) are plotted in Fig. 2. This figure shows all the planets that were formed in 40 simulations, but the reader should be aware that most systems contain 3 to 5 terrestrial planets. Each panel of the figure pertains to a different configuration of Jupiter and Saturn (see Raymond et al. 2009). The terrestrial planets of our Solar System are depicted with filled squares while the masses of the planets formed in the numerical simulations are depicted by grey bullets. Even though the masses of Venus and Earth are reproduced at their current locations, Mercury and Mars are not. From the figure one can see that the masses of the planets and their locations are systematically too high. Apart from the panel labelled 'EEJS', where Jupiter and Saturn are placed on more eccentric orbits than their current values (currently $e_J = 0.05$, $e_S = 0.06$), the small mass of Mars is not reproduced. In many simulations a Mars-mass planet remains on an eccentric orbit, but it is farther from the Sun than Mars is today. Planets at the present location of Mars are almost always more massive, typically 0.2 to 0.6 Earth masses ($M_\oplus$). Since there is no evidence for Jupiter and Saturn ever having had larger eccentricities than currently for a prolonged period of time (Morbidelli et al. 2010), we must conclude that current terrestrial planet formation simulations have great difficulty reproducing the mass of Mars.

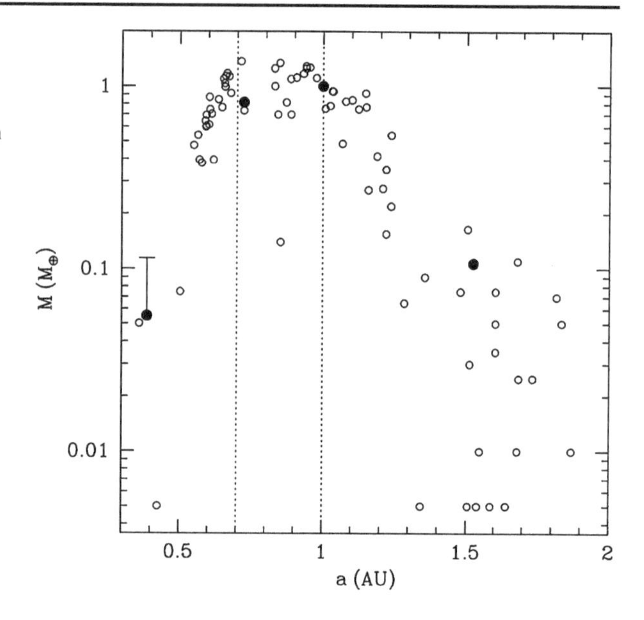

Fig. 3 Mass vs semi-major axis for a range of end states of terrestrial planet formation from a narrow annulus, originally between 0.7 AU and 1 AU. Taken from Hansen (2009). *Black bullets* are the Solar System planets

Chambers (2001) already noted that it was difficult to reproduce the mass of Mars in terrestrial planet formation simulations. He argued that the high mass concentration in Venus and Earth can only be explained in two ways: either in the regions occupied by Mars and Mercury the surface density of the disc was lower than at Venus and the Earth, or mass was lost from the system in these regions during the accretion of the terrestrial planets. Figure 1 above shows that the second option, losing mass from the region that Mars now occupies, does not occur, so we are left with the first: the surface density of the disc of material from which the terrestrial planets formed was lower at the present position of Mars than it was at Earth and Venus.

So, let us examine how terrestrial planet formation would proceed when the source material is confined to the Venus-Earth region.

3 Terrestrial Planet Formation from a Narrow Annulus

Hansen (2009) has examined the formation of the terrestrial planets starting from an annulus of material with constant surface density between 0.7 AU and 1 AU. He examined the evolution of a system of 400 equal-mass planetesimals on essentially circular and coplanar orbits, essentially only studying the third stage of terrestrial planet formation. The total mass in the system is 2 $M_{\oplus}$. The gas giant Jupiter was included in the simulations. After integrating this system for 1 Gyr, Hansen (2009) is left with either 3 or 4 terrestrial planets whose spacing and AMD match the current terrestrial planets and he was also able to reproduce the small masses of Mercury and Mars because these regions were originally depleted in material.

Figure 3, taken from Hansen (2009), is similar to Fig. 2 in that it depicts the mass versus semi-major axis of all the terrestrial planets that were formed from his simulations. The current terrestrial planets are depicted by filled circles while the open circles are fictitious planets from the simulations. As can be seen, the planets at 1.5 AU typically have the same mass as Mars while Mercury is marginally reproduced. Another interesting feature of Hansen's simulations is the fact that the planets on the edges (Mercury and Mars) are

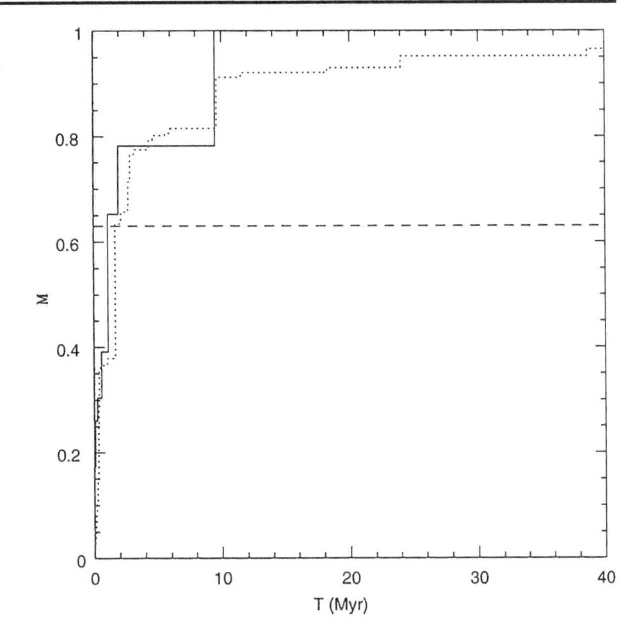

Fig. 4 Mass vs time for a Mars-sized analogue. Taken from Hansen (2009)

systematically more eccentric and inclined than the two central planets (Earth and Venus), similar to the Solar System. This suggests that the former two were scattered out of the disc and essentially remained there without subsequently accreting much more material. Indeed, this is what Hansen (2009) claims to have happened, arguing that both Mars and Mercury are protoplanets that ceased to grow to full-fledged terrestrial planets because of a lack of material to accrete in their direct surroundings. If true, then the growth of Mars (and Mercury) should have occurred very quickly in the beginning and essentially stopped shortly afterwards. Figure 4 depicts the mass of a Mars-sized planet (solid) line, and that of an Earth analogue (dotted line). As one may see, the Mars-sized planet stops growing approximately 10 Myr after the onset of the third stage. Since the second stage usually takes up a fraction of this time, the formation of Mars roughly coincides with the time that oligarchic growth has finished and the third phase of terrestrial planet formation commences. By comparison, in the simulations the Earth was still growing after 40 Myr, and did not reach its final mass until much later (Hansen 2009). What is also interesting to note is that the growth of both planets, while rapid at first, slows down but occurs in discrete steps rather than continuous growth. These sudden increases in mass coincide with collisions with other protoplanets, which tend to dominate the long-term behaviour.

3.1 Caveats

It appears as if terrestrial planet formation from a narrow annulus is able to solve a long-lasting problem in planetary science: the small mass of Mars (and that of Mercury) compared to the Earth. The annulus formation scenario also explains why Mercury and Mars are more inclined and eccentric than the Earth and Venus, and the number of planets, the mass concentration, spacing and AMD of the resulting planets all match the current terrestrial system well. However, there are two problems that need to be addressed. First, the formation of the terrestrial planets from an annulus cannot account for the water on Earth because the material this close to the Sun is essentially dry (e.g. Sasselov and Lecar 2000). Thus the water either had to be delivered later, or the annulus had to be wetter than most theories seem to suggest. Secondly, what caused the disc to be truncated? The existence of an inner edge can be explained by the migration of close-in protoplanets. These formed when the gas was still present in the solar nebula (Ida and Lin 2008). The time scale of the formation of protoplanets is proportional to their distance to the Sun, and thus the protoplanets form quicker closer to the Sun than further out. As the gas is depleted, the protoplanets stop migrating into the star and some of these remain behind. Beyond a threshold distance, the planetesimal disc remains intact because no protoplanets have yet formed and no migration has yet occurred. Ida and Lin (2008) demonstrate that this inner edge is typically around 0.5 AU to 1 AU. It could be argued that this migration scenario can be used to explain the small mass of Mercury, which is always difficult to reproduce in numerical simulations. Recent results from the volatiles at Mercury measured by the MESSENGER mission suggest that it did not suffer a giant impact which stripped its mantle nor could this stripping have occurred through solar heating (Peplowski et al. 2011), so that its formation location is uncertain. That said, the formation of Mercury is not fully understood and discussing it is beyond the scope of this paper.

However, there is no natural explanation for the existence of an outer edge of the planetesimal disc apart from it arising through a dynamical origin. Hansen (2009) concluded that a very sharp, almost discontinuous edge was needed; a Gaussian density distribution was not sufficient. The disc could not have been truncated by photoevaporation because this would have prevented the formation of the giant planets farther out than where the disc was truncated. Likewise, a truncation through a close stellar passage can also be ruled out because this would either have prevented the formation of the giant planets or destabilised an existing giant planet system. Other possibilities include a steeper decline in the surface density of the disc in Mars' region than closer in Chambers and Cassen (2002), particle pile-up through turbulent stirring (Haghighipour and Boss 2003; Johansen et al. 2007), or even a planet trap (Masset et al. 2006). A planet trap is a place in the gas disc where there is a steep surface-density gradient; this steep gradient prevents giant planet migration. However, none of these theories has been particularly favoured nor accepted by the community. Therefore we explore another, more generic possibility: dynamical sculpting by migration of Jupiter and Saturn.

4 The Grand Tack Scenario

Masset and Snellgrove (2001) demonstrated that when Jupiter and Saturn formed in the gaseous protoplanetary disc, well before the formation of the terrestrial planets, both migrated towards the Sun. The migration of the giant planets in a gas disc is a generic outcome of planet formation. The migration is illustrated in Fig. 5. In this simulation, Jupiter is massive enough to open a cavity in the disc, and thus its migration time scale is tied to the viscous evolution of the disc (Type II migration; Papaloizou and Lin 1984). On the other hand, Saturn is not massive enough to open a complete cavity and thus its migration speed is much faster (Type I migration; Goldreich and Tremaine 1980). Both planets move inwards because of the asymmetric Lindblad torques generated from spiral wakes. As can be seen, the migration of Jupiter ceases when Saturn falls into the 2 : 3 mean motion resonance with Jupiter i.e. when Saturn performs two orbits for every three of Jupiter. In fact, the migration even reverses direction! This reversal occurs because the proximity of Saturn to Jupiter causes the cavities they open up in the gas disc to overlap (Morbidelli and Crida 2007), and the asymmetry of the gap cancels and reverses the torques. This reversal of migration is

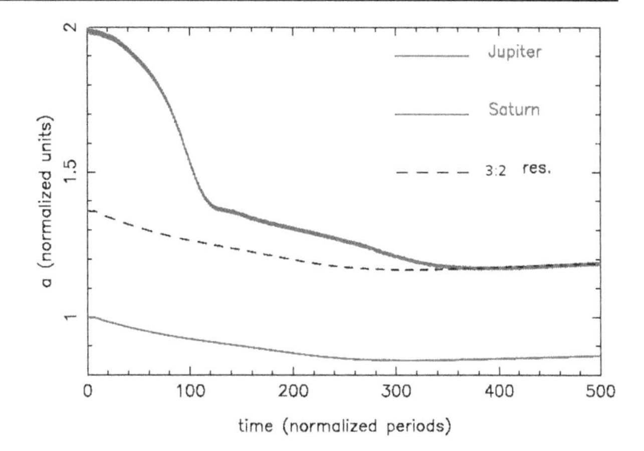

Fig. 5 Migration of Jupiter and Saturn in a gas disc. Taken from Masset and Snellgrove (2001)

once again a generic outcome, since the probability of Saturn getting caught in the 2 : 3 resonance with Jupiter is high (Pierens and Nelson 2008). It should be noted that the slowing of Jupiter's migration and even its reversal is only possible because Saturn is less massive than Jupiter; if the situation was reversed, both planets would migrate into the Sun because the outer, more massive planet would push the inner, less massive one Sunwards. In our system, it is possible that Jupiter and Saturn migrated inwards far enough to truncate the disc of material from which the terrestrial planets formed before migrating out again. All that needs to be specified is the turning point, which is described in Walsh et al. (2011) and is called the 'Grand Tack' scenario. Here I will just give a short summary of its workings.

The Grand Tack scenario relies on two assumptions: Jupiter formed around the snow line, in the range 2.5 AU to 4.5 AU, and it reversed migration at 1.5 AU. Walsh et al. (2011) demonstrate that the asteroid belt can survive this migration of Jupiter and Saturn because scattering by Jupiter initially empties the belt when it migrates inwards, but then repopulates the asteroid belt as it migrates outwards. The reversal at 1.5 AU of Jupiter truncates the disc at 1 AU, effectively creating Hansen's (2009) annulus. Figure 6 shows the migration of the giant planets in the reference simulation of Walsh et al. (2011) as well as the growth of their masses. After 0.1 Myr Saturn catches up with Jupiter, falls into the 2 : 3 resonance and they reverse their migration. Eventually all four giant planets end up in a multi-resonant configuration on nearly-circular, coplanar orbits, similar to the resonant initial conditions of the Nice model (Morbidelli et al. 2007).

The workings of the Grand Tack scenario are illustrated in Fig. 7. Here the planets are outlined by large, filled bullets, with sizes scaling as the mass and the planetary embryos and terrestrial planets are depicted using open circles. The blue dots are water-rich planetesimals from the outer solar system i.e. from beyond the snow line (beyond 3 AU). These are left over from giant planet formation. The red dots are water-poor planetesimals from the inner solar system (interior to 3 AU). Open circles are terrestrial planets and planetary embryos. The sizes of the circles scale with the masses. The axes depict semi-major axis and eccentricity of the bodies.

The top panel depicts the initial conditions. The second panel from the top, at $t = 70$ kyr depicts Jupiter having migrated to approximately 2 AU. Note that the mass in the inner Solar System is squeezed together by the migration of Jupiter, as some of the planetesimals interior to the planet will move with it. Also note that some of the planetesimals in the inner Solar System are blue because they were scattered by Jupiter and Saturn. In other words, in the Grand Tack scenario water delivery to the terrestrial planets is a generic outcome. I shall

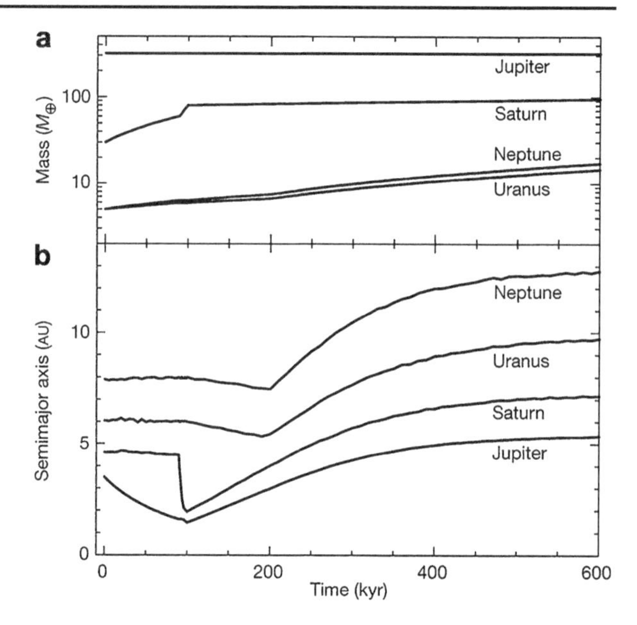

Fig. 6 Growth and migration of the giant planets in a gas disc during the Grand Tack scenario. Taken from Walsh et al. (2011)

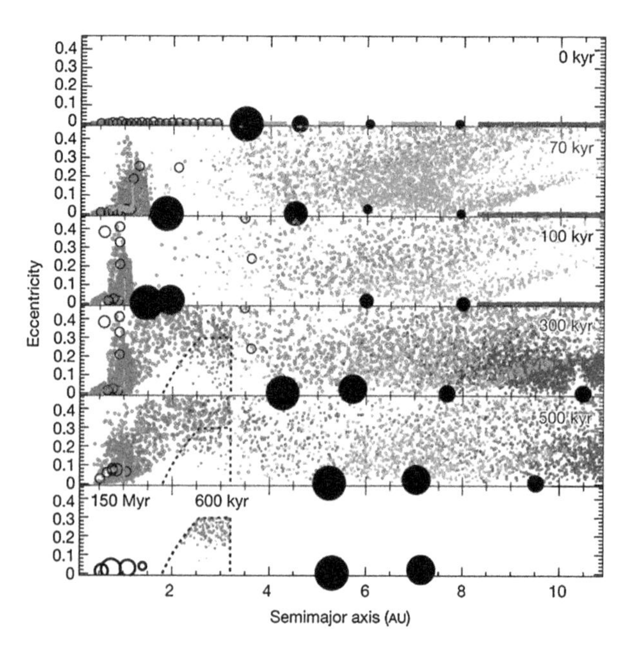

Fig. 7 Migration of the giant planets in a gas disc during the Grand Tack scenario and the evolution of the asteroid belt and inner solar system. Taken from Walsh et al. (2011)

return to this issue later. In the third panel, at $t = 100$ kyr, Jupiter and Saturn are about to reverse their migration. Now the outer Solar System is also filled with red dots, so that a radial mixing of material occurs through the gravitational scattering by Jupiter and Saturn as they migrate through the gas disc. The next two panels depict further outward migration of Jupiter and Saturn. Most of the red dots that originated in the inner solar system form the terrestrial planets and the blue dots that originate in the outer solar system are ejected by the giant planets. Eventually, in the bottom panel, we are left with an asteroid belt, consisting

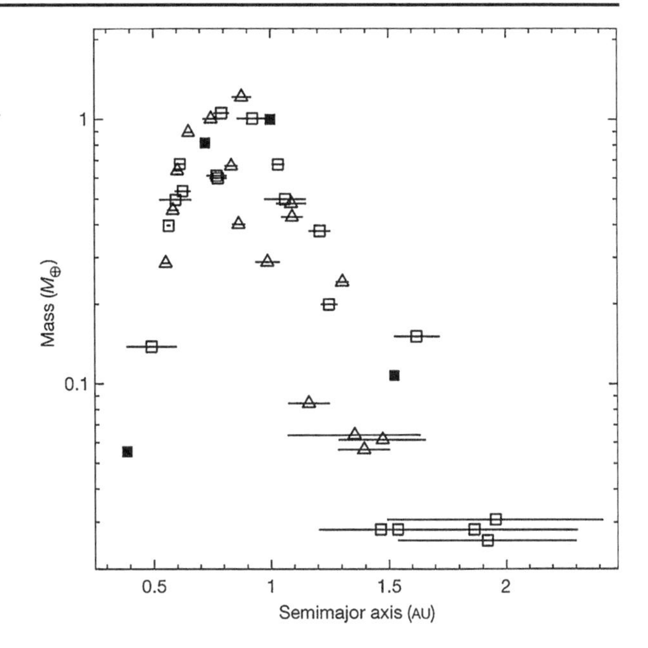

Fig. 8 Masses vs semi-major axis of the simulated terrestrial planets in the Grand Tack scenario. Taken from Walsh et al. (2011)

of both blue and red dots, and the terrestrial planets 150 Myr later. The radial mixing in the asteroid belt is a possible explanation to the compositional gradient of the asteroid belt.

After this short overview of the Grand Tack scenario, I shall turn to the formation of the terrestrial planets.

4.1 Terrestrial Planet Formation During the Grand Tack Scenario

The formation of the terrestrial planets in the Grand Tack scenario proceeds in a similar manner to that described in Hansen (2009). In Fig. 8 I depict the masses of the planets that were formed in several simulations from Walsh et al. (2011) as a function of their semi-major axis. This figure should be compared with Fig. 3, taken from Hansen (2009). The filled squares correspond to the current terrestrial planets. The triangles and squares correspond to simulations with different starting conditions, but the outcome tends to be the same: the masses of the terrestrial planets are well reproduced, which is no surprise because the model basically reproduced the initial conditions of Hansen (2009). Once again, the small mass of Mars is also reproduced, which was the aim of the model, and Mercury and Mars are more inclined and eccentric than Earth and Venus.

I showed above that the Grand Tack scenario brought water to the terrestrial planets. The estimation of the water delivery to the terrestrial planets in the Grand Tack scenario is an ongoing study, but I will present some preliminary results here, based on the work of O'Brien et al. (2010). Table 1 shows the mass, orbital elements and fractional make up of Mars-sized planets from several of the simulations presented in Walsh et al. (2011) and O'Brien et al. (2010). There are several interesting trends. First of all, the majority of the mass of the Mars-sized analogues consist of material from embryos rather than planetesimals, with all cases being formed from a single embryo that accreted some planetesimals. If these simulations are representative of the formation of Mars then it suggests that Mars is a leftover planetary embryo. Second, the orbital elements are more or less consistent with the current orbit of Mars. The last two embryos seem to have most

Table 1 This table shows the mass, fraction of mass from embryos, semi-major axis, eccentricity and inclination of Mars analogues formed in a few of the Grand Tack simulations of Walsh et al. (2011). The remaining columns show, in percent, the fraction of material that originated from the various regions

M ($M_\oplus$)	Frac. emb. (%)	a (AU)	e	i (°)	< 1	1–1.5	1.5–2	2–3	> 3
0.06	82.3	1.50	0.03	11.7	0.0	4.1	86.4	4.1	5.4
0.05	96.2	1.67	0.08	5.1	0.0	0.0	0.0	96.2	3.8
0.07	78.4	1.38	0.07	5.1	3.9	11.8	0.0	82.3	2.1

of their material come from between 2 AU and 3 AU while the first one has most of its material from between 1.5 AU and 2 AU, suggesting that the real Mars is most likely a mixture of both. In terms of water content, if I naively assume that only material beyond 3 AU has a water content (Morbidelli et al. 2000), and when pegging this at 5 % by mass (Morbidelli et al. 2000), then the putative Mars should have approximately 0.1–0.2 % by mass in water. This is most likely an upper limit and carries a large uncertainty, and thus it should only be used as a guidance rather than absolute. In comparison, for the Earth this value is approximately 0.05 % to 0.1 % (Morbidelli et al. 2000; Murakami et al. 2002). In any case, it suggests that Mars, shortly after its formation, could have carried a substantial amount of water. However, whether it could retain it remains to be seen. In any case, while the final outcome in terms of planet semi-major axis and mass in the Grand Tack scenario is very similar to that of Hansen (2009), there is a large difference in the final outcome in terms of planetary composition: Hansen's planets are dry while the Grand Tack planets have a certain percentage of icy material from beyond the snow line.

It appears that Grand Tack scenario can sculpt the disc from which the terrestrial planets form to a small enough size to reproduce Hansen's (2009) initial conditions, and to account for the water content of the Earth. The next question then is whether or not the Grand Tack scenario is compatible with cosmochemical constraints. This is discussed in the next section, focusing on Mars.

5 Comparison with Cosmochemical Constraints

In this section the formation of Mars in the Grand Tack scenario is compared with cosmochemical constraints. Hansen (2009) already discussed some of this when applied to all of the terrestrial planets, and showed that the growth of Mars occurs on a time scale less than 10 Myr (see Fig. 4 above). This growth time scale is consistent with the cosmochemical evidence presented in Nimmo and Kleine (2007), who suggested a Martian formation time between 1 Myr to 10 Myr. In addition, the Grand Tack scenario suggests that the composition of Mars consists of a mixture of materials whose origin span several AU in range, consistent with findings by Lodders (2000). The short formation time scale based on cosmochemical evidence, the small mass of Mars and the results from the formation simulations in the Grand Tack scenario suggest that Mars is a left-over embryo rather than a fully-formed planet. Is this indeed the case?

The best, most recent evidence that Mars is actually a planetary embryo comes from Dauphas and Pourmand (2011). They analyse the 182 Hafnium-Tungsten (^{182}Hf-^{182}W) decay system in the shergottite-nakhlite-chassignite (SNC) meteorites and improve on the currently-known values by analysing the Th/Hf and ^{176}Hf/^{177}Hf ratios in chondrites that reflect remobilisation of Lutetium and Thorium during parent-body processes. The excess abundance of ^{182}W, produced from the radioactive decay of ^{182}Hf, relative to other non-radiogenic isotopes in the Martian mantle, can be used to determine the age of core formation of Mars. See also the chapter by Metzger et al. (Core Formation and Mantle Differentiation on Mars).

During oligarchic growth of protoplanets the mass of the planet can be approximated as $M(t) = M_{\text{Mars}} \tanh^3(t/\tau)$ (Chambers 2006). The value of τ is the accretion time scale. Dauphas and Pourmand (2011) use the excess abundance of ^{182}W in the Martian mantle, the decay of ^{182}Hf, the Hf/W ratios and the typical composition of a chronditic uniform reservoir to compute $\tau = 1.8^{+0.9}_{-1.0}$ Myr. They argue this is a robust upper limit. Thus, most of the Martian accretion took place during the first 4 Myr of the Solar System, even before dissipation of the solar nebula. This time scale is in excellent agreement with the findings of Hansen (2009) and suggests that Mars is indeed a left-over planetary embryo. It is worthy to note that Mars had gained almost its full size while in parts of the disc closer to the Sun planets were still forming, based on the fact that chondrites formed $\gtrsim 3$ Myr after CAIs (Kleine et al. 2009). The idea that Mars is an embryo may also explain the similarities between its atmosphere and that of the Earth because both show isotopically fractionated Xenon (Pepin 1991). The Earth may have inherited the fractionated Xe from the atmosphere of a Mars-sized embryo that collided with it and formed the Moon.

6 Summary and Conclusions

In this chapter I have given an overview of the latest models of terrestrial planet formation, and focused on Mars in particular. The small mass of Mars has been a long-standing problem in terrestrial planet formation because it is very difficult to reproduce in numerical simulations. Indeed, planets at the position of Mars were always a factor of a few too massive.

A breakthrough in this long-standing problem came when Hansen (2009) considered terrestrial planet formation from a narrow annulus of material between 0.7 AU and 1 AU. Hansen noted that Mars finished growing within 10 Myr and from the dynamics suggested that Mars was a protoplanet that got scattered outwards and accreted no additional material. While the coagulation of material from this ring was able to reproduce the current terrestrial system very nicely, it had two shortcomings: the material was dry and there was no physical explanation to truncate the disc at 1 AU. Several reasons were proposed but the most recent one was a dynamical sculpting by the migration of Jupiter and Saturn in the gas disc, called the Grand Tack scenario. This migration could truncate the disc at 1 AU and reproduce the initial conditions of Hansen (2009). The model has the advantage that water-rich material from the outer Solar System got mixed with the water-poor material from the inner Solar System, thereby accounting for the water on Earth and the other terrestrial planets. From this model it is expected that Mars contains up to 0.2 % of water by mass.

When analysing the growth of Mars in the Grand Tack scenario, it appears that Mars is a left-over protoplanet that got stranded outside the main disc of material and stopped growing after a few million years. A recent study by Dauphas and Pourmand (2011), based on Thorium/Tungsten and Thorium/Hafnium ratios in the Martian mantle, confirm that the growth time of Mars was of the order of 2 Myr, well within the upper limit of 10 Myr suggested by Hansen (2009) and once again confirming that Mars is a remnant planetary embryo.

Acknowledgements The author is grateful to Germany's Helmholtz Alliance for partially funding this research through their 'Planetary Evolution and Life' programme while still based in France. I thank Kevin Walsh and Alessandro Morbidelli for their input, David O'Brien and Tilman Spohn for constructive reviews, and Kevin Walsh and David O'Brien again for providing me with data from their numerical simulations.

References

R.M. Canup, Dynamics of lunar formation. Annu. Rev. Astron. Astrophys. **42**, 441–475 (2004)
J.E. Chambers, Making more terrestrial planets. Icarus **152**, 205–224 (2001)
J.E. Chambers, A semi-analytic model for oligarchic growth. Icarus **180**, 496–513 (2006)
J.E. Chambers, P. Cassen, The effects of nebula surface density profile and giant-planet eccentricities on planetary accretion in the inner solar system. Meteoritics and Planetary. Science **37**, 1523–1540 (2002)
J.N. Cuzzi, R.C. Hogan, K. Shariff, Toward planetesimals: dense chondrule clumps in the protoplanetary Nebula. Astrophys. J. **687**, 1432–1447 (2008)
N. Dauphas, A. Pourmand, Hf-W-Th evidence for rapid growth of Mars and its status as a planetary embryo. Nature **473**, 489–492 (2011)
R. Greenberg, W.K. Hartmann, C.R. Chapman, J.F. Wacker, Planetesimals to planets—numerical simulation of collisional evolution. Icarus **35**, 1–26 (1978)
P. Goldreich, S. Tremaine, Disk-satellite interactions. Astrophys. J. **241**, 425–441 (1980)
N. Haghighipour, A.P. Boss, On gas drag-induced rapid migration of solids in a nonuniform solar nebula. Astrophys. J. **598**, 1301–1311 (2003)
B.M.S. Hansen, Formation of the terrestrial planets from a narrow annulus. Astrophys. J. **703**, 1131–1140 (2009)
S. Ida, J. Makino, Scattering of planetesimals by a protoplanet—slowing down of runaway growth. Icarus **106**, 210 (1993)
S. Ida, D.N.C. Lin, Toward a deterministic model of planetary formation. IV. Effects of type I migration. Astrophys. J. **673**, 487–501 (2008)
A. Johansen, J.S. Oishi, M.-M. Mac Low, H. Klahr, T. Henning, A. Youdin, Rapid planetesimal formation in turbulent circumstellar disks. Nature **448**, 1022–1025 (2007)
T. Kleine, M. Touboul, B. Bourdon, F. Nimmo, K. Mezger, H. Palme, S.B. Jacobsen, Q.-Z. Yin, A.N. Halliday, Hf-W chronology of the accretion and early evolution of asteroids and terrestrial planets. Geochim. Cosmochim. Acta **73**, 5150–5188 (2009)
E. Kokubo, S. Ida, Orbital evolution of protoplanets embedded in a swarm of planetesimals. Icarus **114**, 247–257 (1995)
E. Kokubo, S. Ida, Oligarchic growth of protoplanets. Icarus **131**, 171–178 (1998)
E. Kokubo, J. Kominami, S. Ida, Formation of terrestrial planets from protoplanets. I. Statistics of basic dynamical properties. Astrophys. J. **642**, 1131–1139 (2006)
J. Kominami, S. Ida, The effect of tidal interaction with a gas disk on formation of terrestrial planets. Icarus **157**, 43–56 (2002)
J. Kominami, S. Ida, Formation of terrestrial planets in a dissipating gas disk with Jupiter and Saturn. Icarus **167**, 231–243 (2004)
K. Lodders, An oxygen isotope mixing model for the accretion and composition of rocky planets. Space Sci. Rev. **92**, 341–354 (2000)
F. Masset, M. Snellgrove, Reversing type II migration: resonance trapping of a lighter giant protoplanet. Mon. Not. R. Astron. Soc. **320**, L55–L59 (2001)
F.S. Masset, A. Morbidelli, A. Crida, J. Ferreira, Disk surface density transitions as protoplanet traps. Astrophys. J. **642**, 478–487 (2006)
A. Morbidelli, J. Chambers, J.I. Lunine, J.M. Petit, F. Robert, G.B. Valsecchi, K.E. Cyr, Source regions and time scales for the delivery of water to Earth. Meteoritics and Planetary. Science **35**, 1309–1320 (2000)
A. Morbidelli, A. Crida, The dynamics of Jupiter and Saturn in the gaseous protoplanetary disk. Icarus **191**, 158–171 (2007)
A. Morbidelli, K. Tsiganis, A. Crida, H.F. Levison, R. Gomes, Dynamics of the giant planets of the solar system in the gaseous protoplanetary disk and their relationship to the current orbital architecture. Astron. J. **134**, 1790–1798 (2007)
A. Morbidelli, R. Brasser, R. Gomes, H.F. Levison, K. Tsiganis, Evidence from the asteroid belt for a violent past evolution of Jupiter's orbit. Astron. J. **140**, 1391–1401 (2010)
A. Morbidelli, J.I. Lunine, D.P. O'Brien, S.N. Raymond, K.J. Walsh, Building terrestrial planets. Annu. Rev. Earth Planet. Sci. **40**, 251–275 (2012)

M. Murakami, K. Hirose, H. Yurimoto, S. Nakashima, N. Takafuji, Water in Earth's lower mantle. Science **295**, 1885–1887 (2002)

F. Nimmo, T. Kleine, How rapidly did Mars accrete? Uncertainties in the Hf W timing of core formation. Icarus **191**, 497–504 (2007)

D.P. O'Brien, A. Morbidelli, H.F. Levison, Terrestrial planet formation with strong dynamical friction. Icarus **184**, 39–58 (2006)

D.P. O'Brien, K.J. Walsh, A. Morbidelli, S.N. Raymond, A.M. Mandell, J.C. Bond, Early giant planet migration in the solar system: geochemical and cosmochemical implications for terrestrial planet formation. Bull. Am. Astron. Soc. **42**, 948 (2010)

J. Papaloizou, D.N.C. Lin, On the tidal interaction between protoplanets and the primordial solar nebula. I—Linear calculation of the role of angular momentum exchange. Astrophys. J. **285**, 818–834 (1984)

R.O. Pepin, On the origin and early evolution of terrestrial planet atmospheres and meteoritic volatiles. Icarus **92**, 2–79 (1991)

P.N. Peplowski et al., Radioactive elements on Mercury's surface from MESSENGER: implications for the planet's formation and evolution. Science **333**, 1850 (2011)

A. Pierens, R.P. Nelson, Constraints on resonant-trapping for two planets embedded in a protoplanetary disc. Astron. Astrophys. **482**, 333–340 (2008)

S.N. Raymond, T. Quinn, J.I. Lunine, High-resolution simulations of the final assembly of Earth-like planets I. Terrestrial accretion and dynamics. Icarus **183**, 265–282 (2006)

S.N. Raymond, D.P. O'Brien, A. Morbidelli, N.A. Kaib, Building the terrestrial planets: constrained accretion in the inner Solar System. Icarus **203**, 644–662 (2009)

V.S. Safronov, E.V. Zvjagina, Relative sizes of the largest bodies during the accumulation of planets. Icarus **10**, 109 (1969)

D.D. Sasselov, M. Lecar, On the snow line in dusty protoplanetary disks. Astrophys. J. **528**, 995–998 (2000)

K.J. Walsh, A. Morbidelli, S.N. Raymond, D.P. O'Brien, A.M. Mandell, A low mass for mars from Jupiter's early gas-driven migration. Nature **475**, 206–209 (2011)

S.J. Weidenschilling, J.N. Cuzzi, Formation of planetesimals in the solar nebula. Protostars Planets **III**, 1031–1060 (1993)

G. Wurm, J. Blum, J.E. Colwell, NOTE: a new mechanism relevant to the formation of planetesimals in the solar nebula. Icarus **151**, 318–321 (2001)

Space Sci Rev (2013) 174:27–48
DOI 10.1007/s11214-012-9935-8

Core Formation and Mantle Differentiation on Mars

Klaus Mezger · Vinciane Debaille · Thorsten Kleine

Received: 10 February 2012 / Accepted: 17 September 2012 / Published online: 16 October 2012

Abstract Geochemical investigation of Martian meteorites (SNC meteorites) yields important constraints on the chemical and geodynamical evolution of Mars. These samples may not be representative of the whole of Mars; however, they provide constraints on the early differentiation processes on Mars. The bulk composition of Martian samples implies the presence of a metallic core that formed concurrently as the planet accreted. The strong depletion of highly siderophile elements in the Martian mantle is only possible if Mars had a large scale magma ocean early in its history allowing efficient separation of a metallic melt from molten silicate. The solidification of the magma ocean created chemical heterogeneities whose ancient origin is manifested in the heterogeneous ^{142}Nd and ^{182}W abundances observed in different meteorite groups derived from Mars. The isotope anomalies measured in SNC meteorites imply major chemical fractionation within the Martian mantle during the life time of the short-lived isotopes ^{146}Sm and ^{182}Hf. The Hf-W data are consistent with very rapid accretion of Mars within a few million years or, alternatively, a more protracted accretion history involving several large impacts and incomplete metal-silicate equilibration during core formation. In contrast to Earth early-formed chemical heterogeneities are still preserved on Mars, albeit slightly modified by mixing processes. The preservation of such ancient chemical differences is only possible if Mars did not undergo efficient whole mantle convection or vigorous plate tectonic style processes after the first few tens of millions of years of its history.

K. Mezger (✉)
Institut für Geologie, Universität Bern, Baltzerstrasse 1, 3012 Bern, Switzerland
e-mail: klaus.mezger@geo.unibe.ch

V. Debaille
Laboratoire G-Time, Université Libre de Bruxelles CP160/02, Avenue F.D. Roosevelt 50, 1050 Brussels, Belgium

T. Kleine
Institut für Planetologie, Westfälische Wilhelms-Universität Münster, Wilhelm-Klemm-Str. 10, 48149 Münster, Germany
e-mail: thorsten.kleine@uni-muenster.de

Keywords Core formation · Magma ocean · Chemical differentiation · Mantle evolution · Short-lived nuclides

1 Introduction

Mars is the outermost terrestrial planet in the solar system. Its chemical budget compared to the rocky planets of the inner solar system is characterized by a higher abundance of moderately volatile elements, which most likely is due to its larger distance from the Sun. Mars is the second smallest among the terrestrial planets (Mars has about one ninth the mass of Earth), a feature that has puzzled many generations of scientists. It is now proposed that the early migration inwards then outwards of Jupiter and Saturn at the beginning of the solar system inhibited the full formation of Mars (Walsh et al. 2011; Brasser 2012). Alternatively, a Mars-sized body escaped further accretion by rapid outward migration to the outer region of the inner solar system (Minton and Levison 2011). Whatever the exact formation mechanism of Mars, its accretion time appears to have been shorter than that of Earth (e.g., Kleine et al. 2002, 2009; Dauphas and Pourmand 2011). Understanding the accretion and differentiation history of Mars, therefore, is important not just in its own right, but also because it provides essential insights into the accretion history of rocky planets in the inner solar system. In addition, Mars provides an important analogue of what the Earth and other terrestrial planets may have looked like in their earliest stages of accretion. The objective of this article is to review the accretion and early differentiation history of Mars as constrained mainly from geochemical and isotopic observations.

From mineralogical and chemical studies of Martian meteorites, the measurements of its gravity and magnetic field by orbiting satellites and *in-situ* analyses by exploration rovers, it is known that the planet is chemically differentiated, i.e. it has a metallic core, a heterogeneous silicate mantle and mafic to intermediate crust covered with a layer of weathered material and an atmosphere. Its differentiation processes may have many similarities with the chemical differentiation of the Earth. However, due to the much smaller size of Mars compared to Earth its overall geological activity declined earlier after planetary accretion. This is because to a first approximation smaller bodies lose their primordial heat from accretion, heat generated by radioactive decay of short-lived isotopes and core formation by conduction much more efficiently than larger bodies, which lose significant amounts of heat via convection as expressed in volcanic activity on the surface. Whether convection occurs in addition to conduction also depends on the water-content of the planetary interior. Since Mars seems to have outgassed substantially in its early history, the water content of the Martian mantle is now about 2-5 times lower than on Earth (McCubbin et al. 2012), which disfavors convection as well. However, recent crater counting measurements indicate very young (100 Ma) or possibly even near-recent volcanic activity at the surface of Mars (Neukum et al. 2004, 2010), which imply that not all heat is lost by conduction. Thermodynamical modeling of Mars seems to indicate that its mantle may be convecting at the present time (Li and Kiefer 2007). However, despite possible convection in the mantle, the planet is and has been for the major part of its history in a stagnant-lid regime (Lenardic et al. 2004). Compared to Earth the planet showed vigorous geologic activity only during its very early history (e.g. Breuer and Spohn 2003; O'Neill et al. 2007). As a consequence it can be expected that chemical and isotopic signatures of early planetary differentiation processes may be still well preserved on Mars. In contrast, such signatures are largely lost on Earth due to the vigorous convection of the Earth's mantle which results in the formation of new and the destruction of old oceanic and continental crust to the present day. This convection drives homogenization of the mantle after an early differentiation event, like core formation or the crystallization of a magma ocean. However, Mars was able to retain some of its primary heterogeneity implying that over time mantle convection was much less vigorous on Mars than on Earth.

The chemical make-up of Mars and the processes leading to a chemically differentiated planetary body can be investigated in some detail because material for chemical, physical, mineralogical and petrological study is available for analysis in the laboratory. Apart from the Earth, Mars is currently the only planet of the solar system from which material is available for direct analysis in the laboratory. Based on images of the Martian surface and spectral analyses of surface rocks, it is clear that the meteorite material is somewhat limited and does not cover all rock types present on Mars. All meteorites are igneous rocks and none have been found so far that derive from the clastic and chemical sediments that cover large parts of the Martian surface. The study of SNC meteorites in combination with *in-situ* analyses on Mars by the rovers Spirit and Opportunity as well as results from earlier missions (particularly the Viking missions in the 1970's and remote sensing data obtained by several orbiting missions) provides insights into the chemical, mineralogical and petrological composition of part of Mars from the micron to the planetary scale. These studies also provide constraints on the chemical composition of the solar nebula since Mars may have sampled a section of the solar system from which we do not have other material for direct study. Together with meteorites mostly derived from smaller bodies located in the asteroid belt, and the terrestrial and lunar material only a small section of the solar system is currently accessible to direct study.

Remote and *in-situ* studies combined with information derived from Martian meteorites can be used to reconstruct the formation of Mars and its evolution through time. This evolution includes aspects of accretion from planetesimals to the final formation of the planet, its core-mantle segregation, the chemical differentiation of the mantle, the formation of the crust and magmatic activity through time. Core formation in particular is the most dramatic chemical differentiation observed in planets and planetesimals, and has profound consequences for their evolution that range from thermal and chemical processes in the silicate part, to surface evolution and the generation of a magnetic field. Magmatism starting with a magma ocean stage and continuing later with volcanic and subvolcanic activities is a key process that influences the chemical composition of the atmosphere and is the dominant vector for the transfer of volatiles from mantle to the atmosphere. The availability of Martian material in the form of meteorites for direct mineralogical, petrological and chemical study, therefore, provides important insights into the composition and evolution of Mars as a planetary body.

2 Martian Material and the Composition of Mars

Currently about 100 Martian meteorites are available for direct study in the laboratory (The Meteoritical Bulletin Database, http://www.lpi.usra.edu/meteor/metbull.php), but among them, only ~60 are unpaired (http://www.imca.cc/mars/martian-meteorites.htm). Almost all samples are finds from hot or cold deserts and show different degrees of terrestrial weathering. Most Martian meteorites collected so far are quite small and only about 20 of them have a mass of more than 500 g. This limits the number of samples that can be studied in detail with currently available destructive geochemical techniques, particularly high precision isotope measurements or trace elements that require large amounts of sample material.

The Martian meteorites are commonly classified as SNC meteorites named after the prominent representative in each group that also represents an observed fall: Shergotty

(India, 1865), Nakhla (Egypt, 1911) and Chassigny (France, 1815). All known samples are of magmatic origin (e.g., McSween 1994; Treiman et al. 2000; Nyquist et al. 2001; McSween et al. 2009). The gases contained in almost all SNCs match those measured in situ for the Martian atmosphere by the Viking missions and constitute the absolute proof of their Martian origin (e.g., Bogard and Johnson 1983; Treiman et al. 1986, 2000; Bogard et al. 2001) The gases may have been trapped during shock by impacts of other meteorites prior or during ejection of the Martian samples from near the planetary surface. One exception is the sample Chassigny, which is a cumulate and has different gas abundances that may reflect the abundances in the Martian mantle (Ott and Begemann 1985; Ott 1988).

The *shergottites* are the most common type and represent mafic to ultramafic igneous lithologies with phenocrystic olivine and orthopyroxene and only rarely clinopyroxene. Based on their crystal size and modal mineralogy they are subdivided into basaltic, olivine-phyric, and lherzolitic shergottites and one wehrlite. This mineralogical subdivision more or less reflects another subdivision based on their geochemical composition (depleted, intermediate and enriched in light rare earth elements) (e.g., McSween 1994; Borg et al. 2002; Debaille et al. 2007; Debaille et al. 2008). The remarkable characteristic of the shergottites is their "young" crystallization ages, ranging from 165 to 475 Ma (Nyquist et al. 2001), even though Pb-Pb isotope systematics were used to argue for "old" crystallization ages of around 4.1 Ga (Bouvier et al. 2005; Bouvier et al. 2008). The age significance of the Pb-Pb isotope systematics of shergottites, however, has been questioned and are more commonly interpreted to reflect mixing and contamination with terrestrial Pb (e.g., Borg et al. 2005), or disturbance related to shock (Gaffney et al. 2011). A more complete overview about the ages of shergottites is presented by Grott et al. (2012).

Currently there are about ten known *nakhlites* available; the most prominent of the group is the meteorite Nakhla, which fell in El-Nakhla, Alexandria, Egypt, in 1911. It is an unusually large sample with an estimated weight of 10 kg and the only known fall in this group of meteorites. The members of the Nakhlite group represent clinopyroxenites with 70–80 % augite and about 10 % olivine (Treiman 2005). From the group of the *chassignites* only two samples are currently known (Beck et al. 2005), and they are dunites with cumulative texture that are made up of dominantly olivine and contain minor intercumulus pyroxene, feldspar, and oxides. Nakhlites and chassignites show similar ages, within analytical uncertainties, of $\sim$1.3 Ga (Nyquist et al. 2001). Beside the three main groups, a sample in its own group is the meteorite *ALH84001* found in Antarctica. It is a relatively coarse grained orthopyroxenite. This specimen is not only distinct from other Martian meteorites based on its mineralogy, but also due its old age, recently revised from $\sim$4.5 Ga (Nyquist et al. 1995) to 4.091 ± 0.030 Ga (Lapen et al. 2010), consistent with the age of 4.074 ± 0.099 Ga obtained by Bouvier et al. (2009).

In addition to chemical analyses of Martian meteorites, numerous major element with selected trace element analyses are available for Martian material analyzed directly on the surface during the Pathfinder, Spirit and Opportunity missions. A striking observation is that almost all Martian rocks on the surface are broadly of basaltic composition (Fig. 1) but not identical to the samples that are available in the meteorite collections (e.g. McSween et al. 2009; Grotzinger et al. 2011). All in-situ analyses show significantly higher total alkali contents than the meteorites. The samples analyzed by Pathfinder have the highest SiO_2 content (e.g. McSween et al. 2009). This may be due to the fact that they represent surface sediments derived from weathered material rather than fresh igneous rocks. Some analyses may also be compromised by dust that covers the surface of Mars (e.g. Wänke et al. 2001; McSween et al. 2009). During the later Spirit and Opportunity missions the rock surface

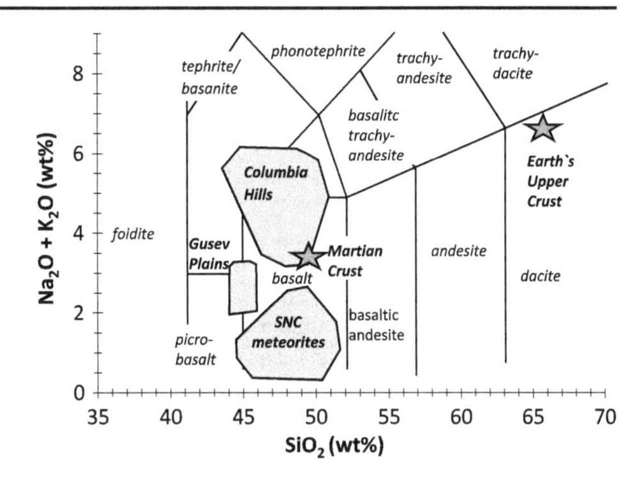

Fig. 1 Composition of Martian meteorites, soils and rocks (after Grotzinger et al. 2011)

was generally cleaned with a brush and analyses were made on the solid rock material. Thus the later analyses can be considered more representative of the primary igneous rocks.

Based on the analyses of SNC meteorites and in-situ analyses of Martian rocks and sediments by space missions different models for the bulk composition of Mars, its core and its silicate reservoir have been proposed (e.g. Dreibus and Wänke 1985; Ohtani and Kamaya 1992; Bertka and Fei 1997). A striking feature of Mars is that its bulk chemical position is higher in moderately volatile elements than that of the Earth (e.g. Dreibus and Wänke 1985). Compared to igneous rocks on Earth, basaltic rocks from Mars have about twice the FeO content at similar SiO_2 content (e.g. Halliday et al. 2001). This implies that the Martian mantle is richer in FeO than the Earth's mantle and Mars has a less massive metallic core than the Earth. Indeed, the radius of Mars is 3,400 km and the radius of its core is between 1,480 km (e.g. Kavner et al. 2001) and $1{,}794 \pm 65$ km (Sohl and Spohn 1997; Rivoldini et al. 2011). The small size of Mars results in internal pressures that are only about one-third of those in the Earth at any given depth which leads to mineral assemblages in the Martian mantle that are different from the Earth (e.g. Bertka and Fei 1997). As a consequence mantle rheology and melt compositions produced by partial melting of the mantle are also different. From these chemical and physical differences it can be expected that differentiation on Mars will have followed paths that are somewhat different from the Earth (see also Grott et al. 2012).

3 Magma Ocean and Core Formation on Mars

Since Mars has a bulk density that is similar to chondrites and its outer parts consist of much lighter silicate rocks it has to have a dense core (e.g., Johnston et al. 1974). Based on results from rotational dynamics of the planet, the size and density distribution as well as the physical state of the core can be determined (e.g. Dehant et al. 2003). Furthermore, chemical analyses of SNC meteorites and *in-situ* analyses during the rover missions on Mars demonstrate that all rocks have Fe contents that are significantly below chondritic abundances. The Martian core, therefore, consists dominantly of metallic Fe, the only abundant element that can produce a large amount of a heavy melt in a planetary body. This metallic melt most likely has dissolved minor amounts of other elements. Due to the high content of moderately volatile elements on Mars and thus its high S-content, it is expected from studies of the partition behavior of elements between silicate and metal melts that the core contains up to 16 % S (Stewart et al. 2007; Rivoldini et al. 2011). This high S-content also reduces the melting point of the metallic liquid considerably and may result in a present-day partially liquid core on Mars (e.g., Sohl and Spohn 1997; Stewart et al. 2007; Rivoldini et al. 2011). If the core is still molten, it does not generate a significant magnetic field since such a field seems absent since ca. 4 Ga (e.g., Schubert and Spohn 1990; Acuna et al. 1999; Connerney et al. 2004; Lillis et al. 2008; Lapen et al. 2010).

A core can form on a planet only if the planet is at least partially molten to allow for a segregation of a heavy metallic melt from a lighter silicate melt. Efficient core formation requires that the silicate melt dominates the system, because only then the immiscible iron melt can segregate quantitatively. The major evidence for an efficient metal melt segregation on Mars is the abundance of the highly siderophile elements (HSE), particularly Re, Os, Ir, Ru, Pt, Rh, Pd and Au, in its silicate rocks. All Martian rocks are strongly depleted in the HSE as are Earth's mantle rocks (Brandon et al. 2012). Such chemical signatures require an efficient metal-silicate separation that can only be achieved by the segregation of a metallic melt from a silicate melt. In contrast, the separation of molten metal from solid silicate by percolation is less efficient and leaves behind some stranded metal (Yoshino et al. 2004). Moreover, percolation requires dihedral angles of less than $\sim 60°$ to allow formation of an interconnected network of metal melt. However, throughout the major part of the Martian mantle dihedral angles appear to be larger than 60° (Terasaki et al. 2007), suggesting that percolation has been limited. This implies that efficient metal segregation and differentiation has been possible only with a large degree of silicate melting.

The energy required for melting and magma ocean formation may have been provided by the decay of short-lived ^{26}Al ($t_{1/2} \approx 0.73$ Ma), if accretion of Mars occurred in the first 1–2 Ma of the solar system (Nimmo and Kleine 2007; Dauphas and Pourmand 2011). Alternatively, large impacts during protracted accretion delivered the energy for widespread melting (e.g., Tonks and Melosh 1993; Elkins-Tanton et al. 2005a). The early loss of a dense atmosphere (Lammer et al. 2008) may be seen as an argument for a large impact. Whether the individual impactors have caused only local heating and the formation of smaller magma oceans (Senshu et al. 2002; Reese and Solomatov 2006) or if alternatively the magma ocean on Mars was global (Elkins-Tanton et al. 2005b) is a matter of current debate, as are the longevity and depth of the magma ocean(s).

Core formation causes almost quantitative removal of the HSE to the core and also leads to the depletion of other siderophile elements in the mantle, including the moderately siderophile W, Cr, P, Co, Ni, Mo, Ga, and V (Kong et al. 1999; Righter and Chabot 2011). The depletion of these moderately siderophile elements can be used to infer the pressure and temperature conditions of metal-silicate equilibration during core formation and, hence, the depth of the Martian magma ocean. Using this approach Righter et al. (1998) suggested a magma ocean depth of 700 to 800 km. However, a more recent study, using new data for the depletion of moderately siderophile elements in the Martian mantle, arrived at a magma ocean depth of about 1300 km (Righter and Chabot 2011). Borg and Draper (2003) used petrogenetic modeling of the lithophile trace element systematics in shergottites to infer a magma ocean depth of at least 1000 km. This depth is also consistent with the requirement that garnet be a liquidus phase. Debaille et al. (2008) used the Sm-Nd and Lu-Hf isotope systematics of depleted shergottites to suggest a magma ocean depth of at least 1350 km. Finally, Elkins-Tanton et al. (2003, 2005a, 2005b) argued on theoretical grounds that Mars could have been entirely molten due to the large amounts of heat present at the beginning of the solar system (see discussion above) and, therefore, proposed the possibility of a magma ocean reaching down all the way to the core-mantle boundary ($\sim$2000 km in Mars). Such a deep magma ocean, however, seems to be inconsistent with the observed depletions of moderately siderophile elements in the Martian mantle (Righter and Chabot 2011), unless metal-silicate equilibration occurred at different depths and models derived from trace element partitioning would thus yield an "average depth" of metal-silicate separation.

The depletion of the highly siderophile elements in the Martian mantle is not only important for constraining the extent of melting during core formation, but also provides essential insights into the temporal relationship of accretion and core formation, and the late accretion history of Mars. The HSE have an extremely high affinity for an iron-melt that is manifested in their metal-silicate partition coefficients that all exceed 10,000 and are different for the different HSEs (e.g., Borisov et al. 1994; Brandon et al. 2000; Holzheid et al. 2000; Brandon et al. 2012). Thus core formation should result in an extreme depletion of these elements and a pattern of HSEs abundances in Martian silicate rocks that are highly fractionated when normalized to chondritic abundances. This is not observed, however. As shown in Brandon et al. (2012) the HSE are depleted by a factor of slightly more than 100 relative to CI chondrites and have an elevated Pd/Pt ratio, but the relative abundances of Re, Os, Ir and Ru are close to chondritic. This element pattern is consistent with a late veneer of ca. 0.08 % (Bottke et al. 2010) of near chondritic material that was added to the Martian mantle after core formation. Thus after the end of core formation there was only minor addition of material and this implies that the time of core formation defines essentially the cessation of planetary growth (see below Sect. 4).

Another important aspect regarding core formation in Mars is that the degree of depletion of some moderately siderophile elements depends on the oxidation state of the magma. From the study of SNC meteorites it is obvious that the Martian mantle is currently more reducing than the Earth's mantle. However, the higher Fe content of the Martian mantle and the smaller metallic core require that bulk Mars is more oxidized than the bulk Earth. This is consistent with accretion models for terrestrial planets that indicate that the early phase of planet formation involved the accretion of more refractory material that is volatile depleted and reduced, followed by a later addition of more volatile rich material that is also more oxidized (e.g. Ringwood 1977; Dreibus and Wänke 1987; Albarède 2009; Nebel et al. 2011; Rubie et al. 2011). Since bulk Mars has a higher abundance of volatile elements (Dreibus and Wänke 1987) it is also more oxidized than the Earth. As a consequence the depletion of the redox-sensitive moderately siderophile elements by core formation is less severe than on Earth. This resulted in a lesser depletion of W in the Martian mantle than in the terrestrial mantle and thus silicate Mars has a lower Hf/W than silicate Earth. The Hf/W is a crucial parameter for the estimation of core formation ages using the ^{182}Hf-^{182}W chronometer, as discussed in the following section.

4 Hf-W Age of Core Formation and the Accretion History of Mars

The ^{182}Hf-^{182}W system is a powerful tool for investigating the timescales of planetary accretion and core formation (e.g., Jacobsen 2005; Kleine et al. 2009). The segregation of the Martian core fractionated the highly lithophile Hf from the moderately siderophile W. If this metal-silicate segregation happened during the life time of the short lived isotope ^{182}Hf, then the Martian mantle should have an excess in ^{182}W and analyses of SNC meteorites confirm that this signal is preserved in Martian meteorites (Lee and Halliday 1997; Kleine et al. 2004b; Foley et al. 2005). Using the W isotope composition of Martian samples to estimate the timing of core formation requires knowledge of the Hf/W and ^{182}W/^{184}W ratios of bulk Mars and the bulk silicate portion of Mars. Furthermore, the degree to which

Table 1 Parameters used for calculating core formation model ages

$(^{182}Hf/^{180}Hf)_0$	$(9.72 \pm 0.44) \times 10^{-5}$	Burkhardt et al. (2008, (2012)
$\varepsilon^{182}W$ Martian mantle	0.45 ± 0.15	Kleine et al. (2009)
$^{180}Hf/^{184}W$ Martian mantle	4.14 ± 0.53	Dauphas and Pourmand (2011)
$\varepsilon^{182}W$ bulk Mars (= carbonaceous chondrites)	-1.9 ± 0.1	Kleine et al. (2002); Schönberg et al. (2002); Yin et al. (2002)
$^{180}Hf/^{184}W$ bulk Mars (= carbonaceous chondrites)	1.23 ± 0.15	Kleine et al. (2009)

$\varepsilon^{182}W = [(^{182}W/^{184}W)_{sample}/(^{182}W/^{184}W)_{terrestrial\ standard} - 1] \times 10^4$

metal-silicate equilibrium was achieved during core formation must be evaluated (e.g., Kleine et al. 2009).

4.1 Hf-W Systematics of the Martian Mantle

Since both Hf and W are refractory it can be assumed that bulk Mars has chondritic relative abundances of Hf and W, and that its $^{182}W/^{184}W$ is similar to that found in chondrites. However, as a result of metal-silicate fractionation in the early solar nebula, the Hf/W ratio is different among the various groups of chondrites. The average $^{180}Hf/^{184}W$ ratio of carbonaceous chondrites is 1.23 ± 0.15, corresponding to a present-day $^{182}W/^{184}W$ ratio of -1.9 ± 0.1 $\varepsilon^{182}W$ (Kleine et al. 2002; Schoenberg et al. 2002; Yin et al. 2002; $\varepsilon^{182}W$ is the deviation of $^{182}W/^{184}W$ from the terrestrial standard value in parts per 10^4, see Table 1). Ordinary chondrites have more variable $^{180}Hf/^{184}W$ ratios ranging from ~0.6 to ~1.1 for H chondrites and from ~1.6 to ~2.1 for LL chondrites, and corresponding $\varepsilon^{182}W$ values between ~ -2.5 and ~ -2.0 and ~ -1.7 and ~ -1.5, respectively (Kleine et al. 2008, 2009). Calculating a Hf-W age of core formation, therefore, requires estimating which of these chondritic Hf-W compositions is most appropriate for Mars.

Estimating the Hf/W and $^{182}W/^{184}W$ ratios of the bulk Martian mantle is more complicated. Unlike the Earth and Moon, the mantle of Mars does not have a uniform $^{182}W/^{184}W$ ratio but exhibits ^{182}W heterogeneities (Lee and Halliday 1997; Kleine et al. 2004b; Foley et al. 2005). The shergottites are characterized by $\varepsilon^{182}W$ values of ~0.2 to ~0.8, whereas the nakhlites and Chassigny have $\varepsilon^{182}W$ values of around ~3 (Fig. 2). These ^{182}W variations most probably reflect early silicate differentiation processes, which led to a fractionation of highly incompatible W from less incompatible Hf (Righter and Shearer 2003). Using the W isotopic data to infer the timescales of core formation in Mars, therefore, requires distinguishing ^{182}W variations caused by silicate differentiation processes from those that reflect Hf/W fractionation due to early core formation. As a consequence of an early silicate differentiation, Martian meteorites also show variations in $^{142}Nd/^{144}Nd$ ratios. While all nakhlites are characterized by large ^{142}Nd excesses of ~0.6 ε-units, shergottites show more variable $^{142}Nd/^{144}Nd$ ranging from ~ -0.2 to $\sim +0.6$ $\varepsilon^{142}Nd$ (Debaille et al. 2007; Caro et al. 2008). The ^{142}Nd variations in shergottites and nakhlites were interpreted to reflect an early silicate differentiation in Mars between ~30 and ~100 Ma after solar system formation (see Sect. 5.1), early enough to also have caused variations in $^{182}W/^{184}W$ ratios.

Kleine et al. (2004b) and Foley et al. (2005) argued that the $^{182}W/^{184}W$ ratio of shergottites best represents the W isotopic composition of bulk silicate Mars. In spite of their variable $^{142}Nd/^{144}Nd$, shergottites exhibit a limited range in $\varepsilon^{182}W$ values from ~0.3 to

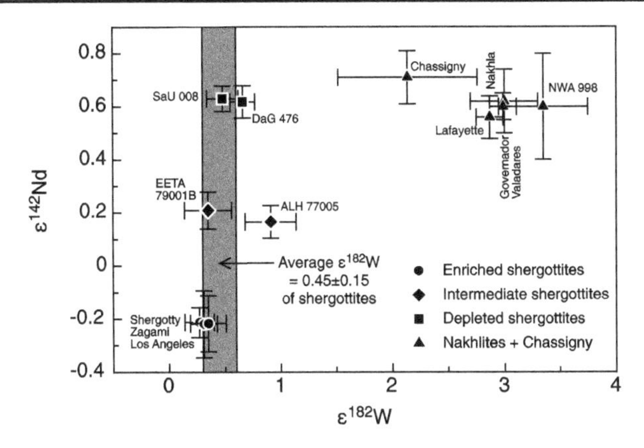

Fig. 2 ε^{182}W-ε^{142}Nd systematics of SNC meteorites (data from Lee and Halliday 1997, Kleine et al. 2004b; Foley et al. 2005; Debaille et al. 2007; Caro et al. 2008). The isotope variation implies the preservation of distinct mantle reservoirs that formed during the initial stages of chemical differentiation on Mars. $\varepsilon^{182}\mathrm{W} = [(^{182}\mathrm{W}/^{184}\mathrm{W})_{\mathrm{sample}}/(^{182}\mathrm{W}/^{184}\mathrm{W})_{\mathrm{terrestrial\ standard}} - 1] \times 10^4$, $\varepsilon^{142}\mathrm{Nd} = [(^{142}\mathrm{Nd}/^{144}\mathrm{Nd})_{\mathrm{sample}}/(^{142}\mathrm{Nd}/^{144}\mathrm{Nd})_{\mathrm{terrestrial\ standard}} - 1] \times 10^4$

$\sim$0.6 (Kleine et al. 2004b; Foley et al. 2005). Only ALH 77005 has a somewhat higher ε^{182}W value of 0.9 ± 0.3 (Foley et al. 2005). The limited range in ^{182}W/^{184}W of the shergottites indicates that the early silicate differentiation processes that led to the ^{142}Nd variations did not result in significant ^{182}W variations. Furthermore, several of the shergottites exhibit ^{142}Nd/^{144}Nd ratios close to the estimated Nd isotopic composition of bulk Mars (i.e., ε^{142}Nd between ~ -0.2 and 0 (Debaille et al. 2007; Caro et al. 2008). The average ^{182}W/^{184}W of shergottites of $\varepsilon^{182}\mathrm{W} = 0.45 \pm 0.15$ (Kleine et al. 2009), therefore, provides the current best estimate for the W isotopic composition of bulk silicate Mars.

The Hf/W ratio of the bulk Martian mantle cannot be measured directly because the Hf/W ratios of available Martian meteorites are the result of complex melting histories (Treiman et al. 1986; Kleine et al. 2004b; Nimmo and Kleine 2007). During silicate melting W is more incompatible than Hf, resulting in large Hf/W fractionations during partial melting. The Hf/W ratio characteristic for the entire Martian mantle must, therefore, be inferred by comparing the W concentrations in Martian meteorites to those of a refractory lithophile element, whose incompatibility is similar to that of W and whose abundance relative to Hf is known. Trace element studies on lunar and terrestrial basalts show that this condition is most closely met by U and Th (Palme and Rammensee 1981; Newsom et al. 1996; Arevalo and McDonough 2008). Owing to the more oxidized nature of the Martian mantle, its Hf/W ratio is much lower than that of the bulk silicate Earth. Nimmo and Kleine (2007) pointed out that the Hf/W ratio is only slightly elevated compared to chondrites and that this ratio, therefore, must be known precisely to obtain useful information on the timescale of Martian core formation. Based on an improved determination of the chondritic Hf/Th ratio, Dauphas and Pourmand (2011) used the constant Th/W ratio of Martian meteorites to estimate a Hf/W ratio of 3.51 ± 0.45 for the bulk mantle of Mars. This estimate is consistent with, but more precise than previous estimates.

4.2 W Isotope Evolution in Different Accretion and Core Formation Models

Table 1 summarizes the Hf-W parameter used to calculate model ages for core formation in Mars. In the simplest model of core formation it is assumed that the entire core segregated instantaneously from a fully formed planet. The instantaneous core formation model age for Mars is $4.2^{+3.0}_{-3.3}$Ma after CAI formation, assuming that the Hf-W systematics of bulk Mars are similar to those of carbonaceous chondrites. If a bulk composition similar to that of H chondrites is used instead, the instantaneous core formation age changes to $3.4^{+2.7}_{-3.0}$Ma after CAI formation, indistinguishable from the model age calculated using a carbonaceous chondrite composition. The particular choice of different bulk compositions for Mars, therefore, does not have a significant effect on the calculated ages. A more severe problem exists if the core did not form instantaneously but rather segregated continuously during accretion of Mars, which may be a more likely core formation scenario. Calculating realistic core formation ages, therefore, requires modeling the W isotopic evolution of the Martian mantle during accretion and concomitant core formation.

Information regarding the accretion history of Mars may be obtained from numerical simulations of terrestrial planet accretion but in these models the small size of Mars has been a persistent problem. However, two different models have been proposed recently that successfully reproduce the size of Mars. Walsh et al. (2011) modeled the small size of Mars by invoking an early migration of Jupiter that removed planetesimals from the inner solar system. In this model the accretion history of Mars was protracted and probably involved some giant impacts. In an alternative model, Minton and Levison (2011) proposed that a Mars-sized body escaped further accretion by rapid outward migration to the outer region of the inner solar system. This model predicts a rapid accretion of Mars within less than 1 Ma. Given the different accretion timescales predicted in the two models, the ^{182}Hf-^{182}W chronometer could provide the necessary constraint to distinguish between the two models.

A widely used model for using W isotopic data to constrain the timescales of planetary accretion and core formation is the exponential growth model (Halliday et al. 1996; Harper and Jacobsen 1996; Jacobsen 2005). This model assumes that accretion occurred at an exponentially decaying rate, such that:

$$M_{\text{Mars}}(t)/M_{\text{Mars}} = 1 - e^{-t/\tau}$$

where $M_{\text{Mars}}(t)/M_{\text{Mars}}$ is the cumulative fractional mass of Mars at time t, and τ is the mean time of accretion, corresponding to the time at which 63 % growth was achieved. The exponential model assumes that accretion occurred by the addition of numerous small planetesimals to the growing Mars. While such a scenario might be appropriate if Mars is a stranded planetary embryo, accretion may have also occurred by a few distinct giant impacts that delivered large masses of material at once. In this case the exponential model nevertheless provides useful age information, because these giant impacts probably also occurred at an approximately exponentially decaying rate.

In the exponential model, the end of accretion and core formation is poorly constrained and so the timescale of core formation is given by τ, the time taken to achieve 63 % growth. In addition, the time of 90 % accretion, $t_{90\%}$, is often used. Using the parameter listed in Table 1, accretion timescales of $\tau = 2.3^{+1.6}_{-1.8}$ Ma and $t_{90\%} = 5.4^{+3.6}_{-4.2}$ Ma are obtained, similar to $\tau = 1.8^{+0.9}_{-1.0}$ Ma obtained by Dauphas and Pourmand (2011) using a slightly different accretion model. Therefore, in the case of exponential accretion Mars may have reached 90 % of its mass as early as ~1.2 Ma or as late as ~9 Ma after CAI formation. The large uncertainty on this age estimate primarily reflects the low Hf/W of the Martian mantle, leading to only slightly elevated ^{182}W/^{184}W relative to chondrites.

An additional variable that has a strong influence on the estimated accretion and core formation timescales is the degree of metal-silicate equilibration during metal segregation (Halliday 2004; Kleine et al. 2004a; Kleine et al. 2009; Nimmo et al. 2010; Rudge et al. 2010; Kleine and Rudge 2011). Figure 3 shows the effect of metal-silicate disequilibrium

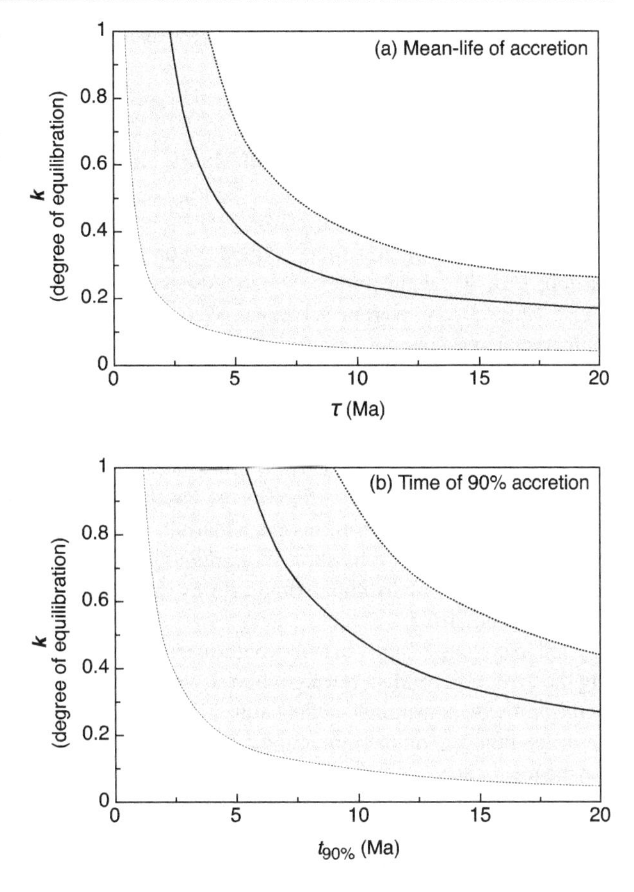

Fig. 3 Dependence of calculated Hf-W ages on the assumed degree of metal-silicate equilibration. For full metal-silicate equilibration $k = 1$. (**a**) Mean life of accretion τ vs. k. In the exponential accretion model τ corresponds to the time of 63 % accretion. (**b**) Same as (**a**) but for the time of 90 % accretion, $t_{90\ \%}$

on the calculated Hf-W ages. For instance, the time of 90 % accretion, $t_{90\%}$, changes from $\sim$ 5.4 Ma for complete equilibration to ~12.2 Ma for 40 % equilibration (corresponding to a shift in τ from ~2.3 Ma to ~5.3 Ma). Thus, precisely defining the accretion timescale for Mars requires knowledge of the degree of equilibration of newly accreted metal cores with the silicate mantle of Mars. However, while there is strong evidence, particularly from the abundances of the HSE, for efficient metal segregation and core formation on Mars (see above Sect. 3), it is less clear if mantle and core have always been in isotopic equilibrium. Full equilibrium can only be assumed if Mars accreted predominantly by collisions with small planetesimals. Collisions in which the impactor is much smaller than the target lead to vaporization of the impactor and subsequent efficient equilibration of the impactor with the mantle of proto-Mars. However, the fate of the impactor material during collisions involving larger bodies is less well understood. During giant impacts (i.e., for target/impactor ratios of ~0.1) fragments of the metal core of the impactor may have rapidly merged with the core of the target, leaving no opportunity for equilibration with the mantle of the target (Dahl and Stevenson 2010; Deguen et al. 2011). Thus, if accretion of Mars involved giant impacts, incomplete metal-silicate equilibration might be expected. We are, therefore, left with the dilemma that the Hf-W data are equally consistent with (i) a very rapid accretion of Mars as a stranded embryo (Dauphas and Pourmand 2011), in which efficient metal-silicate equilibration occurred during metal segregation, as well as with (ii) a more protracted accretion history involving several giant impacts and metal-silicate disequilibrium during core formation. The exact timing of core formation in Mars as well as its accretion history thus remains uncertain.

5 Magma Ocean Solidification and Mantle Differentiation

Silicate differentiation on terrestrial planets is thought to be related to the presence of a magma ocean. While direct evidence for the existence of magma oceans has now disappeared, it can be inferred indirectly on geochemical and theoretical grounds (see above Sect. 3). A magma ocean on Mars was not only important for facilitating efficient metal segregation during core formation, but also played a key role in establishing compositionally distinct mantle source regions that later melted to give rise to Martian basaltic meteorites. For instance, Borg and Draper (2003) showed that the trace element and isotope systematics of shergottites are most easily accounted for by mixing variable proportions of ancient reservoirs formed during crystallization of a magma ocean. The compositional variability of basaltic Martian meteorites can alternatively be explained by assimilation of evolved Martian crust (Jones 1989; Herd et al. 2002). However, this latter model falls short of explaining the isotope systematics of the shergottites, in particular their Os isotope compositions (Brandon et al. 2000; Borg and Draper 2003; Brandon et al. 2012).

Understanding magma ocean evolution and solidification, therefore, is key for constraining the evolution of Mars through time. Isotope studies are particularly useful in this regard as they provide constraints on the nature and timing of magma ocean solidification and mantle differentiation, on the mineralogical composition of the mantle, and on the dynamics of the magma ocean.

5.1 Duration of a Martian Magma Ocean

The duration of the magma ocean stage on Mars largely depends on the presence or absence of an insulating cover, like a thick solid lid or a thick atmosphere (Abe 1997; Elkins-Tanton 2008). In the absence of an insulting cover, the magma ocean would have crystallized rapidly in less than to 10^5 years (Elkins-Tanton 2008). In contrast, the presence of an insulating atmosphere may have increased substantially the lifetime of a magma ocean. For instance, a thick atmosphere rich in H_2O and H_2 could have led to surface temperatures exceeding the silicate solidus, such that the residual shallow portion of a magma ocean may have persisted up to a few hundred Ma (Abe 1997). In the case of a volatile element-rich planet like Mars, such an atmosphere may have formed due to degassing of the mantle during crystallization of the magma ocean. After it was eroded due to solar wind, hydrodynamic escape or later large impact, the planet finally crystallized (e.g., Lammer et al. 2008).

Information on the lifetime of a magma ocean on Mars can be obtained from the study of various long- and short-lived isotope systems. Evidence for an early global differentiation of Mars comes from the fact that Martian meteorites define a ~4.5 Ga Rb-Sr whole-rock isochron (e.g., Shih et al. 1982), and from Pb-Pb isotope systematics of shergottites that define an upper intercept at ~4.5 Ga on the U-Pb Concordia diagram (Chen and Wasserburg 1986). While these isotope systematics provide clear evidence for an early differentiation of the Martian mantle, they cannot constrain the duration of magma ocean solidification precisely. Much tighter age constraints can instead be obtained from short-lived radionuclide systems such as the ^{146}Sm-^{142}Nd decay system [$t_{1/2} \sim 68$ Ma; note that all ages presented in the following were recalculated using the revised decay constant of ^{146}Sm and the revised

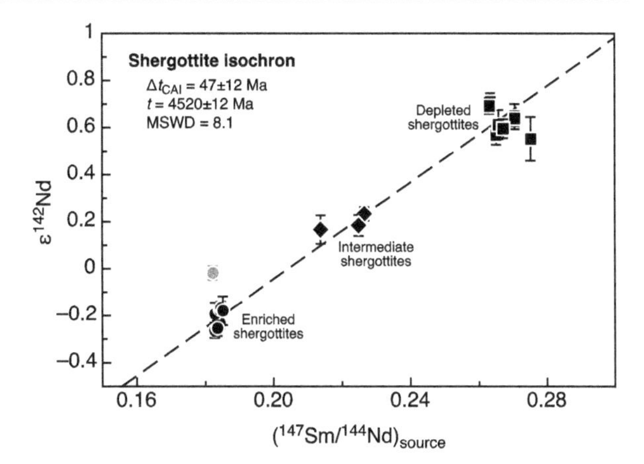

Fig. 4 ^{146}Sm-^{142}Nd isochron for shergottites. The source ^{147}Sm/^{144}Nd are calculated using the approach outlined in Caro et al. (2008). Data are from Debaille et al. (2007) and Caro et al. (2008). The age of differentiation is calculated relative to a solar system initial ^{146}Sm/^{144}Sm of 0.0094 (Kinoshita et al. 2012)

solar system initial ^{146}Sm/^{144}Sm (Kinoshita et al. 2012)]. The radionuclide ^{146}Sm was extant only during the first ~350 Ma of the solar system and so measurable ^{142}Nd variations could only have been produced during this brief period of time. This makes the ^{146}Sm-^{142}Nd system uniquely useful for investigating early silicate differentiation processes. Another strength of this system is that it is coupled to the long-lived ^{147}Sm-^{143}Nd decay system ($t_{1/2}$ ~ 106 Ga), which permits a precise determination of the early Sm/Nd fractionations that led to the ^{142}Nd variations measured today.

Martian meteorites show large variations in ^{142}Nd/^{144}Nd, ranging from ε^{142}Nd values of ~ −0.2 for enriched shergottites to ~ +0.6 for depleted shergottites and nakhlites (Fig. 2) (Borg et al. 1997; Borg et al. 2003; Debaille et al. 2007; Caro et al. 2008). The ^{142}Nd variation in the Martian mantle is thus much larger than that observed in the Earth and Moon, indicating a very early differentiation of the Martian mantle. The coupled ^{142}Nd-^{143}Nd systematics of shergottites suggest that the mantle sources of all shergottites were generated simultaneously by a major fractionation within the Martian mantle at 47 ± 12 Ma after solar system formation (Fig. 4) (Borg et al. 2003; Foley et al. 2005; Caro et al. 2008). However, the identification of small ^{142}Nd deficits in chondrites (Boyet and Carlson 2005) has complicated the interpretation of the ^{146}Sm-^{142}Nd record of Mars. Debaille et al. (2007) observed that chondrites do not plot on the ^{142}Nd-^{143}Nd correlation line defined by the shergottites and on this basis suggested that this correlation line is not an isochron but rather a mixing line (Fig. 5). This would imply that the depleted and enriched shergottites sample two different Martian reservoirs that formed at different times during solidification of a magma ocean characterized by chondritic Sm/Nd and ^{142}Nd/^{144}Nd. Debaille et al. (2007) proposed that the depleted shergottites derive from an upper mantle reservoir that crystallized ~37 Ma after solar system formation, while the enriched shergottites sample an incompatible trace element-enriched component (similar to the lunar KREEP) representing the residual liquid of the Martian magma ocean that crystallized ~90 Ma after solar system formation (Fig. 5). An alternative interpretation of the ^{142}Nd-^{143}Nd systematics of shergottites was advanced by Caro et al. (2008), who argued that bulk Mars (as well as the Earth and Moon) is characterized by super-chondritic Sm/Nd and ^{142}Nd/^{144}Nd ratios. In this case chondrites should not plot on the shergottite isochron and the ^{142}Nd-^{143}Nd systematics of shergottites could still be interpreted as a single differentiation event in the Martian mantle that occurred at 47 ± 12 Ma after solar system formation (Fig. 4) (Borg et al. 2003; Foley et al. 2005; Caro et al. 2008).

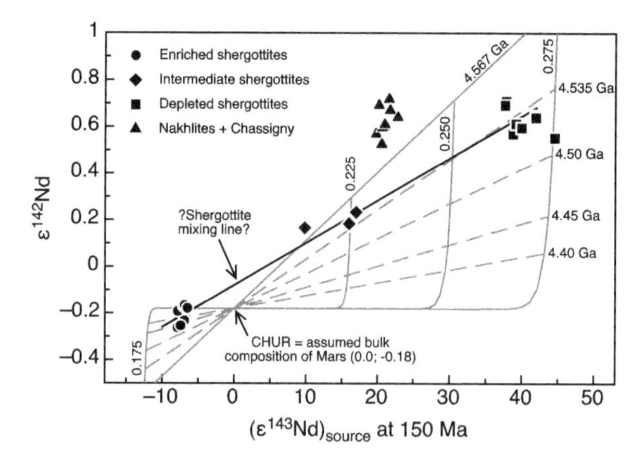

Fig. 5 Two-stage Nd isotope evolution model for the Martian meteorites. Only samples plotting inside the field defined by solid gray lines are consistent with a two-stage evolution. In this model an initially chondritic Martian magma ocean (stage 1) undergoes differentiation (stage 2). *Solid (near vertical) lines* represent constant $^{147}Sm/^{144}Nd$ in the source, *dashed lines* are calculated model isochrons. Note that this two-stage model assumes that bulk Mars has the same Nd isotope composition than ordinary chondrites, as given by the central point of the graph (0.0; −0.18). The graph would shift to a higher central point and the slope of the isochrons would change if Mars is characterized by a non-chondritic Sm/Nd (Caro et al. 2008; Caro and Bourdon 2010). Data from Debaille et al. (2007), Caro et al. (2008), Debaille et al. (2009)

The ^{142}Nd-^{143}Nd record of the nakhlites is even more difficult to interpret than that of the shergottites. The elevated $^{142}Nd/^{144}Nd$ ratio of the nakhlites corresponds to a two-stage model age of ~25 Ma after CAI formation (Harper et al. 1995; Foley et al. 2005). This age is also consistent with that obtained from combined ^{142}Nd-^{182}W systematics for Sm/Nd and Hf/W fractionation in a garnet-bearing shallow mantle (Kleine et al. 2004b; Foley et al. 2005). However, the nakhlites plot outside the two-stage evolutionary field in the ^{142}Nd vs. ^{143}Nd diagram (Fig. 5), indicating that the two-stage model ages might be meaningless and that a more complex model may be required to constrain the age of the nakhlite source (Borg et al. 2003; Debaille et al. 2009). The ^{146}Sm-^{142}Nd systematics of the nakhlites are similar to those of the depleted shergottites, whereas the ^{147}Sm-^{143}Nd systematics reveal differences. This could indicate a disturbance of the Sm-Nd system after extinction of ^{146}Sm, possibly related to the addition of a LREE-enriched fluid (Borg et al. 2003; Foley et al. 2005). Wadhwa and Crozaz (1995) argued that the REE in nakhlites evolved as a closed system, however, making post-crystallization addition of LREE-enriched fluids unlikely. Debaille et al. (2009) proposed that the ^{146}Sm-^{142}Nd, ^{147}Sm-^{143}Nd and ^{182}Hf-^{182}W systematics of the nakhlites are best explained by a multistage process during crystallization of the Martian magma ocean, involving garnet segregation during cumulate overturn soon after magma ocean solidification ~90 Ma after solar system formation (see Sect. 5.2).

The $^{146,147}Sm$-$^{142,143}Nd$ systematics of the SNC meteorites provide clear evidence for differentiation of the Martian mantle in the first few ten to hundred Ma after solar system formation, the exact timescale of magma ocean solidification remains uncertain and depends on the differentiation model and bulk Sm-Nd isotope composition assumed for Mars. Either the shergottites witness a major single-event crystallization of the magma ocean at 47 ± 12 Ma after solar system formation (Borg et al. 2003; Foley et al. 2005; Caro et al. 2008), or a protracted shallow magma ocean that finished crystallizing ~ 90 Ma after solar system formation (Debaille et al. 2007). Clearly, tighter constraints on the timescales and processes involved in the primordial differentiation of Mars will require a better knowledge of the bulk composition of Mars, and the genetic relationships among the different groups of Martian meteorites.

5.2 Magma Ocean Crystallization and Cumulate Overturn

Crystallization of a magma ocean will occur from the bottom up, because the thermal adiabat crosses the liquidus curve first at the base of the mantle (e.g., Elkins-Tanton et al. 2003). Segregation of minerals from the magma will occur only if these crystals are denser than the melt and large enough to overcome the forces generated by the convecting magma ocean. The settling velocity depends on the density difference between mineral and melt, the viscosity of the melt and the velocity of convection (Solomatov 2000). The dominant factors controlling the mineralogical composition of the crystallizing assemblages are the bulk composition of the magma ocean and the depth and temperature dependence of the stability of the minerals. On Mars the crystallization sequence of the magma ocean was roughly as follows: majorite and γ-olivine crystallized from the bottom of the magma ocean, followed by an assemblage of garnet, olivine and pyroxene, which in turn was followed by crystallization of olivine and pyroxene above the low-pressure end of garnet stability (Borg et al. 2003; Borg and Draper 2003; Elkins-Tanton et al. 2003; Elkins-Tanton et al. 2005b; Debaille et al. 2008).

Crystallization of a magma ocean ultimately generates a chemical gradient. The content in incompatible trace elements in the remaining melt will progressively increase with the percentage of solidification, because the first minerals formed at the bottom of the mantle are depleted in incompatible trace elements. Furthermore, the first minerals to crystallize from the magma ocean are Mg-rich, causing a progressive Fe enrichment in the liquid. This ultimately leads to the formation of an unstable density stratification, where early-formed Mg-rich minerals are overlying later-formed, more dense Fe-rich minerals located at the top of the mantle (Elkins-Tanton et al. 2003, 2005a, 2005b). Such a gravitationally unstable stratigraphy will overturn via Rayleigh-Taylor instabilities, where denser cumulates sink to the bottom of the mantle and deeper cumulates rise to shallower levels, resulting in a stable density stratification (Elkins-Tanton et al. 2003, 2005a, 2005b).

The isotope and trace element systematics of Martian meteorites provide important constraints on the crystallization and dynamics of the magma ocean, as well as on its mineralogical composition. For instance, the Rb-Sr and Sm-Nd isotopic systematics of Martian meteorites indicate that Martian mantle sources were generated during crystallization of a magma ocean (Borg et al. 1997, 2003; Borg and Draper 2003). More specifically, the trace element and isotopic variations observed among shergottites is successfully modeled as a mixture between a depleted mantle component, represented by mafic cumulates of the magma ocean, and a strongly enriched component, most likely represented by the late-stage residual liquid of the magma ocean (Borg et al. 2003; Borg and Draper 2003; Elkins-Tanton et al. 2003, 2005b; Debaille et al. 2008). Debaille et al. (2008) observed that equilibrium crystallization is more appropriate to generate the observed chemical compositions of the shergottites, while they acknowledged that the crystallization of a magma ocean is certainly a hybrid process, where some crystals have time to equilibrate with the magma while others sink directly to the bottom of the magma ocean without equilibration. This observation is consistent with the expected dynamics of a turbulent magma ocean, because turbulence can keep the minerals suspended long enough for equilibration to occur. However, as soon as the fraction of residual melt becomes lower than 20 %, fractional crystallization is more appropriate to model the chemical evolution of the remaining magma ocean (Snyder et al. 1992; Borg and Draper 2003). Elkins-Tanton et al. (2003, 2005a) also argued, on theoretical grounds, for some fractional crystallization of the magma ocean. At pressures exceeding ~7.5 GPa (~600 km depth in Mars), olivine and pyroxene become positively buoyant with respect to the coexisting liquid and float, while garnet is denser than the liquid and sinks to the bottom of the magma ocean. Thus, in this pressure regime, garnet is fractionated from the magma ocean and is segregated from olivine and pyroxene (Elkins-Tanton et al. 2003, 2005a).

Combined Hf and Nd isotope studies show that at some stage of their evolution garnet must have been present in the source of the shergottites (Blichert-Toft et al. 1999; Debaille et al. 2007). However, experimental work suggests that garnet is not a liquidus phase of primary shergottite melts. This would be consistent with the garnet segregation model of Elkins-Tanton et al. (2003, 2005a), because fractional crystallization of garnet would impart a strong signature on the relative REE abundances of the remaining melt, which would then later crystallize as the shergottite source region. Nevertheless, based on combined Hf and Nd isotopic data, Debaille et al. (2007) argued that the shergottite magmas formed by partial melting of a garnet-bearing source.

The formation of the source of nakhlites by magma ocean processes has proven more difficult to interpret. This is partly due to the fact that nakhlites plot outside the two-stage evolutionary field in ε^{142}Nd-ε^{143}Nd space (Fig. 5), indicating a more complex evolution (see above, Sect. 5.1). Furthermore, nakhlites show relative large ^{182}W excesses, which are difficult to interpret as a result of core formation alone, because the Hf/W ratio of the Martian mantle appears to be too low to have generated the radiogenic ε^{182}W of $\sim+3$ observed for the nakhlites. The Hf-W systematics of the nakhlites rather seem to indicate an early Hf/W fractionation in a garnet-bearing source (Righter and Shearer 2003; Kleine et al. 2004a, 2004b; Foley et al. 2005). Recently, Debaille et al. (2009) proposed a detailed model for the evolution of the nakhlite source, based on ^{176}Lu-^{176}Hf, ^{146,147}Sm-^{142,143}Nd and ^{182}Hf-^{182}W systematics. These authors argued that the combined Hf-Nd isotope systematics of the nakhlites require segregation of garnet/majoritic garnet from their source and proposed the following model evolution (Fig. 6): after Hf/W fractionation during global metal-silicate separation at t_1, the source of nakhlites differentiated from the magma ocean at some time t_2, involving crystallization of garnet/majoritic garent. This lead to high Sm/Nd, Lu/Hf and Hf/W in the nakhlite source and can account for the radiogenic ε^{142}Nd and ε^{182}W of the nakhlites. Then, at t_3, garnet/majorite was removed from the mantle source of the nakhlites. This event may have occurred when ^{182}Hf was already extinct but while ^{146}Sm was still extant, resulting in a decoupling of ε^{142}Nd and ε^{182}W. Due to its high Sm/Nd ratio, garnet segregation has a significant effect on the Sm/Nd ratio of the nakhlite source. This event, therefore, is consistent with the observation that the combined ε^{142}Nd-ε^{143}Nd systematics of the nakhlites require a more complex history than a two-stage evolution. Segregation of garnet is only possible in a highly molten source at shallow depth, where garnet is negatively buoyant while olivine and pyroxene are positive buoyant (see above). Debaille et al. (2009), therefore, suggested that the episode of re-melting required for garnet segregation that occurred at t_3 and allowed for garnet segregation is related to a major mantle cumulate overturn in the Martian mantle. During such overturn, where hot material is brought up to shallower levels, up to 60 % melting can occur by adiabatic decompression (Elkins-Tanton et al. 2003, 2005a). The generation of large volumes of melt related to the overturn may have created the primary crust of Mars, and possibly participated to the generation of the crustal dichotomy (Elkins-Tanton et al. 2005b).

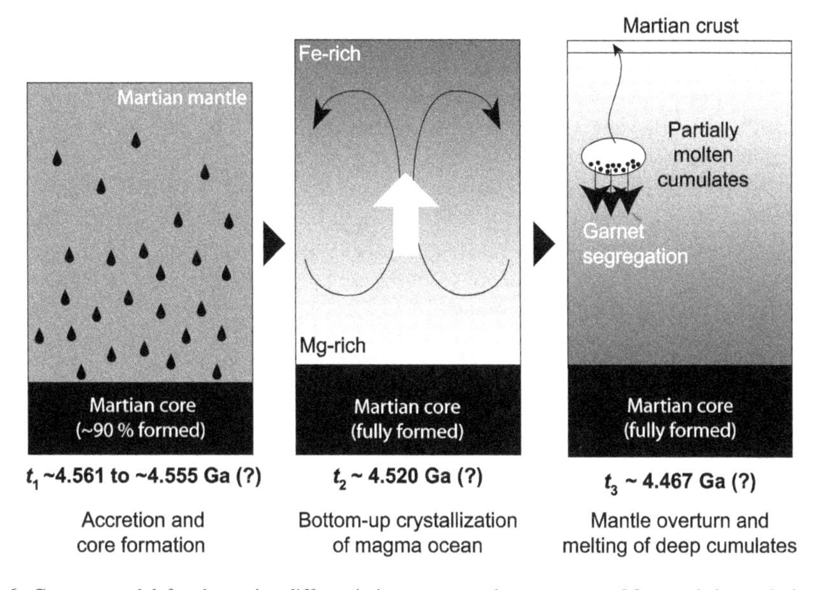

Fig. 6 Cartoon model for the major differentiation events and processes on Mars and the evolution of a Martian magma ocean (after Debaille et al. 2009). The metallic core (in black) segregated from the silicate portion of Mars at t_1. The timing of this event is not precisely known and depends on the assumed degree of metal-silicate equilibration during core formation. Here the time of 90 % accretion for complete metal-silicate equilibration is given. After core formation, the magma ocean crystallized from the bottom up (white arrow) at t_2. The timing and duration of this event can be constrained by combined ^{142}Nd-^{143}Nd systematics, but the interpretation of the Nd isotope data strongly depends on the assumed composition of bulk Mars. Here, a differentiation age of 4.520 Ga (i.e., 47 Ma after solar system formation) based on Nd isotopic for shergottites is shown. Note that the same data have been used to argue for protracted crystallization of the magma ocean between ~30 and ~100 Ma after solar system formation. Note further that the source of the nakhlites may have formed even earlier (see text for details). Crystallization of the magma ocean resulted in an unstable density stratification where dense Fe-rich cumulates are located on top of less dense, Mg-rich cumulates. This resulted in mantle overturn at t_3. Based on Nd and W isotopic data for nakhlites it has been argued that this event occurred at ~100 Ma after solar system formation (see text). During overturn upwelling of deep cumulates caused melting by adiabatic decompression, leading to segregation of garnet. Magmas generated by adiabatic decompression may have also resulted in forming the early Martian crust (Elkins-Tanton et al. 2005a)

6 Conclusions

Due to the broad chemical similarity of Earth and Mars, the two planets show great similarities in their basic building plan. Both have a metallic core, a silicate mantle and crust and an atmosphere. However, in detail there are striking differences in their evolution and current state.

Compared to Earth, Mars has a smaller core, a heterogeneous mantle inherited from differentiation processes in the first few tens of millions of years, only a basaltic crust, no felsic igneous rocks. The relatively small core is due to the higher oxidation state of the whole planet. The preservation of ancient chemical heterogeneities attests to the existence of a magma ocean and is also strong evidence against vigorous whole mantle convection and plate tectonic processes, although some convective mixing immediately following the magma ocean stage is likely. The lack of evolved felsic melts precludes hydrous melts, which are a common feature of terrestrial subduction zone related magmatism. The limitation to mafic and ultramafic volcanic and subvolcanic rocks from the near surface of Mars is more consistent with plume triggered volcanism above hot spots, which in turn is consistent with the observation that Mars is host to individual shield type volcanoes and lacks pronounced liner arrays showing magmatic activity similar to mid-oceanic ridges and subduction zones on Earth that generate the vast majority of mantle derived melts. However, some lateral crustal mobility may have occurred locally as indicated by magnetic lineaments (Connerney et al. 2005). The trace element and isotope systematic of currently available Martian samples show that the chemical heterogeneity of the Martian mantle is directly inherited from the crystallization of a magma ocean with some later, albeit limited, mixing. The existence of a magma ocean step was important, not only for the core segregation, but also for the chemical evolution of the planet and its atmosphere.

Acknowledgements This study was in part supported by the DFG (Leibniz Award) and the Helmholtz Society (Helmholtz Alliance 'Planetary Evolution and Life') to KM, by the FRS-FNRS to VD and a Förderungsprofessur des Schweizerischen Nationalfonds to TK. We are grateful to Francis Nimmo for advice regarding the accretion history of Mars. We thank L.T. Elkins-Tanton and Tilman Spohn for the constructive reviews that helped improving the manuscript.

References

Y. Abe, Thermal and chemical evolution of the terrestrial magma ocean. Phys. Earth Planet. Inter. **100**, 27–39 (1997)

M.H. Acuna, J.E.P. Connerney, N.F. Ness, R.P. Lin, D. Mitchell, C.W. Carlson, J. McFadden, K.A. Anderson, H. Reme, C. Mazelle, D. Vignes, P. Wasilewski, P. Cloutier, Global distribution of crustal magnetization discovered by the Mars Global Surveyor MAG/ER experiment. Science **284**, 790–793 (1999)

F. Albarède, Volatile accretion history of the terrestrial planets and dynamic implications. Nature **461**, 1227–1233 (2009)

J.R. Arevalo, W.F. McDonough, Tungsten geochemistry and implications for understanding the Earth's interior. Earth Planet. Sci. Lett. **272**, 656–665 (2008)

P. Beck, J.A. Barret, P. Gillet et al., The Diderot Meteorite, the second chassignite. Lunar Planet. Sci. **XXXVI**, Abstract #1326 (2005)

C. Bertka, Y. Fei, Mineralogy of the Martian interior up to core-mantle boundary pressures. J. Geophys. Res. **102**, 5251–5264 (1997)

J. Blichert-Toft, J.D. Gleason, P. Telouk, F. Albarède: The Lu–Hf isotope geochemistry of shergottites and the evolution of the Martian mantle-crust system. Earth Planet. Sci. Lett. **173**, 25–39 (1999)

D.D. Bogard, P. Johnson, Martian gases in an Antarctic meteorite. Science **221**, 651–654 (1983)

D.D. Bogard, R. Clayton, K. Marti, T. Owen, G. Turner, Martian volatiles: isotopic composition, origin and evolution. Space Sci. Rev. **96**, 105–164 (2001)

L.E. Borg, D.S. Draper, A petrogenetic model for the origin and compositional variation of the Martian basaltic meteorites. Meteorit. Planet. Sci. **38**, 1713–1731 (2003)

L.E. Borg, J.E. Edmunson, Y. Asmeron, Constraints on the U-Pb isotopic systematics of Mars inferred from a combined U-Pb, Rb-Sr, and Sm-Nd isotopic study of the Martian meteorite Zagami. Geochim. Cosmochim. Acta **69**, 5819–5830 (2005)

L.E. Borg, L.E. Nyquist, H. Wiesmann, Y. Reese, Constraints on the petrogenesis of Martian meteorites from the Rb-Sr and Sm-Nd isotopic systematics of the lherzolitic shergottites ALHA77005 and LEW88516. Geochim. Cosmochim. Acta **66**, 2037–2053 (2002)

L.E. Borg, L.E. Nyquist, L.A. Taylor, H. Wiesmann, C.-Y. Shih, Constraints on Martian differentiation processes from Rb-Sr and Sm-Nd isotopic analyses of the basaltic shergottite QUE 94201. Geochim. Cosmochim. Acta **61**, 4915–4931 (1997)

L.E. Borg, L.E. Nyquist, H. Wiesmann, C.-Y. Shih, Y. Reese, The age of Dar al Gani 476 and the differentiation history of the Martian meteorites inferred from their radiogenic isotopic systematics. Geochim. Cosmochim. Acta **67**, 3519–3536 (2003)

A. Borisov, H. Palme, B. Spettel, Solubility of palladium in siclica melts—implications for core formation on Earth. Geochim. Cosmochim. Acta **58**, 705–716 (1994)

W.F. Bottke, R.J. Walker, J.M.D. Day, D. Nesvorny, L.T. Elkins-Tanton, Stochastic Late Accretion to Earth, the Moon, and Mars. Science **330**, 1527–1530 (2010)

A. Bouvier, J. Blichert-Toft, F. Albarède, Martian meteorite chronology and the evolution of the interior of Mars. Earth Planet. Sci. Lett. **280**, 285–295 (2009)

A. Bouvier, J. Blichert-Toft, J.D. Vervoort, F. Albarède, The age of SNC meteorites and the antiquity of the Martian surface. Earth Planet. Sci. Lett. **240**, 221–233 (2005)

A. Bouvier, J. Blichert-Toft, J.D. Vervoort, P. Gillet, F. Albarède, The case for old shergottites. Earth Planet. Sci. Lett. **266**, 105–124 (2008)

M. Boyet, R.W. Carlson, ^{142}Nd evidence for early (>4.53 Ga) global differentiation of the silicate Earth. Science **309**, 576–581 (2005)

A.D. Brandon, R.J. Walker, J.W. Morgan, G.G. Goles, Re-Os isotopic evidence for early differentiation of the Martian mantle. Geochim. Cosmochim. Acta **64**, 4083–4095 (2000)

A.D. Brandon, I. Puchtel, R.J. Walker, J.M.D. Day, A.J. Irving, L.A. Taylor, Evolution of the martian mantle inferred from the ^{187}Re-^{187}Os isotope and highly siderophile element abundance systematics of shergottite meteorites. Geochim. Cosmochim. Acta **76**, 206–235 (2012)

R. Brasser, The formation of Mars: building blocks and accretion time scale. Space Sci. Rev. (2012). doi:10.1007/s11214-012-9904-2

D. Breuer, T. Spohn, Early plate tectonics versus single-plate tectonics on Mars: evidence from magnetic field history and crust evolution. J. Geophys. Res. **108** (2003)

C. Burkhardt, T. Kleine, H. Palme, B. Bourdon, J. Zipfel, J. Friedrich, D. Ebel, Hf-W mineral isochron for Ca,Al-rich inclusions: Age of the solar system and the timing of core formation in planetesimals. Geochim. Cosmochim. Acta **72**, 6177–6197 (2008)

C. Burkhardt, T. Kleine, N. Dauphas, R. Wieler, Nucleosynthetic tungsten isotope anomalies in acid leachates of the Murchison chondrite: implications for Hf-W chronometry. Astrophys. J. Lett. **753**, L6 (2012). doi:10.1088/2041-8205/753/1/L6

G. Caro, B. Bourdon, Non-chondritic Sm/Nd ratio in the terrestrial planets: consequences for the geochemical evolution of the mantle crust system. Geochim. Cosmochim. Acta **74**, 3333–3349 (2010)

G. Caro, B. Bourdon, A.N. Halliday, G. Quitté, Super-chondritic Sm/Nd ratios in Mars, the Earth and the Moon. Nature **452**, 336–339 (2008)

J.H. Chen, G.J. Wasserburg, Formation ages and evolution of Shergotty and its parent planet from U-Th-Pb systematics. Geochim. Cosmochim. Acta **50**, 955–968 (1986).

J.E.P. Connerney, M.H. Acuna, N.F. Ness, T. Spohn, G. Schubert, Mars crustal magnetism. Space Sci. Rev. **111**, 1–32 (2004)

J.E.P. Connerney, M.H. Acuna, N.F. Ness, G. Kletetschka, D.L. Mitchell, R.P. Lin, H. Reme, Tectonic implications of Mars crustal magnetism. Proc. Natl. Acad. Sci. **102**, 14970–14975 (2005)

T.W. Dahl, D.J. Stevenson, Turbulent mixing of metal and silicate during planet accretion and interpretation of the Hf-W chronometer. Earth Planet. Sci. Lett. **295**, 177–186 (2010)

N. Dauphas, A. Pourmand, Hf-W-Th evidence for rapid growth of Mars and its status as a planetary embryo. Nature **473**, 489–U227 (2011)

V. Debaille, A.D. Brandon, Q.-Z. Yin, B. Jacobsen, Coupled ^{142}Nd-^{143}Nd evidence for a protracted magma ocean in Mars. Nature **450**, 525–528 (2007)

V. Debaille, Q.-Z. Yin, A.D. Brandon, B. Jacobsen, Martian mantle mineralogy investigated by the ^{176}Lu-^{176}Hf and ^{147}Sm-^{143}Nd systematics of shergottites. Earth Planet. Sci. Lett. **269**, 186–199 (2008)

V. Debaille, A.D. Brandon, C. O'Neill, Q.Z. Yin, B. Jacobsen, Early martian mantle overturn inferred from isotopic composition of nakhlite meteorites. Nat. Geosci. **2**, 548–552 (2009)

R. Deguen, P. Olson, P. Cardin, Experiments on turbulent metal-silicate mixing in a magma ocean. Earth Planet. Sci. Lett. **310**, 303–313 (2011)

V. Dehant, T. Van Hoolst, O. de Viron, M. Greff-Lefftz, H. Legros, P. Defraigne, Can a solid inner core of Mars be detected from observations of polar motion and nutation of Mars? J. Geophys. Res. (2003). doi:10.1029/2003JE002140. Planets 108

G. Dreibus, H. Wänke, Mars, a volatile-rich planet. Meteoritics **20**, 367–381 (1985)

G. Dreibus, H. Wänke, Volatiles on Earth and Mars; a comparison. Icarus **71**, 225–240 (1987)

L.T. Elkins-Tanton, Linked magma ocean solidification and atmospheric growth for Earth and Mars. Earth Planet. Sci. Lett. **271**, 181–191 (2008)

L.T. Elkins-Tanton, E.M. Parmentier, P.C. Hess, Magma ocean fractional crystallization and cumulate overturn in terrestrial planets: Implications for Mars. Meteorit. Planet. Sci. **38**, 1753–1771 (2003)

L.T. Elkins-Tanton, P.C. Hess, E.M. Parmentier, Possible formation of ancient crust on Mars through magma ocean processes. J. Geophys. Res. **110**, E12S01 (2005a)

L.T. Elkins-Tanton, S.E. Zaranek, E.M. Parmentier, P.C. Hess, Early magnetic field and magmatic activity on Mars from magma ocean cumulate overturn. Earth Planet. Sci. Lett. **236**, 1–12 (2005b)

N.C. Foley, M. Wadhwa, L.E. Borg, P.E. Janney, R. Hines, T.L. Grove, The early differentiation history of Mars from ^{182}W-^{142}Nd isotope systematics in the SNC meteorites. Geochim. Cosmochim. Acta **69**, 4557–4571 (2005)

A.M. Gaffney, L.E. Borg, Y. Asmeron, C.K. Shearer, P.V. Burger, Disturbance of isotope systematics during experimental shock and thermal metamorphism of a lunar basalt with implications for Martian meteorite chronology. Meteorit. Planet. Sci. **46**, 35–52 (2011)

M. Grott, D. Baratoux, E. Hauber, V. Sautter, J. Mustard, O. Gasnault, S. Ruff, S.-I. Karato, V. Debaille, M. Knapmexer, F. Sohl, T. Van Hoolst, D. Breuer, M.J. Toplis, S.M. McLennan, A. Morschhauser, Long-term Evolution of the Martian Crust-Mantle System, Space Sci. Rev. (2012)

J. Grotzinger, D. Beaty, G. Dromart et al., Mars sedimentary geology: key concepts and outstanding questions. Astrobiology **11**, 77–87 (2011)

A.N. Halliday, M. Rehkämper, D.C. Lee, W. Yi, Early evolution of the Earth and Moon: new constraints from Hf-W isotope geochemistry. Earth Planet. Sci. Lett. **142**, 75–89 (1996)

A.N. Halliday, H. Wänke, J.L. Birck, R.N. Clayton, The accretion, composition and early differentiation of Mars. Space Sci. Rev. **96**, 197–230 (2001)

A.N. Halliday, Mixing, volatile loss and compositional change during impact-driven accretion of the Earth. Nature **427**, 505–509 (2004)

C.L. Harper, S.B. Jacobsen, Evidence for ^{182}Hf in the early Solar System and constraints on the timescale of terrestrial accretion and core formation. Geochim. Cosmochim. Acta **60**, 1131–1153 (1996)

C.L. Harper, L.E. Nyquist, B. Bansal, H. Wiesmann, C.-Y. Shih, Rapid accretion and early differentiation of Mars indicated by ^{142}Nd/^{144}Nd in SNC meteorites. Science **267**, 213–217 (1995)

C.D.K. Herd, L.E. Borg, J.H. Jones, J.J. Papike, Oxygen fugacity and geochemical variations in the martian basalts: Implications for martian basalt petrogenesis and the oxidation state of the upper mantle of Mars. Geochim. Cosmochim. Acta **66**, 2025–2036 (2002)

A. Holzheid, P. Sylvester, H.S.C. O'Neill, D.C. Ruble, H. Palme, Evidence for a late chondritic veneer in the Earth's mantle from high-pressure partitioning of palladium and platinum. Nature **406**, 396–399 (2000)

S.B. Jacobsen, The Hf-W isotopic system and the origin of the Earth and Moon. Annu. Rev. Earth Planet. Sci. **33**, 531–570 (2005)

J.H. Jones, Isotopic relationships among the shergottites, the nakhlites and Chassigny. Proc. Lunar Planet. Sci. Conf. **19**, 465–474 (1989)

D.H. Johnston, T.R. McGetchi, M.N. Toksöz, Thermal state and internal structure of Mars. J. Geophys. Res. **79**, 3959–3971 (1974)

A. Kavner, T.S. Duffy, G.Y. Shen, Phase stability and density of FeS at high pressures and temperatures: implications for the interior structure of Mars. Earth Planet. Sci. Lett. **185**, 25–33 (2001)

N. Kinoshita, M. Paul, Y. Kashiv, P. Collon, C.M. Deibel, B. DiGiovine, J.P. Greene, D.J. Henderson, C.L. Jiang, S.T. Marley, T. Nakanishi, R.C. Pardo, K.E. Rehm, D. Robertson, R. Scott, C. Schmitt, X.D. Tang, R. Vondrasek, A. Yokoyama, A shorter ^{146}Sm half-life measured and implications for ^{146}Sm-^{142}Nd chronology in the solar system. Science **335**, 1614–1617 (2012)

T. Kleine, J.F. Rudge, Chronometry of meteorites and the formation of the Earth and Moon. Elements **7**, 41–46 (2011)

T. Kleine, C. Munker, K. Mezger, H. Palme, Rapid accretion and early core formation on asteroids and the terrestrial planets from Hf-W chronometry. Nature **418**, 952–955 (2002)

T. Kleine, K. Mezger, H. Palme, E. Scherer, C. Münker, The W isotope evolution of the bulk silicate Earth: constraints on the timing and mechanisms of core formation and accretion. Earth Planet. Sci. Lett. **228**, 109–123 (2004a)

T. Kleine, K. Mezger, C. Münker, H. Palme, A. Bischoff, ^{182}Hf-^{182}W isotope systematics of chondrites, eucrites, and Martian meteorites: chronology of core formation and early mantle differentiation in Vesta and Mars. Geochim. Cosmochim. Acta **68**, 2935–2946 (2004b)

T. Kleine, M. Touboul, J.A. Van Orman, B. Bourdon, C. Maden, K. Mezger, A. Halliday, Hf-W thermochronometry: closure temperature and constraints on the accretion and cooling history of the H chondrite parent body. Earth Planet. Sci. Lett. **270**, 106–118 (2008)

T. Kleine, M. Touboul, B. Bourdon et al., Hf-W chronology of the accretion and early evolution of asteroids and terrestrial planets. Geochim. Cosmochim. Acta **73**, 5150–5188 (2009)

P. Kong, M. Ebihara, H. Palme, Siderophile elements in martian meteorites and implications for core formation in Mars. Geochim. Cosmochim. Acta **63**, 1865–1875 (1999)

H. Lammer, J.F. Kasting, E. Chassefiäre, R.E. Johnson, Y.N. Kulikov, F. Tian, Atmospheric escape and evolution of terrestrial planets and satellites. Space Sci. Rev. **139**, 399–436 (2008)

T.J. Lapen, M. Righter, A.D. Brandon, V. Debaille, A.D. Beard, J.T. Shafer, A.H. Peslier, A younger age for ALH84001 and its geochemical link to shergottite sources in Mars. Science **328**, 347–351 (2010)

D.-C. Lee, A.N. Halliday, Core formation on Mars and differentiated asteroids. Nature **388**, 854–857 (1997)

A. Lenardic, F. Nimmo, L. Moresi, Growth of the hemispheric dichotomy and the cessation of plate tectonics on Mars. J. Geophys. Res. (2004). doi:10.1029/2003JE002172

Q.-S. Li, W.S. Kiefer, Mantle convection and magma production on present-day Mars: Effects of temperature-dependent rheology. Geophys. Res. Lett. **34**, L16203 (2007). doi:10.1029/2007GL030544
R.J. Lillis, H.V. Frey, M. Manga, Rapid decrease in Martian crustal magnetization in the Noachian era: Implications for the dynamo and climate of early Mars. Geophys. Res. Lett. (2008). doi:10.1029/2008GL034338
F.M. McCubbin, E.H. Hauri, S.M. Elardo, K.E. Van der Kaaden, J. Wang, C.K. Shearer, Hydrous melting of the martian mantle produced both depleted and enriched shergottites. Geology (2012). doi:10.1130/G33242.1
H.Y. McSween, G.J. Taylor, M.B. Wyatt, Elemental composition of the Martian Crust. Science **324**, 736–739 (2009)
H.Y.J. McSween, What we have learned about Mars from SNC meteorites. Meteoritics **29**, 757–779 (1994)
D.A. Minton, H.F. Levison, Why is Mars small? A new terrestrial planet formation model including planetesimal-driven migration, in *Lunar Planet. Sci. Conf. XLII* (2011), Abstract # 2577
O. Nebel, K. Mezger, W. van Westerenen, Rubidium isotopes in primitive chondrites: Constraints on Earth's volatile element depletion and lead isotope evolution. Earth Planet. Sci. Lett. **305**, 309–316 (2011)
G. Neukum, R. Jaumann, H. Hoffmann, E. Hauber, J.W. Head, A.T. Basilevsky, B.A. Ivanov, S.C. Werner, S. van Gasselt, J.B. Murray, T. McCord, Recent and episodic volcanic and glacial activity on Mars revealed by the High Resolution Stereo Camera. Nature **432**, 971–979 (2004)
G. Neukum, A.T. Basilevsky, T. Kneissl et al., The geologic evolution of Mars: episodicity of resurfacing events and ages from cratering analysis of image data and correlation with radiometric ages of Martian meteorites. Earth Planet. Sci. Lett. **294**, 204–222 (2010)
H.E. Newsom, K.W.W. Sims, P. Noll, W. Jaeger, S. Maehr, T. Beserra, The depletion of W in the bulk silicate Earth: constraints on core formation. Geochim. Cosmochim. Acta **60**, 1155–1169 (1996)
F. Nimmo, T. Kleine, How rapidly did Mars accrete? Uncertainties in the Hf-W timing of core formation. Icarus **191**, 497–504 (2007)
F. Nimmo, D.P. O'Brien, T. Kleine, Tungsten isotopic evolution during late-stage accretion: constraints on Earth-Moon equilibration. Earth Planet. Sci. Lett. **292**, 363–370 (2010)
L.E. Nyquist, B.M. Bansal, H. Wiesmann, C.-Y. Shih, "Martians" Young and old: Zagami and ALH84001', in *Proc. 26th Lunar Planet. Sci. Conf.* (1995), pp. 1065–1066 (abstract)
L.E. Nyquist, D.D. Bogard, C.-Y. Shih, A. Greshake, D. Stoffler, O. Eugster, Ages and geologic histories of Martian meteorites. Space Sci. Rev. **96**, 105–164 (2001). Chronology and Evolution of Mars
C. O'Neill, A.M. Jellinek, A. Lenardic, Conditions for the onset of plate tectonics on terrestrial planets and moons. Earth Planet. Sci. Lett. **261**, 20–32 (2007)
E. Ohtani, N. Kamaya, The geochemical model of Mars: An estimation from the high pressure experiments. Geophys. Res. Lett. (1992). doi:10.1029/1092GL02369
U. Ott, Noble-gases in SNC meteorites—Shergotty, Nakhla, Chassigny. Geochim. Cosmochim. Acta **52**, 1937–1948 (1988)
U. Ott, F. Begemann, Are all the Martian meteorites from Mars? Nature **317**, 509–512 (1985)
H. Palme, W. Rammensee, The significance of W in planetary differentiation processes: evidence from new data on eucrites, in *Proc. 12th Lunar Planet. Sci. Conf.* (1981), pp. 949–964
C.C. Reese, V.S. Solomatov, Fluid dynamics of local martian magma oceans. Icarus **184**, 102–120 (2006)
K. Righter, C.K. Shearer, Magmatic fractionation of Hf and W: Constraints on the timing of core formation and differenciation in the Moon and Mars. Geochim. Cosmochim. Acta **67**, 2497–2507 (2003)
K. Righter, N.L. Chabot, Moderately and slightly siderophile element constraints on the depth and extent of melting in early Mars. Meteorit. Planet. Sci. **46**, 157–176 (2011)
K. Righter, R.L. Hervig, D.A. Kring, Accretion and core formation on Mars: Molybdenum contents of melt inclusion glasses in three SNC meteorites. Geochim. Cosmochim. Acta **62**, 2167–2177 (1998)
A.E. Ringwood, Composition of the core and implications for origin of the Earth. Geochem. J. **11**, 111 135 (1977)
A. Rivoldini, T. Van Hoolst, O. Verhoeven, A. Mocquet, V. Dehant, Geodesy constraints on the interior structure and composition of Mars. Icarus **213**, 451–472 (2011)
D.D. Rubie, D.J. Frost, U. Mann et al., Heterogeneous accretion, composition and core-mantle differentiation of the Earth. Earth Planet. Sci. Lett. **301**, 31–42 (2011)
J.F. Rudge, T. Kleine, B. Bourdon, Broad bounds on Earth's accretion and core formation constrained by geochemical models. Nat. Geosci. **3**, 439–443 (2010)
R. Schoenberg, B.S. Kamber, K.D. Collerson, O. Eugster, New W-isotope evidence for rapid terrestrial accretion and very early core formation. Geochim. Cosmochim. Acta **66**, 3151–3160 (2002)
G. Schubert, T. Spohn, Thermal history of Mars and the sulphur content of its core. J. Geophys. Res. Planets **95**, 14095–14104 (1990)
H. Senshu, K. Kuramoto, T. Matsui, Thermal evolution of a growing Mars. J. Geophys. Res. **107**, 5118 (2002)

C.-Y. Shih, L.E. Nyquist, D.D. Bogard, G.A. McKay, J.L. Wooden, B.M. Bansal, H. Wiesmann, Chronology and petrogenesis of young achondrites, Shergotty, Zagami, and ALHA 77005: late magmatism on a geologically active planet. Geochim. Cosmochim. Acta **46**, 2323–2344 (1982)

G.A. Snyder, L.A. Taylor, C.R. Neal, A chemical model for generating the sources of mare basalts: combined equilibrium and fractional crystallization of the lunar magmasphere. Geochim. Cosmochim. Acta **56**, 3809–3823 (1992)

F. Sohl, T. Spohn, The interior structure of Mars: implications from SNC meteorites. J. Geophys. Res. **102**, 1613–1635 (1997)

V.S. Solomatov, Fluid dynamics of a terrestrial magma ocean, in *Origin of the Earth and Moon*, ed. by R. Canup, K. Righter (University of Arizona Press, Tucson, 2000), pp. 323–338

A.J. Stewart, M.W. Schmidt, W. van Westrenen, C. Liebske, Mars: A new core-crystallization regime. Science **316**, 1323–1325 (2007)

H. Terasaki, D.J. Frost, D.C. Rubie, F. Langenhorst, Interconnectivity of Fe-O-S liquid in polycrystalline silicate perovskite at lower mantle conditions. Phys. Earth Planet. Inter. **161**, 170–176 (2007)

W.B. Tonks, H.J. Melosh, Magma ocean formation due to giant impacts. J. Geophys. Res. **98**, 5319–5333 (1993). Planets

A.H. Treiman, The nakhlite meteorites: Augite-rich igneous rocks from Mars. Chem. Erde **65**, 203–296 (2005)

A.H. Treiman, J.D. Gleason, D.D. Bogard, The SNC meteorites are from Mars. Planet. Space Sci. **48**, 1213–1230 (2000)

A.H. Treiman, M.J. Drake, M.-J. Janssens, R. Wolf, M. Ebihara, Core formation in the Earth and Shergottite parent body (SPB): chemical evidence from basalts. Geochim. Cosmochim. Acta **50**, 1071–1091 (1986)

H. Wänke, J. Bruckner, G. Dreibus, R. Rieder, I. Ryabchikov, Chemical composition of rocks and soils at the Pathfinder site. Space Sci. Rev. **96**, 317–330 (2001)

M. Wadhwa, G. Crozaz, Trace and minor elements in minerals of nakhlites and Chassigny: Clues to their petrogenesis. Geochim. Cosmochim. Acta **59**, 3629–3645 (1995)

K.J. Walsh, A. Morbidelli, S.N. Raymond, D.P. O'Brien, A.M. Mandell, A low mass for Mars from Jupiter's early gas-driven migration. Nature **475**, 206–209 (2011)

Q.-Z. Yin, S.B. Jacobsen, K. Yamashita, J. Blichert-Toft, P. Telouk, F. Albarède, A short timescale for terrestrial planet formation from Hf-W chronometry of meteorites. Nature **418**, 949–951 (2002)

T. Yoshino, M.J. Walter, T. Katsura, Connectivity of molten Fe alloy in peridotite based on in situ electrical conductivity measurements: implications for core formation in terrestrial planets. Earth Planet. Sci. Lett. **222**, 625–643 (2004)

Space Sci Rev (2013) 174:49–111
DOI 10.1007/s11214-012-9948-3

Long-Term Evolution of the Martian Crust-Mantle System

M. Grott · D. Baratoux · E. Hauber · V. Sautter · J. Mustard · O. Gasnault · S.W. Ruff · S.-I. Karato · V. Debaille · M. Knapmeyer · F. Sohl · T. Van Hoolst · D. Breuer · A. Morschhauser · M.J. Toplis

Received: 24 April 2012 / Accepted: 2 November 2012 / Published online: 30 November 2012

Abstract Lacking plate tectonics and crustal recycling, the long-term evolution of the crust-mantle system of Mars is driven by mantle convection, partial melting, and silicate differentiation. Volcanic landforms such as lava flows, shield volcanoes, volcanic cones, pyroclastic deposits, and dikes are observed on the martian surface, and while activity was widespread during the late Noachian and Hesperian, volcanism became more and more restricted to the Tharsis and Elysium provinces in the Amazonian period. Martian igneous rocks are predominantly basaltic in composition, and remote sensing data, in-situ data, and analysis of the SNC meteorites indicate that magma source regions were located at depths between 80

M. Grott (✉) · E. Hauber · M. Knapmeyer · F. Sohl · D. Breuer · A. Morschhauser
Institute of Planetary Research, German Aerospace Center, Rutherfordstr. 2, 12489 Berlin, Germany
e-mail: matthias.grott@dlr.de

D. Baratoux · O. Gasnault · M.J. Toplis
Institut de Recherche en Astrophysique et Planétologie, Université Toulouse III, Toulouse, France

D. Baratoux
e-mail: david.baratoux@irap.omp.eu

V. Sautter
Département Histoire de la Terre Muséum National d'Histoire Naturelle, Paris, France

J. Mustard
Department of Geological Sciences, Brown University, Providence, RI 02912, USA

S.W. Ruff
School of Earth and Space Exploration, Arizona State University, Tempe, AZ, USA

S.-I. Karato
Department of Geology and Geophysics, Yale University, New Haven, CT, USA

V. Debaille
Laboratoire G-Time, Université Libre de Bruxelles, 1050 Brussels, Belgium

T. Van Hoolst
Royal Observatory of Belgium, Brussels, Belgium

and 150 km, with degrees of partial melting ranging from 5 to 15 %. Furthermore, magma storage at depth appears to be of limited importance, and secular cooling rates of 30 to 40 $K\,Gyr^{-1}$ were derived from surface chemistry for the Hesperian and Amazonian periods. These estimates are in general agreement with numerical models of the thermo-chemical evolution of Mars, which predict source region depths of 100 to 200 km, degrees of partial melting between 5 and 20 %, and secular cooling rates of 40 to 50 $K\,Gyr^{-1}$. In addition, these model predictions largely agree with elastic lithosphere thickness estimates derived from gravity and topography data. Major unknowns related to the evolution of the crust-mantle system are the age of the shergottites, the planet's initial bulk mantle water content, and its average crustal thickness. Analysis of the SNC meteorites, estimates of the elastic lithosphere thickness, as well as the fact that tidal dissipation takes place in the martian mantle indicate that rheologically significant amounts of water of a few tens of ppm are still present in the interior. However, the exact amount is controversial and estimates range from only a few to more than 200 ppm. Owing to the uncertain formation age of the shergottites it is unclear whether these water contents correspond to the ancient or present mantle. It therefore remains to be investigated whether petrologically significant amounts of water of more than 100 ppm are or have been present in the deep interior. Although models suggest that about 50 % of the incompatible species (H_2O, K, Th, U) have been removed from the mantle, the amount of mantle differentiation remains uncertain because the average crustal thickness is merely constrained to within a factor of two.

Keywords Mars · Volcanism · Geophysics · Geochemistry

1 Introduction

Following accretion and core formation, Mars may have been initially covered by a magma ocean, which would have crystallized within 50 Myr (Elkins-Tanton et al. 2003, 2005). The gravitationally driven overturn of unstably stratified magma ocean cumulates would then result in the formation of partial melts, thereby forming the 20–30 km thick primary crust of the planet (Halliday et al. 2001; Nyquist et al. 2001; Norman 1999) and establishing the primary geochemical reservoirs of the martian interior. The formation of these reservoirs (primary crust, primordial mantle, and core) is covered elsewhere in this volume (Mezger et al. 2012, this issue), and we will focus here on the evolution of the crust-mantle system on longer time-scales. In the absence of plate tectonics and crustal recycling, reservoir differentiation should be mainly driven by partial melting of the mantle. Indeed, the martian surface shows abundant evidence for the longstanding volcanic history of the planet.

The martian surface record is well preserved, and accessible surface units cover the timespan from the early Noachian around 4.2 Gyr ago to relatively young surfaces (Tanaka 1986; Neukum et al. 2004) which show cratering model ages of less than 10 Myr. While activity was widespread and essentially global in nature during the early evolution, very young volcanic surfaces are only observed in the Tharsis and Elysium volcanic provinces (Werner 2009).

Data to constrain the details of mantle differentiation come from a variety of sources, and the task of unraveling the magmatic history of the planet is truly multidisciplinary. Topography and image data at large scales are essential to identify and map the extent of volcanic provinces, and to estimate magma production rates as a function of time. Images with a spatial resolution of up to decimeter scale (Jaumann et al. 2007; Malin et al. 2007; McEwen et al. 2007) may be used to characterize eruption parameters (e.g., presence of volatiles, lava rheology, effusion rates), and allow for the study of the geomorphological characteristics of volcanically emplaced surfaces in great detail. Mineralogical information based on spectroscopic images is available with resolutions of 100 m/pixel (Christensen et al. 2003; Bibring et al. 2006; Murchie et al. 2007), reaching 18 m/pixel in selected areas (Mustard et al. 2008), while observations in the infrared add a dataset with a resolution of 3×6 km per pixel (Bandfield et al. 2000). Furthermore, gamma ray spectroscopy (Boynton et al. 2007) constrains the elemental abundances of various elements, allowing for an estimation of the SiO_2 content of surface rocks (Boynton et al. 2007). This information is complemented by data obtained in-situ from the Mars exploration rovers, which performed chemical as well as mineralogical investigations (Squyres et al. 2004a, 2004b), as well as Mars Pathfinder (Rieder et al. 1997) and the Phoenix lander (Boynton et al. 2009).

Further information on the lithologies of martian igneous rocks comes from the Shergottite, Nahklite, Chassignite (SNC) clan of meteorites, generally accepted to come from Mars (Bogard and Johnson 1983). Shergottite samples in particular are predominantly basaltic in composition (McSween 1985; Treiman 2003; Bridges and Warren 2006), and their petrological investigations allow for an estimation of the conditions in the magma source regions during melting, including the pressure, degree of partial melting, and water content (McSween et al. 2001). This information can then be tied to numerical models, which aim at describing the thermo-chemical evolution of the planet from a global perspective. One of the main inputs to these models is the bulk silicate composition with respect to heat producing elements, which have been constrained from SNC meteorites (Wänke and Dreibus 1994) and remote sensing data (Taylor et al. 2006).

This review attempts to provide an extensive summary of our current knowledge of volcanic processes on Mars, and we will begin with a description of the volcanic surface record observed in imaging data. We will then summarize the petrological data on martian igneous rocks, which has been gathered from orbit, in-situ, and from the SNC meteorites. Subsequent chapters focus on results from geophysical and geodynamical investigations, and we will conclude by outlining future lines of investigation, which could contribute to further our understanding of the igneous evolution of Mars.

2 History of Martian Volcanism

2.1 Morphological Characteristics of Volcanism over Time

Global mapping on the basis of morphological (Tanaka et al. 1992) and spectral data (e.g., Christensen et al. 2001; Bibring et al. 2005) reveals that large parts of the martian surface are covered by volcanic products. To first order, spectral signatures of basaltic materials and minerals such as pyroxene (e.g., Poulet et al. 2007) and plagioclase (e.g., Rogers and Christensen 2007) are detected wherever dust mantling is absent (Poulet et al. 2007), suggesting an overall basaltic composition. Andesitic compositions have been proposed for the northern lowlands (Bandfield et al. 2000), but may be alternatively interpreted as weathered basalts (Wyatt and McSween 2002). This does not imply, however, that most of the present surface materials of Mars are pristine solidified lava flows or pyroclastic deposits. From a morphological perspective, volcanic landforms are not uniformly distributed across the surface of Mars. Most of them are concentrated at a few large volcanic provinces that

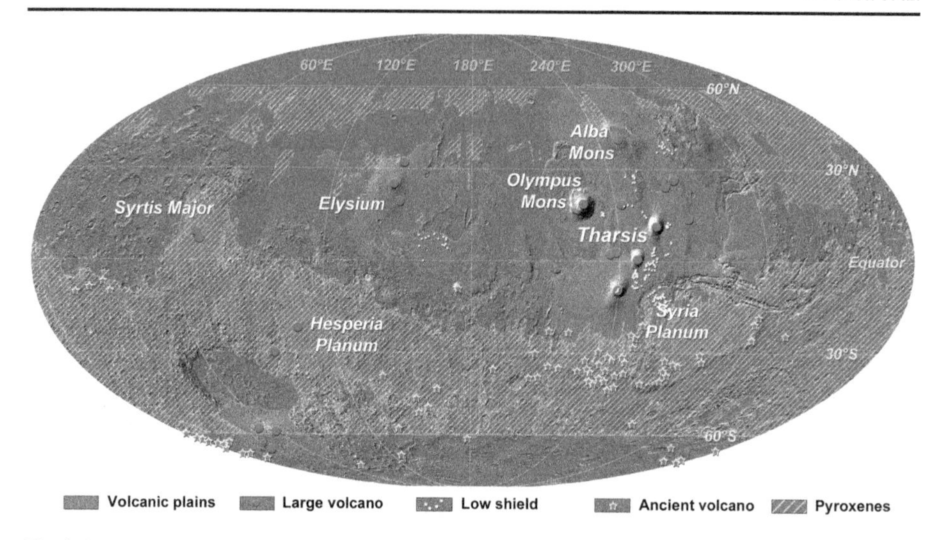

Fig. 1 Map of volcanic landforms on the basis of morphological observations. Background is a shaded digital elevation model derived from MOLA altimetry data in Mollweide projection. Low shields after Hauber et al. (2009), ancient volcanoes after Xiao et al. (2012), pyroxene-bearing surfaces after Ody et al. (2012)

are located at the Tharsis and Elysium rises (Hodges and Moore 1994), around the ancient Hellas impact basin (Williams et al. 2009), and at Syrtis Major (Fig. 1). By far the volumetrically largest of these provinces is the Tharsis bulge ($\sim 3 \times 10^8$ km^3; Phillips et al. 2001), a huge topographic rise centered near the equator which dominates almost the entire western hemisphere. Elysium is the second largest province, but its volume is much smaller than that of Tharsis ($\sim 3.5 \times 10^6$ km^3; Platz et al. 2010). Isolated volcanic edifices outside the large volcanic provinces are sparse and restricted to the cratered highlands (Xiao et al. 2012). The most prominent example is Apollinaris Mons, but there are a few dozen isolated highland massifs that are also interpreted to be of volcanic origin (Scott and Tanaka 1981). The spatial correlation between exposures of mafic minerals and morphological evidence of lava flows is not systematic, but does exist in the case of Syrtis Major and the circum-Hellas-Volcanic province (Fig. 1). However, the Elysium and Tharsis rises are covered by dust, generally precluding the detection of primary mineral phases other than those in the dust. Exposures of mafic materials also exist outside the major volcanic provinces (e.g., Valles Marineris, Noachian terranes). These mafic exposures may be associated with ancient magmatic systems whose surface expressions have been largely eroded, or they may also be attributed to reworked volcanic material that has preserved mafic signatures.

2.2 Styles of Volcanic Eruptions

Volcanic landforms can be subdivided into two broad categories: central volcanoes and volcanic plains. Central volcanoes on Mars have a size range covering at least two orders of magnitude. About 15 large shield volcanoes with diameters greater than 100 km are located in Tharsis and Elysium, the largest of which, Olympus Mons, has a basal diameter of $\sim$600 km and a height of $\sim$23 km. The morphometry of these large shields (Fig. 2(a)) is summarized by Plescia (2004). They share many morphological characteristics with large

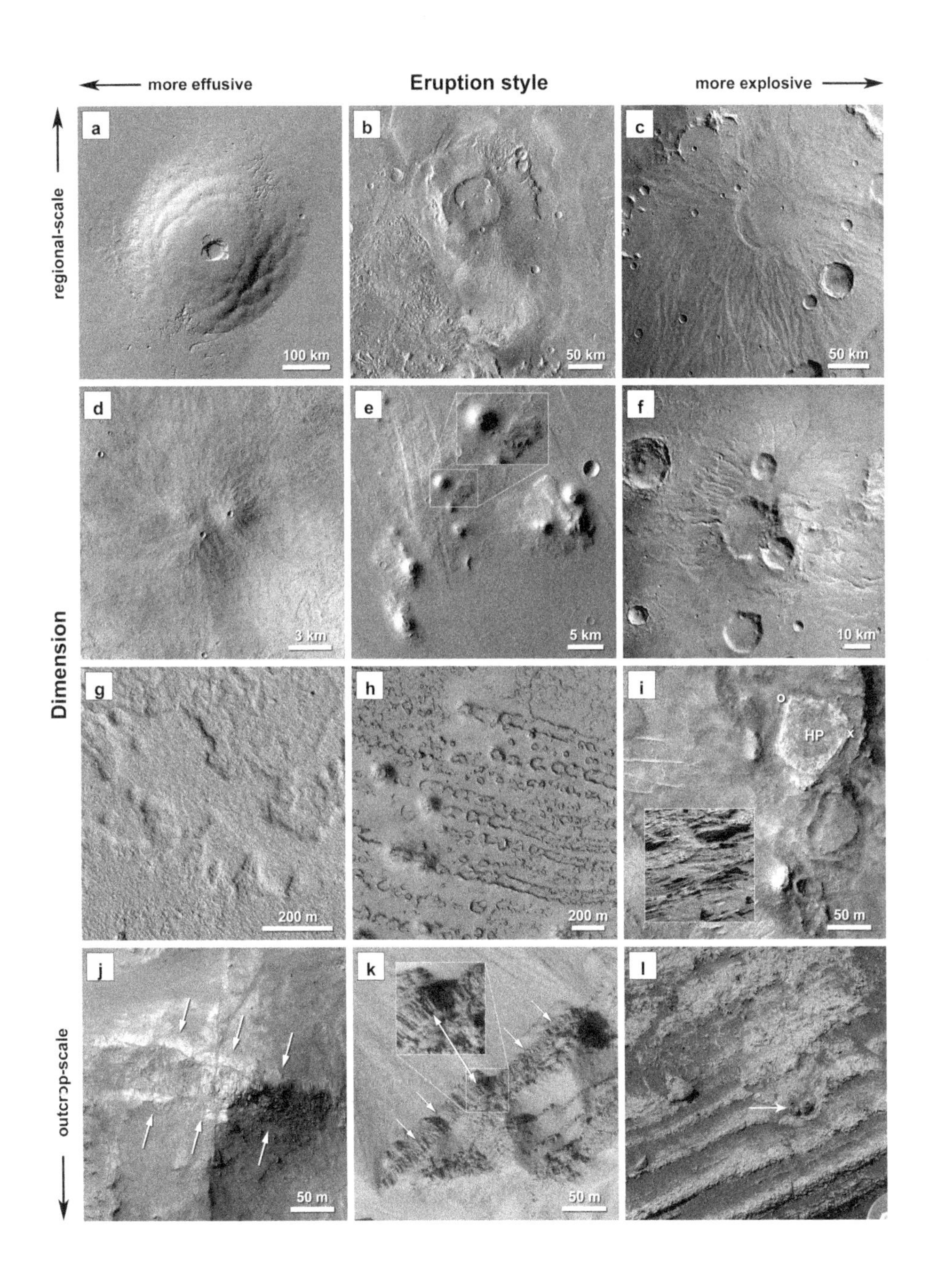
more effusive
Eruption style
more explosive
regional-scale
Dimension
outcrop-scale
a
b
c
d
e
f
g
h
i
j
k
l
100 km
50 km
50 km
3 km
5 km
10 km
200 m
200 m
50 m
HP
o
x
50 m
50 m

◀ **Fig. 2** Range of volcanic surface morphologies on Mars. The trends along the x- and y-axes represent inferred eruption style and scale of the volcanic landforms, respectively. (**a**) One of the large martian shield volcanoes, Ascraeus Mons, in the Tharsis volcanic province (center of caldera at ∼11.2°N/255.6°E). Image: Shaded version of gridded MOLA DEM. (**b**) Apollinaris Patera (8°S/174°E), a large volcano that might have produced effusive lava and pyroclastic material. Image: THEMIS-IR daylight mosaic. (**c**) Hadriaca Patera, a large shield with low relief northeast of the Hellas Basin (center of caldera at −30.18°S/92.8°E). The dissected flanks seem to be composed of easily erodible material, possibly indicating a significant contribution of pyroclastic material (e.g., volcanic ash; Williams et al. 2007). Image: HRSC mosaic of orbits h0528_0000 and h0550_0000. (**d**) Low shields with radial lava flows in Tharsis (2.4°S/252.05°E). Such edifices with very low relief (typically much less than a few hundred meters) and flank slopes ≪1° are thought to be built by lava flows with low viscosities (Hauber et al. 2009). Image: CTX P02_001906_1776. (**e**) Group of volcanic cones with associated lava flows (see *inset*) near the Ulysses Fossae, a set of extensional structures in Tharsis (center of image at 5.8°N/237.2°E). The morphology of this volcanic field is analogous to that of terrestrial fields of cinder cones (Brož and Hauber 2012). Image: HRSC h8396_0009. (**f**) Ancient, possibly volcanic construct in the plains of Thaumasia Planum (18.1°S/300.35°E). Although this and similar edifices elsewhere in the cratered highlands do not display primary volcanic features, their isolated high-standing topography, radial erosion patterns on the flanks, and some putative caldera structures suggest a volcanic origin. Image: HRSC h0449_0009. (**g**) Inflation features in lava flows in the Elysium volcanic province (7.741°N/164.410°E; illumination from the southwest, the feature in the image center has a positive topography). Such inflation features are well known on Earth and are typical for basaltic lava flows. They could be unambiguously identified on Mars in MOC and HiRISE images (Keszthelyi et al. 2008). Image: HiRISE PSP_003241_1880. (**h**) Alignment of rootless cones near southern Arcadia Planitia in the Tartarus Colles (26°N/173.7°E). Image: HiRISE PSP_006959_2060. The chain of cones is interpreted to be the result of successive steam explosions as solidified lava was moving over a fixed source of water (Keszthelyi et al. 2010). (**i**) Putative pyroclastic deposits of Home Plate (HP) in the Columbia Hills (Gusev Crater; center of Home Plate near 14.6°S/175.5°E). *Inset* shows cross-bedded layers (location marked by x in larger image), interpreted as pyroclastic deposits reworked by aeolian activity (Squyres et al. 2007). Image: HiRISE PSP_001513_1655, inset: MER Pancam 2P195076279). (**j**) Exposed dike (*arrows*) in the northern wall of Coprates Chasma (Valles Marineris, near 14.1°S/306.6°E). The dike displays the spectral characteristics of mafic material (Flahaut et al. 2011). Such dikes might have fed eruptions which constructed the km-thick accumulations of volcanic layers around the Valles Marineris region. (**k**) Columnar jointing (*arrows*) could be identified in a large, fresh, unnamed crater near Marte Vallis, centered at 21.52°N/184.35°E (Milazzo et al. 2009). The joints that separate the columns form as the lava contracts and fractures extend perpendicular to the cooling front (HiRISE image PSP_005917_2020). (**l**) Bomb sag-like structure (*arrow*, diameter ∼4 cm) in lower unit of Home Plate (location marked by o in panel (**i**)). The impact of the clast points to an explosive origin, either by volcanism or by an impact. Bomb sags are common in some terrestrial basaltic pyroclastic deposits (e.g., Walker and Croasdale 1971). Image: MER Pancam

terrestrial basaltic shields like those of Hawai'i and the Galapagos Islands, e.g., rift zones and nested caldera complexes (e.g., Crumpler et al. 1996; Mouginis-Mark et al. 2007; Carr and Greeley 1980), very low flank slopes (∼5°), and numerous large individual lava flows that can be channel-fed or tube-fed (Bleacher et al. 2007). Hundreds of much smaller low shield volcanoes, with diameters of a few km to tens of kilometers and heights of tens to hundreds of meters (Fig. 2(d)) are distributed across the wide plains between the large shields in Tharsis (Hauber et al. 2009). They share typical morphological characteristics with low basaltic shields on Earth, such as those in the Snake River Plain (Idaho, USA). To a lesser degree, some small shields are also observed in Elysium Planitiae (Vaucher et al. 2009) and on the Elysium bulge (Platz and Michael 2011). To our knowledge, no such low shields are found outside Tharsis and Elysium on Mars. Lava flows dominate both large and small shields, as well as the huge inter-shield plains in Tharsis and Elysium (Fig. 2(g)), indicating that at least the late stages of volcanism in these provinces were characterized by effusive eruptions.

Some large plains that are located peripherally to Tharsis do not display primary volcanic landforms such as lava flows, but are also thought to be emplaced by effusive volcanic eruptions, perhaps analogous to flood basalt eruptions on Earth. Examples of such plains are Lunae Planum and other plains south of Valles Marineris or around Hellas (e.g., Hesperia Planum). They were typically deformed by contraction, as indicated by sets of wrinkle ridges. The interpretation as volcanic plains (Greeley and Spudis 1981; De Hon 1982) is based on their vicinity to volcanic centers such as Tharsis and by analogy with the basaltic plains of the lunar maria, which also exhibit wrinkle ridges that were formed due to compressional stresses during basin subsidence and cooling. Other evidence for a volcanic origin of these wrinkle-ridged plains comes from inspection of the walls of Valles Marineris, which provide a deep view into the upper crust. The wrinkle-ridged plains there seem to consist of at least 10 km-thick stacks of intact layers of competent material (McEwen et al. 1999). This observation is consistent with lava flows, while such voluminous sedimentary layers are not expected at this regionally high elevation without obvious sources for the sediment.

There may be many more unidentified volcanic plains on Mars. For example, the plains inside Gusev Crater were considered to be sedimentary before the arrival of the Mars Exploration Rover, Spirit, but were subsequently found to consist of low-viscosity basaltic lava flows (Greeley et al. 2005; McSween et al. 2006). Similarly, other smooth-appearing and level crater floors might consist of lava flows. Indirect evidence for igneous processes comes from sets of long and linear grabens radiating outward from Tharsis. One interpretation holds that they are the surface expression of dike swarms analogous to, for example, the Mackenzie dike swarm in the Canadian shield (Mège and Masson 1996; Wilson and Head 2002; Ernst et al. 2001). Further evidence for this idea is provided by direct observations of dikes in the walls of Valles Marineris (Flahaut et al. 2011, Fig. 2(j)).

The evidence for explosive eruptions on Mars is much less obvious than that for effusive eruptions. Several broad edifices with very gentle slopes ($<1^\circ$) display flanks that are heavily eroded and seem to consist of friable material (Greeley and Crown 1990; Crown and Greeley 1993). Examples for such shields, e.g., Hadriaca Patera (Fig. 2(c)), are concentrated in the ancient southern highlands mainly around the Hellas impact basin (Fig. 1). They might consist of accumulations of pyroclastic material which were produced by large explosive eruptions, analogous to the Masaya Caldera Complex (Nicaragua) that produced large volumes of ash-fall deposits, ignimbrites, and surge deposits (Gregg and Williams 1996; Grott and Wieczorek 2012). Another area is the western flank of Hecates Tholus in Elysium, where pyroclastic deposits might have mantled the older substratum, creating a smooth-appearing surface (Mouginis-Mark et al. 1982; Hauber et al. 2005). Widespread layered deposits in equatorial regions, e.g., the Medusae Fossae Formation, may be the result of explosive eruptions (e.g., Hynek et al. 2003; Kerber et al. 2011), but this interpretation is controversial. The isolated volcanic highland massifs (Fig. 2(f)) might also be at least partly constructed from pyroclastic material.

Only a few smaller pyroclastic cones have been detected, and most of them are single and isolated features (e.g., Bleacher et al. 2007; Keszthelyi et al. 2008; Lanz et al. 2010). A notable exception is a field of relatively steep-sided cones with associated flows situated on an ancient, fractured window of older crust in Tharsis (Fig. 2(e)). This cone field is perhaps the morphologically closest analogue to terrestrial cinder cone fields (Brož and Hauber 2012). Evidence for explosive volcanic activity on Mars also comes from ground observations. In the Columbia Hills (Gusev Crater), a bright deposit with a roughly circular outline in plan view shows fine-scale layering and cross-bedding (Squyres et al. 2007) (Fig. 2(i)), similar to what is observed in terrestrial volcanic surge deposits formed by phreatomagmatic explosions. A feature resembling a bomb-sag (Fig. 2(l)) further contributes to the hypothesis that this deposit, termed Home Plate, might have formed by explosive volcanism. Although the vent location could not be unambiguously identified, it seems likely that Home Plate formed by hydrovolcanic activity as a maar or tuff ring (Lewis et al. 2008) when the atmosphere was much denser than today (Manga et al. 2012).

Other volcanic surface features can also be explained by the interaction of magma with near-surface water or ice (Squyres et al. 1987; Smellie and Chapman 2002). A particularly convincing case is provided by pseudocraters or rootless cones that may have formed when lava flowed over a wet substrate, causing the water to vaporize and leading to steam explosions which disrupted the overlying lava blanket creating craters without magma conduits (Greeley and Fagents 2001; Lanagan et al. 2001; Fig. 2(h))

2.3 Environmental Effects on Eruption Style

It is well known that environmental conditions have a strong influence on the style of eruptions (Zimbelman and Gregg 2000), hence magma with a given composition and volatile content will erupt differently on Mars and Earth. Of the various factors controlling the behaviour of magma ascent and eruption (Whitford-Stark 1982), gravity and atmospheric pressure are the physical parameters that account for the most significant differences of volcanic processes on Mars and Earth. The lower surface gravity on Mars (only about 38 % that of the Earth) results in a smaller lithostatic pressure gradient and a lower lithostatic pressure at a given depth. This leads to differences in buoyancy-driven processes such as the ascent of magma diapirs. Since the density contrasts between magma and country rock are considered to be similar on Mars and Earth, magma bodies should rise more slowly on Mars. To avoid cooling and solidification, therefore, magma diapirs are required to be larger than on Earth to reach shallow crustal levels (Wilson and Head 1994). Gravity also influences the depth of neutral buoyancy levels, the vertical extent of magma chambers, the width of dikes and thus effusion rates, and the thickness of lava flows (Table 1). The lower lithostatic pressure leads to volatile oversaturation of magma and the nucleation of bubbles at greater depth (Wilson and Head 1994).

The atmosphere of Mars is currently about two orders of magnitude less dense (∼6–10 mbar) than that of the Earth. This difference has consequences on volcanic eruption styles. The pressure decrease during magma ascent leads to the exsolution of volatiles, which form gas bubbles that nucleate at depth (see above) and grow and coalesce as the magma rises, until the magma will be fragmented and disrupted into a mixture of pyroclasts and gas. The lower atmospheric pressure on Mars implies that after fragmentation the gas expansion is accelerated and the ejection velocities of magma clots and gas will be larger on Mars. Since gravitational acceleration is lower and drag forces on the ejected particles are smaller in the thinner atmosphere, ballistic trajectories will be longer and particles will be dispersed more widely around explosive eruption sites. Cinder cones and hydrovolcanic constructs such as maars should therefore be lower and wider, which would make their identification harder (Wilson and Head 1994). Basaltic plinian eruptions should also be quite common on Mars, and thick and widespread deposits of ash and accretionary lapilli could have formed (Wilson and Head 2007). Extensive blankets of (layered) mantling material (e.g., Mouginis-Mark et al. 1982; Grant and Schultz 1990; Hynek et al. 2003) might represent such products of plinian eruptions. It has to be noted, however, that these theoretical predictions on different environmental effects on eruption styles on Mars and Earth are only valid for a given volatile content. It is unknown if the volatile content of Martian magmas is high enough to favour explosive eruptions. Recent

Table 1 Environmental effects on selected volcanic processes on Mars as compared to Earth (after Wilson and Head 1994; Greeley et al. 2000; Parfitt and Wilson 2008)

Parameter	Physical effect	Result
Magma ascent		
Gravity	Smaller buoyancy forces lead to smaller ascent velocities	Rising magma bodies must be larger on Mars (to avoid cooling and solidification)
	Fluid convective motions, crystal settling, and diapiric ascent rates are slower	Larger diapirs can ascend to shallower depths
	Neutral buoyancy zones and magma reservoirs are deeper	Wider dikes and larger driving pressures are needed ⟶ greater eruption rates and individual eruption volumes
	Lower stresses in magma reservoir of given vertical extent	Greater vertical extents of magma chambers before wall rock strength is exceeded
Role of volatiles		
Gravity	Lower lithostatic pressure at given depth (and lower lithostatic pressure gradient)	Volatile exsolution, bubble nucleation and magma disruption occur at greater depths
Atmospheric pressure	Larger gas expansion and more efficient energy release (see above)	Greater eruption velocities, more thorough magma fragmentation and smaller particle sizes: basaltic Plinian eruptions should be common on Mars
	Reduced aerodynamic drag forces	Long trajectories of pyroclasts, efficient cooling
		Wide dispersion of poorly consolidated fine-grained deposits
		Broader edifices with smaller heights, smaller ratios of height to basal diameter of pyroclastic edifices
		Larger ratio of crater diameter to basal diameter of pyroclastic edifices
Effusive eruptions		
Gravity	Reduced shear stress	Thicker lava flows with greater lengths (with sufficient volume)
	For a given volcanic load, the reduced gravitational acceleration causes less flexure of a lithosphere with given strength and thickness	Higher volcanic edifices possible
	Increase in dike widths by factor of two (see above)	Higher effusion rates by factor of five
Atmospheric pressure	Lower heat loss by natural and forced convection	Tendency towards longer (cooling-limited) flows

results from SNC studies indicate that their water content might be low (e.g., Filiberto and Treiman 2009; see also Sect. 3.3.5). In this respect, we note that unambiguous morphological evidence for explosive volcanism on Mars is sparse (e.g., Brož and Hauber 2012).

2.4 Chronology

Early chronological classifications of martian volcanic surfaces were provided on the basis of low-resolution images (e.g., Neukum and Hiller 1981). More recently, more standardized crater counts using images of higher spatial resolution together with improvements in cratering models (Hartman and Neukum 2001) have led to new compilations of volcanic ages (Fig. 3) (e.g., Werner 2009; Williams et al. 2009; Robbins et al. 2011; Hauber et al. 2011; Platz and Michael 2011; Vaucher et al. 2009).

In general, it appears that the oldest volcanic surfaces are roughly 4 to 3.7 Ga old. They are mainly present on the highland shields such as Hadriaca and Tyrrhena Montes, and on the heavily eroded volcanic highland massifs. Tharsis and Elysium show ancient surfaces too, but to a lesser degree. The volcanism in the highlands seems to have stopped at $\sim$1 Ga at the latest (Williams et al. 2009). More recent volcanism has been limited to Tharsis and to Elysium. The uppermost surfaces in Elysium have a very wide range of ages, spanning more than 3 billion years (Platz and Michael 2011). This diversity implies that the bulk of the Elysium bulge is very old and that the mean eruption rates in Elysium over time were low, because otherwise no old volcanic units should have survived at the surface without being buried by younger units. The youngest lavas in the Elysium province were emplaced in the Cerberus plains as recently as a few million years ago (Vaucher et al. 2009). The bulk of the Tharsis bulge is thought to be $>$3.8 Ga old, as indicated by uplifted old surfaces, the age of tectonic structures, and by ancient valley networks that seem to post-date the formation of the main topographic expression of Tharsis (Phillips et al. 2001). The voluminous volcanic material of the ridged plains is old too, with the latest (i.e. uppermost) units emplaced in the Early Hesperian. This apparent early Hesperian age of the ridged plains has been taken as evidence for a peak of volcanic activity in the late Noachian and early Hesperian, but since the bulk of the ridged plains, which can be $>$10 km thick (McEwen et al. 1999), is buried and only partly usable for age dating (Platz et al. 2010), the true age of ridged plains may be much older. On the other hand, however, the caldera floors and thus the last major phases of activity of the huge Tharsis shields are only a few hundred million years old (Neukum et al. 2004; Robbins et al. 2011). Even younger ages were measured for some lava flows on the large shields (Hartmann et al. 1999) and for many low shields in Tharsis (Hauber et al. 2011). It appears possible, therefore, that a certain level of volcanic activity persisted in Tharsis throughout most of Mars' history, perhaps up to the present day.

Overall, it appears that volcanism was intense in the early history of Mars at least during the late Noachian but that this activity gradually declined over time. This overall decline might have been punctuated by episodic periods of higher intensity (e.g., Wilson 2001; Neukum et al. 2004), as also discussed by Head et al. (2001).

2.5 Eruption Rates

The erupted volumes of volcanic materials are difficult to determine. Unless the subsurface can be accessed, for example by radar measurements, only the surface can be directly observed. In theory, radar sounding instruments can detect interfaces at depth, which might correspond to the basal surface of lava flows. For example, the Shallow Radar (SHARAD) on board the Mars Reconnaissance Orbiter mission is capable of detecting interfaces down to a depth of up to $\sim$1000 m, if there is a sufficient contrast in the dielectric constant of the adjacent layers. In reality, however, the detectability of subsurface echoes in non-icy environments is limited by several factors related to surface roughness and dielectric properties. Where lava flows are stacked without displaying a density contrast (a likely situation

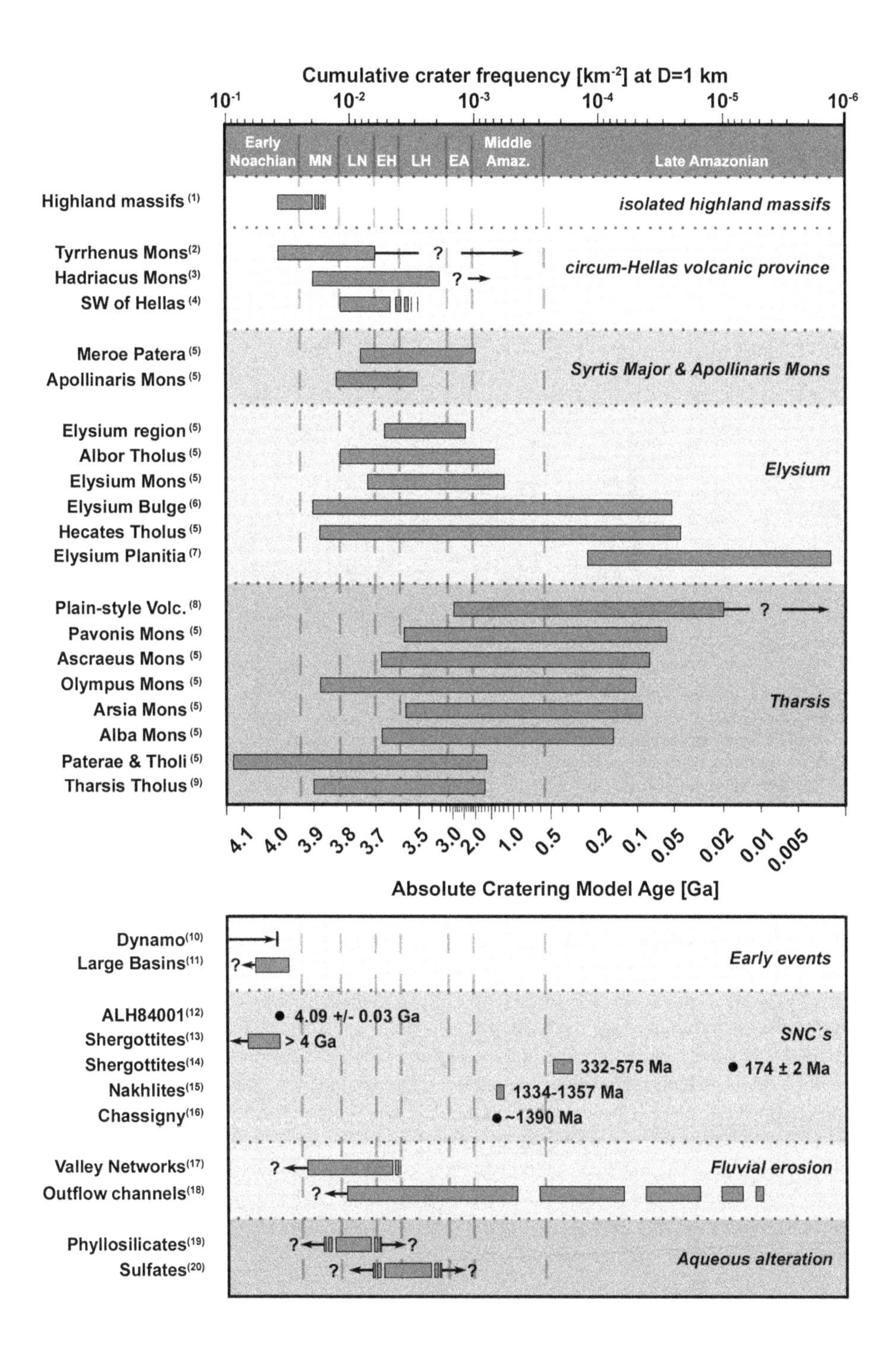
Cumulative crater frequency [km^-2] at D=1 km
10^-1
10^-2
10^-3
10^-4
10^-5
10^-6
Early Noachian
MN
LN
EH
LH
EA
Middle Amaz.
Late Amazonian
Highland massifs (1)
isolated highland massifs
Tyrrhenus Mons(2)
Hadriacus Mons(3)
SW of Hellas (4)
circum-Hellas volcanic province
Meroe Patera (5)
Apollinaris Mons (5)
Syrtis Major & Apollinaris Mons
Elysium region (5)
Albor Tholus (5)
Elysium Mons (5)
Elysium Bulge (6)
Hecates Tholus (5)
Elysium Planitia (7)
Elysium
Plain-style Volc. (8)
Pavonis Mons (5)
Ascraeus Mons (5)
Olympus Mons (5)
Arsia Mons (5)
Alba Mons (5)
Paterae & Tholi (5)
Tharsis Tholus (9)
Tharsis
4.1 4.0 3.9 3.8 3.7 3.5 3.0 2.0 1.0 0.5 0.2 0.1 0.05 0.02 0.01 0.005
Absolute Cratering Model Age [Ga]
Dynamo(10)
Large Basins(11)
Early events
ALH84001(12)
4.09 +/- 0.03 Ga
Shergottites(13)
> 4 Ga
Shergottites(14)
332-575 Ma
174 ± 2 Ma
Nakhlites(15)
1334-1357 Ma
Chassigny(16)
~1390 Ma
SNC´s
Valley Networks(17)
Outflow channels(18)
Fluvial erosion
Phyllosilicates(19)
Sulfates(20)
Aqueous alteration

◂ **Fig. 3** Schematic timeline of Martian volcanism as derived from crater statistics. It is not possible to determine the onset of volcanism for a particular area (i.e. the age of completely buried surfaces), so the start of volcanic activity is uncertain for all regions. Volcanism first ended in the southern highlands (isolated massifs, circum-Hellas province), and focused afterwards at Tharsis and Elysium, where it continued almost until present. The lengths of bars in the upper panel represent diverse ages in the respective volcanic provinces, i.e. they do not represent error bars. *The lower panel* shows the dated ages of Martian meteorites and some key events in Martian history that might have been related to volcanism. The lengths of bars of SNC meteorites represent error bars, since a given meteorite has, theoretically, a unique age. References: (1) (Xiao et al. 2012); (2) (Williams et al. 2008); (3) (Williams et al. 2007); (4) (Williams et al. 2010); (5) (Werner 2009); (6) (Platz and Michael 2011); (7) (Vaucher et al. 2009); (8) (Hauber et al. 2011); (9) (Platz et al. 2011); (10) (Lillis et al. 2008); (11) (Werner 2008); (12) (Lapen et al. 2010); (13) (Bouvier et al. 2008); (14) (Borg et al. 2005); (15) (Misawa et al. 2006); (16) (Park et al. 2009); (17) (Fassett and Head 2008); (18) (Neukum et al. 2010); (19) (Loizeau et al. 2010); (20) (Wray et al. 2009)

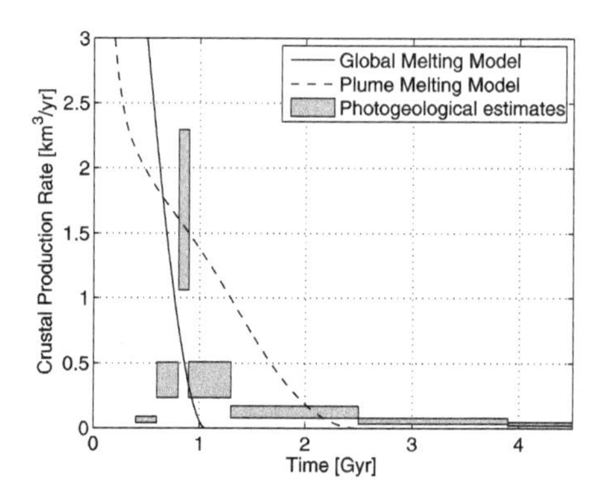

Fig. 4 Crustal production rate as a function of time as obtained from photogeological estimates based on Viking data (Greeley and Schneid 1991). The total amount of produced crust has been calculated using terrestrial ratios of extrusive to intrusive volcanism between 1:5 and 1:12 (White et al. 2006). For comparison, crustal production rates for two numerical models assuming melt generation in a global melt layer (*solid line*) and melt generation in localized plumes (*dashed line*) are also shown (Grott et al. 2011; Morschhauser et al. 2011). Total average thickness of produced crust in these models is ~50 km

in the Tharsis region), radar instruments will not be able to detect any interface, and to date SHARAD has identified only a few subsurface interfaces in volcanic settings (Simon et al. 2012). The Mars Advanced Radar for Subsurface and Ionosphere Sounding (MARSIS) on board Mars Express has a longer wavelength than SHARAD and can probe the subsurface to greater depths, but the strong attenuation of radar signals in basalt prevents the large penetration depths required to see the base of km-thick lava sequences. For large-scale analyses it is therefore necessary to infer the thicknesses and volumes of geologic units from circumstantial evidence, such as the record of ancient crater populations not fully buried by more recent deposits (De Hon 1974). Another difficulty concerns the distinction between intrusive and extrusive materials. In the absence of subsurface knowledge, their volumetric ratio cannot be determined. Geophysical methods such as the analysis of the topography and gravity fields can help to assess magmatic volumes at very large scales (e.g., Phillips et al. 2001; Grott and Wieczorek 2012), but the limited spatial resolution of geophysical data sets prevents their application on more local scales.

The most comprehensive analysis of erupted volumes to date was performed using Viking data, i.e. prior to the detailed global topographic information from MOLA. According to this study (Greeley and Schneid 1991), the total volume of extrusive and intrusive magma generated over the last ~3.8 Ga is about 654×10^6 km^3, or 0.17 km^3 per year (see Table 2 for

Table 2 Martian magma volumes (from Greeley and Schneid 1991; Exp.: Exposed; Extr.: Extruded; Thickn.: Thickness; Topogr.: Topography; LA: Late Amazonian; MA: Middle Amazonian; EA: Early Amazonian; LH:Late Hesperian; EH: Early Hesperian; LN: Late Noachian; MN: Middle Noachian; EN: Early Noachian). The magma volume extruded on plains was determined assuming that some portion of volcanic plains are covered by non-volcanic units, so the total area of volcanic plains is larger than the exposed area of volcanic plains. The total magma volume assumes a ratio of intrusive to extrusive magma of 8.5:1 (which is an average ratio for Earth, Crisp 1984)

Epoch	Area of plains [10^6 km^2]	Exp. plains volume [10^6 km^3]	Thick. of plains [km]	Extr. plains volume [10^6 km^3]	Volume from topogr. [10^6 km^3]	Total extr. volume [10^6 km^3]	Total magma volume [10^6 km^3]
LA	1.06	0.29	0.27	0.33	1.78	2.11	20
MA	3.28	1.04	0.32	1.42	7.07	8.49	81
EA	7.93	2.22	0.28	3.61	12.1	15.76	150
LH	7.63	1.87	0.24	4.54	11.0	15.63	148
EH	22.52	3.95	0.18	10.8	6.82	17.65	168
LN	9.31	1.28	0.14	4.31	3.46	7.77	74
MN	2.85	0.47	0.17	1.39	0.00	1.39	13
EN	?	?	?	?	?	?	?
Total	54.57	11.12	0.20	26.43	42.37	68.80	654

a summary). It has to be noted that the amount of intrusive igneous material was estimated in the study of Greeley and Schneid (1991) on the basis of terrestrial ratios of intrusive to extrusive volumes. Because erosion rates are low on Mars and large-scale exhumation by tectonic events is limited, however, direct observations of intrusive magmatic bodies are rare (Flahaut et al. 2011). Nevertheless, intrusives are expected to be common, since the lower buoyancy forces and the deeper levels of neutral buoyancy would seem to favor stalling of magma diapirs and dikes at depth. Recent analysis of magnetic and spectral data seem to indirectly confirm their existence (Lillis et al. 2009; Farrand et al. 2011). The rate of magma production reported by Greeley and Schneid (1991) is substantially lower than rates for the Earth (26 to 34 km^3 yr^1) or Venus (<20 km^3 yr^{-1}), but much higher than for the Moon (0.025 km^3 yr^{-1}), in qualitative agreement with what would be expected from a planet with intermediate mass. Importantly, this is an average rate over the entire history of Mars. The determination of eruption rates at specific times is complicated by the fact that the lowermost stratigraphic units of large volcanic edifices and provinces are not accessible for age determination. The relatively minor volumes of extruded material in the Noachian (Fig. 4) (Greeley and Schneid 1991) is likely an observational bias, since it is expected from thermal models that the Noachian was a period of intense magmatism (e.g., Morschhauser et al. 2011). Note, however, that it has also been suggested that there might have been a peak in magmatic activity during the Hesperian period (Bibring et al. 2006), which could have been caused by mantle heating following a relatively cold initial state after core formation. Improved methods to determine the volumes of volcanic provinces on the basis of accurate topographic data and crater populations (Vaucher et al. 2009; Platz et al. 2010) together with detailed chronologies (Platz and Michael 2011) will help to refine the rates of magmatism over time.

3 Martian Igneous Petrology

3.1 Chemistry and Mineralogy of Igneous Rocks from Orbit

3.1.1 Methods and Limitations

Several remote sensing methods are now available for characterizing the chemistry and mineralogy of igneous materials at the surface of solid planets. Here we include image-based morphometric analyses of lava flows, which provide clues to composition. Mineralogical information is generally obtained using the visible and infrared range of the electromagnetic spectrum (Bandfield et al. 2000; Poulet et al. 2009a), whereas abundances of chemical elements are principally derived from gamma-rays (Boynton et al. 2007) and neutron spectroscopy (Feldman et al. 2011). These methods offer advantages and disadvantages in terms of spatial resolution, depth of penetration, and sensitivity to the chemical and mineralogical parameters that are important for understanding magmatic processes. Deriving the chemical or mineralogical composition generally involves deconvolution algorithms to separate individual components, and some of these procedures are at different degrees of development. As a consequence, the degree of uncertainty varies between one method and another and is discussed below.

Chemistry from Nuclear Spectroscopy Elements that are both detectable and are major constituents of igneous rocks are Fe, Si, and Ca (Boynton et al. 2007; Boynton et al. 2008); also relevant are the incompatible elements K and Th (Taylor et al. 2006), and mobile elements such as Cl (Keller et al. 2006a). Three other interesting elements are detected, but their derived abundances are still preliminary: Al, S, and U (Evans et al. 2008). The specific case of hydrogen detection is discussed in the chapter by Lasue (2012, this issue). The Gamma Subsystem (GS), onboard the Mars Odyssey orbiter, measured the flux of gamma photons emitted by atomic nuclei with distinguishing energies (Evans et al. 2006). The analysis of these data provided elemental abundances (mass fractions) in the subsurface to decimeter depths (Boynton et al. 2007; Karunatillake et al. 2007). The surface is assumed to have a uniform composition over the sampling depth (several tens of centimeters). Attempts to detect layering signatures, by monitoring gamma-ray lines from a given element at different energies, did not reveal any obvious effect (Keller et al. 2006b), except for hydrogen (Maurice et al. 2011; Feldman et al. 2011) and chlorine (Diez et al. 2009). Because these observations were conducted from orbit and at high-energy (above 100 keV), the signal was not collimated and the spatial resolution is limited to about 500 km (corresponding to 50 % of the photon flux received from the surface). The data are usually reported on a cylindrical grid made of 5-deg. bins, based on cumulative spectra processed with a mean filter (Karunatillake et al. 2011). At these scales, the surface of Mars is seen as a mixture of materials of regional origin with variable amounts of the local bedrock, including mantling materials (fragments of rocks, sediments, and dust) (Newsom et al. 2007).

The intrinsic properties of the GS detection are appropriate to study the weakly altered basaltic materials that make up most of the surface of Mars. Five to eight broad homogeneous provinces can be defined, primarily driven by their chlorine and hydrogen contents, although some diversity is found in both mantled and less-mantled provinces (Gasnault et al. 2010; Taylor et al. 2010). In particular, the GS data reveal some variability in the spectrally uniform high-albedo areas (Newsom et al. 2007). The compositions detected in the volcanic regions are compatible with primary melts of the martian mantle and may be used to infer the conditions of partial melting (El Maarry et al. 2009). However, secondary surficial processes must be invoked to explain the elemental composition in other areas (Karunatillake et al. 2009), which is consistent with the extensive alteration and sedimentary processes and the associated mineralogical diversity revealed by other instruments (Grotzinger and Milliken 2011).

Visible and Near-Infrared Spectroscopy Visible and Near-Infrared (VNIR) spectroscopy takes advantage of electromagnetic radiation in the wavelength region from 0.35 to 5.0 μm. The precise wavelength where VNIR ends and the thermal IR begins depends to some extent on the temperature of the surface, but the transition occurs where the fundamental electromagnetic radiance becomes dominated by reflected radiation from the sun versus the radiance emitted from a heated surface. For cold surfaces the transition is at longer wavelengths and for warmer surfaces it is at shorter wavelengths. Processes that lead to absorptions in the VNIR range are dominated by two categories: electronic absorptions and molecular vibrations (e.g., Clark 1999).

Electronic absorptions are first and foremost dominated by crystal field absorptions in iron-bearing mafic minerals. Electronic transitions between energy levels in the d-orbitals of iron in distorted octahedral sites in minerals such as olivine and pyroxene (Burns 1993a) lead to broad and diagnostic absorptions. For example, pyroxene shows two electronic transition absorptions near 1 and 2 μm, the exact wavelengths of which vary with Fe, Mg, and Ca cation concentration (Adams 1974; Cloutis et al. 1986). The systematics of how these absorptions vary with composition have been characterized empirically, allowing discrimination of low calcium pyroxene from high calcium pyroxene, their broad iron contents, and the iron content of olivine (e.g., Sunshine et al. 1990; Sunshine and Pieters 1993; Kanner and Mustard 2007; Isaacson and Pieters 2009). Charge transfer, where electrons transition between anions and cations or oxidation states (e.g., between Fe^{2+}–O^{2-} or Fe^{2+}–Fe^{3+}) have diagnostic absorptions in the 0.2–1.1 μm region, and are particularly useful for recognizing ferric oxides (Burns 1993b; Morris et al. 1985; Bell et al. 1990).

VNIR data contain a host of narrow, diagnostic absorptions due to overtones and combination tones of fundamental molecular absorptions. The most common of these are due to water and hydroxyl in crystal structures. For example, smectite clays show a diagnostic absorption near 1.9 μm due to the combination of the fundamental OH stretch and H-O-H bend in water. The position, shape and character of this absorption depends on water abundance and the particular mineral site it is held in. In the 2.0–2.5 μm region are numerous narrow absorptions due to the combination tones of hydroxyl band modes. The molecular absorptions are diagnostic of the common clays, phyllosilicates, carbonates, and sulfates.

VNIR absorption features are maximized when the particle size of the material being measured is in the size range of 45–125 μm. With larger particles there is less penetration of light into the material and thus the path length of photons is reduced leading to weaker absorptions. Conversely, with smaller particles there is more scattering of light because of increased first-surface Fresnel reflection leading to, again, less penetration of photons into the material and weaker absorptions. In general the penetration depth of photons is several hundred microns so the detection of minerals with VNIR is very much a surface measurement.

The majority of the analyses with VNIR data have been directed to the detection of mineral phases based on the molecular vibration bands and their characteristics and this has been very successful (e.g., Poulet et al. 2005; Bibring et al. 2005; Mustard et al. 2008). There has recently been an effort to calculate the phase assemblages from VNIR data (e.g., Poulet et al. 2009a, 2009b) which has the potential of modeling mineral abundances to ±10 % with optical constants appropriate to the phases modeled. In addition, several studies have used the Modified Gaussian Model (MGM, Sunshine and Pieters 1993) to characterize the mafic mineral phases in martian VNIR data (Mustard et al. 1997; Baratoux et al. 2007; Skok et al. 2010).

There have been three VNIR imaging spectrometers on orbital platforms that have acquired data at Mars: (1) Imaging Spectrometer for Mars (ISM, Bibring et al. 1989), (2) Observatoire pour la Minéralogie, l'Eau, les Glaces, et l'Activité (OMEGA, Bibring et al. 2004) on the Mars Express spacecraft and (3) the Compact Reconnaissance Imaging Spectrometer for Mars (CRISM, Murchie et al. 2007) on the Mars Reconnaissance Orbiter. ISM was on the Phobos-II spacecraft and this instrument acquired a number of pioneering data sets for the martian surface at 22 km spatial resolution in the wavelength range 0.76–3.16 μm. OMEGA acquired data across three detectors from 352 contiguous channels covering 0.35 to 5.1 μm and the spectral sampling varied from 7 nm in the visible to 20 nm at 2.5 to 5.1 μm. Because Mars Express was in an elliptical orbit, the spatial resolution for OMEGA varied from 300 m to 4.8 km and it has acquired near global coverage at the lower resolution. CRISM's data covers the spectral range 0.36 to 3.92 μm in 522 channels. The spatial resolution varies from 16 m at the highest resolution, targeted mode, to 200 m in the multispectral survey mode.

Thermal Infrared Spectroscopy Spanning the wavelength range of ~5 to 50 μm (2000 to 200 cm^{-1}), thermal infrared (TIR) spectroscopy offers the capability to identify and quantify the abundance of all primary igneous phases, including glasses. The technique is sensitive to the fundamental vibrational modes of molecules bound in crystal lattices; in short-range ordered molecules in amorphous phases; and unbound molecules in gases and liquids. Orbital measurements rely on the passive emission of TIR radiance produced by solar insolation. Radiance spectra are converted to temperature-independent emissivity spectra that typically display a range of broad overlapping bands and narrow discrete bands that are diagnostic of composition.

Because the vibrational frequencies of molecules are a function of both the anion and cation components and their relationship to adjacent molecules in a crystal lattice, TIR spectra are very sensitive to variations in composition and crystal structure. This leads to the ability to distinguish among different varieties of major rock-forming minerals like pyroxenes, olivines, and feldspars. Members of solid-solution series can be distinguished from endmember compositions in all cases. For example, members of the olivine series can be identified with a precision of Fo ±10 (e.g., Koeppen and Hamilton 2008).

In addition to compositional information, TIR spectra present features that are a function of particle size. As the size of unbound particles (e.g., regolith) approaches TIR wavelengths (<100 μm), the effect of multiple scattering is manifested in the spectra as a loss of contrast of the primary bands known as reststrahlen features and the introduction of secondary bands known as transparency features (e.g., Salisbury and Eastes 2008). On Mars, the relatively high albedo reddish dust is an example of fine particulate material that displays particle size effects in its TIR spectrum. This dust also obscures the underlying substrate as its thickness approaches TIR wavelength, such that as little as 100 μm precludes characterization of a dust-coated surface (e.g., Johnson et al. 2002).

The abundance of component phases present in mixtures can be determined from deconvolution (unmixing) of TIR spectra. Unlike at VNIR wavelengths, TIR spectra of intimate mixtures of mineral grains in a rock or coarse particles (>100 μm) in regolith have been shown to represent a linear combination of the individual components. A linear least squares algorithm and a library of spectral endmembers can be applied to mixed spectra in order to identify components and their abundance (Ramsey and Christensen 1998). Non-linear effects arise from intimate mixtures of fine ($\ll$100 μm) and coarse particles in the field of view, which confounds deconvolution. Detection limits for component phases are variable because of the varying spectral contrast of different minerals, but are in the range of ~5 to 15 % by area fraction (e.g., Christensen et al. 2001).

Two orbital TIR instruments have been sent to Mars to investigate surface mineralogy. The Thermal Emission Spectrometer (TES) on the Mars Global Surveyor spacecraft was a hyperspectral instrument (~6–50 μm; 1670–200 cm^{-1}; ~10 cm^{-1} sampling) with a 3 × 2 array detector with a spatial resolution of ~3 × 6 km that operated from 1996 to 2006 Christensen et al. 2001). The Thermal Emission Imaging System (THEMIS) on the Mars Odyssey spacecraft is a multispectral (9 unique bands from ~7 to 15 μm) imager capable of 100 m spatial resolution (Christensen et al. 2003). THEMIS spectral data typically are used in a complementary fashion with TES spectra (e.g., Christensen et al. 2005; Michalski and Fergason 2009).

3.1.2 Morphometric Analysis of Lava Flows

The morphology of lava flows can provide clues to their rheology, which in turn provides information on lava composition and eruption conditions. Temperature, effusion rate, magma composition and volatile content together with external parameters such as the surface atmospheric conditions and the slope of the underlying terrain all combine to give a lava flow a final morphological shape. The study of this shape and of the inferred rheological properties of a lava flow provides insights into these parameters. Even for terrestrial lava flows, however, it is far from straightforward to determine their rheological properties. On Mars it is more difficult yet, since there are no observations of active lava flows, no samples of known source regions, and, due to the ubiquitous dust cover, very little compositional information of the well-preserved lava flows of Tharsis and Elysium. Therefore, the only way to assess the rheology and nature of martian lava flows is the study of their morphometry. The foundations of these techniques were laid by Jefffreys (1925), who studied the purely viscous behaviour of water flowing on an inclined surface, and Hulme (1974), who described a method to determine the yield strength of a Bingham fluid, recognizing that lava typically displays a non-Newtonian behavior. A compilation of rheological interpretations of morphometric measurements in lava flows is given in Table 3. As pristine morphologies are required to apply the above methods, results are mostly confined to the Amazonian period.

It needs to be emphasized, however, that there are many caveats associated with studies of martian lava flow rheology from morphometry. An example is the process by which a lava flow is delivered to the surface. It can be fed directly from a vent that is linked to a magma chamber, but it can also be fed by a lava lake that originates from accumulation during a Hawaiian-style eruption of magma clots ejected in a fire fountain that are still sufficiently hot (i.e. molten) upon landing. In the latter case, the initial temperature of a lava flow will be lower. Rheology depends on temperature, and care must be taken to avoid misinterpretations of such temperature differences as compositional differences (Wilson and Head 1994). Moreover, most empirical models that link morphology to rheology are based on laboratory results, not on field studies, and they apply to terrestrial lava compositions, not to the Fe-rich compositions that are found at the surface of Mars.

3.1.3 Young Volcanism

Evidence for recent volcanism (see Sect. 2.4) is now well accepted and activity is concentrated in the Tharsis and Elysium provinces (Neukum et al. 2004; Jaeger et al. 2007;

Table 3 Viscosity η and yield strength τ for lava flows in different regions on Mars. Methods: a: The effusion rate is determined from Graetz and Jeffrey equations modified by Sakimoto et al. (1997) for a rectangular flow. b: Yield strengths determined from bingham flow equations (Hulme 1974), c: Effusion rate from the Graetz equation and viscosity from Wilson and Head (1983), d: Effusion rate from the Graetz equation and viscosity from Fink, Griffiths and Fink (1978, 1992). References: [1]Baptista et al. (2008), [2]Basilevskaya and Neukum (2006), [3]Hiesinger et al. (2007), [4]Vaucher et al. (2009), [5]Warner and Gregg (2003), Sakimoto et al. (1997), [6]Sakimoto et al. (1997), [7]Zimbelman (1985)

Region	Age	Methods	Results
Syria Planum[1]	Hesperian	a	$\eta = 6.7 \times 10^5$–4.2×10^6 Pa s
		b	$\tau = 7.9 \times 10^2$–1.7×10^3 Pa
Olympus Mons[2]	Amazonian	c	$\eta = 1.4 \times 10^3$–2.8×10^7 Pa s
		b	$\tau = 0.9 \times 10^3$–3.6×10^4 Pa
Ascraeus Mons[3]	Amazonian	c	$\eta = 1.8 \times 10^4$–4.2×10^7 Pa s
		d	$\tau = 2.0 \times 10^4$–4.2×10^7 Pa
Central Elysium Planitia[4]	Amazonian	a	$\eta = 1.0$–2.5×10^5 Pa s
		b	$\tau = 100$–500 Pa
Arsia Mons[5]	Amazonian	a	$\eta = 10^4$–10^6 Pa s
		b	$\tau = 10^3$ Pa
Alba Patera[6]	Amazonian	a	$\eta = 10^2$–10^5 Pa s
Ascraeus Mons[7]	Amazonian	a	$\eta = 6.4 \times 10^5$–2.1×10^8 Pa s
		b	$\tau = 3.3 \times 10^3$–8.4×10^4 Pa

Vaucher et al. 2009; Hauber et al. 2011). Unfortunately, volcanic regions that were active during the Amazonian are often covered by dust, affecting the remote sensing analysis of the composition of the underlying igneous rocks. Morphologic studies of Martian lava flows in such regions indicate that they are characterized by low viscosities (Table 3) with rheologic properties that are most often similar to terrestrial basaltic flows, whereas exceptionally fluid flows, compatible with ultramafic compositions, have been also reported. GRS data have revealed substantial chemical differences between terrains mantled by dust, which suggests that superficial material has a local origin, rather than being homogenized at the global scale (Newsom et al. 2007). In this case, recent volcanic regions may have a chemical signature consistent with the partial melting of the mantle. This hypothesis has been tested for the case of the Tharsis region and the mantle composition proposed by Dreibus and Wänke (1985), leading to the conclusion that magma was likely produced from 3–10 % of partial melt at pressures of 1.5–2 GPa (corresponding to a depth of 125–170 km) (El Maarry et al. 2009). Based on GRS data, the chemistry of young volcanic regions is characterized by a Si-poor (42–45 wt%) composition, iron oxide being relatively variable (16–22 wt%) (Baratoux et al. 2011a).

At the resolution of CRISM and THEMIS, a few young and dust-free exposures of volcanic materials exist, such as Amazonis Planitia (Stockstill-Cahill et al. 2008), Noctis Labyrinthus, and Echus Chasma (Mangold et al. 2010). In addition, several dozens of domes in Acidalia Planitia have been interpreted as recent volcanic features or cryptodomes associated with volcanic intrusions (Farrand et al. 2005). This hypothesis is now supported by their association with mafic signatures in visible and near-infrared spectroscopy (Farrand et al. 2011). In general, the mineralogy associated with young volcanic material is dominated by a high-calcium pyroxene and plagioclase. Olivine has been detected as the dominant phase in low-albedo deposits at Amazonis Planitia, implying the occurrence of mafic to ultramafic magmas (SiO_2 ~40–52 wt%) (Stockstill-Cahill et al. 2008). However, the silica abundance is higher than 45 wt% at the scale of GRS data, suggesting that ultramafic magmas do not spatially dominate. Examination of the morphology of the olivine-rich outcrops reveals that they are often associated with sandy surfaces, or sand dunes. An apparent olivine enrichment in the sand relative to the bedrock may be explained by aeolian sorting, as observed from basaltic sands on the Earth (Baratoux et al. 2011b) and on Mars (Sullivan et al. 2008), questioning the real occurrence of ultramafic magma. Although the modal abundance of mineral species may be explained by aeolian activity, it remains that the chemical composition of these melts must be compatible with the crystallization of olivine.

3.1.4 Hesperian Volcanism

In contrast to young volcanism, volcanic material of the Hesperian era is well exposed, and is typically represented by Syrtis Major and the Circum-Hellas volcanic province. Based on GRS data, Hesperian volcanic rocks appear to be richer in silica (SiO_2 ranging from 45 wt%–48 wt%), poorer in thorium (0.3 to 0.5 wt ppm), and have iron-rich compositions around 18–20 wt% (Baratoux et al. 2011a; Gasnault et al. 2010). Variations in silica, thorium, and iron content indicate variable degrees of partial melting (Baratoux et al. 2011a), and it has been shown that these trends in composition may be associated with global cooling of the Martian mantle. Results by Baratoux et al. (2011a), who investigated 12 major volcanic provinces of various ages, are compatible with a cooling rate of 30–40 K/Gyr, corresponding to a lithosphere thickening rate of 17–25 km/Gyr and a heat flow decrease from 44 to 32 mW m^{-2} from the Hesperian to the Amazonian period. Inferred values for the mantle temperature and lithospheric thickness are consistent with estimates from experimental studies of Adirondack-class basalt compositions measured in-situ at Gusev crater using the multiple-saturation strategy (Monders et al. 2007).

Spectroscopic observations systematically report the occurrence of two pyroxenes over Hesperian terrains (Bishop et al. 2009; Salvatore et al. 2010; Baratoux et al. 2007; Williams et al. 2010; Mustard et al. 2009) in agreement with the occurrence of clinopyroxene augite and pigeonite in the basaltic shergottites (McSween 1994). Plagioclase is present, as indicated by TES observations (Bandfield 2002) and accounts for the neutral component involved in the deconvolution of visible and near-infrared spectra. Olivine is locally present, but never dominates the mineral assemblage (Koeppen and Hamilton 2008). The pyroxenes are dominated by the calcium-rich end member, with a typical low-calcium pyroxene (LCP) to high-calcium pyroxene (HCP) ratio of $LCP/(HCP + LCP) = 0.2–0.3$ (Poulet et al. 2009a, 2009b). The composition of olivine is difficult to assess from visible and near-infrared spectroscopy as the spectrum of forsterite with large grains and that of fayalite with smaller grains can be similar (Mustard et al. 2005). Nevertheless, Poulet et al. (2009a) have defined spectral indices for both end-members and found that Hesperian pyroxene-rich regions are associated with Mg-rich olivine, consistent with TES observations (Koeppen and Hamilton 2008).

Syrtis Major is a broad 1100 km wide Hesperian-aged volcanic complex with an estimated flow thickness of 0.5–1 km (Schaber et al. 1981; Hiesinger and Head 2004). It has long been a target for studying the volcanic composition of Mars because of its large size and the relative lack of dust cover compared to other regions (e.g., Singer et al. 1982; Mustard et al. 1993). Using the near-infrared Imaging Spectrometer for Mars (ISM), Bibring et al. (1989) documented the presence of both low-calcium and high-calcium pyroxenes (Mustard and Sunshine 1995). Data from the Thermal Emission Spectrometer (TES)

(Christensen and Moore 1992) instrument yielded a regional modal mineralogy of 31 % plagioclase, 29 % HCP, 12 % high-silica phases, 7 % olivine, 4 % LCP, and 17 % other minerals for the low albedo regions of Syrtis Major (Rogers and Christensen 2007). Data from the near-infrared OMEGA instrument are consistent with these results. Poulet et al. (2009c) applied a radiative transfer model based on the Skuratov formulation and derived a modal mineralogy of 34 % HCP, 9 % LCP, 48 % plagioclase, <5 % olivine, and 8 % other. The results are broadly similar for Fe-bearing mafic minerals but differ for the other constituents due to varying sensitivities of the corresponding wavelength region.

Evidence for magmatic differentiation is found in the Nili Patera caldera of Syrtis Major. TES and THEMIS measurements of a small lava flow from a volcanic construct on the caldera floor indicate a dacite composition, which is significantly enriched in SiO_2 relative to the basaltic composition that dominates Syrtis Major (Christensen et al. 2005). The appearance of an eruptive sequence that evolved in composition from basaltic to dacitic suggests that fractional crystallization occurred in the magma chamber associated with the Hesperian Syrtis Major volcano. Although this appears to be an isolated example, it nevertheless makes the case that magmatic differentiation has occurred on Mars.

More extreme magmatic differentiation is suggested by TES and THEMIS observations of what appears to be quartz-rich materials found in craters in the Syrtis Major region (Bandfield et al. 2004; Christensen et al. 2005; Bandfield 2006). Although these materials spectrally resemble granitoid rocks, the possibility remains that secondary processes produced the identified quartz. On the other hand, the association with central peaks in at least two cases suggests excavation from depth and the possibility of a highly differentiated intrusive body.

3.1.5 The Noachian Crust

The originally igneous nature of the Noachian crust has been significantly modified and reworked by impact, erosion, and tectonic processes as well as chemical alteration forming phyllosilicate minerals (Bibring et al. 2006). It is therefore challenging to find unaltered remnants of the early crust. However, at the scale of orbiting spectrometers, it is possible to find significant sections of well-exposed Noachian crust, such as the surroundings of the Isidis basin, characterized by a strong signature of mafic silicates (Hoefen et al. 2003; Hamilton and Christensen 2005; Mustard et al. 2005; Tornabene et al. 2008) associated with phyllosilicates (Bibring et al. 2006; Poulet et al. 2005; Mangold et al. 2007; Mustard et al. 2008; Ehlmann et al. 2009). Concerning the mafic signatures, analysis of spectroscopic data has revealed a clear increase of the LCP/(HCP + LCP) ratio relative to younger volcanic terrains with values up to 0.5 (Poulet et al. 2009a, 2009b; Flahaut et al. 2011). Plagioclase is also present, and olivine-rich outcrops have been reported (Flahaut et al. 2011).

3.2 Chemistry and Mineralogy from in-situ Analysis of Igneous Rocks

3.2.1 Viking Landers

The two Viking landers, which operated from 1976 to 1980 (Viking Lander 2) and to 1982 (Viking Lander 1) in Utopia Planitia and Chryse Planitia, were the first spacecraft to measure the martian regolith in situ. Each was equipped with an X-ray fluorescence spectrometer that provided the first geochemical lander data from the surface of Mars (e.g., Toulmin et al. 1977). However, the capability of the arm-mounted scoop limited sampling to loose regolith rather than the local rocks. Typically described as martian fines or soil, this material is dominated by mafic silicates indurated to varying degree by sulfate-rich cement. Direct measurements of igneous rocks in situ would have to await subsequent rovers.

3.2.2 Mars Pathfinder

The Mars Pathfinder mission, which arrived in Ares Vallis in 1997 and operated for nearly 3 Earth months, combined an immobile lander with a small (<1 m long) mobile rover capable of driving a few 10s of meters. The Sojourner rover was equipped with an alpha proton X-ray spectrometer (APXS) that could be deployed directly onto rock surfaces (e.g., Rieder et al. 1997). Five different rocks were measured with the APXS. Although the measured compositions have significant differences, this is attributed to varying amounts of soil coating the rocks. If these rocks are of igneous origin, as is generally accepted, then their relatively high measured sulfur content (up to 2 wt%) can be attributed to a S-rich soil coating. A soil-free rock composition was determined using linear regressions of rock and soil compositions and the assumption of 0.3 wt% sulfur as the end point (e.g., Foley et al. 2003; Brückner et al. 2003). The result is a rock composition much more felsic (57 % SiO_2) than the mafic soils measured at the Viking and Pathfinder sites. This appears to be consistent with the more andesitic composition of materials concentrated in Acidalia Planitia first recognized from orbit with TES data (Bandfield et al. 2000). However, the possibility that surface alteration on the rocks is responsible for the elevated values of SiO_2 remains a viable alternative (McSween et al. 2009; Horgan and Bell 2012).

3.2.3 Mars Exploration Rovers

The two Mars Exploration Rovers Spirit and Opportunity landed on Mars in 2004, Spirit in Gusev crater, and Opportunity in Meridiani Planum. Both sites were chosen because of evidence in orbital data for the presence of liquid water at the surface during some time in martian geologic history (Squyres et al. 2003). Spirit ceased operating in 2010, while at the time of writing Opportunity continues to operate, although with diminished capabilities. Although neither landing site was chosen to address igneous geology, both have yielded chemical and mineralogical information on martian igneous rocks, most notably in Gusev crater.

Spirit encountered olivine-rich basaltic rocks, subsequently named Adirondack class, which are interpreted to be part of impact-disrupted lava flows on the plains of Gusev crater (Squyres et al. 2004a) and dated by morphological data and crater counts to be 3.65 Ga old (Greeley et al. 2005). These represent the largest occurrence of a single rock type observed by Spirit, but additional igneous rocks were encountered throughout the traverse of the adjacent Columbia Hills, which are embayed by the flows of Adirondack-class basalt. Of the hundreds of rocks observed by the Miniature Thermal Emission Spectrometer (Mini-TES) and 10s of rocks measured with the rover's arm-mounted instruments, five more classes are considered minimally-altered volcanic or volcaniclastic rocks, some in outcrop, others as float rocks. They are mostly basaltic in composition as determined from the rover's APXS instrument, but examples of picrobasalt, trachybasalt, and basanite tephrite have been observed (McSween et al. 2008). Three of the rock classes (Wishstone, Backstay, and Irvine) have alkalic compositions as determined by APXS, constituting the first definitive alkaline rocks on Mars and perhaps indicative of fractional crystallization at various depths of oxidized, hydrous basaltic magmas similar in composition to Adirondack-class rocks (McSween et al. 2006). The basalt compositions measured by Spirit have been used as the basis for laboratory studies of mantle melting conditions on early Mars, such as melting temperature, pressure range, and melt fraction (Monders et al. 2007; Filiberto et al. 2008, 2010).

Mineralogically, the igneous rocks in Gusev crater can be grouped according to the dominant silicate phase as identified with Mini-TES spectra (Arvidson et al. 2008). Olivine-rich rocks abound, represented by the Adirondack-class basalts and Algonquin-class picrobasalts. Pyroxene dominates the spectra of Backstay and Irvine class alkaline rocks. In a notable departure from any martian meteorites, plagioclase is the dominant phase in the Al-rich Wishstone-class tephrites found as cobbles and boulders covering the north side of Husband Hill (Ruff et al. 2006). These results are consistent with measurements by Spirit's Mössbauer spectrometer (MB), although only Fe-bearing phases are sensed with this technique. Among these, hematite typically is observed in much lower abundance relative to magnetite, a reflection of the generally low oxidation state of Gusev's minimally altered igneous rocks (Morris et al. 2008).

The Meridiani Planum landing site encountered by Opportunity along its >30 km traverse is dominated by sulfate-rich sedimentary outcrops covered by a lag of basaltic sand and gray crystalline hematite spherules that are likely to be groundwater-derived concretions (e.g., Squyres et al. 2004b; Grotzinger et al. 2005). Igneous rocks along the traverse are limited to various cobbles that were likely delivered by impact processes or are themselves meteorites (e.g., Schröder et al. 2008). A well-studied example of likely impact ejecta known as Bounce Rock is rich in pyroxene to the exclusion of all other Fe-bearing phases as determined from MB data, despite its basaltic composition (Squyres et al. 2004b). In both its chemistry and mineralogy, Bounce Rock is similar to the martian shergottite meteorite EETA79001 Lithology B, although not a direct match as determined by a comparison of Mini-TES and laboratory spectra (Christensen et al. 2004; Zipfel et al. 2011). There is evidence to suggest that Bounce Rock was ejected from a relatively fresh $\sim$25 km crater located 75 km southwest of Opportunity's landing site, which exposed volcanic material different from the basaltic sands that partially cover Meridiani Planum (Squyres et al. 2004b).

3.3 Martian Meteorites

Martian meteorites are called the SNC group after three of its end members, the shergottites, nakhlites, and chassignites. In the absence of other samples, these rocks are the only pieces of Mars that may be analyzed in the laboratory on Earth to study in detail the magmatic processes affecting the crust-mantle system, from its differentiation up to the most recent events represented in the present collection. At the time of writing, the SNC group is a set of 104 igneous rocks (including pairs) (Meyer 2009), all of which have oxygen isotope compositions related to each other by a common mass fractionation trend (Clayton and Mayeda 1996), indicating an origin from the same planetary body. The definitive link to Mars was made from the isotopic composition of trapped atmospheric gases extracted from shock melt pockets of shergottites, which is identical to the martian atmospheric composition as determined by the Viking landers (Bogard and Johnson 1983). Relatively young crystallization ages (Nyquist et al. 2001) of all the SNCs (shergottites between 450 and 175 Ma, nakhlites and chassignites 1.3–1.4 Ga, but see Sect. 3.3.6 for a discussion of shergottite ages) have been considered as indirect evidence for a martian origin. The exception is ALH 84001, for which age estimates range from 3.8–4 Ga (Ash et al. 1996) to 4.56–4.5 Ga (Jagoutz et al. 1994; Nyquist et al. 1995), with most recent studies narrowing the age to 4.091 ± 0.030 Gyr (Lapen et al. 2010).

The time of ejection of the SNC meteorites from the martian surface is estimated from Cosmic Ray Exposure data obtained from cosmogenic nucleide measurements (Christen et al. 2005), and exposure times between 1 and 20 Myr have been obtained. The entire collection of martian meteorites appears to sample six different as yet undetermined localities (Meyer 2009). The most representative samples are the shergottites, which originate from four distinct sites and constitute 78 % of the SNCs discovered so far. However, despite the general assumption that the SNC meteorites have young crystallization ages young volcanic surfaces are remarkably sparse, representing less than 20 % of the planet's exposed surfaces (Hartman and Neukum 2001). Sampling of the martian crust by the SNCs would thus be highly biased towards young igneous rocks. Furthermore, it should be noted that spectral analysis of the martian surface does not provide a perfect match to the spectral signature of the SNC meteorites (Hamilton et al. 2003; Lang et al. 2009). However, this may be due in part to the fact that young volcanic rocks such as those found in Tharsis and Elysium are hidden by dust.

3.3.1 Petrology

The unpaired shergottites are sub classified on mineralogical grounds into three classes (McSween and Treiman 1998), and 46 different specimen have been discovered so far: 16 basaltic (Shergotty, Zagami, Los Angeles, 2 from Antarctica, 11 from Sahara); 10 lherzolitic (8 from Antarctica and 2 from Sahara) and 20 picritic (Tissint, 14 from Sahara, 2 from Antarctica, 2 from Oman and 1 from Libya).

Basaltic Shergottites have a tholeitic basaltic composition. They consist predominantly of coarse pyroxene crystals (augite and pigeonite) with lesser amounts of maskelynite (diaplectic glass formed from plagioclase during impact). The pyroxenes are typically zoned with Fe-rich rims. The proportion of pyroxene varies from 70 % in Shergotty to 44 % in QUE 94201, while that of maskelynite is 22–47 % (Meyer 2009). In more detail, pyroxenes are pigeonite 36–45 % and augite 10–34 %. Minor phases are oxides (titanomagnetite, ilmenite), sulfides (pyrrhotite) and accessory apatite, quartz, baddeleyite, and fayalite. Based on textures (coarse grain size) and high pyroxene/plagioclase ratio that cannot represent cotectic melt, most basaltic shergottites are considered as basaltic cumulates, which crystallized from fractionated melt (Usui et al. 2009). QUE 94201 is an exception as pyroxene core compositions are consistent with formation from a melt with the composition of the bulk sample (Kring et al. 2003). The shergottites range in oxygen fugacity from 3-log units below the Quartz-Fayalite-Magnetite (QFM) oxygen buffer to one log unit below QFM (QUE94201 $<$ EET 790001, Zagami, Los Angeles, and Shergotty). QUE 94201 is the least oxidized shergottite (FMQ-3) (Herd et al. 2002; McSween 1994).

Picritic Shergottites are basalts containing up to 7–29 % large (1 to 3 mm across) olivine grains. Picritic shergottites as a whole seem to be the least affected by crystal accumulation processes. Goodrich (2002) described large zoned olivine grains ($Fo_{60–76}$) as phenocrysts, and pyroxene in picritic shergottites is richer in Mg compared to basaltic shergottites. At least 4 picritic shergottites (Y 980459, NWA 5789, NWA 2990, and NWA 6234) represent magma compositions (Musselwhite et al. 2006; Gross et al. 2012; Filiberto and Dasgupta 2011). LAR 06319, which has been suggested to represent a near magma composition, actually contains $\sim$11 wt% excess olivine (Filiberto and Dasgupta 2011). Picritic shergottites contain chromites rather than Ti-magnetite and ilmenite as the main oxide phases. The sulfides are coarse blebs of pyrrhotite-pentlandite association (Lorand et al. 2005). DAG 476 is the least oxidized picritic shergottite (QFM-2.47 $\pm$ 0.73, Herd et al. 2002)

Lherzolitic Shergottites consist of medium grained olivine (40–60 %) and chromite, poikilitically enclosed by large (up to 5 mm) pigeonite / orthopyroxene crystals (9–25 %). Minor phases include maskelynite (less than 10 %), augite, whitlockite, and chromite filling the interstices. They are interpreted as olivine and pyroxene ultramafic cumulates possibly formed at depth.

Nakhlites and Chassignites are clinopyroxenite and dunite cumulates, respectively, and only represent 20 % of the meteoritic falls. The 8 nakhlites (Nakhla, Lafayette, Governador Valadares, 2 from antarticca, 3 from Sahara) are all olivine-bearing clinopyroxenites with cumulate textures defined by coarse augite and olivine crystals set in fine-grained mesostasis of silica-rich phases, oxides, sulfides, and phosphates. The coarse crystals hint at a rather slow growth over a relatively long period of time (>100 yr) within a single subsurface magma chamber or series of related but distinct flows (Lentz et al. 1999; Day et al. 2006), whilst the mesostasis corresponds to faster cooling as crystal mushes were extruded to the surface. Assuming a single thick lava flow or magma chamber environment, subtle mineralogical and chemical differences have been used to classify the 8 nakhlites in a stratigraphic order (Mikouchi and Miyamoto 2002). The nakhlites contain very small proportions of Cu-Ni-Fe sulfides. They thus show low modal sulfide abundances about five times lower than shergottite lavas (Lorand et al. 2005). The two Chassignites (Chassigny, Sahara) are olivine-rich cumulates with minor pigeonite, augite, alkali feldspar, chromite, troilite, pentlandite, chlorapatite, ilmenite, rutile, and baddeleyite (Floran et al. 1978; Beck et al. 2006). In the type specimen Chassigny, olivine is Fo_{68} whilst it is Fo_{78} in the only other known chassignite NWA 2737.

AllanHills (ALH 84001) is a unique sample that was added to the SNC classification (Mittlefehldt 1994). It is a coarse grained, cataclastic orthopyroxenite, consisting of 97 % orthopyroxene up to 5 mm long, 2 % chromite, 1 % maskelynite, and 0.15 % phosphate, with minor augite, olivine, and secondary Fe-Mg-Ca carbonates. Orthopyroxene is $En_{70}Fs_{27}Wo_3$, olivine is Fo_{65}, and maskelynite is $An_{35}Ab_{62}Or_3$.

3.3.2 Geochemistry

Shergottites Abundant geochemical data are available for the shergottites (see, e.g., McSween and Treiman 1998 and Meyer 2009) including trace element data (Treiman et al. 1986) and various isotopic ratios (Borg et al. 2002). The variation of mineralogy between the three groups translates into variable magnesium numbers Mg#, defined as $100 \times Mg/(Mg + Fe)$, where Mg and Fe are the magnesium and iron mole fractions. Basaltic shergottites vary from Mg# 23 (Los Angeles) to Mg# 52 (Zagami); picritic shergottites vary from Mg# 58 to Mg# 68, and lherzolitic shergottites have Mg# > 70. The Mg# of a rock is a function of both magmatic differentiation (liquids becoming more iron-rich as crystallization proceeds) and crystal content (mafic minerals having higher Mg# than their coexisting liquids). In the case of the Shergottites, the fact that Mg# correlates with decreasing SiO_2 concentration indicates significant olivine accumulation. Oxygen fugacities vary by 3 log units between the most and least reduced samples, the former of which have fO_2 equal to -3 log QFM units (close to the IW buffer) (Herd et al. 2002). All shergottites have low total alkali and Al_2O_3. A super-chondritic ratio of CaO/Al_2O_3 reflects a high pyroxene/plagioclase ratio. Sulfur concentrations in shergottites are high (from 500 to 3000 ppm) (Lorand et al. 2005 and references therein). The parallel decrease in pentlandite/pyrrhotite ratio, Ni concentration in pyrrhotite, and bulk rock Mg# indicates that: (1) sulfides are comagmatic; and (2) sulfide melts segregated from progressively more evolved (i.e. Mg- and Ni-depleted) silicate melts.

Shergottites may be sub classified into three groups based on Rare Earth Elements (REE) geochemistry: The Light REE (LREE)/Heavy REE (HREE) ratio varies from LREE highly depleted (HD) shergottites[1] (QUE 94201, DAG 476, Y 980459, NWA 5789, and NWA 6234) to moderately depleted (MD) shergottites and slightly depleted (SD) shergottites. Here, depletion refers to the incompatible behavior of Rare Earth Elements with respect to major minerals formed during igneous differentiation. Most HD shergottites are picritic, but QUE 94201 is a basaltic shergottite (Kring et al. 2003). Most MD are lherzolitic shergottites, and most SD shergottites are basaltic. This geochemical classification correlates only weakly with the mineralogical subclasses, but geochemical subclasses correlate with fO_2: less oxidized shergottites such as QUE 94201 and DAG 476 are HD, while the more oxidized ones are SD (Herd et al. 2002).

During igneous differentiation, the incompatible alkalis and LREE concentrate in the primary crust, leaving behind a depleted mantle. SD shergottites with flatter REE profiles correspond to an enriched reservoir that may reside inside the crust. The fO_2 variation would be due to mixing between reduced mantle-derived basalts and an oxidized crust-like reservoir. The correlation between the degree of geochemical contamination and magmatic oxidation state could be consistent with crustal assimilation (Herd et al. 2002), although Re/Os isotopic data argue for contamination in the mantle, (Brandon et al. 2000, 2012; Debaille et al. 2008). The geochemical characteristics of the shergottites have been modified by fractionation and/or assimilation during ascent, except for the HD shergottites QUE 94201, DAG 476, Y 980459, NWA 5789, and NWA 6234. These rocks are unfractionated and uncontaminated by enriched components. Therefore, the compositions of these samples represent equilibrium crystallization from the most primitive primary mantle melt, which could have equilibrated at a depth of 80–100 km in the martian mantle (Musselwhite et al. 2006).

Exposure ages (Nyquist et al. 2001) define at least four different groups of shergottites, excluding the interpretation that they would have been extracted by a single impact event. Rather, they likely correspond to four distinct sites on Mars.

Nakhlites, Chassignites and ALH84001 Nakhlites and chassignites represent a rather homogeneous set of meteorites. Sharing the same cosmic exposure age (11 ± 1.5 Ma), they likely originate from the same locality on Mars (Nyquist et al. 2001; Treiman 2005). Their Mg# is quite constant and around 50 for nakhlites, between 68 and 78 for chassignites, and 72 for ALH 84001. They are cumulate rocks which are not representative of the primary melt. Trapped melt inclusions in olivine and augite cumulus crystals are the only way to reconstruct the most primitive melt from which these cumulates crystallized (Treiman 2005 and references therein). In nakhlites, the recalculated parental melt has a Mg# of 29, much more iron rich than liquids produced by partial melting of the (Dreibus and Wänke 1985) mantle composition (DW85) at reasonable depths and degrees of partial melting (for which Mg# is typically 50) (Sautter et al. 2012). In contrast to the shergottites, nakhlites and chassignites are LREE-enriched (Nakamura et al. 1982), and they are depleted in sulfide components. ALH 84001 has an unusually old cosmic ray exposure age (15 Ma) for a martian meteorite, and a mineral composition similar to those of lherzolitic shergottites. It is LREE depleted, but its absolute abundances of REE are lower than in lherzolitic shergottites.

3.3.3 Comparison between SNCs and in-situ Analysis

The Gusev magmatic province is now the most thoroughly characterized igneous site on Mars. The Mars Exploration Rover Spirit has encountered mainly unaltered basalts, first

[1] It is of note that the following terminology is also often used: Depleted shergottites (equivalent to HD), Intermediate shergottites (equivalent to MD) and Enriched shergottites (equivalent to SD).

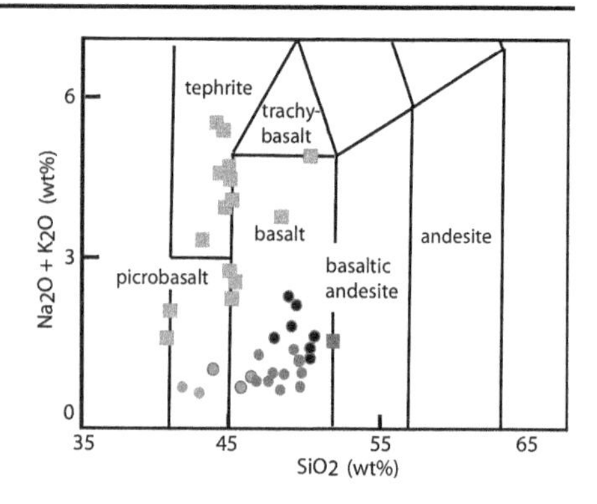

Fig. 5 Total alkali vs. silica (TAS) diagram used for the classification of volcanic rocks. Squares refer to rocks analyzed in situ by the Spirit and Opportunity rovers (Gusev rocks are indicated in *yellow*, while Bounce rock at Meridiani Planum is indicated in *red*). Dots refer to SNC meteorites, where *black dots* represent basaltic shergottites, *blue dots* represent picritic shergottites, and *green dots* represent lherzolitic shergottites

in the Gusev plain, and subsequently in the Columbia Hills. Major and minor elements were determined by APXS, Mini-TES investigated the mineralogy of rocks remotely in the thermal infrared, and Fe-bearing minerals were identified by the Mossbauer spectrometer. Most Gusev basalts are fine-grained with few olivine phenocrysts visible in the microscopic images down to about 100 microns resolution. They range from picritic basalts to alkali basalts (Fig. 5) and four classes are distinguished by element composition (McSween et al. 2008): (1) Adirondack-class basalts which consist of olivine and pyroxene bearing blocks on the Gusev plain (McSween et al. 2004). (2) Backstay-class basalts that include fine-grained olivine-, pyroxene-, ilmenite-, and magnetite bearing float rocks near the crest of Husband Hill. (3) Irvine-class rocks that include massive to vesicular pyroxene magnetite bearing rocks found on the flank of Husband Hill and in the inner basin of the Columbia Hills. (4) Algonquin-class rocks that are picritic basalt containing variable amounts of coarse crystals of cumulus olivine.

Gusev alkali and picritic basalts differ from the shergottites as they are all fairly silica-poor and generally have higher total alkali abundance as well as Al_2O_3 and Na_2O, and lower Al_2O_3/CaO relative to the shergottites. They are all fairly oxidized iron-rich basalts with a narrow range of Mg# from 51–55 (McSween et al. 2006) consistent with that calculated from primary melts in equilibrium with a DW 85 mantle (Mg# 75). All rocks from Gusev, except for Wishstone-class of P and Ti-rich basalts interpreted as pyroclastic rocks, can be related through fractional crystallization of a common parental magma at depths ranging from the mantle source at 85 km depth to the near surface. Adirondack class rocks only represent unfractionated picritic basalt while Algonquin class rocks accumulated olivine crystals on their way to the surface (McSween et al. 2006).

The Picritic shergottite that looks the most like Adirondack class rock with approximately the same proportion of olivine phenocrysts (10 to 20 %, Meyer 2009) is DAG 476. Olivine core composition (Fo76) is consistent with that expected from olivine crystallizing from primary magma derived from 15–20 % partial melt of a DW85 mantle source at 85 km depth (Bertka and Fei 1997; Schmidt and McCoy 2010; Baratoux et al. 2011a). Therefore, picritic basalts of Gusev (the Adirondack class) and picritic shergottite could be derived from the same source. Consequently olivine-phyric shergottites provide the most direct analogue to Gusev picritic basalt. The two basaltic shergottites EET 79001B (Steele and Smith 1982) and QUE 94201 are fine-grained with sub-ophitic textures. They are pyroxene-rich rocks dominated by low-Ca pyroxene (pigeonite) with subordinate high-Ca pyroxene (augite) similar to Bounce Rock.

All SNCs, except for the picritic shergottites such as DAG 476 and some of the basaltic shergottites (QUE 94201) are cumulates. Thus they are hypo-volcanic rocks (i.e., not surface rocks) as opposed to Gusev Adirondack-class rocks, which are extrusive olivine-rich basalts very similar to the picritic shergottites. All SNCs, except for the chassignites, are pyroxene-rich and plagioclase-poor while the martian surface is dominated by true basaltic rocks in which pyroxene and plagioclase proportions are similar. This indicates some degree of differentiation at depth by fractional crystallization. Such a subsurface origin for the shergottites would explain why no spectral signature related to these lithologies has been found from orbital measurements. This implies limited erosional or tectonic exposure of the subsurface and sampling at depth during strong impacts. Moreover, cosmic ray exposure ages are short (a few million years) with respect to the age of the SNCs, indicating that the specimens were shielded from interaction with cosmic rays, and buried at some depth below the surface.

3.3.4 Magnetic Properties

Ultramafic cumulates (i.e. Chassigny, ALH 84001) and lherzolitic shergottites are too weakly magnetized to account for the observed crustal magnetization (Acuña et al. 2008), assuming that magnetization took place in an Earth like magnetic field (Rochette et al. 2005). Basaltic shergottites and nakhlites are the most strongly magnetized SNCs, and in nakhlites the magnetic carrier is titanomagnetite. However, magnetization in nakhlites may be shock induced or linked to their shallow emplacement, and therefore they may not be representative of deeper unshocked rocks. In most basaltic shergottites pyrrhotite—a magmatic phase composed of iron sulfide (Lorand et al. 2005)—is the carrier of magnetization (Rochette et al. 2001). However, pyrrhotite magnetization was likely acquired during post-impact pressure release, and not from thermal remanence (Rochette et al. 2001). Furthermore, it should be noted that apart from ALH 84001 the SNCs are apparently much younger (<1.2 Ga) than the inferred 4 Ga lifetime of the martian dynamo (Langlais et al. 2004).

3.3.5 Water Content

The bulk water content of the SNC meteorites is one of the few constraints available to estimate the amount of water in the Martian mantle. This is of importance as the solidus (Sect. 5.1) and rheology (Sect. 4.2) of mantle rocks strongly depend on water content. Even just a few tens of ppm water can significantly reduce viscosity, and water content above $\sim$100 ppm will notably reduce the mantle solidus. The bulk water content of the SNC meteorites is relatively small and ranges from 50 to 150 ppm (Leshin et al. 1996), indicating low mantle water contents in the source regions of the SNC meteorites between only 5 and 15 ppm (Filiberto and Treiman 2009). However, it has been argued that magmatic degassing could be responsible for the low bulk water content in the SNC meteorites (McSween et al. 2001), and the water content of the martian mantle could be considerably larger. As of now, the water content of the martian mantle remains the subject of ongoing debate, and while some studies argue that water contents could be as low as 1–36 ppm H_2O (Mysen et al. 1998; Filiberto and Treiman 2009), making the martian mantle dryer than that of the Moon (Hauri et al. 2011), others derive water contents of 55 to 220 ppm H_2O (McCubbin et al. 2010; McCubbin et al. 2012), values similar to that of the Earth's mantle.

Evidence for a high mantle water content comes from studies of the depleted shergottites, which have an unambiguous mantle origin, and which are derived from partial melt that has not been mixed with an enriched component (either crustal or metasomatic). Olivine and pyroxene hosted primary melt inclusions, which contain trapped apatite and amphibole crystals, directly sample the water reservoir of the martian mantle, provided they have not been altered by terrestrial or martian weathering. Apatites in basaltic shergottites span a range of H_2O contents similar to those found in terrestrial mafic minerals (McCubbin et al. 2012), and based on apatite-melt partition coefficients for water, shergottite magmas should have contained 1.5 to 2.85 wt% H_2O at the time of apatite crystallization (McCubbin et al. 2012). This implies a much higher bulk water content in the parental melt than what is currently present in the shergottites, consistent with significant degassing. Taken together, apatite data indicate that water contents in the source regions of the depleted shergottites range from approximately 55 to 160 ppm H_2O (Watson et al. 1994; Leshin 2000; McCubbin et al. 2012), very similar to values estimated for the martian mantle from kaersutite in the Chassigny meteorite (McCubbin et al. 2010). For the enriched shergottites, a water content of 120–220 ppm H_2O has been estimated, although it remains controversial whether the source of this water is from a mantle or crustal reservoir (McCubbin et al. 2012). Taken together, evidence from the SNC meteorites indicates that rheologically significant amounts of water are likely present in the martian mantle, while it remains to be seen whether water content is also petrologically significant.

3.3.6 Dating the SNC Meteorites

The time of crystallization of the shergottites has been a controversial subject because discordant values have been obtained using different chronometric methods. As the mineralogy and chemistry of the shergottites have been used to infer the composition of their respective source regions as well as the conditions during partial melting, their crystallization age is highly relevant to the reconstruction of the composition and evolution of the crust-mantle system. If crystallization occurred relatively recently, derived water contents and melting conditions could represent a reservoir in the present-day martian mantle, while old ages would indicate the almost primordial state of the mantle shortly after core formation.

Young ages for the shergottites (late Amazonian period, 160–650 Ma) are obtained using U-Pb, Sm-Nd, Rb-Sr, Ar-Ar and Lu-Hf chronometers (Nyquist et al. 2001 and references therein), but an old late-Noachian to early Hesperian age of 4 Ga has been inferred from Pb isotope systematics in earlier studies (Chen and Wasserburg 1986; Jagoutz and Wänke 1986). Old crystallization ages have been confirmed by Bouvier et al. (2005, 2008, 2009) based on the same Pb geochronometer, and it has been argued that the young age isochrons for the shergottites date the last isotopic resetting due to impacts or acidic hydrothermal alterations (Bouvier et al. 2005, 2008, 2009), rather than the crystallization age. Different but still young Rb-Sr and Sm-Nd ages have been obtained for QUE 94201 (Borg et al. 1997) and DAG 476 (474 ± 11 Ma, Borg et al. 2000) interpreted as indicating that more than one perturbation event such as hydrothermal acidic leaching occurred on the martian surface (Bouvier et al. 2005).

The difference between old ^{206}Pb–^{207}Pb ages and young ages from other chronometers is not simple to interpret. Usually, a discordance between different chronometers is related to the robustness of their respective host-phases to contamination and disturbance events such as (shock) metamorphism. Bouvier et al. (2005, 2008, 2009) justified their hypothesis of the shergottites being old by arguing that Pb is mainly located in maskelynite that is robust to Pb diffusion during metasomatic events, while REE are mainly held in phosphates, a phase highly prone to disturbance. However, it is noteworthy that their very high content in REE makes metasomatic fluids less efficient for a complete resetting of their trace element contents. Also, other robust mineral phases, such as high-Ca pyroxene, can contain a substantial amount of REE (Wadhwa et al. 1994; Hsu et al. 2004). On the other hand, maskelynite is a glassy mineral resulting from the melting of plagioclase after a high-pressure shock event (Chen and El Goresy 2000), and glassy structures in which crystallographic bounds have been broken would be more prone to disturbance following metasomatic and/or metamorphic events (Ostertag and Stoffler 1982; Gaffnee et al. 2011).

There is good agreement between Rb-Sr, Sm-Nd, and Lu-Hf chronometers, despite their being hosted by different minerals: For Rb-Sr, maskelynite and to a lesser extent low and high-Ca pyroxene indicate that minerals cannot have been randomly reset (e.g., Borg et al. 2002). Similarly, phosphates and to a lesser extent high-Ca pyroxene indicate the same for the Sm-Nd and Lu chronometers (e.g., Wadhwa et al. 1994), while maskelynite shows the same for the Hf chronometer (e.g., Bouvier et al. 2005, 2008). Given the different robustness of each host-phase, a shock event followed by metasomatism would have deeply perturbed the different chronometers, and no consistency would be observed between the derived ages. This is confirmed by shock and heating experiments in which Rb-Sr and Sm-Nd remain preserved after heating the samples to 1000 °C and exposing them to shocks of 55 GPa (corresponding to the inferred conditions suffered by the shergottites), while U-Pb and even more ^{207}Pb–^{206}Pb are perturbed (Gaffnee et al. 2011). Thus the geochronological evidence suggests a young age for the shergottites, but the old ^{206}Pb–^{207}Pb systematic still needs to be explained.

Apart from geochronological arguments, aluminum depletion of the SNCs relative to terrestrial rocks as well as the Hesperian rocks of Gusev crater has been used as a geochemical discriminant to argue that the SNCs result from an Al-depleted evolved mantle source region left behind after early extraction of a primitive aluminum-rich crust (McSween 2002). This would support a young age of the shergottites relative to Gusev rocks. However, this super-chondritic ratio in the shergottites could be a mineralogical effect due to clinopyroxene fractionation and accumulation rather than a geochemical feature inherited from the deep mantle source. On the other hand, several indirect lines of evidence may suggest that shergottites are late Noachian to early Hesperian. First of all, shergottites are magnetic and the magnetic crust on Mars is exclusively Noachian (Rochette et al. 2005 and references therein). Furthermore, being S-rich, shergottites are not devolatilized, whilst younger nakhlites and chassignites are S-poor (Lorand et al. 2005; Chevrier et al. 2011). Finally, the most primitive uncontaminated and unfractionated shergottites (QUE 94201 and DAG 476) appear to be derived from partial melt of a primitive mantle source with DW85 composition.

4 Rheology and Tides

4.1 Lithospheric Deformation

The lithosphere comprises the crust and upper mantle, and its susceptibility to deformation can be described by strength-depth profiles of the corresponding crustal and mantle materials, which indicate the amount of stress that can be supported. In the cold (shallow) regions, the strength is controlled by the fracture strength, whereas in the deep (hot) regions the strength is controlled by the resistance to ductile flow. Such general strength profiles were first proposed by Goetze and Evans (1979), and a schematic profile is shown in Fig. 6. In this model, the strength of materials in the brittle regime is controlled by the frictional strength along preexisting faults and is generally insensitive to the specific materials (except for sheet silicates that have a smaller resistance to friction).

Using this so-called yield-strength envelope model, the mechanical and elastic thickness of the lithosphere can be defined, where the former corresponds to the depth up to which the

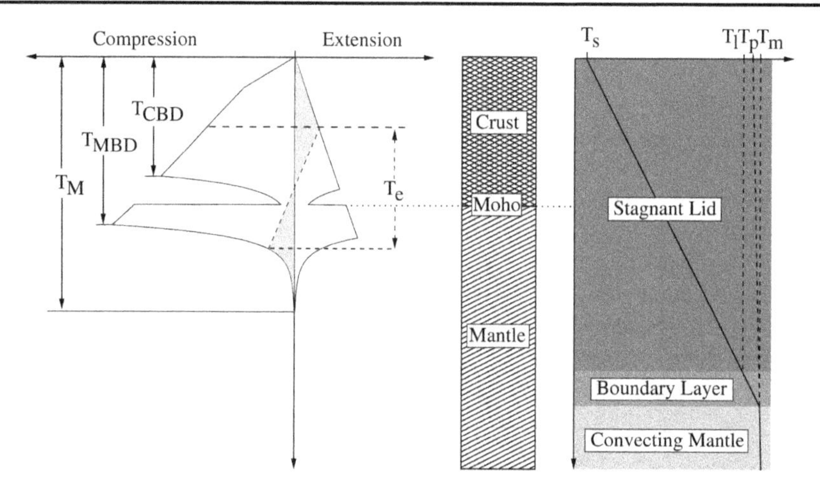

Fig. 6 Schematic diagram showing the relationship between elastic lithosphere thickness T_e, mechanical lithosphere thickness T_M, and the depth to the brittle-ductile transition in the crust (T_{CBD}) and mantle (T_{MBD}). T_e is defined by the depth range over which stresses are supported elastically, here indicated by shades. The chemical layering of the lithosphere is shown in the mid-column, while dynamical denominations are given in the right panel along with their corresponding temperatures: T_s is surface temperature, T_l is the temperature at the base of the stagnant lid, and T_m is upper mantle temperature. T_p is the mantle potential temperature, which is calculated by extrapolating T_m to the surface using the adiabatic temperature gradient in the mantle

lithosphere carries significant mechanical strength (usually assumed to be 10 to 50 MPa). Deformation of the lithospheric plate will then result in the generation of bending stresses, and brittle as well as ductile yielding will take place wherever stresses exceed the limits defined by the yield strength envelope. The elastic thickness of the lithosphere T_e can then be defined as the thickness of a purely elastic plate which carries the same bending moment as the real elasto-plastic plate defined by T_M.

For the two-layer rheology considered in Fig. 6, the yield strength envelope consists of two domains, each corresponding to one homogeneous chemical layer. Usually, crust and upper mantle are both part of the elastic and mechanical lithospheres, and both are generally part of the stagnant lid, in which heat is transported by conduction. Only deeper parts of the mantle located at depths larger than the stagnant lid thickness will convect, and temperatures in the convecting region will increase along the adiabatic temperature gradient.

Due to the reduced gravity on Mars and the associated reduced lithostatic pressure, the brittle strength of the lithosphere is only $\sim 1/3$ that of the Earth. Ductile deformation is governed by plastic flow laws of the corresponding material and is sensitive to temperature, but water content also plays an important role. Consequently, the transition from the strong lithosphere to a weak asthenosphere depends both on the temperature-depth and the water content-depth profiles, as indicated in Fig. 7.

The asthenosphere is situated directly below the lithosphere and comprises a region in the upper mantle which deforms plastically on geological timescales. Partial melting predominantly occurs at depths below the lithosphere, and the location of the lithosphere—asthenosphere boundary can be identified with an isotherm. The depth of the lithosphere—asthenosphere boundary is the depth below which deformation is dominated by viscous flow, and this region is characterized by a Maxwell time

$$\tau_M = \frac{\eta_{\text{eff}}(z)}{E} \tag{1}$$

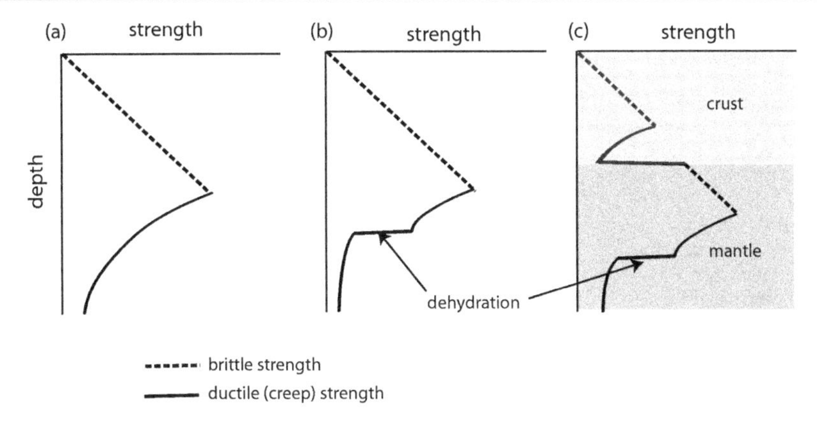

Fig. 7 Schematic diagrams showing lithospheric strength-depth profiles. (**a**) Homogeneous lithospheric composition with only temperature and pressure dependent creep strength. (**b**) Same as (**a**), but including the water content dependence of the creep strength. A water-depleted (dehydrated) layer above the indicated depth is assumed. (**c**) Layered lithospheric composition including crust and mantle. A layered water content similar to (**b**) is assumed

which becomes shorter than the time scale of deformation τ. Here, η_{eff} is the effective viscosity, and E is the elastic modulus. For two-dimensional flow, the effective viscosity is defined as (e.g., Karato 2008)

$$\eta_{\text{eff}} = \frac{\sigma}{2\dot{\epsilon}} \tag{2}$$

where σ is stress and $\dot{\epsilon}$ is strain-rate. The latter may be expressed as a power-law relationship, and if the grain-size dependence is ignored for simplicity, $\dot{\epsilon}$ is given by

$$\dot{\epsilon} = \left(A_{\text{dry}} + A_0 C_W^r\right) \exp\left(-\frac{H^*}{RT}\right)\sigma^n \tag{3}$$

where A_{dry} and A_0 are deformation constants, T is temperature, R is the gas constant, C_W is water content, r is a non-dimensional parameter representing the water sensitivity of deformation (close to 1–1.2 for high temperature creep in olivine), H^* is activation enthalpy, and n is the stress exponent. Consequently, the lithosphere will appear thick for deformation at a low stress level and thin for deformation at high stress levels.

The dependence of the effective viscosity on water content is caused by hydrogen, which enhances transport properties if dissolved in silicates such as olivine. The precise mechanisms of enhancement are not well understood, but enhancement is partly due to the increased population of point defects such as atomic vacancies created by hydrogen (e.g., Chapter 10 of Karato 2008). In many cases, even a small amount of hydrogen can significantly affect transport properties, and the critical water content above which viscosity is significantly reduced can be estimated by comparing the relative importance of A_{dry} and $A_0 C_W^r$ in Eq. (3). For high-temperature (dislocation) creep in olivine, the critical water content at which $A_{\text{dry}} = A_0 C_W^r$ is only 2 ppm wt (~30 ppm H/Si) (Mei and Kohlstedt 2000a). This implies that if the water content is close to 100 ppm wt, then the viscosity is reduced by a factor of ~50 from the completely dry situation. Therefore, a few tens of ppm water can be rheologically significant, but the exact value depends on the operating deformation mechanism. As limiting cases, viscosities of 10^{19} and 10^{21} Pa s at a temperature of 1600 K are generally assumed to represent wet and dry mantle rheologies.

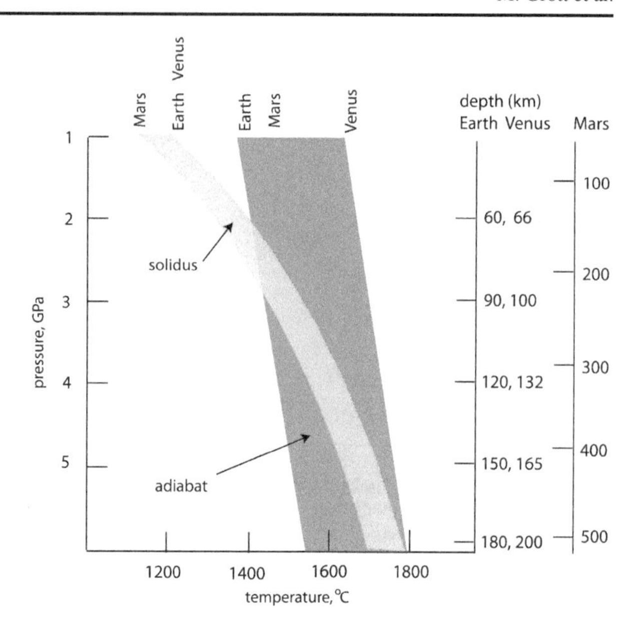

Fig. 8 A diagram illustrating the extent of a potentially dehydrated zone in the mantles of Earth, Venus, and Mars. Substantial dehydration will occur if mantle materials undergo a large degree of partial melting, and assuming a partitioning coefficient of 0.01, the residual mantle will be more than 90 % dehydrated at melt fractions larger than 10 %. The depth of partial melting corresponds to the depth at which the solidus intersects the adiabat. Due to the iron-rich composition, the dry solidus of the martian mantle is lower than that for the Earth

The water content of the martian mantle is poorly constrained, but it seems likely that the minimum estimates of water concentrations (see Sect. 3.3.5) are sufficient for rheological weakening. In addition, the water content in the martian mantle may be stratified, as has been proposed for Earth's upper mantle (e.g., Karato 1986; Hirth and Kohlstedt 1996), and this stratification would be associated with a large change in the effective viscosity as a function of depth. A potential stratification in water content on Mars could be the consequence of partial melting, which results in dehydration of the mantle minerals. The depth range of melting depends on the relative positions of the adiabat and the solidus, and the depth to which a potential stratification in water content would occur on Mars is expected to be deeper than on either Earth or Venus (Fig. 8). Furthermore, dehydration stiffening only becomes efficient near the dry solidus, as even small amounts of water can significantly reduce the viscosity (Hirth and Kohlstedt 1996; Karato and Jung 1998).

4.2 Mantle Rheology

Below the lithosphere, much of the deformation occurs via plastic flow through thermally activated motion of atoms, and the rheological properties of the martian mantle play a key role for the thermal and chemical evolution of the planet. Rheology influences processes like crustal production and possibly also the history of the magnetic field (Breuer and Spohn 2006). The interpretation of lithosphere thickness (Sect. 4.4) also requires some knowledge of rheological properties. The rheological properties of minerals in the ductile regime depend on temperature, pressure, water content, grain-size and mineralogy (major element chemistry and crystal structure) (Karato 2008), with water having a significant influence on the strength of mantle minerals.

Mantle mineralogy is similar on Mars and Earth, with a modest increase in Fe content (e.g., Ohtani and Kamaya 1992; Bertka and Fei 1997), which reduces the viscosity at a given stress by a factor of ~3–5 (see Fig. 4 of Zhao et al. 2009). In addition, a higher Fe content reduces the pressure at which a phase transformation from olivine to wadsleyite or ringwoodite occurs by up to ~0.5 GPa. Consequently, the rheological properties of the martian mantle likely follow a trend similar to the Earth's mantle, with minor modifications caused by the influence of Fe content and potentially larger modifications caused by differences in water content.

For mantle minerals of the Earth, detailed data sets on olivine (and olivine-dominated rocks) exist that can be applied to a depth of ~1000 km on Mars. From a depth of ~1000 to ~1800 km, the martian mantle is mainly composed of wadsleyite, ringwoodite and majorite, for which experimental data on rheological properties are sparse (e.g., Nishihara et al. 2008; Kawazoe et al. 2010). At the very bottom of the martian mantle (from ~1800 to ~2000 km), a thin layer made mostly of perovskite and (Mg,Fe)O may exist, depending on core size, and the rheological properties of materials below ~1000 km must be viewed as highly uncertain. In addition, a possible difference between the rheological behaviour of the martian and terrestrial mantle could be related to the larger fraction of the martian mantle that is composed of wadsleyite and ringwoodite, which can store a large quantity of water. In such a case, a large regional variation in rheological properties may occur associated with the regional variation in water content.

4.3 Tidal Dissipation

The tidal environment of Mars allows to infer the rheological properties of the martian interior from the orbital evolution of the planet's two tiny moons, Phobos and Deimos, which are rotationally locked into a 1:1 spin-orbit resonance. The satellites' orbital evolution is closely connected to the dissipation of tidal energy within the martian interior, and a phase lag between the tidal-bulge direction and the direction to the tide-generating satellite causes exchange of angular momentum between planetary rotation and satellite orbital motion. The mean specific dissipation factor of Mars Q_p is then related to the tidal lag angle γ by $Q_p = \cot 2\gamma$.

While Deimos' almost circular orbit is close to synchronous with Mars' present rotation rate and subject to less secular deceleration, Phobos's highly eccentric orbit is well within the synchronous distance and, therefore, the satellite undergoes notable secular acceleration. The orbital evolution of Phobos is mainly affected by the degree-two semidiurnal tide, while the third-order tide is expected to contribute less than 10 % to the secular acceleration of Phobos (Zharkov and Gudkova 1997). The latter is now known with high accuracy due to long-term monitoring of Phobos' orbital longitude (see Jacobson 2010 for a review). From long-term observations of the secular accelerations of the martian satellites $\dot{n}_i/n_i$ or equivalent time rates of change of their semi-major axes, $\dot{a}_i/a_i = -2\dot{n}_i/3n_i$, the ratio k_2/Q_p of the elastic Love number k_2 and the planet's mean specific dissipation factor Q_p can be inferred according to

$$\dot{a}_i = 3m_i \frac{k_2}{Q_p}\left(\frac{G}{a_i(M_p + m_i)}\right)^{1/2}\left(\frac{r_p}{a_i}\right)^5, \tag{4}$$

where G is the gravitational constant, M_p is the planet's mass, and m_i, a_i, and n_i are mass, semimajor axis, and mean motion of each satellite, respectively. As a measure for Mars' gravitational response to tides exerted by the Sun, k_2 is dependent on the planet's interior structure, rheological layering, and tidal forcing period. Together with the mean density and axial moment-of-inertia of Mars, k_2 imposes additional constraints on the size and physical state of the martian core (e.g., Sohl et al. 2005; Verhoeven et al. 2005; Rivoldini et al. 2011).

Based on the analysis of radial weighting functions for the distribution of elastic strain energy, Zharkov and Gudkova (1997) conclude that the tidal phase lag is not sensitive to the martian crust and uppermost mantle. Dissipation of tidal energy mainly occurs at greater depth in the middle or lower mantle. Furthermore, dissipation in the middle or lower mantle supports the view that the martian mantle cannot be completely dry.

This analysis is consistent with recent radar sounder observations of lithospheric deflection beneath the north polar units of Mars. These measurements indicate that the geodynamic response of the martian lid, i.e., the lithosphere and uppermost mantle to the loading history of the north polar layered deposits (NPLD) is less than previously expected (Phillips et al. 2008). Deposition of the NPLD is believed to be driven by climate variations on timescales of several 10,000 of years. This is sufficiently short that the corresponding geodynamic creep response of the martian mantle could be dominated by time-dependent or transient rheological behavior (e.g., Faul and Jackson 2005).

4.4 Elastic Lithosphere Thickness

In the absence of direct heat flow measurements (see Sect. 6.2), the evolution of the elastic lithosphere thickness T_e is one of the few clues we have to reconstruct the thermal history of the planet. The elastic thickness is a measure for the stiffness of the lithospheric plates and can be estimated by analyzing the response of the lithosphere to loading. Given a rheological model, the mechanical thickness of the lithosphere can be derived from the elastic thickness, and mechanical thickness can in turn be identified with an isotherm (McNutt 1984). In this way, elastic thickness estimates can be used to test the predictions of thermal evolution models (e.g., McGovern et al. 2002; Montési and Zuber 2003; Grott and Breuer 2008, 2010; Ruiz et al. 2011).

Most T_e estimates have been derived from gravity and topography data (McGovern et al. 2004; Kiefer 2004; Belleguic et al. 2005; Hoogenboom and Smrekar 2006; Wieczorek 2008; Grott et al. 2011), but some geological features allow for more direct approaches. Phillips et al. (2008) have modeled the lithospheric deflection due to polar cap loading, and analysis of rift flank uplift has been used by Barnett and Nimmo (2002), Grott et al. (2005), and Kronberg et al. (2007) to constrain T_e at the Tempe Terra, Coracis Fossae, and Acheron Fossae rift systems, respectively, while the position of Nili Fossae has been modeled by assuming these circumferential graben to be a tectonic feature caused by mascon loading at the Isidis basin (Comer et al. 1985; Ritzer and Hauck 2009). Furthermore, T_e has been constrained from the seismogenic layer thickness T_s, which is defined as that part of the lithosphere which deforms in a brittle fashion when exposed to tectonic stresses. T_s usually is of the same order as T_e or slightly smaller, and T_s has been determined at thrust faults by Schultz and Watters (2001), Grott et al. (2007a), and Ruiz et al. (2009), who modeled lobate scarp topography by varying depth of faulting and slip distribution on the fault surface. Other more indirect methods to estimate T_e include modeling of fault spacing using localization instabilities (Montési and Zuber 2003), and modeling of the thermal state in chaotic terrain (Schumacher and Zegers 2011).

The times corresponding to the observed paleo-flexure are usually assumed to be given by the age of the deformed surfaces, but this assumption may not always be valid. Age assignments at the Tharsis volcanoes are particularly difficult, as these have been active throughout the history of the planet (see Sect. 2.4). While the Tharsis volcanoes exhibit Amazonian surface ages, loading was likely mainly finished by the end of the Hesperian, and the observed paleo-flexure might correspond to either of these periods, resulting in age uncertainties of up to 3 Gyr. Further complications arise due to the fact that although paleo-flexure is generally assumed to be frozen-in at the time of loading, stresses in the lithosphere will decay as a function of time due to viscous relaxation. As a result, the elastic thickness T_e will also decrease as a function of time, and the time corresponding to the observed paleo-flexure is determined by a competition between loading rate, lithospheric cooling rate, and stress relaxation rate (Albert and Phillips 2000). While stress relaxation is of limited importance for single-layer rheologies, it may become important for multi-layer configurations (Brown and Phillips 2000). Furthermore, while the strength of the Noachian lithosphere was almost exclusively carried by the crust, mantle contributions to lithospheric strength became important in the cooler Hesperian and Amazonian lithospheres. Therefore, care must be taken when interpreting mantle temperatures derived from loading models, as these may not easily be associated with a given point in time.

An overview of published elastic thickness estimates is compiled in Fig. 9, where the spatial distribution of analyzed features is given in the top panel, while T_e is given as a function of time in the bottom panel. Elastic thicknesses were small and between 0 and 20 km during the Noachian period, but quickly increased to values above 50 km during the Hesperian. Values in the Amazonian range from 40 to 150 km, and best estimates for the present day elastic thickness are above 150 km. In particular, lithospheric deflection due to loading at the north polar cap locally constrains present day T_e to values greater than 300 km at this location (Phillips et al. 2008).

This evolution is in general agreement with planetary cooling models (McGovern et al. 2004; Grott and Breuer 2008), and small elastic thicknesses during the earliest evolution are consistent with higher than present-day mantle temperatures and an associated high Noachian heat flux (McGovern et al. 2002; Grott and Breuer 2010). Thermal evolution models typically predict Noachian heat fluxes above 60 $\mathrm{mW\,m^{-2}}$, with slightly lower values around 50 $\mathrm{mW\,m^{-2}}$ during the Hesperian period. Heat flows then steadily decline to around 20 $\mathrm{mW\,m^{-2}}$ today (Morschhauser et al. 2011). However, higher heat fluxes alone are insufficient to explain the small elastic thicknesses derived for the Noachian period, and rheological layering also seems to play an important role (Grott and Breuer 2008, 2009). Decoupling of the crust and mantle lithosphere as indicated by reduced strength at the base of the crust in Fig. 7(c) can significantly reduce the overall lithospheric thickness, and the disappearance of this layer due to planetary cooling then results in a sudden increase of T_e (Burov and Diament 1995; Grott and Breuer 2008).

In order for an incompetent layer to exist, a weak crustal rheology is required (Grott and Breuer 2008, 2010), and given the absence of large-scale felsic deposits, it seems likely that water is responsible for rheological weakening (see Sect. 4.2). Furthermore, small T_e require relatively thin thermal boundary layers in the upper mantle, indicating that mantle convection was active and vigorous during this time period. Therefore, any stable mantle stratification following magma ocean solidification and mantle overturn (Elkins-Tanton et al. 2003) could not have survived longer than 300–500 Myr. Thin thermal boundary layers also require small mantle viscosities, indicating that the convecting mantle is best described by a wet or only partially dehydrated rheology with viscosity around 10^{19} Pa s at 1600 K.

An open question concerns the large elastic thickness values derived from polar cap loading, which require lithospheric temperatures to be colder than expected. This could either imply a sub-chondritic bulk composition in terms of heat producing elements (Phillips et al. 2008), or a large degree of spatial heterogeneity of the mantle heat flow (Phillips et al. 2008; Grott and Breuer 2009, 2010; Kiefer and Li 2009), which—potentially lacking upwellings in the polar regions—could be below the average there. These hypotheses are difficult to test by models, but in-situ heat flow measurements could help to discriminate between the two (see Sect. 6.2).

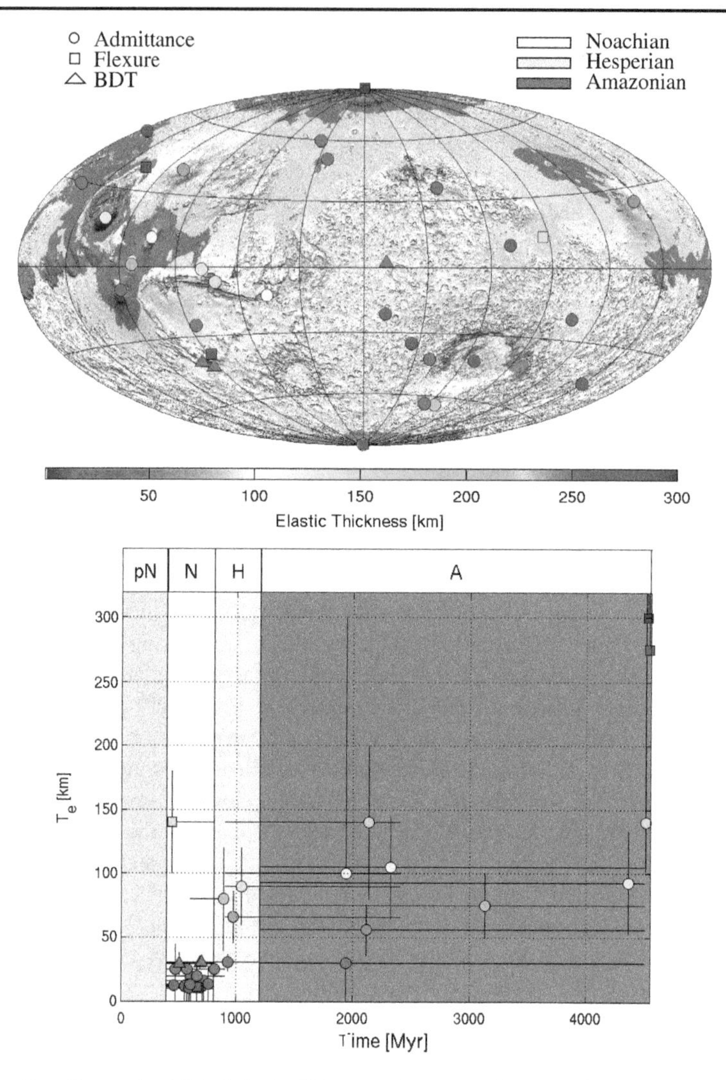

Fig. 9 *Top*: Compilation of published elastic thickness estimates overlaid over a MOLA shaded relief topographic map. The major martian geological epochs derived from the maps of Scott et al. (1986, 1987a, 1987b) are indicated by color. Elastic thicknesses are given in color code, and the methods used to estimate T_e are indicated by symbols (see text for details). *Bottom*: Same dataset as above, but shown as a function of surface age for the corresponding geological units. Errorbars for the individual data points are indicated. Times of loading were assigned based on the surface ages of the corresponding features

5 Thermo-Chemical Evolution

5.1 Partial Melting

Partial melting is one of the most important processes for the chemical evolution of terrestrial planets, controlling the redistribution of various elements among molten materials and residual solids, chemical differentiation occurring via melt transport and element diffusion. The composition of melts and residual solids has been thoroughly studied under shallow

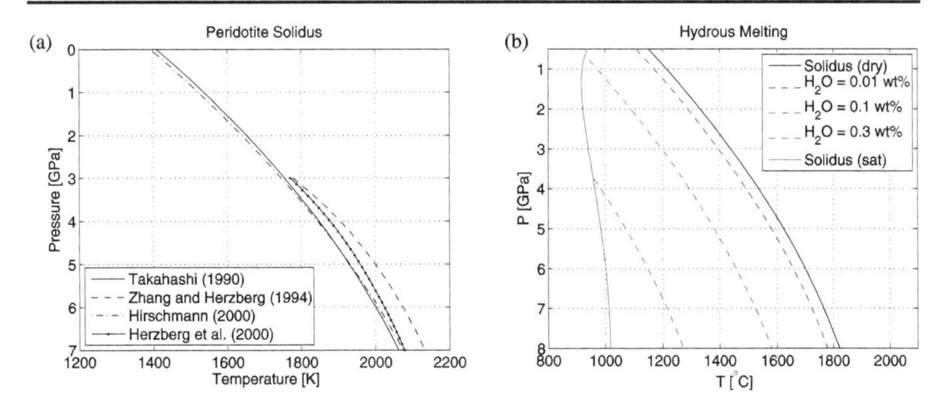

Fig. 10 (**a**) Different parameterizations of the peridotite solidus as a function of pressure. Note that the parameterizations by Zhang and Herzberg (1994) and Herzberg et al. (2000) are only valid above pressures of 3 GPa. (**b**) Influence of water content on the solidus of peridotite as a function of pressure. Cf. Fig. 3 in Katz et al. (2003)

mantle conditions (e.g., Kushiro 2001), and an extensive literature on the melting behaviour of terrestrial KLB-1 peridotite exists (e.g., Hirschmann 2000 and references therein). Apart from the parameterization by McKenzie and Bickle (1988), current solidus parameterizations (Takahashi 1990; Zhang and Herzberg 1994; Herzberg et al. 2000; Hirschmann 2000; Katz et al. 2003) agree to within 50 K, a difference attributed to varying experimental techniques and differences in sample compositions (Hirschmann 2000). Fits to the experimentally derived terrestrial solidi are shown in Fig. 10(a), where curves by Herzberg et al. (2000) and Zhang and Herzberg (1994) are only valid at pressures greater than 3 GPa.

Water can have a significant influence on the solidus of mantle materials (Hirth and Kohlstedt 1996; Asimow and Langmuir 2003; Katz et al. 2003; Hirschmann 2006), and a bulk water content of 250 ppm (0.025 wt%) can lower the peridotite solidus by 100 K. The effect of water on the solidus of the terrestrial mantle is summarized in Fig. 10(b), where the pressure dependent water saturation curve is indicated to the left of the diagram. As discussed above, the bulk water content of the martian mantle is still a matter of debate, and while dry scenarios would not result in a significant reduction of the mantle solidus, the presence of petrologically significant amounts of water of a few hundred ppm cannot be ruled out.

In the absence of a broad experimental database for melting of martian mantle materials, the KLB-1 peridotite solidus derived for the Earth's mantle is usually also applied to Mars. However, it should be pointed out that alkali content and, to a lesser extent Mg# too, influence the mantle solidus (Herzberg et al. 2000; Hirschmann 2000). The solidus decreases by about 100 K when increasing the total $Na_2O + K_2O$ content from 0.2 to 0.75 wt%, while the solidus is increased by a similar amount when increasing Mg# from 85 to 91. Therefore, a firmer experimental basis to parametrize mantle melting on Mars is clearly desirable.

The composition of melts produced under deep mantle conditions ($P > 14$ GPa) is less well constrained than shallow melting (Bertka and Holloway 1994a, 1994b) and melt compositions can be distinctly different in the deep mantle (e.g., Litasov and Ohtani 2002). High pressure melts tend to be Fe-rich and silica-poor, and hence heavier than the surrounding minerals. In such a case, melts may sink to the core-mantle boundary rather than ascend to the surface. These melts are dense and water-rich and can play an important role as a water reservoir in the deep interior.

During partial melting, incompatible elements like water (hydrogen), or the heat producing elements K, Th, and U will preferentially partition into the liquid, thus being extracted from the mantle along with the melt. The degree to which hydrogen and other elements are removed from the minerals and rocks by partial melting depends on the process of melting, and two modes of melt-solid separation are often considered: (1) Fractional melting is the melting process in which chemical equilibrium between the melt and the mineral is attained only at the interfaces between the two. The majority of the melt is assumed to be removed quickly from the source, such that a majority of the melt does not reach equilibrium with the residual solids. This mode of melting occurs when the time scale of melt extraction is faster than the time scale of element diffusion. (2) If all the melt produced by partial melting is in equilibrium with the solids, a batch melting model is appropriate to describe the melting process. In this case, the time scale of melt extraction is slower than the element's diffusion time scale, such that chemical equilibrium is reached for the entire melt volume.

The mode of melt-solid separation can be evaluated using the non-dimensional number ξ (e.g., Iwamori 1992, 1993)

$$\xi = 2\sqrt{\frac{DL}{d^2 v}} \tag{5}$$

where D is the diffusion coefficient of a relevant atomic species, L is the characteristic length scale of the melting region, d is the grain-size, and v is the velocity of mantle convection. The diffusion coefficients D differ significantly among different species, and D is close to 10^{-10} m^2/s for hydrogen (Kohlstedt and Mackwell 1998), while it is about 10^{-20} m^2/s for Th and U (Van Orman et al. 1998). Choosing $L \sim 100$ km, $d \sim 1$ cm, $v \sim 1$ cm/yr, we find that for hydrogen equilibrium melting (batch melting) is likely, whereas Th and U are better described by disequilibrium (fractional) melting. In numerical models of the martian thermo-chemical evolution, both melting models have been applied irrespective of the different diffusion coefficients (Hauck and Phillips 2002; Fraeman and Korenaga 2010; Morschhauser et al. 2011). While this has only a minor influence on the bulk extraction of K, Th, and U, the reduced concentration of residual hydrogen and the associated dehydration stiffening of the mantle rheology may play an important role in spatially resolved convection models. However, experimental data on fractional melting is sparse, because melting phase relationships change continuously as melt is extracted. Therefore, fractional melting experiments are usually limited to a few discrete jumps in composition. It is also worth noting that melt extracted at depth by fractional melting might equilibrate with its surroundings at shallower depth, resulting in an overall process that is closer to batch melting.

5.2 Parameterized Thermo-Chemical Evolution Models

The thermal evolution of Mars is intimately related to the history of martian volcanism and some insight into the evolution of the crust-mantle system may be gained from parameterized convection models. These models calculate the thermo-chemical evolution starting from an initial state after core formation and solve the energy balance equations for core, mantle, and lithosphere as a function of time (see Fig. 11) (Spohn 1991; Hauck and Phillips 2002; Breuer and Spohn 2006; Parmentier and Zuber 2007). Mantle differentiation is treated by comparing mantle temperatures to the peridotite solidus and the degree of partial melting is calculated by a linear interpolation between solidus and liquidus (Hauck and Phillips 2002; Schumacher and Breuer 2006; Morschhauser et al. 2011). Although the degree of partial melting is a non-linear function of temperature, the linear approximation is in satisfactory agreement with the experimental data for melt fractions below

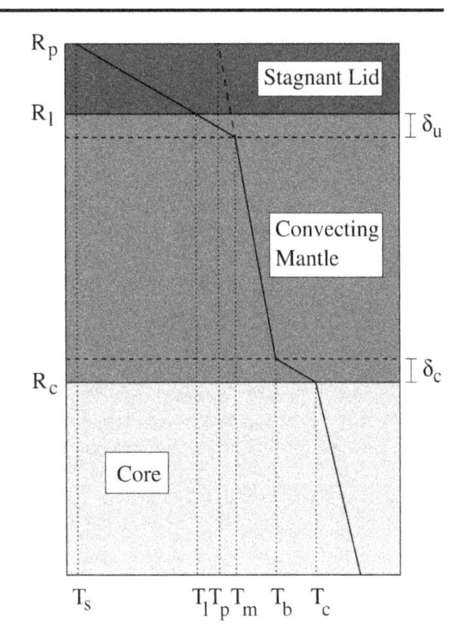

Fig. 11 Schematic diagram showing the reservoirs considered in parameterized thermal evolution models, which comprise the stagnant lid, the convecting mantle including the thermal boundary layers, and the core. The planetary radius R_p, stagnant lid radius R_l, and core radius R_c are indicated. δ_u and δ_c are the upper and CMB boundary layer thickness, respectively. Temperatures shown are the surface temperature T_s, the stagnant lid temperature T_l, the potential mantle temperature T_p, the upper mantle temperature T_m, the lower mantle temperature T_b, and the core temperature T_c. Together, these quantities define the thermal state of the planet

30 % (McKenzie and Bickle 1988; de Smet et al. 1999). In this way the degree of partial melting as well as the amount of partial melt extracted from the mantle can be calculated.

Thermo-chemical evolution models parametrize planetary heat loss in terms of the mantle Rayleigh number and while earliest models used scaling laws derived from boundary layer theory (Stevenson et al. 1983; Schubert and Spohn 1990; Spohn 1991), it was later recognized that the temperature dependence of the mantle viscosity can result in the development of a stagnant lid (Solomatov and Moresi 1997; Grasset and Parmentier 1998; Reese et al. 1998). Current parameterizations either treat the stagnant lid separately and consider the mantle to convect like an isoviscous fluid (Grasset and Parmentier 1998), or they parametrize surface heat flow in terms of the whole mantle Rayleigh number including the lid (Solomatov and Moresi 1997; Reese et al. 1998). While the former approach enables the modeling of feedback-mechanisms between lithosphere and mantle dynamics, the latter approach has been extended to treat non-Newtonian rheologies. In addition, latest models also try to parametrize the influence of mantle melting on the efficiency of mantle energy transport (Korenaga 2009; Fraeman and Korenaga 2010).

It has been noted early on that the single most important parameter in these calculations is the mantle viscosity (Schubert et al. 1979), but few a priori constraints exist from mineral physics. Major uncertainties are related to the unknown water content of the martian mantle (see Sects. 3.3.5 and 4.2), and plausible mantle viscosities span at least two orders of magnitude. As a consequence, a large parameter space needs to be studied. A second poorly constrained parameter is the initial temperature profile, but a feedback mechanism between temperature and mantle viscosity reduces the range of admissible present day thermal scenarios: While high temperatures result in low mantle viscosities and fast cooling, low temperatures result in high viscosities and heat accumulation. Consequently, mantle temperatures will tend to similar present day values and heat transport today will be equally efficient almost irrespective of the initial temperature profile. On the other hand, reconstructing the thermal history of the planet from present day measurements is very challenging for the same reasons. Also note that different mantle temperatures during the early evolution have important implications for mantle melting and crustal production.

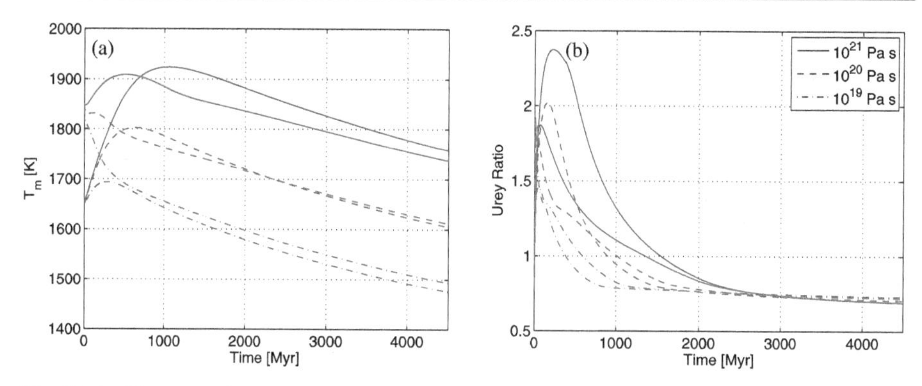

Fig. 12 (**a**) Upper mantle temperature as a function of time for parameterized thermo-chemical evolution models using three different mantle reference viscosities and two different initial upper mantle temperatures. Cold models (*blue curves*) correspond to an initial upper mantle temperature of 1650 K, while hot models (*red curves*) correspond to initial mantle temperatures of 1850 K. (**b**) Urey ratio, i.e., the ratio between the radiogenically generated heat flow and the surface heat flow for the same models

In addition to viscosity and initial temperature, energy transport is influenced by the mode of mantle convection. While Mars is currently in the stagnant lid mode of mantle convection (Solomatov and Moresi 1997; Grasset and Parmentier 1998; Reese et al. 1998), plate tectonics might have been operating during its earliest evolution. The thin crust of the northern lowlands has been taken as an indication for seafloor spreading (Sleep 1994), and the growth of the southern highlands could have then inhibited subduction, resulting in a transition to the stagnant lid mode of mantle convection (Lenardic et al. 2004). Parallel faults offsetting magnetic field contours in a way similar to transform faults on Earth have been taken as support for the plate tectonics hypothesis (Connerney et al. 2000, 2005), and planetary cooling driven by plate tectonics could have powered a core dynamo (Nimmo and Stevenson 2000; Breuer and Spohn 2003). In addition, folds and thrusts in the Thaumasia Plateau, Daedalia Planum, and Aonia Terra have been interpreted as orogenic belts (Anguita et al. 2006), a possible remnant of convergent plate margins, and the lineations in Terra Cimmeria have also been hypothesized to have been formed by convergent plates through collision and accretion of terranes (Fairén et al. 2002).

On the other hand, thermal evolution models have shown that the transition from the plate tectonics to the stagnant lid mode of mantle convection would have resulted in considerable heating and melting of the mantle (Breuer and Spohn 2003), leading to a late peak in crustal production. As a result, a substantial part of the martian crust would have been produced around 2 Gyr, contrary to observations (Norman 1999; Halliday et al. 2001; Nyquist et al. 2001; Phillips et al. 2001, see also Sect. 2.4). This, together with the persistent absence of evidence for subduction zones, argues against the plate tectonics hypothesis.

The unresolved issue of the existence or absence of plate tectonics on early Mars notwithstanding, the thermo-chemical evolution of Mars is generally modeled assuming that Mars has been in the stagnant lid mode of mantle convection ever since the onset of mantle convection (Spohn 1991; Hauck and Phillips 2002; Breuer and Spohn 2006; Parmentier and Zuber 2007; Fraeman and Korenaga 2010; Morschhauser et al. 2011). Typical cooling models are shown in Fig. 12(a), where reference viscosities spanning the range from wet (10^{19} Pa s) to dry (10^{21} Pa s) olivine have been considered. As pointed out above, the exponential dependence of mantle viscosity on temperature acts as a thermostat, and temperature profiles are similar irrespective of the initial temperature. Models predict

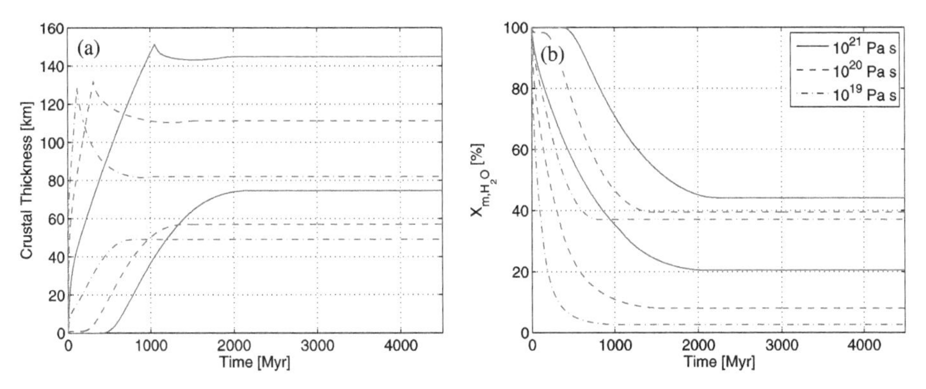

Fig. 13 (**a**) Crustal thickness as a function of time for the same models as in Fig. 12. Models showing crustal erosion are usually considered to be incompatible with observations (see text). (**b**) The bulk mantle water content as a percentage of the initial water content

present day mantle cooling rates of 40 to 50 K Gyr^{-1}, irrespective of mantle viscosity. These values are in general agreement with geochemical observations from which cooling rates of 30–40 K Gyr^{-1} have been derived (Baratoux et al. 2011a).

The rate of mantle cooling can also be described by the Urey ratio, which is given by the proportion of the radioactively produced energy output to the total energy loss of the planet. Thermal insulation by the stagnant lid makes heat loss inefficient and results in Urey ratios above unity during the early evolution, but Urey ratios tend towards their present day value close to 0.7 after the self-adjustment of mantle temperature and viscosity (Fig. 12(b)). This value is typical for stagnant lid planets, but much larger than the Urey ratio of the Earth, which is estimated to be 0.3–0.5 (Korenaga 2008; Nakagawa and Tackley 2012).

The chemical evolution of the mantle is characterized by the removal of crustal components and the extraction of incompatible elements during partial melting. Thermo-chemical evolution models with high initial temperatures possess voluminous melt zones, and crustal thicknesses quickly reach values in excess of 120 km (Fig. 13(a)). Due to the crustal enrichment of heat producing elements and the relatively low thermal conductivity of the crustal layer, these models accumulate heat in the stagnant lid, which can result in an erosion of the lid from below. In this way, crustal material can be delaminated and recycled back into the mantle (Morschhauser et al. 2011). However, isotopic anomalies in the martian meteorites (Chen and Wasserburg 1986; Harper et al. 1995; Borg et al. 1997; Brandon et al. 2000; Halliday et al. 2001) indicate an early mantle differentiation event and the formation of distinct geochemical reservoirs about 4.5 Ga ago, and insignificant reservoir mixing thereafter (Mezger et al. 2012, this issue). Therefore, crustal recycling is generally believed to be incompatible with this observation, and models showing crustal recycling are usually not considered (Morschhauser et al. 2011; Grott et al. 2011).

Crustal recycling is absent in models with low initial temperatures, and extracted crustal volumes are generally smaller. Models typically predict crustal thicknesses between 40 and 80 km (Hauck and Phillips 2002; Breuer and Spohn 2006; Morschhauser et al. 2011; Grott et al. 2011), and small crustal thicknesses are compatible with estimates derived from gravity and topography data, which predict the crust to be 57 ± 24 km thick on average (Wieczorek and Zuber 2004). The timing of crustal production is mainly a function of the reference viscosity. For large viscosities, convection is relatively sluggish and melt zones are pushed to greater depth, such that the mantle needs to heat up before partial melt can be produced.

On the other hand, low mantle viscosities allow for the production of partial melt during the early evolution. As most of the martian crust probably formed before 4 Gyr (Nimmo and Tanaka 2005), wet rheologies resulting from bulk mantle water contents of at least a few tens of ppm H_2O are generally preferred (Hauck and Phillips 2002; Breuer and Spohn 2006; Fraeman and Korenaga 2010; Morschhauser et al. 2011; Grott et al. 2011).

Apart from the bulk crustal thickness, parameterized models can give a first order estimate of the depth and degree of partial melting, the latter of which has a large influence on the efficiency of incompatible element removal. For small melt fractions, heat producing elements and water are efficiently removed from the deep interior, and melt fractions predicted by parameterized models are between 10 and 20 % during the early evolution, but drop below 10 % between 1 and 1.5 Gyr after accretion (cf. Figs. 12 and 13). Thereafter, large scale crustal production ceases, and present day melting is not predicted by parameterized models as melting is inhibited by the thick stagnant lid. Present day magmatism would be driven by a locally thickened crust (Schumacher and Breuer 2007), or the penetration of mantle plumes into the stagnant lid (Li and Kiefer 2007; Grott and Breuer 2009), but these processes are not considered in parameterized models. Ruling out models which show crustal recycling, about 50 % of the total inventory of incompatible elements is typically extracted from the interior (Fig. 13(b)). Successive depletion of the mantle is consistent with surface thorium abundances of the Hesperian and Amazonian volcanic provinces, which indicate that Hesperian volcanism was fed from a non-depleted mantle source, while Amazonian volcanics point towards a depleted mantle source during this time period (Baratoux et al. 2011a).

The link between planetary cooling models and the depth of partial melting is not straight-forward to make, as stagnant lid planets primarily cool by growth of the stagnant lid, while at the same time keeping the interior relatively warm. Depending on the initial temperature profile and mantle viscosity, stagnant lid planets can experience phases of mantle heating during the early evolution, and stagnant lid thickness need not increase monotonously (Breuer and Spohn 2003, 2006; Schumacher and Breuer 2006; Morschhauser et al. 2011). Therefore, there is no one-to-one correspondence between depth of melting and crystallization age. However, the situation is less complex during the later evolution following the Hesperian, when planetary cooling and growth of the stagnant lid push the zone of melting to ever increasing depths.

Taken together, results of thermo-chemical evolution calculations indicate that a wet mantle rheology is consistent with the martian crustal and elastic thickness evolution, where it should be noted that as little as a few tens of ppm water are rheologically significant. Hydrous melting becomes important if the mantle water content exceeds a few hundred ppm and is currently not considered in most models. Evolution models with low initial upper mantle temperatures are generally preferred, as these do not give rise to crustal erosion. The produced crustal volumes are in general agreement with estimates from gravity and topography studies (Wieczorek and Zuber 2004). Initial upper mantle temperatures on Mars were probably close to 1700 K, and secular cooling of 40–50 $\mathrm{K\,Gy^{-1}}$ resulted in average present day upper mantle temperatures of 1500 K, inhibiting global magmatic activity, that is furthermore hindered by the presence of a stagnant lid. However, localized activity may still be feasible in regions of a locally thickened crust or in the heads of uprising mantle plumes. Furthermore, approximately 50 % of the incompatible elements were likely removed from the mantle in the course of its entire evolution.

5.3 The Crustal Dichotomy

The surface of Mars shows specific features such as the prominent crustal dichotomy, whose formation cannot be addressed using one-dimensional models. The southern highlands and northern lowlands of Mars differ markedly in average elevation (Smith et al. 1999) and crustal thickness (Zuber et al. 2000; Neumann et al. 2004), the crust being about 25 km thicker in the highlands than in the lowlands. It is generally accepted that this crustal dichotomy is one of the oldest features on Mars, but the exact timing of the dichotomy formation is still debated. Estimates vary between the first 50 Ma after solar system formation (Solomon et al. 2005) and the Early Noachian period (>3.9 Ga) (Frey et al. 2002; Nimmo and Tanaka 2005). Possible explanations of the origin of the dichotomy invoke either exogenic (Wilhelms and Squyres 1984; Frey and Schultz 1988) or endogenic processes (Wise et al. 1979), but none of the proposed formation mechanisms appears to be fully convincing, which is in part due to the uncertainty in the timing of the dichotomy formation.

Proposed exogenic processes include formation by one (Wilhelms and Squyres 1984) or several (Frey and Schultz 1988) large impacts, and this idea has recently been revived in a series of articles explaining the observed elliptical shape of the lowlands (Andrews-Hanna et al. 2008; Nimmo et al. 2008; Marinova et al. 2008). However, an endogenic origin of the dichotomy cannot be ruled out.

In this respect, different mechanisms associated with (1) an early phase of plate tectonics (Sleep 1994); (2) an episode of degree-one (one-ridge) mantle convection (Schubert and Lingenfelter 1973; Wise et al. 1979; Zhong and Zuber 2001; Keller and Tackley 2009); (3) the evolution of an early magma ocean (Hess and Parmentier 2001; Elkins-Tanton et al. 2003); and (4) a thermal anomaly following a giant impact in the southern hemisphere (Reese et al. 2010; Golabek et al. 2011) have been proposed. As the plate tectonics hypothesis has been discussed in the previous chapter, we will focus the discussion here on points 2–4.

Low-degree mantle convection could be responsible for the formation of the dichotomy if a single large upwelling plume were present underneath the southern hemisphere. Such a flow structure can be obtained for particular mantle viscosity profiles, and an episode of long-wavelength mantle convection has been suggested to be the consequence of a layered mantle viscosity including a viscosity jump of more than a factor of 25 in the mid-mantle and a relatively low viscosity in the upper mantle (Zhong and Zuber 2001; Roberts and Zhong 2006). Alternatively, a mid-mantle viscosity jump linked to the martian mineralogical transition zone below 1000 km depth also promotes large scale (low-degree) flow (Keller and Tackley 2009). The time required to develop degree-one or low-ridge convection ranges from 100 Ma to several hundred Ma and is consistent with most assumptions about the timing of the dichotomy formation. However, these models result in continuous magmatic resurfacing of the southern hemisphere, thereby destroying the initially dichotomous crustal structure in the long-term.

Furthermore, such models implicitly assume the martian mantle to be homogeneous, but the existence of distinct geochemical reservoirs is suggested by isotopic anomalies in the martian meteorites (Chen and Wasserburg 1986; Harper et al. 1995; Borg et al. 1997; Brandon et al. 2000; Halliday et al. 2001). These data indicate an early mantle differentiation event about 4.5 Ga ago, and insignificant reservoir mixing thereafter. It seems unlikely that long-wavelength flow could maintain the separation of reservoirs to the present day, and dynamic models still need to be improved to be consistent with all observations.

One possible mechanism to form distinct and separate geochemical reservoirs is magma ocean solidification by fractional crystallization, which likely results in an unstable density stratification in the martian mantle. Late-stage silicates that crystallize at shallow depths are denser than earlier cumulates that crystallize near the base of the magma ocean (Hess and Parmentier 2001; Elkins-Tanton et al. 2003, 2005), and a potential long-wavelength overturn of this unstable situation may account for the crustal dichotomy (Hess and Parmentier 2001).

Although such a process would form distinct geochemical reservoirs, models of the interior evolution after mantle overturn indicate that different reservoirs remain dynamically strictly separated, because the density differences between different layers are large. This contradicts the small elastic lithosphere thicknesses derived for the Noachian period (Sect. 4.4), which require the lithosphere to be warm and the upper thermal boundary layer to be thin. Furthermore, a potential reservoir in the deep mantle would have no surface expression and thereby would remain'invisible', contrary to observations. Therefore, alternative explanations for the early formation of reservoirs that did not mix significantly during the subsequent evolution are needed. One such model includes the rapid formation of a harzburgite layer (Schott et al. 2001; Ogawa and Yanagisawa 2011) by early crustal formation, which would be less dense than the primordial mantle, and would remain stable as a consequence of dehydration and subsequent stiffening.

5.4 The Tharsis and Elysium Volcanic Provinces

Other surface features which cannot be understood using one dimensional models are the two large volcanic provinces of Tharsis and Elysium, which show longstanding and stable volcanic activity. Their presence argues for only a limited number of mantle plumes and the operation of low degree mantle convection. While activity in Tharsis and Elysium took place throughout martian history, these provinces likely acquired their present shape relatively early (e.g., Phillips et al. 2001; Solomon et al. 2005; Nimmo and Tanaka 2005), and as both Tharsis and Elysium are located near or at the dichotomy boundary, a dynamic link between these two features has been suggested (Wenzel et al. 2004; Zhong 2009; Šrámek and Zhong 2010, 2012). Taking the presence of a hemispheric lithospheric keel representing a depleted mantle reservoir into account, Šrámek and Zhong (2012) demonstrated that a stiff keel can induce a rotation of the entire lithosphere relative to the underlying mantle, until the upwelling is stabilized near the edge of the keel. This edge corresponds to the dichotomy boundary, thus accounting for the location of the Tharsis rise.

In addition to the aforementioned viscosity structure, low-degree mantle convection may also be caused by mantle phase changes like the olivine to spinel and spinel to perovskite transitions (e.g., Weinstein 1995; Harder and Christensen 1996; Breuer et al. 1998; Harder 2000; Buske 2006). In particular, the endothermic spinel to perovskite phase transition that might be located close to the core-mantle boundary (Sohl and Spohn 1997) would stabilize upwelling plumes and reduces their number to one or two. Although the long timescale necessary to form a single upwelling in most models is inconsistent with an early formation of either the dichotomy or the volcanic provinces (Roberts and Zhong 2006), more recent three dimensional convection models with strongly temperature dependent rheology suggest a shorter formation time scale, consistent with the observations (Buske 2006). Other models capable of generating longstanding plumes consider chemical layering (Schott et al. 2001; Wenzel et al. 2004), but these models do not predict volcanic provinces to be at the dichotomy boundary. Rather, volcanic activity is predicted to be focused in the center of regions with a thickened crust, contrary to observations.

6 Future Investigations

The combined analysis of data from orbiting and landed spacecraft, laboratory investigations of the martian meteorites, and theoretical modeling, have revealed a complex evolution of the crust-mantle system and much has been learned about the geological history of the planet. With the presently ongoing debates concerning the age of the shergottites and the water content of the martian mantle, a considerable range of scenarios for the magmatic history of Mars seems plausible, and the link between thermal models and petrological investigations remains non-unique. To make significant progress in quantifying the martian geochemical reservoirs, a new suite of space missions focusing on the geophysical exploration of Mars is needed. Furthermore, the return of carefully selected samples would be extremely valuable. In the following, we outline what observations are needed and what may be expected from future exploration and modeling efforts.

6.1 Mars Science Laboratory

The Mars Science Laboratory (MSL) called Curiosity has landed at the Hesperian aged Gale Crater, in which sediments showing evidence of sulfates and phyllosilicates can be accessed by the rover (Grotzinger 2011). Although the mission is dedicated to the search for possible traces of biological activity and the characterization of the surface environment, new in-situ observations of ancient igneous materials are possible. For example, the rover began the mission by traversing an alluvial fan and is at the time of writing providing new observations of igneous rocks. The payload, which includes an APXS instrument similar to that of the MERs, a remote laser-induced breakdown spectrometer (ChemCam) for chemical analyses, and an X-ray diffractometer (CheMin) for mineralogical analyses, is well suited to providing new information to address magmatic processes from these new samples.

6.2 Heat Flow

The average heat flow from a planet reflects the bulk abundance of heat producing elements in the planetary interior. In the absence of direct measurements, martian heat flow has been estimated from the deformation of the lithosphere as discussed above (e.g., Schultz and Watters 2001; McGovern et al. 2004; Grott et al. 2005; Ruiz et al. 2009 amongst others), yet large uncertainties are associated with this method. Heat flow estimates are then related to the time of deformation, which is again poorly known and may not be represented by the crater retention age of the corresponding surfaces (Beuthe et al. 2012).

In order to constrain the bulk abundance of heat producing elements in the martian interior, a suite of direct heat flow measurements is needed (Grott et al. 2007a; Dehant et al. 2012a). While a large number of measurements was needed to constrain the heat flow of the Earth (e.g., Pollack et al. 1993) the situation on Mars is probably less complex, and average surface heat flow can likely be constrained from measurements at only a few well chosen sites. Being a one-plate planet, heat flow variations on Mars are expected to be smaller than those on the Earth, although some geographical variation depending on factors like local concentration of radioactive elements and crustal thickness can be expected (Grott and Breuer 2010, see also Fig. 14). As such, heat flow around Terra Sirenum is expected to be elevated due to a local enrichment of heat producing elements in the crust, whereas heat flow in large impact basins is probably lower due to small crustal thicknesses. In order to quantify the average heat flow, measurements in three to four geologically representative regions (e.g., the northern lowlands, southern highlands, Tharsis, and Hellas) are needed, and given the average heat flow it is then straightforward to estimate the average heat production rate in the martian interior. This is due to the fact that the present day Urey ratio is almost independent of the planet's cooling history and close to 0.7 (see Sect. 5.2), and the bulk abundance of radioactive elements in the planetary interior can thus be linked to the surface heat flow.

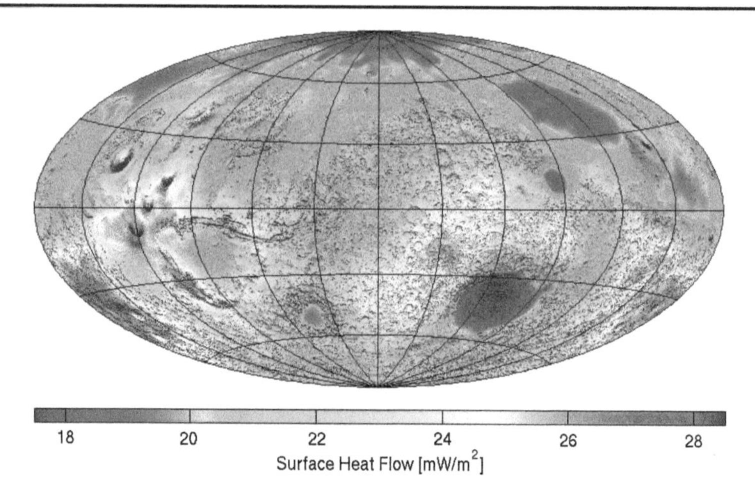

Fig. 14 Expected present day surface heat flow on Mars assuming bulk radioactive heat source abundances corresponding to the compositional model of Wänke and Dreibus (1994), crustal heat source abundances derived from gamma ray spectroscopy (Taylor et al. 2006), and crustal thicknesses derived from gravity and topography data (Neumann et al. 2004). Surface abundances of heat producing elements have been assumed to be representative for the entire crustal column (Taylor et al. 2006), and mantle heat flow has been calculated from a thermal evolution model resulting in a Urey ratio of 0.7 (Grott and Breuer 2010)

A comparison of heat flow in regions with locally thin and thick crust would furthermore indicate the crustal enrichment of heat producing elements with respect to the mantle. It is currently assumed that the crustal thickness inside the Hellas basin is extremely small and probably close to 5 km (Zuber et al. 2000; Neumann et al. 2004), such that heat flow inside the basin should primarily reflect its mantle value. By comparing heat flow values obtained inside the basin to others obtained in, for example, a representative highland area, much could be learned about processes like mantle melting and differentiation.

Other considerations which need to be taken into account include the possibility of active mantle plumes beneath the Tharsis and Elysium volcanic provinces. Crater counts suggest that some areas of Tharsis may have been volcanically active within the last tens of millions of years (Neukum et al. 2004; Hauber et al. 2011), and if plumes indeed exist underneath these regions, local heat flow would be expected to be elevated by a factor of two (Kiefer and Li 2009; Grott and Breuer 2010). Thus, a plume signature would be clearly visible in the data and heat flow measurements could help constrain the state of mantle convection, adding to the discussion of large scale mixing in the martian interior. The recently selected Discovery-class mission InSight (Banerdt et al. 2012) scheduled to launch in 2016 will carry a heat flow probe (Spohn et al. 2012) to the martian surface. In this way, an important baseline measurement will be conducted to constrain the heat flow from the martian interior.

6.3 Electrical Conductivity

A measurement of the electrical conductivity of the martian mantle would be a valuable indicator of the physical and chemical conditions in the planetary interior, in particular in light of the possible sensitivity to the presence of water (hydrogen) in the mantle (see, e.g., Poe et al. 2010; Karato and Wang 2011). To a lesser extent, electrical conductivity is also sensitive to other parameters such as temperature and iron content (Mg#) but these parameters can be estimated independently with a certain amount of confidence. As the distribution

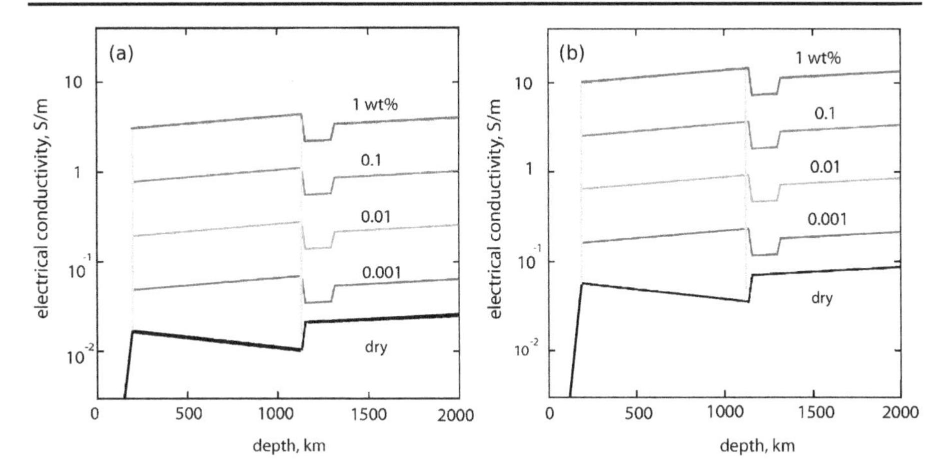

Fig. 15 (**a**) Electrical conductivity as a function of depth for different mantle water contents. The mantle transition zone representing phase changes in olivine is indicated by a kink in the profiles. An upper mantle temperature of 1700 K has been assumed. (**b**) Same as (**a**), but for an upper mantle temperature of 2000 K

of water reflects the history of the chemical evolution of Mars, knowledge of the electrical conductivity has the potential to place constraints on chemical evolution models. Electrical conductivity in the deep interior can be inferred from electromagnetic induction experiments (e.g., Grimm 2002), that can be conducted using a surface magnetometer, as has been done for the Moon (e.g., Sonett et al. 1971).

Electrical conductivity in minerals and rocks is determined by transfer of charged species, and in most minerals the dominant charge carriers are ferric iron- and/or hydrogen-related point defects (Karato and Wang 2011). Extensive laboratory studies have been carried out to determine the dependence of electrical conductivity on temperature, pressure, major element chemistry, water content and oxygen fugacity, and nearly complete data sets are now available for the electrical conductivity of most dominant minerals in the crust and mantle of Mars (olivine, orthopyroxene, clinopyroxene, garnet, wadsleyite and ringwoodite).

Using laboratory data on the influence of temperature, pressure, and water content on the electrical conductivity of mantle minerals (Romano et al. 2006, 2009; Xu et al. 1998, 2000), models of the conductivity-depth profile can be constructed if the temperature-depth profile is known. Figure 15 shows some synthetic profiles as a function of water content using the data compiled in Karato and Wang (2011). A small effect of higher Fe content has been taken into account. A comparison of Figs. 15(a) and (b) demonstrates that the uncertainties introduced by the imperfect knowledge of the mantle temperature are modest. In Fig. 15, water content was assumed to be constant as a function of depth, but layering could in principle be taken into account.

The way in which electrical conductivity depends on mantle water content is the subject of ongoing debate (Yoshino et al. 2006, 2009; Poe et al. 2010; Karato and Wang 2011; Yang 2012), but it is generally accepted that water plays an important role for the mantle conductivity structure. Note, however, that alternative models include the effects of carbonatite melts (Gaillard et al. 2008), which also show high electrical conductivities, indicating that the relationship between conductivity and water content may not be straightforward. Depending on the exact sensitivity to water content compared to other parameters, a measurement of the electrical conductivity in the mantle could yield valuable constraints on the

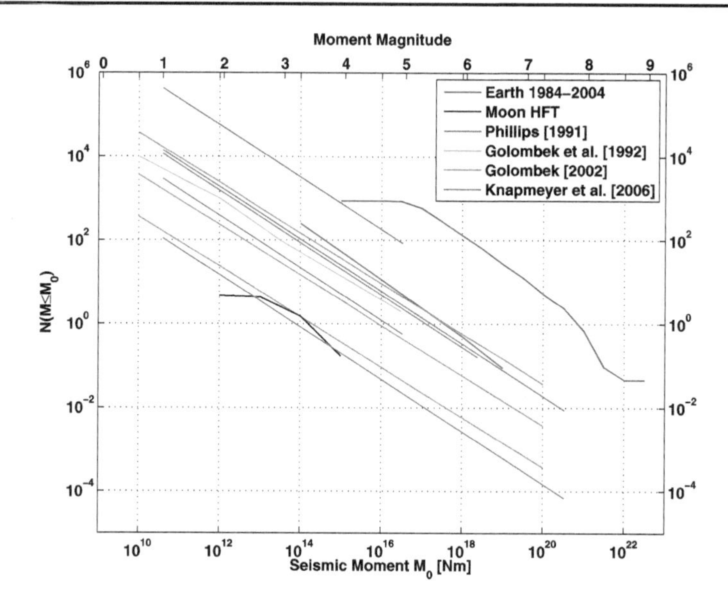

Fig. 16 Number of events per year that exceed a given seismic moment, according to different estimates (see *legend*). Data for the Earth is from the Harvard CMT project (Dziewonski et al. 1983 and annual follow-ups), which provides a source mechanism catalog complete for magnitudes larger than about 5. Data for the Moon (Oberst 1992) is for shallow moonquakes only, the contribution of the deep moonquakes is negligible

water content in the martian interior. In this way, one could start to quantify this fundamental geochemical reservoir, which is otherwise only indirectly constrained.

6.4 Seismology

Seismological experiments have been conducted on Mars using the seismometers onboard the two Viking landers (Anderson et al. 1977), but the experiment did not return a unique detection of any seismic event. This non-detection only supported the conclusion that Mars is likely seismically less active than Earth. Subsequent analysis has shown that the instrument characteristics, the mounting of the instrument on top of the lander bus, and the elastic attenuation within Mars would have prevented the detection of even a magnitude 9 event on Tharsis by the Viking 2 lander in Utopia Planitia (Goins and Lazarewicz 1979), such that the information gathered by the Viking experiments is of limited value.

On the other hand, several authors have shown that a significant seismicity is to be expected on Mars. Phillips (1991) evaluated different sources of elastic deformations and concluded that secular cooling of the martian lithosphere is the most effective one, resulting in a seismic moment release of 4.8×10^{18} Nm/yr (see Fig. 16 for a conversion of seismic moment into magnitudes). Golombek et al. (1992) estimated from an assessment of the total slip of visible faults that the annual moment release is about 10^{18} Nm. Knapmeyer et al. (2006) combined the approach of Phillips (1991) with new results on planetary cooling rates, a magnitude-frequency relation, and a geographic source distribution model. Based on the uncertainties of all input parameters, Knapmeyer et al. (2006) conclude that the moment release is between 3.4×10^{16} Nm/yr and 4.8×10^{18} Nm/yr (Fig. 16). Since Mars lacks the microseismic noise generated by oceans, even magnitude 4 events will be globally detectable by well-deployed modern instrumentation (e.g., Dehant et al. 2012a). If the Tharsis volcanoes are still active today, as suggested by Neukum et al. (2004), a spatial concentration of seismic activity beyond the distribution model proposed by Knapmeyer et al. (2006) may result. Further indications for seismic activity come from observations of fallen boulder populations in Cerberus Fossae (Roberts et al. 2012), as well as from faults crosscutting young lava flows (Vaucher et al. 2009).

The endogenous seismicity of Mars is complemented by meteorite influx. While Davis (1993) estimated from an upscaled meteorite flux of the Earth that an Apollo-type seismometer on Mars could expect to detect about 116 meteorite impacts per year, Teanby and Wookey (2011) estimated from the rate of direct cratering observations that globally detectable impacts occur only once every few years. The differences between these two estimates and also the flux that could be expected according to the "meteoritic hum" model of Lognonné et al. (2009) reflect the uncertainties of flux scaling between different planets and also of the estimated seismic efficiency of impacts.

Gravity-based models of the crustal thickness (Zuber et al. 2000) would benefit from one or more anchor points of seismological measurements, not only to resolve the thickness-density trade-off, but also to account for lateral variations of crustal density. Using a network of at least three stations, deployed in a seismically active region, this can be achieved using the classical inversion of travel time curves especially for the Moho head wave. A modern standard method that can be applied with teleseismic data recorded at a single three-component station, is the receiver function inversion of Zhu and Kanamori (2000). This approach yields a best fitting value not only for the Moho depth, but also for the Poisson ratio of the crustal rock. It does not necessarily need precise source locations if the crustal scattering is not too strong and thus the slowness of the incoming P wave can be estimated from the computation of the receiver functions themselves (Knapmeyer and Harjes 2000).

Core size is one of the main parameters controlling the existence of the perovskite phase transition in the deep martian mantle (Bertka and Fei 1998; Sohl and Spohn 1997), and knowledge of the core size would thus have important consequences for models of martian mantle dynamics (see Sect. 5.3). Furthermore, knowledge of the core size would constrain the size of the mantle reservoir. If seismic source location and time were known, a single station would suffice to determine the size of the martian core by exploiting the almost linear relationship between core size and the extent of the P wave core shadow (Oldham 1906; Knapmeyer 2011). Source localization could be achieved by monitoring meteorite impacts by camera observation of the flashes that are caused by thermal radiation of vaporized target material (Bouley et al. 2012). Observations of lunar impacts can be conducted with Earth based telescopes (Suggs et al. 2008), and it has been proposed to use impact flashes to localize seismic sources by Mimoun et al. (2012b). In the case of Mars, an orbiter could be parked in the Sun-Mars Lagrange point L2, from which it would always see the martian night side, or in a sun-synchronous orbit, which would be much closer to the martian surface but would also allow to view the surface at a specific local time.

Another approach that can be applied to data from only a single station is the interpretation of free oscillations of the planet. These allow inferring not only the seismic velocities, but also the density structure of the deep interior and the core size. Usually known for being excited by strong earthquakes, it has been found that Earth's free oscillations are continuously excited by dynamic pressures exerted by the atmosphere, and that the atmospheric activity of Mars is also likely to excite free oscillations of the planet (Kobayashi and Nishida 1998). The InSight mission (Banerdt et al. 2012) will carry a seismometer (Mimoun et al. 2012a) to the surface of Mars and establish a first single-station geophysical observatory.

6.5 Radio Science

The mean moment of inertia of Mars provides a global constraint on the radial density profile of Mars and as such contains information on the crust and mantle structure. The moment of inertia with respect to the polar axis has been obtained from radio science measurements in which two-way Doppler and ranging measurements are performed on radio links between ground stations on the Earth and the Mars Pathfinder and Viking landers on the surface of Mars. In addition, radio links between the Earth and orbiters around Mars have been used. From these measurements, the precession rate of Mars, which is inversely proportional to the polar moment of inertia, can be derived. By combining precession estimates with the accurately known degree-two gravitational field of Mars, the mean moment of inertia has been determined with a relative precision of about 0.1 % (Konopliv et al. 2011). Although this result is rather precise, no accurate direct inferences can be made on the size and composition of the mantle and crust because the moment of inertia of Mars also depends on the properties of the core. In particular, the core composition, temperature, and radius are not sufficiently well known to allow for an accurate determination of the core contribution to the moment of inertia. From indirect measurements of the tides on Mars raised by the Sun (Konopliv et al. 2011), the currently most accurate estimate of the core radius is 1794 ± 65 km (Rivoldini et al. 2011), corresponding to 53 % of the planetary radius.

Using radio tracking of a single lander over an extended period of time (one Martian year or more), the error on the precession estimate could be decreased by at least a factor of five (Dehant et al. 2009; Le Maistre et al. 2012). Recent analyses of several months of tracking data of the Mars Exploration Rover Opportunity already lead to an improved estimate of the precession rate (Le Maistre et al. 2010; Kuchynka et al. 2012). Further improvements are expected with multiple landers (Yseboodt et al. 2003) and interferometric techniques (Iess et al. 2012; Dehant et al. 2012b). This would set a stronger global constraint on the radial density profile in Mars, but more importantly the nutation (the periodic changes in the orientation) of Mars could be measured with such a radio science experiment. The main nutations of Mars have annual, semi-annual, and ter-annual periods in inertial space with amplitudes of up to about 500 milliarcsecond, corresponding to a displacement at the surface of Mars of about 8 m (Roosbeek 2000; Dehant et al. 2000). Nutation is particularly interesting since it depends on the moment of inertia of the core and some of the main nutations are amplified by a resonance with a free mode, the Free Core Nutation, representing a rigid rotation of the core around the rotation axis of the mantle. With a single equatorial lander, and one Martian year of Doppler data, the core moment of inertia could be determined with a precision of 30 % (Le Maistre et al. 2012) and significantly better precisions are expected with more advanced radio tracking experiments. The advantage of nutation observations in addition to precession is that the moments of inertia of the core and the silicate mantle and crust can be determined independently, allowing to set improved constraints on the density profile in both the core and the silicate outer part of Mars.

6.6 Sample Return

Returned samples of igneous martian rocks would dramatically increase our understanding of the planet‘s thermal evolution. Although the 104 martian meteorites provide important insights into critical topics such as mantle mineralogy and water content, the timing of differentiation and the thermal state of the interior (see, e.g., McLennan et al. 2011 and references therein), they have significant shortcomings. Despite several attempts, for example, the source regions of the SNC meteorites are unknown. Therefore, they lack field context, which limits their use as ground truth for orbital measurements (see Table 2 in McLennan et al. 2011 for a list of main shortcomings of SNCs as representative samples of Mars).

The ability to determine absolute model ages of the martian surface by analysis of crater size-frequency distributions is limited by the lack of radiometrically dated samples of a known origin, which are required to calibrate the crater dating method. For the case of the Moon, this critical knowledge comes from the samples collected and returned by the Apollo missions (e.g., Neukum and Ivanov 1994). For the case of Mars, a fresh igneous sample of early Amazonian age (e.g., 2 Ga) from a surface unit with a one-stage emplacement history would provide an important calibration measurement in the range of the cratering curve that is least constrained (Neukum 1988). Thus, sample return (or in situ dating by landers or rovers) would dramatically improve global dating from orbital imaging (Hartman and Neukum 2001).

Even younger volcanic rocks would also bear information on the range of volcanic processes and the role of water on recent Mars. On the other hand, samples of ancient igneous rocks have the potential to inform on early interior and surface processes. Early Noachian rocks might reveal details on silicate differentiation, on the crystallization of a magma ocean and its possible overturn, the formation of initial reservoirs and a primitive crust (McLennan et al. 2011). Preferred lithologies would include (a) cumulate-free basalts, from the chilled flow margin to the flow interior, to infer the properties of the parent melt, (b) ultramafic rocks with xenoliths from the mantle or lower crust, to study the mantle mineralogy and the thermal state of the martian interior, and (c) evolved igneous rocks, to study the diversity of igneous compositions and test hypotheses on the existence of felsic (granitic) rocks on Mars (e.g., Bandfield et al. 2004).

Paleomagnetic measurements of ancient martian rocks, both in situ or using returned samples, would be a major step towards a better knowledge of the intensity and timing of the magnetic field (Weiss et al. 2008). Such knowledge is essential to understand the thermal evolution of the martian core, the possibility of early plate tectonics, and the atmospheric loss on early Mars (Weiss et al. 2008). Ideally, strategies would focus on the acquisition of oriented samples from coherent bedrock (e.g., bedded basalts) with paleohorizonal indicators (Weiss et al. 2008), containing primary igneous magnetic minerals such as titanomagnetite and pyrrhotite (McLennan et al. 2011).

6.7 Theoretical and Laboratory Studies

The surface of Mars preserves a long-standing record of the planet's thermal evolution and this situation may be unique in the solar system. Estimates of lithospheric thicknesses, numerical modeling of mantle convection, mineralogical and chemical observations of igneous rocks (in-situ, from orbit, and in the lab) help reconstruct the planet's volcanic history, and geomorphological observations provide complementary insight into the magmatic evolution of the crust-mantle system. However, integrating the different observations into a consistent picture remains a challenging goal, which requires discrimination between global trends and local diversity, which naturally results from mantle heterogeneities and small-scale processes.

A major challenge for integrating different approaches is related to age determinations and there may be considerable timing discrepancies when characterizing a site using geophysical and petrological methods. While estimates of the local elastic thickness and heat flow usually date the time the bulk of a load was emplaced onto the surface (e.g., McGovern et al. 2002; Belleguic et al. 2005; Grott et al. 2005, 2011; Ruiz et al.

2009, 2011) petrological (Monders et al. 2007; Filiberto et al. 2010, 2011; Baratoux et al. 2011a) and geomorphological (e.g., Werner 2009; Robbins et al. 2011; Vaucher et al. 2009; Hauber et al. 2011) studies tend to date the last resurfacing events. Therefore, it is difficult to link the derived thermal state to the observed petrology, and a better understanding of the timing for the construction of volcanic surfaces is required.

Problems encountered when modeling the thermal evolution of the martian mantle include a lack of knowledge of the involved transport parameters, which generally have relatively large uncertainties. In particular, the thermal conductivity of the crust and lithospheric mantle are only known to within a factor of 2 for the deeper part (Breuer and Moore 2009), and due to their dependence on porosity, pore size, and cementation (Piqueux and Christensen 2009a, 2009b) their values are even more uncertain for the upper crust (<1 km). This range may be narrowed down by experimental studies of the thermal properties of iron-rich materials, analogous to the martian crust and mantle, and a determination of the regolith thickness. Thermodynamic relations and the process of melting in the iron-rich martian mantle also need to be addressed and validated by experimental investigations.

A major shortcoming when trying to link morphological observations to the nature of the lavas is the lack of experimental data on extra-terrestrial silicate melts such as the iron- and phosphorous-rich martian magmas (Chevrel et al. 2011a, 2011b). Lava rheology also bears on the process of magma ascent, which influences the chemistry of primitive melts by crustal contamination and fractional crystallization. Although geochemical observations suggest that a large proportion of the volcanically emplaced surfaces is in appearance consistent with primary melts of the primitive martian mantle (Baratoux et al. 2011a), the SNCs emphasize that details of the petrological processes need to be considered to explain the large range of Mg# and REE content (Meyer 2009). In addition, real primary melts are rare on Earth, which might indicate that spatial averaging on the regional scale might also mask the petrological complexity of crustal sections on Mars as inferred from orbit.

Evidence from remote sensing, in-situ analysis, and the analysis of the SNC meteorites suggests that primary and evolved liquids, crustal contamination, and cumulate rocks are present on Mars. The resolution of remote sensing instruments is now able to reveal, for instance, the magmatic signature of dikes along with that of their host rocks in the walls of Valles Marineris (Flahaut et al. 2011), and this opens up the possibility to explore the process of intrusive magma emplacement in more detail. In addition, the mineralogical analysis of a wide range of magmatic ages is important to confirm mineralogical trends, such as the change in pyroxene compositions at the Noachian/Hesperian boundary. Global changes in mineral chemistry may then be linked to the evolution of conditions for partial melting, and thus, to the thermal evolution of the martian mantle.

7 Summary and Conclusions

Present-day Mars operates in the stagnant lid mode of mantle convection, and geological evidence suggests that this has been the case since the planet's earliest evolution. In addition, isotopic ratios in the martian meteorites indicate that no significant mixing of early established geochemical reservoirs occurred (Chen and Wasserburg 1986; Harper et al. 1995; Borg et al. 1997; Brandon et al. 2000; Halliday et al. 2001), implying that crustal rock was not recycled back into the mantle. Therefore, the evolution of the crust-mantle system was driven by partial mantle melting, silicate differentiation, and volcanism. The martian surface bears witness to a rich history of volcanic processes and volcanism appears to have been a global process during the early evolution of the planet (Werner 2009). As planetary cooling progressed, activity became more and more focused, and the youngest volcanic surfaces are found exclusively in the Tharsis and Elysium volcanic provinces. Crater retention ages of these surfaces indicate a relatively recent emplacement, a possible indication that the process of mantle differentiation is ongoing today.

The total amount of mantle differentiation is poorly constrained and geophysical studies estimate the martian crust to be 57 ± 24 km thick on average (Wieczorek and Zuber 2004). Other information to constrain the amount of mantle differentiation is derived from photogeological estimates of crustal production rates (Greeley and Schneid 1991), but these estimates suffer from the fact that old surfaces are often covered by younger flows and are thus not accessible to investigation. In this sense, the scarcity of observed Noachian volcanic material is almost certainly an observational bias. In addition, only a minor part of the magma is expected to be extruded onto the surface, and intruded volumes are generally believed to be 5 to 10 times larger than extruded volumes (Lillis et al. 2009). Therefore, the crustal thickness must be viewed as uncertain to within a factor of two, a situation which could be improved by future seismological investigations (Dehant et al. 2012a).

The thickness of the martian crust is an indicator for the total amount of mantle differentiation, which remains equally uncertain. As the amount of incompatible elements removed from the mantle is proportional to the amount of produced crust, this uncertainty also affects the distribution of H_2O, K, Th, and U between crust and mantle. Numerical models indicate that about 50 % of these elements were removed from the mantle to the crust if an average crustal thickness of 50 km is assumed (Kiefer 2003; Morschhauser et al. 2011; Grott et al. 2011), but this value might need to be revised when better estimates of the total crustal thickness become available.

Information on the differentiation products, i.e., igneous rocks, comes from a variety of sources including remote sensing, in-situ analysis, and laboratory investigations. Martian igneous rocks are predominantly mafic in composition, and SiO_2 contents of 41–44 wt% and 44–47 wt% have been reported on the basis of gamma ray data for Amazonian and Hesperian surface units, respectively (Baratoux et al. 2011a; Gasnault et al. 2010). The composition of the Noachian crust is less well constrained, as it has been heavily altered. Unaltered representative units detectable at the scale of the gamma ray data resolution have not been unambiguously identified, and the composition of the Noachian crust is difficult to quantify from orbit. In-situ data generally agree with a mafic composition of the martian crust, and few more felsic compositions have been identified (Foley et al. 2003; Brückner et al. 2003).

Geochemical analysis of the SNC meteorites as well as remote sensing data indicate that igneous rocks on Mars may have crystallized from primary melts, with little magma storage at depth (Bertka and Fei 1997; Schmidt and McCoy 2010; Baratoux et al. 2011a). Although evidence for magmatic differentiation exists, it appears to be volumetrically minor. TES and THEMIS measurements of Nili Patera indicate a dacitic composition, enriched in SiO_2 relative to the basaltic composition that dominates Syrtis Major (Christensen et al. 2005), and even more extreme magmatic differentiation products rich in quartz may be present in other areas of the Syrtis Major region (Bandfield et al. 2004; Christensen et al. 2005; Bandfield 2006). Although these examples appear to be isolated, they nevertheless indicate that magmatic differentiation has occurred on Mars.

The composition of the martian crust as observed by gamma ray spectroscopy indicates that mantle melting during the Hesperian likely took place at temperatures between 1650 and 1700 K, and that magma source regions were located at depths between 90 and 110 km (Baratoux et al. 2011a). The degree of partial melting is estimated to have been 10–15 %, and a general trend from old, high temperature, high melt fraction melting to young, lower temperature, small melt fraction melting is observed. Accordingly, Amazonian volcanism is associated with temperatures between 1600 and 1670 K, source region depths around 150 km, and melt fractions between 5 and 12 % (Baratoux et al. 2011a). These estimates are consistent with depths derived from the Gusev crater basalts, which indicate source region depths around 85 km for these likely Hesperian aged rocks. Estimates of mantle temperature, source region depth, and melt fraction are in general agreement with numerical models of the thermo-chemical evolution of Mars, which predict magma source region depths around 100–200 km and melt fractions between 20 and 5 % during the Hesperian and Amazonian periods (Morschhauser et al. 2011; Grott et al. 2011).

Our understanding of the process of mantle differentiation is hampered by our lack of knowledge of the mantle water content, which has a large influence on the efficiency of mantle convection (Karato 2011) and could strongly affect the mantle solidus (e.g., Hirschmann 2000). Estimates of the mantle water content range from relatively dry (1–36 ppm H_2O) (Wänke and Dreibus 1994; Mysen et al. 1998; Filiberto and Treiman 2009) to moderately wet conditions (55–220 ppm H_2O) (McSween et al. 2001; McCubbin et al. 2010), but given that only small amounts of water are needed to be rheologically significant (Mei and Kohlstedt 2000a; Mei and Kohlstedt 2000b), the martian mantle is best described by a wet or only partially dehydrated rheology. Independent evidence for rheologically significant amounts of water in the martian crust and mantle comes from elastic thickness studies (Grott and Breuer 2010), and wet mantle rheologies are also preferred by thermo-chemical evolution models (Hauck and Phillips 2002; Grott et al. 2011; Morschhauser et al. 2011; Fraeman and Korenaga 2010). Geochemical analysis of the SNC meteorites indicates an upper limit of 220 ppm water in the martian mantle (McSween et al. 2001; McCubbin et al. 2010), which could result in a solidus reduction of $\sim$100 K (Katz et al. 2003). It therefore remains to be investigated whether petrologically significant amounts of water (in excess of $\sim$100 ppm) are present in the martian interior. While the bulk mantle water content remains poorly constrained, electro-magnetic sounding experiments would help to better quantify this value (Grimm 2002).

The thermo-chemical evolution of Mars is largely driven by the heat liberated in its interior, and the bulk abundance of heat producing elements has been estimated from the SNC meteorites. The most widely accepted compositional model (WD94, Wänke and Dreibus 1994) predicts abundances of 16 ppb U, 56 ppb Th, and 305 ppm K, and the predicted abundance ratio of K/Th closely matches the observed ratio of K/Th on the martian surface (Taylor et al. 2006). Radioactive elements in the WD94 model produce heat at a rate comparable to heat production in chondritic meteorites, and numerical models using these abundances are in good agreement with the inferred thermo-chemical evolution of the planet (Hauck and Phillips 2002; Grott et al. 2011; Morschhauser et al. 2011; Fraeman and Korenaga 2010). However, recent estimates of the elastic lithosphere thickness at the polar caps indicate a very cold lithosphere (Phillips et al. 2008), which could either imply a spatially heterogeneous mantle heat flow (Kiefer and Li 2009; Grott and Breuer 2010), or a sub-chondritic heat production rate in the martian interior. If the latter were true, models for the thermo-chemical evolution of Mars would need to be revisited, and in-situ heat flow measurements are needed to directly address this question (Kömle et al. 2011; Dehant et al. 2012a).

Acknowledgements M. Grott, E. Hauber, D. Breuer, and F. Sohl acknowledge funding by the Helmholtz Association through the research alliance "Planetary Evolution and Life". D. Baratoux and M.J. Toplis acknowledge funding from the Programme National de Planétologie of INSU. T. Van Hoolst acknowledges funding from the Belgian PRODEX program managed by the European Space Agency in collaboration with the Belgian Federal Science Policy Office.

References

M.H. Acuña, G. Kletetschka, J.E.P. Connerney, in *The Martian Surface: Composition, Mineralogy, and Physical Properties*, ed. by J.F. Bell III (Cambridge University Press, Cambridge, 2008), pp. 242–262, Chap. 11
J.B. Adams, J. Geophys. Res. **79**, 4829–4836 (1974)
R.A. Albert, R.J. Phillips, Geophys. Res. Lett. **27**(16), 2385–2388 (2000)
D.L. Anderson, W.F. Miller, G.V. Latham et al., J. Geophys. Res. **82**, 4524–4546 (1977)
J.C. Andrews-Hanna, M.T. Zuber, W.B. Banerdt, Nature **453**(7199), 1212–1215 (2008)
F. Anguita, C. Fernández, G. Cordero et al., Icarus **185**(2), 331–357 (2006)
R.E. Arvidson, S.W. Ruff, R.V. Morris et al., J. Geophys. Res. **113**, E12S33 (2008)
R.D. Ash, S.F. Knott, G. Turner, Nature **380**(6569), 57–59 (1996)
P.D. Asimow, C.H. Langmuir, Nature **421**, 815–820 (2003)
J.L. Bandfield, J. Geophys. Res. **107**, E6 (2002)
J.L. Bandfield, Geophys. Res. Lett. **33**, L06203 (2006)
J.L. Bandfield, V.E. Hamilton, P.R. Christensen, Science **287**, 1626–1630 (2000)
J.L. Bandfield, V.E. Hamilton, P.R. Christensen, H.Y. McSween, J. Geophys. Res. **109**(10), E10009 (2004)
W.B. Banerdt, S.E. Smrekar, L. Alkalai et al., *43rd Lunar and Planetary Science Conference*, abstract 2838 (2012)
A.R. Baptista, N. Mangold, V. Ansan et al., J. Geophys. Res. **113**(E12), E09010 (2008)
D. Baratoux, P. Pinet, A. Gendrin et al., J. Geophys. Res. **112**(E11), E08S05 (2007)
D. Baratoux, M.J. Toplis, M. Monnereau et al., Nature **472**, 338–341 (2011a)
D. Baratoux, N. Mangold, O. Arnalds et al., Earth Surf. Process. Landf. **36**, 1789–1808 (2011b)
D.N. Barnett, F. Nimmo, Icarus **157**(1), 34–42 (2002)
E.A. Basilevskaya, G. Neukum, Sol. Syst. Res. **40**, 375–383 (2006)
P. Beck, J.-A. Barrat, P. Gillet et al., Geochim. Cosmochim. Acta **70**, 2127–2139 (2006)
J.F. Bell III, T.B. McCord, P.D. Owensby, J. Geophys. Res. **95**, 14447–14461 (1990)
V. Belleguic, P. Lognonne, M. Wieczorek, J. Geophys. Res. **110**, E11005 (2005)
C.M. Bertka, Y. Fei, J. Geophys. Res. **107**, 5251–5264 (1997)
C.M. Bertka, Y. Fei, Science **281**, 1838–1840 (1998)
C.M. Bertka, J.R. Holloway, Contrib. Mineral. Petrol. **115**(3), 313–322 (1994a)
C.M. Bertka, J.R. Holloway, Contrib. Mineral. Petrol. **115**(3), 323–338 (1994b)
M. Beuthe, S. Le Maistre, P. Rosenblatt et al., J. Geophys. Res. **117**, E04002 (2012)
J.-P. Bibring, M. Combers, Y. Langevin et al., Nature **341**, 591–593 (1989)
J.-P. Bibring, A. Soufflot, M. Berthé et al., in *Mars Express: the Scientific Payload*, ed. by A. Wilson (ESA Publications, Noordwijk, 2004), pp. 37–49. ISBN:92-9092-556-6
J.-P. Bibring, Y. Langevin, A. Gendrin et al., Science **307**, 1576–1581 (2005)
J.-P. Bibring, Y. Langevin, J.F. Mustard et al., Science **312**, 400–404 (2006)
J.L. Bishop, M. Parente, C.M. Weitz et al., J. Geophys. Res. **114**(E13), E00D09 (2009)
J.E. Bleacher, R. Greeley, D.A. Williams et al., J. Geophys. Res. **112**, 9005 (2007)
D.D. Bogard, P. Johnson, Science **221**, 651–654 (1983)
L.E. Borg, L.E. Nyquist, L.A. Taylor et al., Geochim. Cosmochim. Acta **61**, 4915 (1997)
L.E. Borg, L.E. Nyquist, H. Wiesmann et al., *31 Lunar and Planetary Institute Science Conference*, abstract 1036 (2000)
L.E. Borg, L.E. Nyquist, H. Wiesmann, Y. Reese, Geochim. Cosmochim. Acta **66**, 2037–2053 (2002)
L.E. Borg, J.E. Edmunson, Y. Asmeron, Geochim. Cosmochim. Acta **69**, 5819–5830 (2005)
S. Bouley, D. Baratoux, J. Vaubaillon, Icarus **218**, 115–124 (2012)
A. Bouvier, J. Blichert-Toft, J.D. Vervoort, F. Albarède, Earth Planet. Sci. Lett. **240**, 221–233 (2005)
A. Bouvier, J. Blichert-Toft, J.D. Vervoort et al., Earth Planet. Sci. Lett. **266**, 105–124 (2008)
A. Bouvier, J. Blichert-Toft, F. Albarède, Earth Planet. Sci. Lett. **280**, 285–295 (2009)
W.V. Boynton, G.J. Taylor, L.G. Evans et al., J. Geophys. Res. **112**, E12S99 (2007)
W.V. Boynton, G.J. Taylor, S. Karunatillake et al., Elemental abundances determined via the Mars Odyssey GRS, in *The Martian Surface: Composition, Mineralogy, and Physical Properties*, ed. by J.F. Bell III (Cambridge University Press, Cambridge, 2008), pp. 105–124
W.V. Boynton, D.W. Ming, S.P. Kounavas et al., Science **325**(5936), 61–64 (2009)
A.D. Brandon, R.J. Walker, J.W. Morgan et al., Geochim. Cosmochim. Acta **64**, 4083–4095 (2000)
A.D. Brandon, I.S. Puchtel, R.J. Walker et al., Geochim. Cosmochim. Acta **76**, 206–235 (2012)
D. Breuer, W.B. Moore, in *Treatise of Geophysics*, vol. 10, (2009), pp. 299–341
D. Breuer, T. Spohn, J. Geophys. Res. **108**(E7), 507 (2003)
D. Breuer, T. Spohn, Planet. Space Sci. **54**, 153–169 (2006)
D. Breuer, D.A. Yuen, T. Spohn et al., Geophys. Res. Lett. **25**(3), 229–232 (1998)

J.C. Bridges, P.H. Warren, J. Geol. Soc. **163**, 229–251 (2006)
C.D. Brown, R.J. Phillips, J. Geophys. Res. **105**(B6), 13221–13238 (2000)
P. Brož, E. Hauber, Icarus **218**, 88–99 (2012)
J. Brückner, G. Dreibus, R. Rieder et al., J. Geophys. Res. **108**(E12), 8094 (2003)
R.G. Burns, *Mineralogical Applications of Crystal Field Theory*, 2nd edn. (Cambridge University Press, Cambridge, 1993a), 551 pp.
R.G. Burns, Origin of electronic spectra of minerals in the visible-near infrared region, in *Remote Geochemical Analysis: Elemental and Mineralogical Composition*, ed. by C. Pieters, P. Englert (Cambridge University Press, Cambridge, 1993b), pp. 3–29
E.B. Burov, M. Diament, J. Geophys. Res. **100**(B3), 3905–3927 (1995)
M. Buske, *Three-dimensional thermal evolution models for the interior of Mars and Mercury*, Ph.D. Thesis, University of Göttingen (2006)
M.H. Carr, R. Greeley, *Volcanic features of Hawaii*, NASA pub. SP-403 (1980), 211 pp.
M. Chen, A. El Goresy, Earth Planet. Sci. Lett. **179**, 489–502 (2000)
J.H. Chen, G.J. Wasserburg, Geochim. Cosmochim. Acta **50**, 955–968 (1986)
M. Chevrel, D. Dingwell, D. Baratoux, D. Giordano, *42nd International Union for Geodesy and Geophysics Conference*, Melbourne, Australia, abstract 4031 (2011a)
M. Chevrel, T. Platz, E. Hauber, D. Dingwell, *42nd International Union for Geodesy and Geophysics Conference*, Melbourne, Australia, abstract 4029 (2011b)
V. Chevrier, J.-P. Lorand, V. Sautter, Meteorit. Planet. Sci. **46**, 769–784 (2011)
F. Christen, O. Eugster, H. Buseman, Antarct. Meteor. Res. **18**, 117 (2005)
P.R. Christensen, H.J. Moore in *Mars*, ed. by H.H. Kieffer et al.(University of Arizona Press, Tucson, 1992) pp. 686–729
P.R. Christensen, J.L. Bandfield, V.E. Hamilton et al., J. Geophys. Res. **106**(E10), 23823–23872 (2001)
P.R. Christensen, J.L. Bandfield, J.F. Bell et al., Science **300**, 2056–2061 (2003)
P.R. Christensen, M.B. Wyatt, T.D. Glotch et al., Science **306**(5702), 1733–1739 (2004)
P.R. Christensen, H.Y. McSween, J.L. Bandfield et al., Nature **436**(7050), 504–509 (2005)
R.N. Clark, in *Spectroscopy of rocks and minerals and principles of spectroscopy, in Manual of Remote Sensing*, ed. by A.N. Rencz (Wiley, New York, 1999), pp. 3–58
R.N. Clayton, T.K. Mayeda, Geochim. Cosmochim. Acta **60**, 1999–2017 (1996)
E.A. Cloutis, M.J. Gaffey, T.L. Jackowski, J. Geophys. Res. **91**, 11641–11653 (1986)
R.P. Comer, S.C. Solomon, J.W. Head, Rev. Geophys. **23**, 61–92 (1985)
J.E.P. Connerney, M.H. Acuña, P.J. Wasilewski et al., Science **284**(5415), 794–798 (2000)
J.E.P. Connerney, M.H. Acuña, N.F. Ness et al., Proc. Natl. Acad. Sci. USA **102**(42), 14970–14975 (2005)
J.A. Crisp, J. Volcanol. Geotherm. Res. **20**, 177–211 (1984)
D.A. Crown, R. Greeley, J. Geophys. Res. **98**, 3431–3451 (1993)
L.S. Crumpler, J.W. Head, J.C. Aubele, Calderas on Mars: Characteristics, Structure, and Associated Flank Deformation, in *Volcano Instability on the Earth and Other Planets*, vol. 110, ed. by W.C. McGuire, A.P. Jones, J. Neuberg (Geological Society Special Publication, London, 1996), pp. 307–347
P.M. Davis, Icarus **105**, 469 (1993)
J.M.D. Day, L.A. Taylor, C. Floss et al., Meteorit. Planet. Sci. **41**, 581–606 (2006)
R.A. De Hon, in *5th Proc. Lunar Planet. Sci. Conf.*, (1974), pp. 2553–2561
R.A. De Hon, J. Geophys. Res. **87**, 9821–9828 (1982)
J.H. de Smet, A.P. van den Berg, N.J. Vlaar, Lithos **48**, 153–170 (1999)
V. Debaille, Q.-Z. Yin, A.D. Brandon, Earth Planet. Sci. Lett. **269**, 186–199 (2008)
V. Dehant, P. Defraigne, T. Van Hoolst, Phys. Earth Planet. Inter. **117**, 385–395 (2000)
V. Dehant, W. Folkner, E. Renotte et al., Planet. Space Sci. **57**, 1050–1067 (2009)
V. Dehant, B.W. Banerdt, P. Lognonné et al., Planet. Space Sci. (2012a). doi:10.1016/j.pss.2011.10.016
V. Dehant, M. Yseboodt, M. Mitrovic et al, Abstract EGU (2012b)
B. Diez, W.C. Feldman, N. Mangold et al., Icarus **200**, 19–29 (2009)
G. Dreibus, H. Wänke, Meteoritics **20**, 367–381 (1985)
A.M. Dziewonski, A. Friedman, D. Giardini et al., Phys. Earth Planet. Inter. **33**, 76–90 (1983)
B.L. Ehlmann, J.F. Mustard, G.A. Swayze et al., J. Geophys. Res. **114**(E13), E00D08 (2009)
M.R. El Maarry, O. Gasnault, M.J. Toplis et al., J. Volcanol. Geotherm. Res. **185** (2009)
L.T. Elkins-Tanton, E.M. Parmentier, P.C. Hess, Meteorit. Planet. Sci. **38**(12), 1753–1771 (2003)
L.T. Elkins-Tanton, P.C. Hess, E.M. Parmentier, J. Geophys. Res. **110**, E12S01 (2005)
R.E. Ernst, E.B. Grosfils, D. Mège, Annu. Rev. Earth Planet. Sci. **29**, 489–534 (2001)
L.G. Evans, R.C. Reedy, R.D. Starr et al., J. Geophys. Res. **111**, E03S04 (2006)
L.G. Evans, R.C. Reedy, R.D. Starr et al., *Lunar and Planetary Science Conference*, abstract 1875 (2008)
A.G. Fairén, J. Ruiz, F. Anguita, Icarus **160**(1), 220–223 (2002)
W.H. Farrand, L.R. Gaddis, L. Keszthelyi, J. Geophys. Res. **110**(E9), E05005 (2005)

W.H. Farrand, M.D. Lane, B.R. Edwards et al., Icarus **211**, 139–156 (2011)
C.I. Fassett, J.W. Head, Icarus **195**, 61–89 (2008)
U.H. Faul, I. Jackson, Earth Planet. Sci. Lett. **234**, 119–134 (2005)
W.C. Feldman, A. Pathare, S. Maurice et al., J. Geophys. Res. **116**(E11), E11009 (2011)
J.R. Filiberto, R. Dasgupta, Earth Planet. Sci. Lett. **304**, 527–537 (2011)
J.R. Filiberto, A.H. Treiman, Geology **37**(12), 1087–1090 (2009)
J.R. Filiberto, A.H. Treiman, L. Le, Meteorit. Planet. Sci. **43**, 1137–1146 (2008)
J.R. Filiberto, R. Dasgupta, W.S. Kiefer, A.H. Treiman, Geophys. Res. Lett. **37**, L13201 (2010)
J. Fink, J. Volcanol. Geotherm. Res. **4**, 151–170 (1978)
J. Flahaut, J.F. Mustard, C. Quantin et al., Geophys. Res. Lett. **38**, L15202 (2011)
R.J. Floran, M. Prinz, P.F. Hlava et al., Geochim. Cosmochim. Acta **42**, 1213–1229 (1978)
C.N. Foley, T. Economou, R.N. Clayton, J. Geophys. Res. **108**(E12), 8096 (2003)
A.A. Fraeman, J. Korenaga, Icarus **210**, 43–57 (2010)
H.V. Frey, R.A. Schultz, Geophys. Res. Lett. **15**, 229–232 (1988)
H.V. Frey, J.H. Roark, K.M. Shockey et al., Geophys. Res. Lett. **29**(10), 1384 (2002)
A.M. Gaffnee, L.E. Borg, Y. Asmeron et al., Meteorit. Planet. Sci. **46**, 35–52 (2011)
F. Gaillard, M. Malki, G. Iacono-Marziano et al., Science **322**, 1363–1365 (2008)
O. Gasnault, G.J. Taylor, S. Karunatillake et al., Icarus **207**, 226–247 (2010)
C. Goetze, B. Evans, Geophys. J. R. Astron. Soc. **59**, 463–478 (1979)
N.R. Goins, A.R. Lazarewicz, Geophys. Res. Lett. **6**, 368–370 (1979)
G.J. Golabek, T. Keller, T.V. Gerya et al., Icarus **215**(1), 346–357 (2011)
M.P. Golombek, W.B. Banerdt, K.L. Tanaka et al., Science **258**(5084), 979–981 (1992)
C.A. Goodrich, Meteorit. Planet. Sci. **37**, 31 (2002)
J.A. Grant, P.H. Schultz, Icarus **84**, 166–195 (1990)
O. Grasset, E.M. Parmentier, J. Geophys. Res. **103**, 18171–18181 (1998)
R. Greeley, D.A. Crown, J. Geophys. Res. **95**, 7133–7149 (1990)
R. Greeley, S.A. Fagents, J. Geophys. Res. **106**, 20527–20546 (2001)
R. Greeley, B.D. Schneid, Science **254**, 996–998 (1991)
R. Greeley, P.D. Spudis, Rev. Geophys. Space Phys. **19**, 13–41 (1981)
R. Greeley, N.T. Bridges, D.A. Crown et al., Volcanism on the Red Planet: Mars, in *Environmental Effects on Volcanic Eruptions: from Deep Oceans to Deep Space*, ed. by J. Zimbelman, T. Gregg (Kluwer Academic/Plenum Publishers, New York, 2000), pp. 75–112
R. Greeley, B.H. Foing, H.Y. McSween et al., J. Geophys. Res. **110**, E05008 (2005)
T.K.P. Gregg, S.N. Williams, Icarus **122**, 397–405 (1996)
R.W. Griffiths, J.H. Fink, J. Geophys. Res. **97**, 19739–19748 (1992)
R.E. Grimm, J. Geophys. Res. **107**(E2), 5006 (2002)
J. Gross, J. Filiberto, A.H. Treiman et al. *43rd LPSC*, abstract 2693 (2012)
M. Grott, D. Breuer, Icarus **193**(2), 503–515 (2008)
M. Grott, D. Breuer, Icarus **201**(2), 540–548 (2009)
M. Grott, D. Breuer, J. Geophys. Res. **115**, E03005 (2010)
M. Grott, M.A. Wieczorek, Icarus **221**, 43–52 (2012)
M. Grott, E. Hauber, S.C. Werner et al., Geophys. Res. Lett. **32**, L21201 (2005)
M. Grott, E. Hauber, S.C. Werner et al., Icarus **186**, 517–526 (2007a)
M. Grott, A. Morschhauser, D. Breuer, E. Hauber, Earth Planet. Sci. Lett. **308**(3–4), 391–400 (2011)
J.P. Grotzinger, The Mars Science Laboratory Mission. Abstr. Program - Geol. Soc. Am. **43**(5), 411 (2011)
J.P. Grotzinger, R.E. Milliken, The Sedimentary Rock Record of Mars: Distribution, Origins, and Global Stratigraphy, in *Sedimentary Geology of Mars*, ed. by J.P. Grotzinger, R.E. Milliken (SEPM Special Publication, Tulsa, 2011)
J.P. Grotzinger, R.E. Arvidson, J.F. Bell III et al., Earth Planet. Sci. Lett. **240**, 11–72 (2005)
A.N. Halliday, H. Wanke, J.L. Birck, R.N. Clayton, Space Sci. Rev. **96**, 197–230 (2001)
V.E. Hamilton, P.R. Christensen, Geology **33**, 433 (2005)
V.E. Hamilton, P.R. Christensen, H.Y. McSween et al., Meteorit. Planet. Sci. **38**, 871–885 (2003)
H. Harder, Geophys. Res. Lett. **27**(3), 301–304 (2000)
H. Harder, U.R. Christensen, Nature **380**, 507–509 (1996)
C.L. Harper, L.E. Nyquist, B. Bansal et al., Science **267**(5195), 213–217 (1995)
W.K. Hartman, G. Neukum, Space Sci. Rev. **96**, 165–194 (2001)
W.K. Hartmann, M.C. Malin, A.S. McEwen et al., Nature **397**, 586–589 (1999)
E. Hauber, S. van Gasselt, B. Ivanov et al., Nature **434**, 356–361 (2005)
E. Hauber, J. Bleacher, K. Gwinner et al., J. Volcanol. Geotherm. Res. **185**, 69–95 (2009)
E. Hauber, P. Brož, F. Jagert et al., Geophys. Res. Lett. **38**, L10201 (2011)
S.A. Hauck, R.J. Phillips, J. Geophys. Res. **107**(E7), 5052 (2002)

E.H. Hauri, T. Weinreich, A.E. Saal et al., Science **333**, 213–215 (2011)
J.W. Head, R. Greeley, M.P. Golombek et al., Space Sci. Rev. **96**, 263–292 (2001)
C.D.K. Herd, L.E. Borg, J.H. Jones, J.J. Papike, Geochim. Cosmochim. Acta **66**, 2025–2036 (2002)
C. Herzberg, P. Raterron, J. Zhang, Geochem. Geophys. Geosyst. **1**, 1051 (2000)
P.C. Hess, E.M. Parmentier, 32nd Annual Lunar and Planetary Science Conference, abstract 1319 (2001)
H. Hiesinger, J.W. Head, J. Geophys. Res. **109**, E01004 (2004)
H. Hiesinger, J.W. Head, G. Neukum, J. Geophys. Res. **112**, E05011 (2007)
M.M. Hirschmann, Annu. Rev. Earth Planet. Sci. **34**, 629–653 (2006)
M.M. Hirschmann, Geochem. Geophys. Geosyst. **1**, 1042 (2000)
G. Hirth, D.L. Kohlstedt, Earth Planet. Sci. Lett. **144**(1–2), 93–108 (1996)
C. Hodges, H. Moore, *Atlas of Volcanic Landforms on Mars*, US Geological Survey (1994)
T.M. Hoefen, R.N. Clark, J.L. Bandfield et al., Science **302**, 627–630 (2003)
T. Hoogenboom, S.E. Smrekar, Earth Planet. Sci. Lett. **248**(3–4), 830–839 (2006)
B.H. Horgan, J.F. Bell III, Geology **40**, 391–394 (2012)
W. Hsu, Y. Guan, H. Wang, Meteorit. Planet. Sci. **39**, 701–709 (2004)
G. Hulme, Geophys. J. R. Astron. Soc. **39**, 361–383 (1974)
B.M. Hynek, R.J. Phillips, R.E. Arvidson, J. Geophys. Res. **108**, 5111 (2003)
L. Iess, S. Giuliani, V. Dehant, Abstract EGU (2012)
P.J. Isaacson, C.M. Pieters, Icarus **210**, 8–13 (2009)
H. Iwamori, Geophys. Res. Lett. **19**, 309–312 (1992)
H. Iwamori, Earth Planet. Sci. Lett. **114**, 301–313 (1993)
R.A. Jacobson, Astron. J. **139**, 668–679 (2010)
W.L. Jaeger, L.P. Keszthelyi, A.S. McEwen et al., Science **317**, 1709 (2007)
E. Jagoutz, H. Wänke, Geochim. Cosmochim. Acta **50**, 939–953 (1986)
E. Jagoutz, A. Sorowka, J.D. Vogel, H. Wänke, Meteoritics **290**, 478–479 (1994)
R. Jaumann, G. Neukum, T. Behnke et al., Planet. Space Sci. **55**, 7–8, 928–952 (2007)
H.J. Jefffreys, Philos. Mag. **49**(6), 793–807 (1925)
J.R. Johnson, P.R. Christensen, P.G. Lucey, J. Geophys. Res. **107**(E6), 5035 (2002)
L.C. Kanner, J.F. Mustard, Icarus **187**, 442–456 (2007)
S. Karato, Nature **319**, 309–310 (1986)
S. Karato, *Deformation of Earth Materials: Introduction to the Rheology of the Solid Earth* (Cambridge University Press, Cambridge, 2008)
S. Karato, Rheological properties of minerals and rocks, in *Physics and Chemistry of the Deep Earth*, ed. by S. Karato (Wiley-Blackwell, New York, 2011)
S. Karato, H. Jung, Earth Planet. Sci. Lett. **157**, 193–207 (1998)
S. Karato, D. Wang, Electrical conductivity of minerals and rocks, in *Physics and Chemistry of the Deep Earth*, ed. by S. Karato (Wiley-Blackwell, New York, 2011)
S. Karunatillake, J.M. Keller, S.W. Squyres et al., J. Geophys. Res. **112**, E08S90 (2007)
S. Karunatillake, J.J. Wray, S.W. Squyres et al., J. Geophys. Res. **114**(E12), E12001 (2009)
S. Karunatillake, S.W. Squyres, O. Gasnault et al., J. Sci. Comput. **46**, 439–451 (2011)
R.F. Katz, M. Spiegelman, C.H. Langmuir, Geochem. Geophys. Geosyst. **4**(9), 1073 (2003)
T. Kawazoe, S. Karato, J. Ando et al., J. Geophys. Res. **115**, B08208 (2010). doi:10.1029/2009JB007096
T. Keller, P.J. Tackley, Icarus **202**(2), 429–443 (2009)
J.M. Keller et al., J. Geophys. Res. **111**, E03S08 (2006a)
J.M. Keller, W.V. Boynton, R.M.S. Williams et al., *Lunar and Planetary Science Conference*, abstract 2343 (2006b)
L. Kerber, J.W. Head, J.-B. Madelein et al., Icarus **216**, 212–220 (2011)
L. Keszthelyi, W. Jeager, A. McEwen et al., J. Geophys. Res. **113**, 4005 (2008)
L.P. Keszthelyi, W.L. Jaeger, C.M. Dundas et al., Icarus **205**, 211–229 (2010)
W.S. Kiefer, Meteorit. Planet. Sci. **38**(12), 1815–1832 (2003)
W.S. Kiefer, Earth Planet. Sci. Lett. **222**, 349–361 (2004)
W.S. Kiefer, Q. Li, Geophys. Res. Lett. **36**, L18203 (2009)
M. Knapmeyer, Planet. Space Sci. **59**, 1062–1068 (2011)
M. Knapmeyer, H.-P. Harjes, Geophys. J. Int. **143**, 1–21 (2000)
M. Knapmeyer, J. Oberst, E. Hauber et al., J. Geophys. Res. **111**(E11), E11006 (2006)
N. Kobayashi, K. Nishida, Nature **395**, 357–360 (1998)
W.C. Koeppen, V.E. Hamilton, J. Geophys. Res. **113**, E05001 (2008)
D.L. Kohlstedt, S.J. Mackwell, Z. Phys. Chem. **207**, 147–162 (1998)
N.I. Kömle, E.S. Hütter, W. Macher et al., Planet. Space Sci. **59**(8), 639–660 (2011)
A.S. Konopliv, S.W. Asmar, W.M. Folkner et al., Icarus **211**, 401–428 (2011)
J. Korenaga, Rev. Geophys. **46**, RG2007 (2008)

J. Korenaga, Geophys. J. Int. **179**, 154–170 (2009)
D.A. Kring, J.D. Gleason, T.D. Swindle et al., Meteorit. Planet. Sci. **38**, 1833–1848 (2003)
P. Kronberg, E. Hauber, M. Grott et al., J. Geophys. Res. **112**, E04005 (2007)
P. Kuchynka, W.M. Folkner, R.S. Park et al. DPS abstract 12-RC-598-AAS-DPS, DPS (2012)
I. Kushiro, Annu. Rev. Earth Planet. Sci. **29**, 71–107 (2001)
P.D. Lanagan, A.S. McEwen, L.P. Keszthelyi, T. Thodarson, Geophys. Res. Lett. **28**, 2365–2368 (2001)
N.P. Lang, L.L. Tornabene, H.Y. McSween, P.R. Christensen, J. Volcanol. Geotherm. Res. **185**, 103–115 (2009)
B. Langlais, M.E. Purucker, M. Mandea, J. Geophys. Res. **109**, E2 (2004)
J.K. Lanz, R. Wagner, U. Wolf et al., J. Geophys. Res. **115**, 12019 (2010)
T.J. Lapen, M. Righter, A.D. Brandon et al., Science **328**, 347–350 (2010)
J. Lasue et al., Space Sci. Rev. (2012, this issue). doi:10.1007/s11214-012-9946-5
S. Le Maistre, W.M. Folkner, P. Rosenblatt et al., *European Planetary Science Congress*. Extended abstract EPSC2010-191 (2010)
S. Le Maistre, P. Rosenblatt, A. Rivoldini et al., Planet. Space Sci. **68**(1), 105–122 (2012). doi:10.1016/j.pss.2011.12.020
A. Lenardic, F. Nimmo, L. Moresi, J. Geophys. Res. **109**, E02003 (2004)
R.C.F. Lentz, G.J. Taylor, A.H. Treiman, Meteorit. Planet. Sci. **34**, 919–932 (1999)
L.A. Leshin, Geophys. Res. Lett. **27**, 14 (2000)
L.A. Leshin, S. Epstein, E.M. Stolper, Geochim. Cosmochim. Acta **60**(14), 2635–2650 (1996)
K. Lewis, O. Aharonson, J.P. Grotzinger et al., J. Geophys. Res. **113**, E12S36 (2008)
Q. Li, W.S. Kiefer, Geophys. Res. Lett. **34**(16), L16203 (2007)
R.I. Lillis, H.V. Frey, M. Manga et al., Icarus **194**, 575–596 (2008)
R.J. Lillis, J. Dufek, J.E. Bleacher et al., J. Volcanol. Geotherm. Res. **185**, 123–138 (2009)
K. Litasov, E. Ohtani, Phys. Earth Planet. Inter. **134**, 105–127 (2002)
P. Lognonné, M. Le Feuvre, C.L. Johnson et al., J. Geophys. Res. **114**, E12003 (2009)
D. Loizeau, M. Mangold, F. Poulet, Icarus **205**, 396–418 (2010)
J.-P. Lorand, V. Chevrier, V. Sautter, Meteorit. Planet. Sci. **40**, 1257 (2005)
M.C. Malin, J.F. Bell, B.A. Cantor et al., J. Geophys. Res. **112**(E5), E05S04 (2007)
M. Manga, A. Patel, J. Dufek, E.S. Kyte, Geophys. Res. Lett. **39**, 1202 (2012)
N. Mangold, F. Poulet, J.F. Mustard et al., J. Geophys. Res. **112**(E11), E08S04 (2007)
N. Mangold, D. Loizeau, F. Poulet et al., Earth Planet. Sci. Lett. **294**, 440–450 (2010)
M.M. Marinova, O. Aharonson, E. Asphaug, Nature **453**(7199), 1216–1219 (2008)
S. Maurice, W. Feldman, B. Diez et al., J. Geophys. Res. **116**, E11008 (2011)
F.M. McCubbin, A. Smirnov, H. Nekvasil et al., Earth Planet. Sci. Lett. **292**(1–2), 132–138 (2010)
F.M. McCubbin, E.H. Hauri, S.M. Elardo et al., *LPSC 43*, abstract 1121 (2012)
A.S. McEwen, M. Malin, M. Carr, Nature **397**, 584–586 (1999)
A.S. McEwen, E.M. Eliason, J.W. Bergstrom et al., J. Geophys. Res. **112**(E5), E05S02 (2007)
P.J. McGovern, S.C. Solomon, D.E. Smith et al., J. Geophys. Res. **107**, E12 (2002)
P.J. McGovern, S.C. Solomon, D.E. Smith et al., J. Geophys. Res. **109**, E07007 (2004)
D. McKenzie, M.J. Bickle, J. Pet. **29**, 625–679 (1988)
S.M. McLennan, M.A. Sephton, C. Allen et al., *Planning for Mars Returned Sample Science: Final report of the MSR End-to-End International Science Analysis Group (E2E-iSAG)* (2011). The Mars Exploration Program Analysis Group (MEPAG) at http://mepag.jpl.nasa.gov/reports/, 101 pp.
M. McNutt, J. Geophys. Res. **89**, 11180–11194 (1984)
H.Y. McSween, Rev. Geophys. **23**, 391–416 (1985)
H.Y. McSween, Meteoritics **29**, 757–779 (1994)
H.Y. McSween, Meteorit. Planet. Sci. **36**, 7–25 (2002)
H.Y. McSween, A.H. Treiman, Martian meteorites, in *Planetary Materials*, ed. by J.J. Papike. Revs. Mineral., vol. 36, (1998), pp. 6–53
H.Y. McSween, T.L. Grove, R.C.F. Lentz et al., Nature **409**, 487–490 (2001)
H.Y. McSween et al., Science **305**, 842–845 (2004)
H.Y. McSween, S.W. Ruff, R.V. Morris et al., J. Geophys. Res. **111**, E09S91 (2006)
H.Y. McSween, S.W. Ruff, R.V. Morris et al., J. Geophys. Res. **113**, E06S04 (2008)
H.Y. McSween, G.J. Taylor, M.B. Wyatt, Science **324**, 736–739 (2009)
D. Mège, P. Masson, Planet. Space Sci. **44**, 1499–1546 (1996)
S. Mei, D.L. Kohlstedt, J. Geophys. Res. **105**(B9), 21471–21482 (2000a)
S. Mei, D.L. Kohlstedt, J. Geophys. Res. **105**(B9), 21457–21470 (2000b)
C. Meyer, *The Mars Meteorite Compendium*, Johnson Space Center contribution 27672. Houston, Texas, USA (2009). http://curator.jsc.nasa.gov/antmet/mmc/index.cfm
K. Mezger, V. Debaille, T. Kleine, Space Sci. Rev. (2012, this issue). doif:10.1007/s11214-012-9935-8

J.R. Michalski, R.L. Fergason, Icarus **199**, 24–48 (2009)
T. Mikouchi, M. Miyamoto, in *38th Lunar and Planetary Science Conference*, vol. 1343 (2002)
M.P. Milazzo, L.P. Keszthelyi, W.L. Jaeger et al., Geology **37**, 171–174 (2009)
D. Mimoun, P. Lognonné, W.B. Banerdt et al., *43rd Lunar and Planetary Science Conference*, abstract 1493 (2012a)
D. Mimoun, M. Wieczorek, L. Alkalai et al., Exp. Astron. (2012b). doi:10.1007/s10686-011-9252-3
K. Misawa, C.-Y. Shih, Y. Reese et al., Earth Planet. Sci. Lett. **246**(1–2), 90–101 (2006)
D.W. Mittlefehldt, Meteoritics **29**(2), 214–221 (1994)
A.G. Monders, E. Médart, T.L. Grove, Meteorit. Planet. Sci. **42**, 131–148 (2007)
L.G.J. Montési, M.T. Zuber, J. Geophys. Res. **108**, E6 (2003)
R.V. Morris, H.V. Lauer Jr., C.A. Lawson et al., J. Geophys. Res. **90**, 3126–3144 (1985)
R.V. Morris, G. Klingelhofer, C. Schroder et al., J. Geophys. Res. **113**, E12S42 (2008)
A. Morschhauser, M. Grott, D. Breuer, Icarus **212**(2), 541–558 (2011)
P.J. Mouginis-Mark, J.W. Head, L. Wilson, J. Geophys. Res. **87**, 9890–9904 (1982)
P.J. Mouginis-Mark, A.J.L. Harris, S.K. Rowland, M.G. Chapman (eds.), *Terrestrial Analogs to the Calderas of the Tharsis Volcanoes on Mars the Geology of Mars: Evidence from Earth-Based Analog* (Cambridge University Press, Cambridge, 2007), p. 71
S. Murchie, R. Arvidson, P. Bedini et al., J. Geophys. Res. **112**, E05S03 (2007)
D.S. Musselwhite, H.A. Dalton, W.S. Kiefer, A.H. Treiman, Meteorit. Planet. Sci. **41**(9), 1271–1419 (2006)
J.F. Mustard, J.M. Sunshine, Science **267**, 1623–1626 (1995)
J.F. Mustard, S. Erard, J.-P. Bibring et al., J. Geophys. Res. **98**, 3387–3400 (1993)
J.F. Mustard, S.L. Murchie, S. Erard, J.M. Sunshine, J. Geophys. Res. **102**, 25605–25615 (1997)
J.F. Mustard, F. Poulet, A. Gendrin et al., Science **307**, 1595–1597 (2005)
J.F. Mustard, S.L. Murchie, S.M. Pelkey et al., Nature **454**, 305–309 (2008)
J.F. Mustard, B.L. Ehlmann, S.L. Murchie et al., J. Geophys. Res. **114**(E13), E00D12 (2009)
B.O. Mysen, D. Virgo, R.K. Popp, C.M. Bertka, Am. Mineral. **83**, 942–946 (1998)
T. Nakagawa, P.J. Tackley, Earth Planet. Sci. Lett. **329–330**, 1–10 (2012)
N. Nakamura, D.M. Unruh, M. Tatsumoto, R. Hutchuson, Geochim. Cosmochim. Acta **46**, 155–1573 (1982)
G. Neukum, *Workshop on Mars Sample Return Science* (Lunar Planet. Inst., Houston, 1988), pp. 128-129
G. Neukum, K. Hiller, J. Geophys. Res. **86**, 3097–3121 (1981)
G. Neukum, B.A. Ivanov, in *Hazards due to Comets and Asteroids*, ed. by T. Gehrels et al.(University of Arizona Press, Tucson, 1994), pp. 359–416
G. Neukum, R. Jaumann, H. Hoffmann et al., Nature **432**, 971–979 (2004)
G. Neukum, A.T. Basilevsky, T. Kneissl et al., Earth Planet. Sci. Lett. **294**, 204–222 (2010)
G.A. Neumann, M.T. Zuber, M.A. Wieczorek et al., J. Geophys. Res. **109**, E08002 (2004)
H.E. Newsom, L.S. Crumpler, R.C. Reedy et al., J. Geophys. Res. **112**(E11), E03S12 (2007)
F. Nimmo, D. Stevenson, J. Geophys. Res. **105**(E5), 11969–11979 (2000)
F. Nimmo, K. Tanaka, Annu. Rev. Earth Planet. Sci. **33**, 133–161 (2005)
F. Nimmo, S.D. Hart, D.G. Korycansky et al., Nature **453**(7199), 1220–1223 (2008)
Y. Nishihara, D. Tinker, T. Kawazoe et al., Phys. Earth Planet. Inter. **170**, 156–169 (2008)
M.D. Norman, Meteorit. Planet. Sci. **34**, 439–449 (1999)
L.E. Nyquist, B.M. Bansal, H. Wiesmann, C.-Y. Shih, *26th Lunar and Planetary Institute Science Conference*, abstracts 1065 (1995)
L.E. Nyquist, D.D. Bogard, C.Y. Shih et al., Space Sci. Rev. **96**, 105–164 (2001)
J. Oberst, J. Geophys. Res. **92**(B2), 1397–1405 (1992)
A. Ody, F. Poulet, Y. Langevin et al., J. Geophys. Res. **117**, E00J14 (2012)
M. Ogawa, T. Yanagisawa, J. Geophys. Res. **116**, E08008 (2011)
E. Ohtani, N. Kamaya, Geophys. Res. Lett. **19**, 2239–2242 (1992)
R.D. Oldham, Q. J. Geol. Soc. Lond. **62**, 456–475 (1906)
R. Ostertag, D. Stoffler, in *Proceedings of 13th Lunar and Planetary Science Conference*, (1982), pp. 457–463
E.A. Parfitt, L. Wilson, *Fundamentals of Physical Volcanology* (Blackwell, Oxford, 2008), 230 pp.
J. Park, D.H. Garrison, D.D. Bogard, Geochim. Cosmochim. Acta **73**, 2177–2189 (2009)
E.M. Parmentier, M.T. Zuber, J. Geophys. Res. **112**, E02007 (2007)
R.J. Phillips, LPI Tech. Rep. 91-02 LPI/TR-91-02, pp. 35–38, Lunar Planet. Inst., Houston (1991)
R.J. Phillips, M.T. Zuber, S.C. Solomon et al., Science **291**, 2587–2591 (2001)
R.J. Phillips, M.T. Zuber, S.E. Smrekar et al., Science **320**(5880), 1182–1185 (2008)
S. Piqueux, P.R. Christensen, J. Geophys. Res. **114**, E09005 (2009a)
S. Piqueux, P.R. Christensen, J. Geophys. Res. **114**, E09006 (2009b)
T. Platz, G. Michael, Earth Planet. Sci. Lett. **312**, 140–151 (2011)
T. Platz, G. Michael, G. Neukum, Earth Planet. Sci. Lett. **293**, 388–395 (2010)

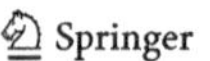

T. Platz, S. Münn, T.R. Walter et al., Earth Planet. Sci. Lett. **305**, 445–455 (2011)
J. Plescia, J. Geophys. Res. **109**, E03003 (2004)
B.T. Poe, C. Romano, F. Nestola, J.R. Smyth, Phys. Earth Planet. Inter. **181**, 3–4 (2010)
H.N. Pollack, S.J. Hurter, J.R. Johnson, Rev. Geophys. **31**(3), 267–280 (1993)
F. Poulet, J.-P. Bibring, J.F. Mustard et al., Nature **438**, 623–627 (2005)
F. Poulet, C. Gomez, J.-P. Bibring et al., J. Geophys. Res. **112**, E08S02 (2007)
F. Poulet, J.-P. Bibring, Y. Langevin et al., Icarus **201**, 69–83 (2009a)
F. Poulet, N. Mangold, B. Platevoet et al., Icarus **201**, 84–101 (2009b)
F. Poulet, D.W. Beaty, J.-P. Bibring et al., Astrobiology **9**, 3 (2009c)
M.S. Ramsey, P.R. Christensen, J. Geophys. Res. **103**, 577–596 (1998)
C.C. Reese, V.S. Solomatov, L.N. Moresi, J. Geophys. Res. **103**(E6), 13643–13658 (1998)
C.C. Reese, C.P. Orth, V.S. Solomatov, J. Geophys. Res. **115**(E5), E05004 (2010)
R. Rieder, T. Economou, H. Wänke et al., Science **278**, 1771–1774 (1997)
J.A. Ritzer, S.A. Hauck, Icarus **201**(2), 528–539 (2009)
A. Rivoldini, T.V. Hoolst, O. Verhoeven et al., Icarus **213**, 451–472 (2011)
S.J. Robbins, G.D. Achille, B.M. Hynek, Icarus **211**, 1179–1203 (2011)
J.H. Roberts, S. Zhong, J. Geophys. Res. **111**(E6), E06013 (2006)
G.P. Roberts, B. Matthews, C. Bristow et al., J. Geophys. Res. **117**, E02009 (2012)
P. Rochette, J.P. Lorand, G. Fillion et al., Meteorit. Planet. Sci. **36**, 176 (2001)
P. Rochette, J. Gattacceca, V. Chevrier et al., Meteorit. Planet. Sci. **40**, 529 (2005)
A.D. Rogers, P.R. Christensen, J. Geophys. Res. **112**, E01003 (2007)
C. Romano, B.T. Poe, N. Kreidie et al., Am. Mineral. **91**(8–9), 1371–1377 (2006)
C. Romano, B.T. Poe, J. Tyburczy et al., Eur. J. Mineral. **21**(3), 615–622 (2009)
F. Roosbeek, Celest. Mech. Dyn. Astron. **75**, 285–300 (2000)
S.W. Ruff, P.R. Christensen, D.L. Blaney et al., J. Geophys. Res. **111**, E12S18 (2006)
J. Ruiz, C. Fernández, D. Gomez-Ortiz et al., Earth Planet. Sci. Lett. **270**(1-2), 1–12 (2009)
J. Ruiz, P.J. McGovern, A. Jiménez-Díaz et al., Icarus **215**(2), 508–517 (2011)
S.E.H. Sakimoto, J. Crisp, S.M. Baloga, J. Geophys. Res. **102**, 6597–6614 (1997)
J.W. Salisbury, J.W. Eastes, Icarus **64**, 586–588 (2008)
M.R. Salvatore, J.F. Mustard, M.B. Wyatt et al., J. Geophys. Res. **115**(E14), E07005 (2010)
V. Sautter, M.J. Toplis, J.P. Lorand, M. Macri, Meteorit. Planet. Sci. (2012)
G.G. Schaber, K.L. Tanaka, J.K. Harmon, in *Lunar and Planetary Science Conference*. LPI Contribution, vol. 441, Houston, TX (1981)
M.E. Schmidt, T.J. McCoy, Earth Planet. Sci. Lett. **296**(1–2), 67–77 (2010)
B. Schott, A.P. van den Berg, D.A. Yuen, Geophys. Res. Lett. **28**(22), 4271–4274 (2001)
C. Schröder, D.S. Rodionov, T.J. McCoy et al., J. Geophys. Res. **113**, E06S22 (2008)
G. Schubert, R.E. Lingenfelter, Nature **242**(5395), 251–252 (1973)
G. Schubert, T. Spohn, J. Geophys. Res. **95**(B9), 14095–14104 (1990)
G. Schubert, P. Cassen, R.E. Young, Icarus **38**, 192–211 (1979)
R.A. Schultz, T.R. Watters, Geophys. Res. Lett. **28**, 4659–4662 (2001)
S. Schumacher, D. Breuer, J. Geophys. Res. **111**(E2), E02006 (2006)
S. Schumacher, D. Breuer, Geophys. Res. Lett. **34**(14), L14202 (2007)
S. Schumacher, T.E. Zegers, Icarus **211**, 305–315 (2011)
D. Scott, K. Tanaka, Icarus **45**, 304–319 (1981)
D.H. Scott, K.L. Tanaka, R. Greeley, J.E. Guest, US Geol. Surv. Misc. Invest. Ser. Map, I-1802-A (1986)
D.H. Scott, K.L. Tanaka, R. Greeley, J.E. Guest. US Geol. Surv. Misc. Invest. Ser. Map, I-1802-B (1987a)
D.H. Scott, K.L. Tanaka, R. Greeley, J.E. Guest. US Geol. Surv. Misc. Invest. Ser. Map, I-1802-C (1987b)
M.N. Simon, L.M. Carter, B.A. Campbell et al., *Lunar Planet. Sci., XLIII*, abstract 1595 (2012)
R.B. Singer, T.B. McCord, R.N. Clark, J. Geophys. Res. **84**, 8415–8426 (1982)
J.R. Skok, J.F. Mustard, S.L. Murchie et al., J. Geophys. Res. **115**, E00D14 (2010)
N.H. Sleep, J. Geophys. Res. **99**(E3), 5639–5655 (1994)
J.L. Smellie, L.G. Chapman (eds.), *Volcano-Ice Interaction on Earth and Mars*, vol. 202 (Geological Society Special Publication, London, 2002)
D.E. Smith, M.T. Zuber, S.C. Solomon et al., Science **284**(5419), 1495 (1999)
F. Sohl, T. Spohn, J. Geophys. Res. **102**, 1613–1635 (1997)
F. Sohl, G. Schubert, T. Spohn, J. Geophys. Res. **110**, E12008 (2005)
V.S. Solomatov, L.N. Moresi, Geophys. Res. Lett. **24**(15), 1907–1910 (1997)
S.C. Solomon, O. Aharonson, J.M. Aurnou et al., Science **307**(5713), 1214–1220 (2005)
C.P. Sonett, D.S. Colburn, P. Dyal et al., Nature **230**, 359–362 (1971)
T. Spohn, Icarus **90**, 222–236 (1991)
T. Spohn, M. Grott, J. Knollenberg et al., *43rd Lunar and Planetary Science Conference*, abstract 1445 (2012)

S.W. Squyres, D.E. Wilhelms, A.C. Moosman, Icarus **70**, 385–408 (1987)
S.W. Squyres, R.E. Arvidson, E.T. Baumgartner et al., J. Geophys. Res. **108**(E12), 8062 (2003)
S.W. Squyres, R.E. Arvidson, J.F. Bell III et al., Science **305**(5685), 794–799 (2004a)
S.W. Squyres, R.E. Arvidson, J.F. Bell III et al., Science **306**(5702), 1698–1703 (2004b)
S.W. Squyres, O. Aharonson, B.C. Clark et al., Science **316**, 5825, 738 (2007)
O. Šrámek, S. Zhong, J. Geophys. Res. **115**(E9), E09010 (2010)
O. Šrámek, S. Zhong, J. Geophys. Res. **117**, E01005 (2012)
I.M. Steele, J.V. Smith, in *13th Lunar Plan. Sci. Conf.*, Houston, TX, March 15–19 (1982), pp. A375–A384
D.J. Stevenson, T. Spohn, G. Schubert, Icarus **54**, 466–489 (1983)
K.R. Stockstill-Cahill, F.S. Anderson, V.E. Hamilton, J. Geophys. Res. **113**(E12), E07008 (2008)
R.M. Suggs, W.J. Cooke, R.J. Suggs et al., Earth Moon Planets **102**, 293–298 (2008)
R. Sullivan, R. Arvidson, J.F. Bell III et al., J. Geophys. Res. **113**, E06S07 (2008)
J.M. Sunshine, C.M. Pieters, J. Geophys. Res. **98**, 9075–9087 (1993)
J.M. Sunshine, C.M. Pieters, S.F. Pratt, J. Geophys. Res. **95**, 6955–6966 (1990)
E. Takahashi, J. Geophys. Res. **95**(B10), 15941–15954 (1990)
K.L. Tanaka, J. Geophys. Res. **91**, E139–E158 (1986)
K.L. Tanaka, D.H. Scott, R. Greeley, in *Mars* (University of Arizona Press, Tucson, 1992), pp. 345–382
G.J. Taylor, W. Boynton, J. Brückner et al., J. Geophys. Res. **111**(E3), E03S10 (2006)
G.J. Taylor, L.M.V. Martel, S. Karunatillake et al., Geology **38**, 183–186 (2010)
N.A. Teanby, J. Wookey, Phys. Earth Planet. Inter. **186**, 70–80 (2011)
L.L. Tornabene, J.E. Moersch, H.Y. McSween et al., J. Geophys. Res. **113**(E12), E10001 (2008)
P. Toulmin III, A.K. Baird, B.C. Clark et al., J. Geophys. Res. **82**(28), 4625–4634 (1977)
A.H. Treiman, Meteorit. Planet. Sci. **38**(12), 1849–1864 (2003)
A. Treiman, Chem. Erde **65**, 203–296 (2005)
A.H. Treiman, M.J. Drake, M.-J. Janssens et al., Geochim. Cosmochim. Acta **50**, 1071–1091 (1986)
T. Usui, H.Y. McSween, C. Floss, Geochim. Cosmochim. Acta **72**(6), 1711–1730 (2009)
J.A. Van Orman, T.L. Grove, N. Shimizu, Earth Planet. Sci. Lett. **160**, 505–519 (1998)
J. Vaucher, D. Baratoux, N. Mangold et al., Icarus **204**, 418–442 (2009)
O. Verhoeven, A. Rivoldini, P. Vacher et al., J. Geophys. Res. **110**, E04009 (2005)
M. Wadhwa, H.Y.J. McSween, G. Crozaz, Geochim. Cosmochim. Acta **58**, 4213–4229 (1994)
G.P.L. Walker, R. Croasdale, Bull. Volcanol. **35**, 303–317 (1971)
H. Wänke, G. Dreibus, Philos. Trans. R. Soc. Lond. A **349**, 285–293 (1994)
N.H. Warner, T.K.P. Gregg, J. Geophys. Res. **108**, 5112 (2003)
L.L. Watson, I.D. Hutcheon, S. Epstein, E.M. Stolper, Science **265**(5158), 86–90 (1994)
S.A. Weinstein, J. Geophys. Res. **100**(E6), 11719–11728 (1995)
B.P. Weiss, I. Garrick-Bethell, J.L. Kirschvink, *Workshop on Ground Truth from Mars: Science Payoff from a Sample Return Mission*. LPI Contribution No. 1401, abstract #4024 (2008)
M.J. Wenzel, M. Manga, A.M. Jellinek, Geophys. Res. Lett. **31**(4), L04702 (2004)
S.C. Werner, Icarus **195**, 45–60 (2008)
S.C. Werner, Icarus **201**, 44–68 (2009)
S.M. White, J.A. Crisp, F.J. Spera, Geochem. Geophys. Geosyst. **7**, Q03010 (2006)
J.L. Whitford-Stark, Earth-Sci. Rev. **18**, 109–168 (1982)
M.A. Wieczorek, Icarus **196**, 506–517 (2008)
M.A. Wieczorek, M.T. Zuber, J. Geophys. Res. **109**, E01009 (2004)
D.E. Wilhelms, S.W. Squyres, Nature **309**, 138–140 (1984)
D.A. Williams, R. Greeley, W. Zuschneid et al., J. Geophys. Res. **112**, E10004 (2007)
D.A. Williams, R. Greeley, S.C. Werner et al., J. Geophys. Res. **113**, E11005 (2008)
D.A. Williams, R. Greeley, R. Fergason et al., Planet. Space Sci. **57**, 895–916 (2009)
D.A. Williams, R. Greeley, L. Manfredi et al., Earth Planet. Sci. Lett. **294**, 451–465 (2010)
L. Wilson, J. Geophys. Res. **106**, 1423–1434 (2001)
L. Wilson, J.W. Head, Nature **302**, 663–668 (1983)
L. Wilson, J.W. Head, Rev. Geophys. **32**, 221–263 (1994)
L. Wilson, J.W. Head, J. Geophys. Res. **107**, 5057 (2002)
L. Wilson, J.W. Head, J. Volcanol. Geotherm. Res. **163**, 83–97 (2007)
D.U. Wise, M.P. Golombek, G.E. McGill, J. Geophys. Res. **84**, 7934–7939 (1979)
J.J. Wray, E.Z. Noe Dobrea, R.E. Arvidson et al., Geophys. Res. Lett. **36**, L21201 (2009)
M.B. Wyatt, H.Y. McSween, Nature **417**, 263–266 (2002)
L. Xiao, J. Huang, P.R. Christensen et al., Earth Planet. Sci. Lett. **323–324**, 9–18 (2012)
Y.S. Xu, B.T. Poe, T.J. Shankland et al., Science **280**(5368), 1415–1418 (1998)
Y.S. Xu, T.J. Shankland, B.T. Poe, J. Geophys. Res. **105**(B12), 27865–27875 (2000)
X. Yang, Earth Planet. Sci. Lett. **317–318**, 241–250 (2012)

T. Yoshino, T. Matsuzaki, S. Yamashita, T. Katsura, Nature **443**, 973–976 (2006)
T. Yoshino, T. Matsuzaki, A. Shatskiy, T. Katsura, Earth Planet. Sci. Lett. **288**, 291–300 (2009)
M. Yseboodt, J.-P. Barriot, V. Dehant, J. Geophys. Res. **108**, 5076 (2003)
J. Zhang, C. Herzberg, J. Geophys. Res. **99**(B9), 17729–17742 (1994)
Y.-H. Zhao, M.E. Zimmerman, D.L. Kohlstedt, Earth Planet. Sci. Lett. **287**, 229–240 (2009)
V.N. Zharkov, T.V. Gudkova, Planet. Space Sci. **45**, 401–407 (1997)
S. Zhong, Nat. Geosci. **2**, 19–23 (2009)
S. Zhong, M.T. Zuber, Earth Planet. Sci. Lett. **189**(1–2), 75–84 (2001)
L. Zhu, H. Kanamori, J. Geophys. Res. **105**, 2969–2980 (2000)
J.R. Zimbelman, J. Geophys. Res. **90**, 157–162 (1985)
J.R. Zimbelman, T.K.P. Gregg (eds.), *Environmental Effects on Volcanic Eruptions* (Kluwer Academic/Plenum, New York, 2000), 206 pp.
J. Zipfel, C. Schröder, B.L. Jolliff et al., Meteorit. Planet. Sci. **46**, 1–20 (2011)
M.T. Zuber, S.C. Solomon, R.J. Phillips et al., Science **287**, 1788–1793 (2000)

Space Sci Rev (2013) 174:113–154
DOI 10.1007/s11214-012-9943-8

Outgassing History and Escape of the Martian Atmosphere and Water Inventory

Helmut Lammer · Eric Chassefière · Özgür Karatekin · Achim Morschhauser · Paul B. Niles · Olivier Mousis · Petra Odert · Ute V. Möstl · Doris Breuer · Véronique Dehant · Matthias Grott · Hannes Gröller · Ernst Hauber · Lê Binh San Pham

Received: 6 December 2011 / Accepted: 16 October 2012 / Published online: 30 November 2012

Abstract The evolution and escape of the martian atmosphere and the planet's water inventory can be separated into an early and late evolutionary epoch. The first epoch started from the planet's origin and lasted ~500 Myr. Because of the high EUV flux of the young Sun and Mars' low gravity it was accompanied by hydrodynamic blow-off of hydrogen and strong thermal escape rates of dragged heavier species such as O and C atoms. After the main part of the protoatmosphere was lost, impact-related volatiles and mantle outgassing may have resulted in accumulation of a secondary CO_2 atmosphere of a few tens to a few hundred mbar around ~4–4.3 Gyr ago. The evolution of the atmospheric surface pressure and water inventory of such a secondary atmosphere during the second epoch which lasted

H. Lammer (✉) · H. Gröller · P. Odert
Space Research Institute, Austrian Academy of Sciences, Schmiedlstr. 6, 8042 Graz, Austria
e-mail: helmut.lammer@oeaw.ac.at

E. Chassefière
Laboratoire IDES, CNRS, UMR8148, Univ. Paris-Sud, Orsay, 91405, France

Ö. Karatekin · V. Dehant · L.B.S. Pham
Royal Observatory of Belgium, Brussels, Belgium

A. Morschhauser · D. Breuer · M. Grott · E. Hauber
German Aerospace Center, Institute of Planetary Research, Rutherfordstr. 2, 12489 Berlin, Germany

P.B. Niles
Astromaterials Research and Exploration Science Johnson Space Center, NASA, Houston, TX, USA

O. Mousis
Observatoire de Besançon, 41 bis, avenue de l'Observatoire, B.P. 1615, 25010 Besançon, France

O. Mousis
UPS-OMP; CNRS-INSU; IRAP, Université de Toulouse, 14 Avenue Edouard Belin, 31400 Toulouse, France

P. Odert · U.V. Möstl
Institute for Physics/IGAM, University of Graz, Universitätsplatz 5, 8010, Graz, Austria

from the end of the Noachian until today was most likely determined by a complex interplay of various nonthermal atmospheric escape processes, impacts, carbonate precipitation, and serpentinization during the Hesperian and Amazonian epochs which led to the present day surface pressure.

Keywords Early Mars · Young Sun · Magma ocean · Volcanic outgassing · Impacts · Thermal escape · Nonthermal escape · Atmospheric evolution

1 Introduction

The present martian atmosphere is the result of numerous interacting processes. On the one hand, these include atmospheric sinks such as erosion by impacts, thermal and nonthermal escape, extreme ultraviolet (EUV) radiation, as well as solar wind forcing (e.g. Lundin et al. 2007). On the other hand, atmospheric sources such as volcanic outgassing or delivery of volatiles by impacts have also to be taken into account for understanding atmospheric evolution. Furthermore, the atmosphere can interact with crustal reservoirs by CO_2 weathering and hydration processes, which occur at the surface and/or in the crust (e.g., Zent and Quinn 1995; Bandfield et al. 2003; Becker et al. 2003; Lundin et al. 2007; Lammer et al. 2008; Pham et al. 2009; Tian et al. 2009; Phillips et al. 2010). A schematic synopsis of these interactions is presented in Fig. 1.

The aim of this work is to review the latest knowledge on the evolution of the martian atmosphere since the planet's origin ~4.55 Gyr ago. In Sect. 2 we discuss the delivery of volatiles, the planet's early hydrogen-rich protoatmosphere, and point out possible reasons why there is an apparent deficiency of noble gases in the present atmosphere. In Sect. 3 we discuss different approaches to constrain volcanic outgassing rates of CO_2 and H_2O. In Sect. 4 we consider the role of atmospheric impact erosion and delivery in the early martian environment. In Sect. 5, we discuss the efficiency of EUV-powered escape during the early Noachian and its influence on the growth of a secondary CO_2 atmosphere. In Sect. 6 we briefly address consequences of the late heavy bombardment (LHB) on the martian atmosphere and its climate, ~3.7–4 Gyr ago. Finally, Sects. 7 and 8 focus on nonthermal atmospheric escape to space and on possible surface sinks of CO_2 and H_2O allowing the surface pressure to reach its present-day value.

2 Origin and Delivery of Volatiles to Mars

The sources and evolutionary histories of volatiles composing the martian atmosphere are poorly understood. They are related to the sources that delivered significant amounts of water to early Mars, which have implications for the formation of the planet's protoatmosphere. Furthermore, isotope variations in volatiles have the potential to provide insights into the origin and atmosphere modification processes in terrestrial planets, possibly related to the observation that the noble gases appear strongly depleted in the martian atmosphere compared to those of Earth and Venus.

2.1 Water Delivery and Formation of the Martian Protoatmosphere

Four main processes are responsible for the early formation of an atmosphere:

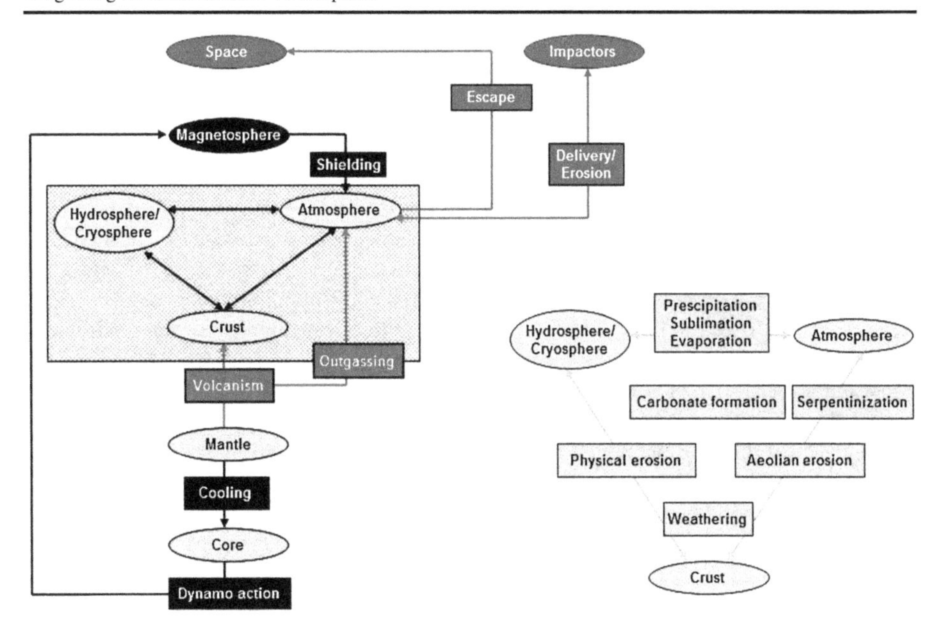

Fig. 1 Sketch showing important interactions between the main reservoirs, i.e. atmosphere, hydrosphere/cryosphere, crust, mantle and core (*ovals* in *light blue*), that have been addressed in the present work. Volcanism results in the formation of the crust and the associated degassing of the mantle produces an atmosphere with time (*red boxes*). Dynamo action in the core, which is triggered by efficient heat transfer in the mantle, and the subsequent shielding of the atmosphere prevents or reduces atmospheric erosion by non-thermal processes (*black boxes*). The erosion of the atmosphere to space can be caused by solar influx or by impacts (*blue boxes*). The latter may also deliver volatiles to the atmosphere depending on the impactors size and composition. The complex interactions between atmosphere, hydrosphere/cryosphere and crust (*green dotted area*) are shown in more detail on the right side (*green boxes*)

- capture and accumulation of gasses from the planetary nebula,
- catastrophic outgassing due to magma ocean solidification,
- impacts,

and

- later degassing by volcanic processes.

As long as nebula gas is present, growing protoplanets can capture hydrogen and He which form gaseous envelopes around the rocky core (e.g., Hayashi et al. 1979; Rafikov 2006). Depending on the host star's radiation and plasma outflow, the nebula dissipation time, the planet's orbital location and the number and orbital location of additional planets in the system, according to Hayashi et al. (1979), planetary embryos with the mass of $\sim 0.1 M_{\mathrm{Earth}}$ can capture hydrogen and other nebula gas from the nebula during $\sim$3 Myr with an equivalent amount of up to $\sim$55 times the hydrogen which is present in the Earth's present day ocean. Furthermore, noble gases delivered by comets accreted during this period were mixed with volatiles remaining after an episode of strong atmospheric escape.

The initial water inventory of a planet is acquired from colliding planetesimals, growing planetary embryos, impacting asteroids and comets (e.g., Lunine et al. 2003; Brasser 2012). Lunine et al. (2003) estimated the cumulative collision probability between small bodies and Mars and found that Mars' initial water inventory may have been equivalent to $\sim$0.06–0.27 times that of an Earth ocean (EO), corresponding to a martian surface pressure of ~10–100 bar. Other simulations which considered different impact regimes suggest that Mars could also have been drier (Horner et al. 2009). In a more recent study, Walsh et al. (2011) argues that the small mass of Mars indicates that the terrestrial planets in the Solar System have formed from a narrow material annulus, rather than a disc extending to Jupiter. In such a scenario the truncation of the outer edge of the disc was the result of the migration of the gas giants, which kept the martian mass small. From cosmochemical constraints one can argue that Mars formed in a couple of Myr and can be considered in agreement with the latest dynamical models as a planetary embryo that never grew to a real planet. In such a case most of Mars' materials consisted of building blocks that formed in a region at ~2–3 AU, and therefore, were more H_2O-rich compared to the materials which formed Earth and Venus. From these arguments Brasser (2012) suggests that Mars may have consisted of ~0.1–0.2 wt.% of water.

A substantial part of the initial inventory of volatiles could have been outgassed as a consequence of the solidification of an early magma ocean (Elkins-Tanton 2008). Water and carbon dioxide enter solidifying minerals in only small quantities and are enriched in magma ocean liquids as solidification proceeds. Close to the surface at low pressure these volatiles degas into the growing atmosphere. Depending on the initial water/volatile content, which was built-in the planetary body during its growth and the depth of the possible magma ocean z_{mag}, steam atmospheres with a surface pressure between ~30 (0.05 wt.% H_2O, 0.01 wt.% CO_2, $z_{mag} \sim 500$ km) to ~800 bar (0.5 wt.% H_2O, 0.1 wt.% CO_2, $z_{mag} \sim 2000$ km) (see Table 3, Elkins-Tanton 2008) could have been catastrophically outgassed. If early Mars consisted of ~0.1–0.2 wt.% water (Brasser 2012) then a steam atmosphere with a surface pressure of more than ~60 bar could have been catastrophically outgassed (Elkins-Tanton 2008). Although it is assumend that most volatiles are degassed into the early atmosphere, a geodynamically significant quantity is still sequestered in the solid cumulates. The amount is estimated to be as much as 750 ppm by weight OH for an initial water content of 0.5 wt.%, and a minimum of 10 ppm by weight in the driest cumulates of models beginning with just 0.05 wt.% water (Elkins-Tanton 2008). Even more water in the martian interior can be expected after the magma ocean solidification phase in the case of a shallow magma ocean in particular in the deep unmolten primordial mantle. In any case, these small water contents significantly lower the viscosity and possibly the melting temperature of mantle materials, facilitating later volcanism, as discussed below.

The early steam atmosphere could have remained stable for a few tens of Myr. During this early stage, environmental conditions were determined by a high surface temperature and frequent impacts, which could have reached up to ~1500 K due to thermal blanketing and frequent impacts (e.g., Matsui and Abe 1986). If such a steam atmosphere is not lost upon cooling, the remaining H_2O vapor can condense and produces liquid water on the surface or ice in case of a cold climate (e.g., Chassefière 1996).

If Mars originated with ~0.1–0.2 wt.% H_2O, as long as the planet was surrounded by a captured dense nebula-based hydrogen envelope, magma ocean related outgassed greenhouse gases (H_2O, CO_2, CH_4, NH_3) would have been protected against dissociation because these heavy molecules would remain closer to the planet's surface compared to the lighter hydrogen in the upper atmosphere. Depending on the amount and the lifetime of accumulated nebula gas and its evaporation time, a combination of a possible H_2 greenhouse (Pierrehumbert and Gaidos 2011; Wordsworth 2012) and the outgassed greenhouse gases may have provided warm and wet conditions on the martian surface for a few tens of Myr.

Finally, it is important to note that the previous investigations of planetary formation suffer from several unknowns including the sources of impactors across the inner Solar System. Such work would require far more detailed model populations for the cometary and asteroidal sources, and would have to include a study of the effects of Oort cloud comets. Because the results of Lunine et al. (2003) and Brasser (2012) are different from those of Horner et al. (2009), it is obvious that our knowledge of terrestrial planet formation and hydration is currently insufficient because it is not possible to predict the real initial deuteration level on each of the planets considered. This piece of evidence, combined with the fact that the D/H ratio in H_2O in comets is not homogeneous (Hartogh et al. 2011), indicate that the water delivery mechanisms to the terrestrial planets can only be established within an uncertainty range.

2.2 The Apparent Noble Gas Deficiency of the Martian Atmosphere

The difference between the measured atmospheric abundances of non-radiogenic noble gases in Venus, Earth, and Mars is striking. It is well known that these abundances decline dramatically as one moves outward from Venus to Mars within the inner Solar System, with these two planets differing in abundance by up to two orders of magnitude (see Fig. 2). Therefore, understanding this variation is a key issue in understanding how the initial atmospheres of the terrestrial planets evolved to their current composition, and requires to study the different delivery mechanisms of the volatiles accreted by these planets (Pepin 1991, 2006; Owen et al. 1992; Owen and Bar-Nun 1995; Dauphas 2003; Marty and Meibom 2007).

In this context, recent n-body simulations have been performed by Horner et al. (2009) in order to study the impact rates experienced by the terrestrial planets as a result of diverse populations of potential impactors. These authors considered a wide range of plausible planetary formation scenarios for the terrestrial planets, and found that the different impact regimes experienced by Venus, Earth, and Mars could have resulted in significant differences between their individual hydration states over the course of their formation and evolution. Horner et al. (2009) found that, on average, the Earth most likely received a flux of impacting comets which is $\sim$3.4 times higher than that experienced by Mars. Assuming that the mass of noble gases delivered by comets to the terrestrial planets is proportional to the rate at which they impacted upon them, it is possible to derive X_E/X_M (the ratio of the noble gas abundances (as a fraction of the total mass of the planet) between Earth and Mars)

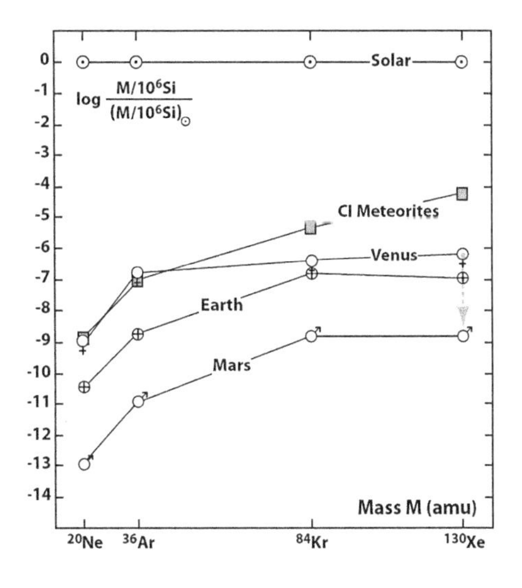

Fig. 2 Measured abundances of Ne, Ar, Kr, and Xe in the atmospheres of the terrestrial planets and primitive CI meteorites. The values shown for these gases are presented relative to their solar abundances, in units of atoms per 10^6 Si atoms (adapted from Fig. 2 of Pepin 1991). The *vertical arrow pointing down* indicates that the Venus atmospheric abundance of Xe is only an upper limit

from the ratio of the number of comets impacting upon those two planets N_E/N_M (~3.4), through the following relation (Mousis et al. 2010)

$$\frac{X_{\mathrm{E}}}{X_{\mathrm{M}}} = \frac{N_{\mathrm{E}}}{N_{\mathrm{M}}} \frac{M_{\mathrm{M}}}{M_{\mathrm{E}}}, \tag{1}$$

where M_{E} and M_{M} are the masses of Earth and Mars, respectively. From this relation, one can infer that the average noble gas abundance on Earth should be ~0.37 times the martian noble gas abundances if these volatiles were solely delivered by comets. This result differs significantly from that inferred from measurements of noble gas abundances, which are observed to be approximately two orders of magnitude larger for the Earth compared to Mars. As a result, subsequent processes that occurred preferably during the post-impact period of Mars are required in order to explain its present-day atmospheric composition. It has been proposed that atmospheric escape could have strongly altered the composition of the atmospheres of terrestrial planets (Pepin 1991, 1997; Dauphas 2003; Jakosky et al. 1994; Chassefière and Leblanc 2004). This hypothesis is supported by both Mars (SNC meteorites) and Earth, which show substantial fractionation of Xe isotopes compared to the plausible primitive sources of noble gases, i.e., solar wind (SW–Xe), meteorites (Q–Xe), or the hypothetical U–Xe source (Pepin 2006). This fractionation then suggests important losses of Xe and other noble gases from the early atmospheres of the Earth and Mars. Impact related loss processes might have been more important for the Earth and Mars than Venus because the latter planet would have escaped impacts of the magnitude that formed the Moon (Canup and Asphaug 2001) or created the largest basins on Mars (Andrews-Hanna et al. 2008). In the case of the Earth, the noble gas fractionation episode could have also been driven by impacts (Pepin 1991, 1997; 2003) in combination with the high EUV radiation of the young Sun (e.g., Ribas et al. 2005; Lammer et al. 2008).

Thus, in the case of Mars, the combination of impacts, EUV-powered hydrodynamic escape, planetary degassing, and fractionation by nonthermal atmospheric escape processes (Jakosky et al. 1994; Luhmann et al. 1992; Carr 1999; Chassefière and Leblanc 2004; Lammer et al. 2008) might have played an important role in sculpting the pattern of the noble gas abundances observed today. An alternative hypothesis proposed to explain the Kr and Xe abundance differences between Earth and Mars is the presence of large amounts of CO_2-dominated clathrates in the martian soil that would have efficiently sequestered these noble gases (Mousis et al. 2010, 2012). In this scenario, these noble gases would have been trapped in clathrates ~4 Gyr b.p. when the CO_2 surface pressure was expected to be of the order of a few tens to a few hundred mbar (Mousis et al. 2012). This scenario implies that the ^{36}Ar, ^{84}Kr, and ^{130}Xe abundances measured in the planet's atmosphere are not representative of its global noble gas budget. Depending on the amount of existing clathrates, the volume of noble gases trapped in these crystalline structures could be much larger than those measured in the atmosphere. In this context, two different scenarios have been proposed by Mousis et al. (2010) to explain the differences between the Ne and Ar abundances of the terrestrial planets.

In the first scenario, cometary bombardment of the planets would have occurred at epochs contemporary with the existence of their primary atmospheres. Comets would have been the carriers of Ar, Kr, and Xe, while Ne would have been gravitationally captured by the terrestrial planets (Owen et al. 1992). Only Ne and Ar would have been fractionated due to thermal and nonthermal atmospheric escape, while the abundances of the heavier noble gases would have been poorly affected by such losses. In this scenario, the combination of processes, such as escape of Ne and Ar, cometary bombardment at the epochs of existence of primary planetary atmospheres, and the sequestration of krypton and xenon in the martian clathrates, would then explain the observed noble gas abundance differences between the Earth and Mars. However, this scenario leads to an important chronological issue because depending on a the hydrogen/He amount of the captured nebula-based protoatmosphere it existed most likely only during the first few to several tens of Myr (Halliday 2003; Pepin 2006; Lammer et al. 2012).

On the other hand one should also note that heavy noble gases could have been supplied during the LHB (Marty and Meibom 2007). In such a second scenario, Mousis et al. (2010) considered impacting comets that contained significantly smaller amounts of Ar, an idea supported by predictions of noble gas abundances in these bodies, provided that they are formed from clathrates in the solar nebula (Iro et al. 2003). Here, Ne and Ar would have been supplied to the terrestrial planets via the gravitational capture of their primary atmospheres and comets would have been the carriers of Kr and Xe only. In this case, the cometary bombardment of the terrestrial planets could have occurred after the formation of their protoatmospheres because only the neon and argon abundances observed today would have been engendered by the escape-fractionation processes in these atmospheres.

Both scenarios preclude the possibility that material with a CI chondrite-like composition could be the main source of noble gases in terrestrial planets because the trend described by the chondritic noble gas abundances as a function of their atomic mass does not reflect those observed on Venus, the Earth, and Mars (Pepin 1992; Owen and Bar-Nun 1995). If the composition is similar to that of CI chondrites, this then excludes the hypothesis of noble gas outgassing from the interior of Mars and also the scenario of asteroidal bombardment. Irrespective of the scenario envisaged, this work does not preclude the possibility that a fraction of the heavy noble gases could have been captured by the Earth and Mars during the acquisition of nebula-based protoatmospheres. In the first scenario, the fraction of Kr and Xe accreted in this way should be low compared to the amount supplied by comets since these noble gases are not expected to have been strongly fractionated by atmospheric escape. In the second scenario, the fraction of Kr and Xe captured gravitationally by the terrestrial planets could be large if escape was efficient.

3 Outgassing and Growth of a Secondary Atmosphere

Volcanic outgassing is one of the main sources of volatiles for the Martian atmosphere and provides an important link between mantle and atmospheric geochemical reservoirs. Information on exchange processes between the different reservoirs is contained in the atmospheric isotopic ratios R of elements such as hydrogen, carbon, and the noble gases. R may change as a function of time as lighter isotopes can escape to space more efficiently than their heavier counterparts. Overall, the efficiency of isotopic fractionation depends on the size S of the considered reservoir, the total escape flux ϕ, and the relative efficiency of isotopic escape which may be expressed by the fractionation factor f (Donahue 2004).

3.1 Estimation of the Martian Water-Ice Reservoir by the Atmospheric D/H Ratio

The ratio of the sizes of the past and present reservoirs in isotopic equilibrium with the atmosphere can be calculated if the respective isotopic ratios are known and is given by Donahue (1995)

$$\frac{S_t}{S_p} = \left(\frac{R_p}{R_t}\right)^{1/(1-f)}, \qquad (2)$$

where variables with index p and t refer to the present and past values, respectively. If an initial isotopic ratio is assumed for R_t, the size of the reservoir when it has last been

Table 1 D/H ratio in the martian atmosphere, the terrestrial sea water, comets and various martian meteorites. The crystallization ages and ejection ages of the meteorites are taken from Nyquist et al. (2001) and for ALH 84001 from Turner et al. (1997). The measured D/H values are taken from Leshin et al. (1996) and are also given in units of terrestrial sea water D/H. The ejection age refers to the estimated time of ejection from the Martian surface

Reservoir	D/H [1×10^{-4}]	D/H [SMOW]	Cryst. age [Myr]	Ejection age [Myr]
Mars atmosphere	8.0	5.13		
Terrestrial sea water (SMOW)	1.56	1.00		
Comets	~3.2	~2.05		
Martian meteorites				
AH 84001	2.45	1.57	3920 ± 40	15.0 ± 0.8
Chassigny	1.49–1.6	0.96–1.03	1340 ± 50	11.3 ± 0.6
Nakhla	2.24–2.73	1.44–1.75	1270 ± 50	10.75 ± 0.4
Lafayette	2.47–2.8	1.58–1.79	1320 ± 20	11.9 ± 2.2
Governador Valadares	1.97	1.26	1330 ± 10	10.0 ± 2.1
Zagami	3.28	2.10	177 ± 3	2.92 ± 0.15
Shergotty	3.39	2.17	165 ± 4	2.73 ± 0.2
Elephant Moraine 79001	3.86	2.47	173 ± 3	0.73 ± 0.15

reset to this value will be obtained. This reset may have happened due to strong volcanic outgassing, delivery of additional material by impacts, or a sudden exchange with other reservoirs not in isotopic equilibrium with the atmosphere. In this way, isotopic ratios found in Martian meteorites can be used for R_t to obtain the size of the reservoir at the time of their crystallization.

In the following, we will consider the size of the water reservoir and use the isotopic ratio of deuterium (D) and atomic hydrogen (H) (D_0/H_0). A compilation of different D/H isotopic ratios in the martian atmosphere, terrestrial sea water as well as comets and martian meteorites is given in Table 1. The initial D/H ratio was modified from its initial ratio to the present one by atmospheric escape processes. One way to estimate the initial ratio on Mars is to assume that isotopic ratios on Earth and Mars were identical following accretion. Given that the amount of water present in Earth's oceans is very large, isotopic ratios have probably changed by less than 0.2 % since accretion (Donahue 2001) and D_0/H_0 can be approximated by its present day value of the Standard Mean Ocean Water (SMOW). Another way to constrain D_0/H_0 is to calculate the Martian primordial composition from dynamical accretion models, which result in a primordial isotopic ratio D_0/H_0 between 1.2 and 1.6 times that of the SMOW (Lunine et al. 2003), indicating the range of uncertainty associated with this value. The current atmospheric D/H ratio on Mars was measured with a 4 m reflecting telescope in combination with a Fourier transform spectrometer at Kitt Peak Observatory providing a significantly fractionated value of $R_p = 5.5 \pm 2$ SMOW (Krasnopolsky et al. 1997). From high resolution spectroscopic observations of D and H Lyman-α emissions of the martian hydrogen corona with the Hubble Space Telescope, the fractionation factor f for D and H was estimated to be ~0.016–0.02 (Krasnopolsky et al. 1998, Krasnopolsky 2000). These values are significantly lower than the theoretically calculated value of 0.32 by Yung et al. (1988), resulting in larger reservoirs than previously assumed. Using $f = 0.02$, the size S_0 of the water reservoir at the time when the isotopic ratio in the Martian atmosphere was last reset can be estimated from Eq. (2) and is shown in Fig. 3 as a function of the initial D/H ratio with $D_0/H_0 = 1$ to 1.6. Reservoir size is given in terms of its present-day size for

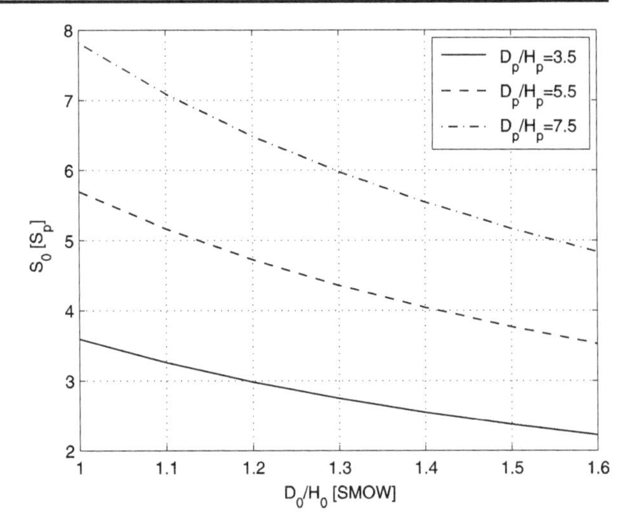

Fig. 3 Size of the martian water reservoir in isotopic equilibrium with the atmosphere as a function of the initial deuterium to hydrogen ratio D_0/H_0 for three different present-day isotopic ratios D_p/H_p. Reservoir size is given in terms of its present-day size H_p and corresponds to the time when the isotopic ratio in the atmosphere was last reset to the initial ratio. The assumed fractionation factor f is 0.02, as measured with the Hubble Space Telescope (Krasnopolsky et al. 1998, Krasnopolsky 2000)

three different values of the current D/H ratio. This calculation implies that the past water reservoir was 2–8 times larger than today, indicating that 50 to 88 % of the past reservoir was lost.

In addition, the absolute size of the past water reservoir can be estimated from the total atmospheric escape flux (Donahue 2004). Results are sensitive to the assumed escape flux and early estimates arrived at a total past reservoir size corresponding to an equivalent global layer (EGL) of water between 0.2 m (Yung et al. 1988) and 30–80 m 1991. More recently, Krasnopolsky and Feldman (2001) estimated a total reservoir size of 65–120 m EGL, Lammer et al. (2003) obtained a value of ∼17–61 m EGL, while Donahue (2004) used isotopic ratios in the Zagami meteorite as an additional constraint to arrive at 100–800 m EGL. A scenario where serpentinization in the crust stored most of the ancient water reservoir has recently been proposed by Chassefière and Leblanc (2011c). They used the present D/H ratio to conclude that up to ∼400 m EGL of free water could hypothetically be stored in crustal serpentine, based on the assumption that D and H atoms released into the atmosphere during serpentinization have escaped and fractionated.

As the initial martian water inventory was most likely affected by hydrodynamic blow-off due to the young Sun's high EUV flux (cf. Sect. 5), H and D will have escaped unfractionated to space between ∼4.0–4.5 Gyr ago. Therefore, these values may give only an estimate of the amount of volatiles delivered by impacts and volcanic outgassing until ∼4 Gyr ago.

3.2 Volcanic Outgassing of CO_2 and H_2O

One way to quantify the rate of volcanic outgassing is to estimate the amount of crustal production as a function of time and to multiply this volume by the magma volatile content. In this case, the outgassing rate can be obtained from

$$\frac{dM_{\mathrm{i}}^{\mathrm{atm}}}{dt} \propto \frac{dM_{\mathrm{cr}}}{dt}\eta X_{\mathrm{i}}^{\mathrm{melt}} \tag{3}$$

where $M_{\mathrm{i}}^{\mathrm{atm}}$ is the outgassed mass of volatile species i, M_{cr} is the amount of extracted magma, X^{i}_{melt} is the concentration of volatile species i in the melt, and η is an outgassing efficiency.

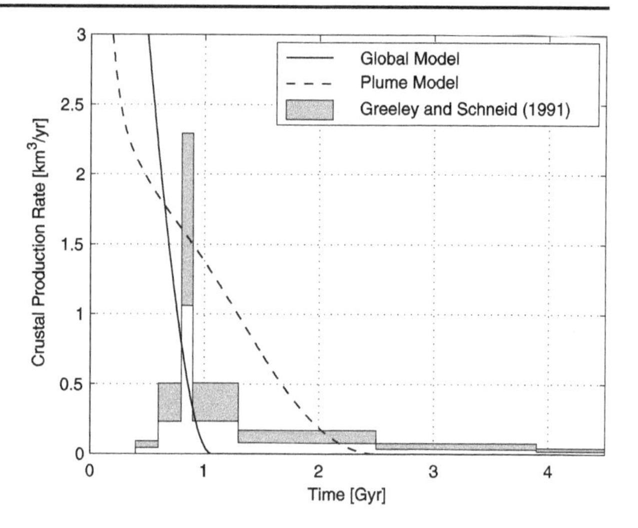

Fig. 4 Different estimates of crustal production rates as a function of time. The *solid* and *dashed lines* show the rates obtained with parameterized thermochemical evolution models assuming a global melt layer and melt generation in localized, hot mantle plumes, respectively (Grott et al. 2011; Morschhauser et al. 2011). The *shaded area* corresponds to the values obtained from photogeological estimates by Greeley and Schneid (1991) with ratios of extrusive to intrusive volcanism ranging from 1:5–1:12

As the solubility of volatiles in magmas at surface pressure is low, essentially all dissolved volatiles will be released when erupting extrusively. For intrusive volcanism, it can be assumed that plutons do not contribute to volcanic outgassing (O'Neill et al. 2007) and R_i will depend on the ratio of intrusive to extrusive volcanism. However, as volatiles will be enriched in the remaining liquid during solidification, it is also possible that dissolved volatiles will outgas at depth and reach the atmosphere (Hirschmann and Withers 2008). An intermediate approach is to assume that volatiles can be delivered to the surface as long as some crustal porosity is present at the depth of the intrusion (Grott et al. 2011). The volume of crustal production has been estimated from the photogeological record and crater counting has been used to age-date the corresponding surfaces. In this way, the rate of crustal production, i.e., the amount of crust produced as a function of time, has been determined. Lava volumes were obtained from the topology of partially filled impact craters by comparing their actual depth to the theoretical values obtained from depth-to-diameter scaling relations (Greeley and Schneid 1991). As only extrusions can be assessed, estimates of the ratio of extrusive to intrusive volcanism are necessary to obtain the total volume of produced crust. Also, older deposits may be covered by later extrusions and therefore early crust production rates may be underestimated.

Alternatively, numerical models of the thermochemical evolution of Mars can be used to calculate the globally averaged crustal production rates (Hauck and Phillips 2002; Breuer and Spohn 2006; Fraeman and Korenaga 2010; Morschhauser et al. 2011). Note, however, that parameterized models cannot account for lateral variations in crustal production, and fully two- or three dimensional models need to be applied in order to resolve young, localized volcanism. Figure 4 compares the photogeological estimates of Greeley and Schneid (1991) with the results of the numerical models by Grott et al. (2011) and Morschhauser et al. (2011), who assume melt production in a global melt layer and melt production in localized, hot mantle plumes, respectively. Within uncertainties of the ratio of extrusive to intrusive volcanism, these approaches show satisfactory agreement at intermediate epochs, but the photogeological approach underestimates crustal production rates in the Noachian, whereas the numerical models cannot provide estimates for younger volcanism. Therefore, both approaches complement each other and are necessary for an overall picture of Mars' volcanic history.

As Mars is in the stagnant-lid mode of mantle convection, volatile contents of magmas associated with intra-plate volcanism on Earth have been considered to be comparable to volatile contents on Mars. However, as Mars may have a different volatile content and mantle oxidation state compared to Earth, these values are at best first-order estimates. However, for lack of better data at the time, terrestrial values have been assumed in several studies. Today, a better understanding of the differences of the magma volatile content of Earth and Mars exist and a more sophisticated approach will be elaborated at the end of this section. For Mauna Loa and Kilauea on Hawaii, the concentration of CO_2 and H_2O in the magma is 0.65 wt.% and 0.30 wt.%, respectively (Gerlach and Graeber 1985; Greenland 1987a, 1987b). Subglacial volcanism on Iceland with a CO_2 content of $\sim$500 ppm has also been used as an analogue for Mars (O'Neill et al. 2007), although it may differ significantly from the type of volcanism expected in the stagnant-lid regime of mantle convection. Using Hawai'ian volcanism as an analogue for Mars and by neglecting atmospheric escape, Phillips et al. (2001) concluded that 1.5 bar of CO_2 have been outgassed during the formation of the Tharsis bulge including intrusions, probably leading to a strong greenhouse effect and climate transition at the end of the Noachian. Since the mid-Noachian, extrusive volcanism may have outgassed an atmosphere of 800 mbar, consisting of 400 mbar CO_2, 8 m EGL H_2O and 6 other minor species as has been estimated from the photogeological record and Hawai'ian volatile concentrations (Craddock and Greeley 2009).

Being of igneous origin, the volatile content of the Martian meteorites may also serve as a proxy for magma volatile contents. An analysis of melt inclusions and a reconstruction of SNC solidification history results in a magma water content of 1.4–1.8 wt.% prior to degassing (McSween and Harvey 1993; McSween et al. 2001; Johnson et al. 1991). As an upper bound, the formation of Tharsis could have outgassed 120 m EGL H_2O in this way, if a magma volatile content of 2 wt.% H_2O is assumed (Phillips et al. 2001). However, water content may have been overestimated and values change to less than 0.3 wt.% if the high chlorine content in Martian meteorites is taken into account (Filiberto and Treiman 2009). These lower water concentrations are also supported by direct measurements in kaersutitic and biotitic melt inclusions (Watson et al. 1994) and would decrease the amount of water outgassed by Tharsis to 18 m EGL.

It may be argued that the magma water content of the SNC meteorites, which are believed to be younger than $\sim$1.3 Gyr, do not represent the typical magma water content at the time of Tharsis formation $\sim$3 Gyr ago. As numerical models predict a total mantle water loss of $\sim$50 % due to volcanic outgassing, the uncertainty associated in using SNC magma water contents for Tharsis outgassing is around a factor of two. Compared with other uncertainties, e.g. the volume of Tharsis or the debate on the water magma concentration of the SNCs, this uncertainty is not significant.

In addition, partitioning of water into the melt can be calculated from a melting model. Accumulated fractional melting may be more appropriate for mantle melting on Mars (Grott et al. 2012, this issue), but batch melting is also widely applied (Hauck and Phillips 2002; Morschhauser et al. 2011). However, the difference between the two approaches is comparatively small and results are not significantly affected. Within the frameworks of batch- and fractional melting, the partioning coefficient, melt fraction, and bulk water content determine the concentration in the melt. The partition coefficient of water is most likely close to 0.01 (Katz et al. 2003), and melt fractions obtained from numerical models average around 5–10 % (Hauck and Phillips 2002; Morschhauser et al. 2011). These values are consistent with melt fractions determined from trace-element analysis of shergottites, which range from 2 % to 10 % (Norman 1999; Borg and Draper 2003) and result in magma water concentrations of 10 to 16 times the bulk water content. The bulk mantle water content of Mars is poorly constrained, and a large range of concentrations have been obtained from different methods: Analysing water content in melt inclusions of Martian meteorites, a bulk mantle water concentration of 1400 ppm was calculated (McSween and Harvey 1993), while numerical accretion models (Lunine et al. 2003) arrive at maximum concentrations of 800 ppm. In contrast, meteoritic mixing models constrained from element ratios in SNC meteorites predict bulk water concentrations of only 36 ppm (Wänke and Dreibus 1994). It should be noted that, even though the mantle is dehydrating with time and SNCs are believed to be geologically young, the inferred bulk mantle water content of the SNCs is larger than that predicted by the other methods. This may be due to the large uncertainties associated with each of these methods.

The solubility of CO_2 in Martian magma can be calculated by considering the underlying chemistry. For CO_2, solubility depends on the form in which graphite is stable in the Martian mantle, which in turn depends on oxygen fugacity (Hirschmann and Withers 2008). The Shergottites, which most likely reflect conditions at the magma source region (Hirschmann and Withers 2008), have oxygen fugacities between the iron-wustite (IW) buffer and one $\log_{10}$ unit above it (IW + 1) (Herd et al. 2002; Shearer et al. 2006). The oldest Martian meteorite, ALH84001, indicates even more reducing conditions around IW-1 (Warren and Kallmeyen 1996). Under these reducing conditions, carbon is stable in the form of graphite (Hirschmann and Withers 2008), and a chemical model for CO_2 solubility under graphite saturated conditions (Holloway et al. 1992; Holloway 1998) can be applied (Hirschmann and Withers 2008). Chemical equilibrium constants controlling CO_2 solubility have been calibrated using terrestrial basaltic magmas (Holloway et al. 1992) and Martian-basalt analogue material (Stanley et al. 2011), as a function of oxygen fugacity. At relatively oxidizing conditions (IW + 1) and melt fractions typically encountered in the Martian mantle (5–10 %), a maximum of $\sim$1000 ppm CO_2 can be dissolved, which is significantly less than the 0.65 wt.% obtained for Kilauea basalts.

Combining the chemical model for CO_2 solubility (Hirschmann and Withers 2008) with parameterized thermal evolution models (Morschhauser et al. 2011), the amount of outgassed CO_2 can be calculated self-consistently (Grott et al. 2011). In order to cover the range of expected mantle dynamics in a one-dimensional model, two end member melting models may be considered: Melting in a global melt channel is likely representative for early martian evolution, whereas melting in localized mantle plumes may be more appropriate for the later evolution. Outgassing for both models is shown in Fig. 5. The degree of partial melting encountered in the plume model is generally higher than that for the melt channel model, and volatile concentrations in the melt are therefore lower (cf. Fig. 5). As approximately the same amount of crust is extracted from the mantle in both models, this results in reduced outgassing efficiencies. In both cases, a total of $\sim$1 bar CO_2 can be outgassed if comparatively oxidizing conditions (IW + 1) are assumed, and for an initial mantle water concentration of 100 ppm a total of 61 and 18 m EGL of H_2O can be outgassed in the global melt channel and plume model, respectively. While the rate of outgassing is lower for the plume model, outgassing in this model persists for $\sim$1 Gyr after melt generation ceases in the global melt channel model. A parameterization of CO_2 and H_2O outgassing rates as a function of oxygen fugacity, initial mantle water content, and outgassing efficiency may be found in Grott et al. (2011).

From these considerations one can see that, depending on geochemical and geological constraints, early Mars could have accumulated a secondary CO_2 atmosphere by volcanic outgassing of $\leq$1 bar $\sim$4 Gyr ago. However, large impacts and atmospheric escape processes should have modified the growth of this secondary atmosphere. In the following sections

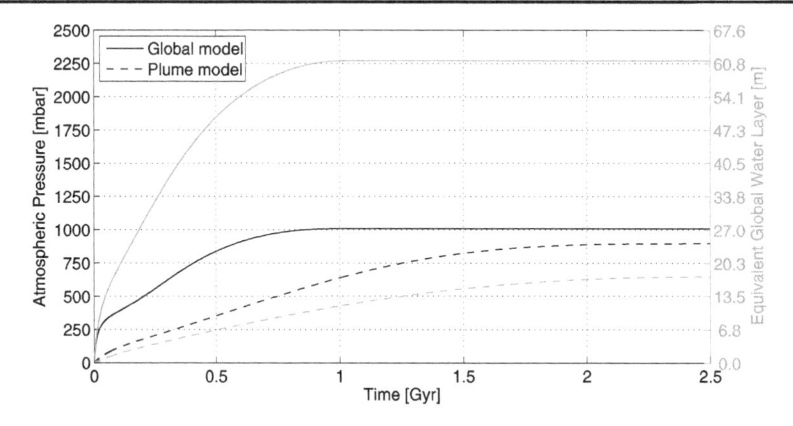

Fig. 5 Modeled cumulative volcanic outgassing of CO_2 given as partial surface pressure in mbar (*black lines*) and of H_2O given as equivalent global water layer (*green lines*) as a function of time. Mantle oxygen fugacity was assumed to be one order of magnitude above the iron-wustite buffer, resulting in an upper limit on CO_2 pressure. Initial mantle water concentration was assumed to be 100 ppm. *Solid curves* correspond to a model considering mantle melting in a global melt channel, whereas *dashed curves* correspond to a model considering melting in mantle plumes covering only a small fraction of the planetary surface. The outgassing efficiency η was set to 0.4 for all models and atmospheric escape is neglected

we will investigate possible changes of the secondary atmosphere in relation to losses and sources caused by large impacts, as well as various atmospheric escape processes which are connected to the change in solar activity.

4 Atmosphere Erosion and Delivery by Large Impacts

Atmospheric erosion and delivery by impactors can be studied with the help of hydrocode simulations which essentially simulate the flow field and dynamic response of materials by taking into account material strength and rheology (Ahrens 1993; Pierazzo and Collins 2003; Shuvalov and Artemieva 2001; Svetsov 2007; Melosh and Vickery 1989). Previous atmospheric erosion studies by hydrocodes have not always provided similar results, mainly due to differences in the physical models such as the choice of an appropriate equation of state, or a proper model of vapor cloud dynamics (Pham et al. 2009). In addition, these simulations require very large computer resources and can not be used directly to simulate long term atmospheric evolution. Therefore, the influence of the major parameters on atmospheric erosion and delivery has been parameterized. Models using the parameterization of the major mechanisms affecting the atmospheric erosion and delivery by the impacts can be instead used to study the evolution of the atmosphere. Many studies applied the so-called "tangent plane model" which has been developed by Melosh and Vickery (1989).

The tangent plane model of Melosh and Vickery (1989) is based on their hydrocode simulation results. Their model has been modified to take into account other simulations as well as additional parameters, and has been used to obtain a global view of the atmospheric mass evolution (Zahnle et al. 1992; Zahnle 1993; Manning et al. 2006, 2010; Pham et al. 2009). The advantage of using analytical models is that they can represent basic aspects of impact erosion and delivery, while reducing computation time since they only use a reduced number of parameters, scaled with numerical hydrocode simulation results.

The principle of the tangent plane model is that, when an impactor above a critical mass, m_{crit}, strikes the planet, the total mass above the plane tangent to the surface at the impact point, m_{tan}, escapes.

The critical mass is the minimal impactor mass that can eject m_{tan}, and it is proportional to m_{tan} through a factor n which represents the impact efficiency. The atmospheric mass above the plane tangent of the impact surface is approximated by $m_{tan} = m_{atm}H/2R_{pl}$, assuming an isothermal atmosphere in hydrostatic equilibrium, where m_{atm} is the total atmospheric mass, H the atmospheric scale height, and R_{pl} the radius of the planet. Note that the tangent plane model is only an approximation of erosion and delivery processes and that small impactors can still remove atmosphere (Zahnle 1993). While the model doesn't reproduce the physics of impact erosion or delivery, it can, with a suitable parameterization of the critical mass, give a global view of the atmospheric mass evolution upon impacts with a minimum set of variables (related to the critical mass value) and a much smaller computation time. The total mass evolution M_{atm} is controlled by the difference between the rates of change of atmospheric erosion, M_{esc} and the delivered volatile mass, M_{del}

$$\frac{dM_{atm}}{dt} = \frac{dM_{del}}{dt} - \frac{dM_{esc}}{dt}, \tag{4}$$

with

$$\frac{dM_{esc}}{dt} = \frac{\partial N_{cum}[>m_{crit}(t), t]}{\partial t} 4\pi R^2 m_{tan}(t) f_{vel} f_{obl}, \tag{5}$$

and

$$\begin{aligned} \frac{dM_{del}}{dt} &= \frac{\partial N_{cum}[>m_{crit}(t), t]}{\partial t} 4\pi R^2 \frac{b}{1-b} m_{crit}(t) y_{imp} f_{vap} \\ &+ \frac{\partial N_{cum}[>m_{crit}(t), t]}{\partial t} 4\pi R^2 m_{crit}(t) y_{imp}(1 - f_{vel} f_{obl} g_{vap}), \end{aligned} \tag{6}$$

where N_{cum} is the cumulative number of impacts with mass larger than m_{crit} at a time t, $\partial N_{cum}(>m_{crit}(t), t)/\partial t$ represents the flux of these impactors and b characterizes the mass distribution of the impactor flux, $b < 1$. We also assumed an exponentially decaying impact flux (e.g., Neukum and Wise 1976; Ivanov 2001; Neukum et al. 2001).

In the above equations the original "tangent plane model" is modified by the additional terms f_{vel} and f_{obl} in Eq. (5) and y_{imp} f_{vap}, g_{vap}, f_{vel} and f_{obl} in Eq. (6) which were not considered in previous studies (Pham et al. 2009, 2011). The fraction of impactors which are fast enough to erode the planet (f_{vel}) as well as the enhancement factor of the erosion due to impact obliquity (f_{obl}) are taken account. The volatile content (y_{imp}) is different for asteroids and comets. We assume a volatile content of $y_{imp} = 0.03$ for comets, and $y_{imp} = 0.01$ wt. for asteroids. The ratio of the vaporized mass to the impactor mass averaged over impact velocities is considered through the parameters f_{vap} or g_{vap} depending on whether the impactor mass is below or above the critical mass, respectively. In the delivery equation, the first term on the right hand side represents the delivered mass rate for $m_{imp} < m_{crit}$, and the second term is the delivered mass rate for the fraction of impactors with $m_{imp} > m_{crit}$ that was not removed by impacts. The relative amount of comets and asteroids in the total impact flux on Mars is assumed to be 6 % and 94 %, respectively, as suggested by Olsson-Steel (1987). In addition comets are differentiated between short-period (SP) comets (~4 %) and long-period comets (LP) from the Oort cloud (~2 %) when the impact velocity is taken into account in the simulations. The values corresponding to the factors given in Eqs. (5) and (6) are shown in Table 2.

Table 2 Values of the factors given in Eqs. (6) and (7)

Factor	Asteroids	Comets	
		SP comets	LP comets
f_{vel}	0.08	0.83	0.99
f_{obl}	7.54	2.16	
y_{imp}	0.01	0.3	
f_{vap}	0.34	1	
g_{vap}	0.21	1	

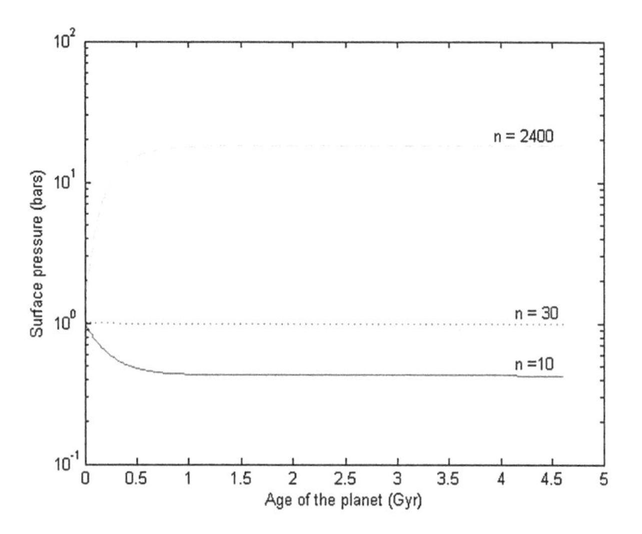

Fig. 6 Maximum diameter of the impactors hitting Mars as a function of time, for an assumed initial atmospheric surface pressure of 300 mbar. The *horizontal lines* show the critical diameter d_{crit} for different values of $n = m_{crit}/m_{tan}$ (details in the main text). Only impactors with diameter larger than d_{crit} can erode the atmosphere. The lower limit of $n = 10$ represents an extreme case corresponding to the upper limit of atmospheric erosion

The efficiency of atmospheric erosion and delivery is determined by the factor n which is given by $n = m_{crit}/m_{tan}$. The exact value of n in the tangent plane model is uncertain. The value suggested initially by Melosh and Vickery (1989), $n = 1$, has been revised by more recent studies to $n = 10$ (Vickery 1990; Manning et al. 2006). On the other hand, the more recent hydrocode simulations performed by Shuvalov and Artemieva (2001), Ivanov et al. (2002), and Svetsov (2007) yield results compatible with much larger values of the critical mass, on the order of $50 < n < 2000$. Note that n is a function of atmospheric pressure (Hamano and Abe 2006; Svetsov 2007). Pham et al. (2009) considered the tangent plane model with different impact erosion efficiencies, using constant as well as pressure-dependent values of n.

The total atmospheric mass evolution calculated by using Eqs. (5) and (6) is plotted in Fig. 6 for three values of n. Depending on the value of n impacts constitute either a factor of erosion or a source of volatiles on Mars. The most recent hydrocode simulations tend to favor larger values, $n > 30$, for which impacts are a source of volatiles (Svetsov 2007). Although impacts can remove atmosphere for smaller values of n the process is not very efficient, since even for the most favorable values of $n = 10$, a 1 bar protoatmosphere can not be eroded to present values over the age of the Solar System. In Sect. 6 we will consider the most favorable parameters for atmospheric loss considering a large impact flux during the late heavy bombardment with lower limit of n to yield an upper limit of atmospheric erosion. Although impact erosion may not have been a very relevant loss process, the question remains if the impact flux for accumulation of a secondary atmosphere was higher than the expected EUV-powered thermal escape flux which will be discussed in Sect. 5.

5 The EUV-Powered Blow-Off of the Protoatmosphere and the Change of a Secondary CO_2 Atmosphere by Escape Processes

From observations of young solar-type G stars it is known that despite a weaker total luminosity, stars with a young age are a much stronger source of X-rays and EUV electromagnetic radiation (e.g., Newkirk 1980; Skumanich and Eddy 1981; Zahnle and Walker 1982; Güdel et al. 1997). Since the 90ies the evolution of UV fluxes of a sample of solar analogue stars, so-called proxies of the Sun have been studied in detail by spectral measurements from the IUE satellite (Dorren et al. 1995). This research was extended by Güdel et al. (1997), Ribas et al. (2005) and recently by Claire et al. 2012 to X-rays and EUV. The wavelength range $\lambda \leq 1000$ Å is relevant for ionization, dissociation and thermospheric heating (e.g., Hunten et al. 1987; Hunten 1993).

Because X-rays dominate at young stellar ages, Owen and Jackson (2012) studied the contribution of harder X-ray's to the heating of hydrogen-rich upper atmospheres. If we compare the X-ray luminosities of solar proxies with younger age (e.g. Ribas et al. 2005; Claire et al. 2012) at Mars' orbit, with the values necessary for having a dominating X-ray driven atmospheric escape (Owen and Jackson 2012, their Fig. 11), one finds that this process is only relevant for hydrogen-rich "Hot Jupiter"-type exoplanets but can be neglected on early Mars which orbits further away from the Sun. Therefore, for early Mars, EUV radiation should be the main heating process in the thermosphere.

Güdel et al. (1997), Ribas et al. (2005) and Claire et al. (2012) analyzed multi-wavelength EUV observations by the ASCA, ROSAT, EUVE, FUSE and IUE satellites of solar proxies with ages <4.6 Gyr and found that the EUV flux is saturated during the first 100 Myr at a value $\sim$100 times that of the present Sun. As shown in Fig. 7 this early active period of the young Sun decreases according to an EUV enhancement factor power law after the first 100 Myr (Ribas et al. 2005)

$$S_{\mathrm{EUV}} = \left(\frac{t_{\mathrm{Gyr}}}{t_0}\right)^{-1.23}, \tag{7}$$

where t_0 is the age of the present Sun and t_{Gyr} the younger or older age of the Sun in time in units of Gyr.

Due to the lack of accurate astrophysical observations from solar proxies with different ages, previous pioneering studies on EUV-driven hydrodynamic escape of primitive atmospheres were based only on rough EUV enhancement scaling factors which were assumed to be similar or up to only $\sim$5–25 times higher than the present value (e.g., Watson et al. 1981;

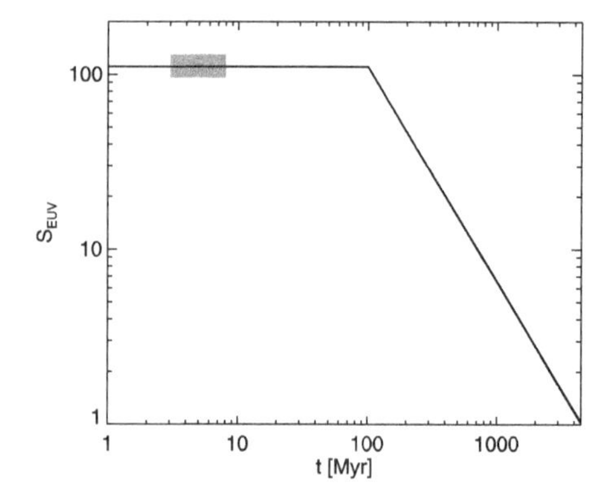

Fig. 7 Solar EUV flux enhancement factor S_{EUV} as obtained from observations of solar proxies. The average nebula evaporation time is $\sim$3 Myr. Since that time period planetary embryos and protoplanets are exposed to the saturated EUV flux value which is $\sim$100 times larger compared to the present solar value for about 90 Myr. The *shaded area* marks the expected time when Mars ended its accretion

Kasting and Pollack 1983; Chassefière 1996). Furthermore, most of the previous studies are based on terrestrial planet formation models where the accretion for Mars occurred quite late at ~100 Myr (Wetherill 1986).

Because Mars can be considered as a planetary embryo that did not collide or merge with other planetary embryos, it developed most likely within ~2–4 Myr after the birth of the Solar System (Dauphas and Pourmand 2011; Brasser 2012). This age agrees with average planetary nebula evaporation time scales of ~3 Myr. Because nebulae life times are <10 Myr (e.g., Lunine et al. 2011), the nebula-based hydrogen-rich martian protoatmosphere was most likely exposed during several tens of Myr or up to ~150 Myr to an EUV flux which was ~50–100 times higher compared to today's Sun (Ribas et al. 2005; Claire et al. 2012). Because of the high EUV flux of the young Sun H_2, H_2O and most CO_2 molecules in the thermosphere are dissociated and H atoms should dominate the upper atmosphere until they escaped to space.

Depending on the composition of the upper atmosphere and the planet's mean density, when the solar EUV flux in the wavelength range $\lambda \approx 2$–120 nm overcomes a critical value, the outward flow of the bulk thermosphere cools due to adiabatic expansion (Tian et al. 2005, 2008). According to studies of Watson et al. (1981), Kasting and Pollack (1983) and Tian et al. (2005), if hydrogen populates the upper atmosphere of a terrestrial planet, its exobase level can expand several planetary radii if the EUV flux is only a few times higher compared to that of today's Sun. A hydrogen-rich upper atmosphere of a martian-type body which is exposed to a EUV flux which is 5–100 times higher compared to the present solar value is, therefore, certainly in the blow-off regime. Under such conditions the exosphere evaporates as long as enough hydrogen is present.

By applying a blow-off formula which is derived from the energy-limited equation (e.g., Hunten et al. 1987; Hunten 1993) the atmospheric mass loss $dM_{\rm esc}/dt$ can be written as

$$\frac{dM_{\rm esc}}{dt} = \frac{3\eta S_{\rm EUV} F_{\rm EUV}}{4G\rho_{\rm pl}}, \tag{8}$$

where the heating efficiency η is the ratio of the net heating rate to the rate of solar EUV energy absorption of ~15–40 % (Chassefière 1996; Lammer et al. 2009; Koskinen et al. 2012), the gravitational constant G, the mean planet density $\rho_{\rm pl}$ and the present time EUV flux $F_{\rm EUV}$ in Mars' orbit. Using this relation one can estimate the atmospheric escape as long as hydrodynamic blow-off conditions occur which means that the thermal energy of the gas kinetic motion overcomes the gravitational energy.

Figure 8a shows the upper limit of a EUV-driven hydrogen-dominated protoatmosphere, which may have been captured from the nebula, during ~1 Gyr after the planet's origin. The loss is estimated from Eq. (8) with a heating efficiency η of 40 % (Koskinen et al. 2012). One can see that Mars could have lost an equivalent hydrogen content as available in ~14 Earth oceans (EO_H). However, if one assumes higher heating efficiency values the upper limit of hydrogen escape from early Mars during the first Gyr would be at ~30 EO_H. Due to the low gravity of Mars and EUV fluxes on the order of ≥50 times that of the present Sun, the blow-off condition was 100 % fulfilled for light hydrogen atoms and most likely also for heavier atomic species such as O, or C if they populated the upper atmosphere (e.g., Tian et al. 2009).

After the dissociation of the H_2O molecules and a fraction of CO_2 one can expect that O atoms are the major form of escaping oxygen. The escape flux of the heavier atoms $F_{\rm heavy}$ which can be dragged by the dynamically outward flowing hydrogen atoms with flux $F_{\rm H}$ can be written as (e.g., Hunten et al. 1987; Chassefière 1996)

$$F_{\rm heavy} = \frac{X_{\rm heavy}}{X_{\rm H}} F_{\rm H}\left(\frac{m_{\rm c} - m_{\rm heavy}}{m_{\rm c} - m_{\rm H}}\right), \tag{9}$$

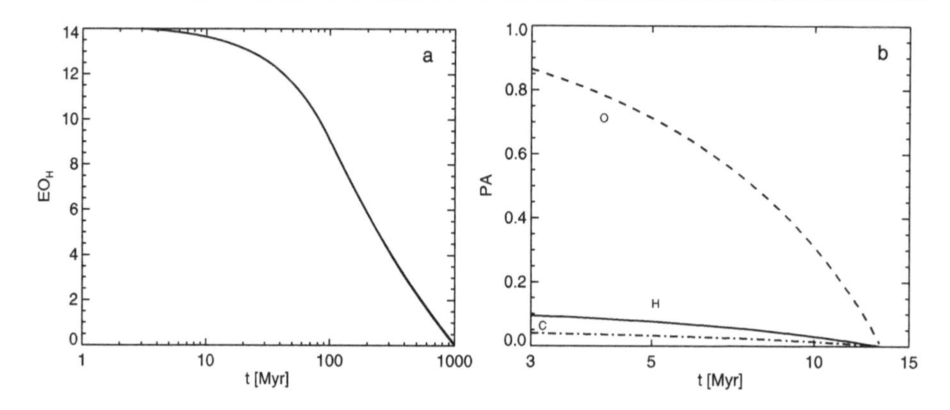

Fig. 8 (**a**) Upper escape value of atomic hydrogen in units of Earth ocean equivalent amounts (EO_H) of atomic hydrogen, with a heating efficiency of 40 % from a martian protoatmosphere between 3 Myr to 1 Gyr after the origin of the Solar System. (**b**) Calculated normalized loss of an outgassed 70 bar water vapor and 12 bar CO_2 steam atmosphere as a function of time. PA = 1 corresponds to the total pressure of 82 bar, while the *solid line* corresponds to loss of the hydrogen content. The *dashed* and *dash-dotted lines* corresponds to dragged oxygen and carbon atoms which originate from dissociation of H_2O and CO_2

where X_H and X_{heavy} are the mole mixing ratios. m_H and m_{heavy} are the masses of the hydrogen atom and the heavy species. m_c is the so called cross over mass

$$m_c = m_H + (kTF_H)(bgX_H), \tag{10}$$

which depends on F_H, a molecular diffusion parameter b (Zahnle and Kasting 1986; Chassefière 1996), the gravity acceleration g, Boltzmann constant k and an average upper atmosphere temperature T, which can be assumed under such conditions for hydrogen to be on the order of ~500 K (Zahnle and Kasting 1986; Chassefière 1996). By applying Eqs. (1) and (2) and assuming that Mars finished its accretion within the EUV-saturated epoch of the young Sun, Fig. 8b shows the atmosphere loss estimation of an outgassed water vapour dominated by a 70 bar H_2O and 12 bar CO_2 steam atmosphere. The loss is normalized to the total outgassed surface pressure of 82 bar, where PA = 1 which corresponds to the total pressure of 82 bar. One can see that under such conditions early Mars could easily lose its initial atmosphere in ~10 Myr. Thus, because Mars accreted early (Dauphas and Pourmand 2011; Brasser 2012), even if the planet would have obtained its volatile inventory later, the high EUV flux of the young Sun would have blown the atmosphere away. One can also see that under these extreme conditions the outgassed CO_2 would be lost in dissociated form as C and O as shown in Fig. 8b. Therefore, a dense CO_2 atmosphere could have been lost very early and could not have accumulated during the early Noachian. On the other hand, if early Mars was surrounded by a nebula-based hydrogen envelope, the outgassed heavier volatiles may have been protected against atmospheric escape until the captured hydrogen was lost and did not dominate the upper atmosphere anymore.

For these time scales the outgassed H_2O/CO_2 atmosphere remained most likely in steam form because the time scale where the surface temperatures may reach the point that H_2O can condense is comparable (Elkins-Tanton 2008). Furthermore, these time scales also agree with studies by several researchers who investigated the early stages of accretion and impacts and expect that due to thermal blanketing hot temperatures could keep the volatiles in vapor phase for several tens of Myr or even up to ~100 Myr (e.g., Hayashi et al. 1979; Mizuno et al. 1980; Matsui and Abe 1986; Zahnle et al. 1988; Abe 1997; Albarède and

Blichert-Toft 2007). One can also see from Figs. 5 and 6 and the discussions in Sect. 4 that during this early evolutionary period most of the impacts occurred. Although the loss effect of these impacts may have been less efficient when Mars had a dense outgassed H_2O/CO_2 steam atmosphere. As mentioned in Sect. 4 impacts have contributed to a permanent heating of the atmosphere. On the other hand volatiles which were brought in to the atmosphere by large impacts should also have been lost due to the strong hydrogen escape.

These results are in agreement with the non-detection of carbonates by the OMEGA instrument on board of ESA's Mars Express spacecraft (Bibring et al. 2005). Mars Express mapped a variety of units based on areas exhibiting hydrated minerals, layered deposits, fluvial floors, and ejecta of deep craters within Vastitas Borealis with a surface resolution in the ~1–3-km range. Besides CO_2-ice in the perennial southern polar cap, no carbonates were reported. Bibring et al. (2005) concluded that the non-detection of carbonates would indicate that no major surface sink of CO_2 is present and the initial CO_2, if it represented a much higher content, would then have been lost from Mars early rather than stored in surface reservoirs after having been dissolved in long-standing bodies of water.

However, so far it is not clear when the outgassing flux from Mars' interior exceeded the expected escape flux so that a secondary CO_2 atmosphere could grow during the later Noachian. It should be noted that the accumulation of both the secondary outgassed atmosphere and volatiles which were possibly delivered by later impacts is highly dependent on atmospheric escape after the strong early hydrodynamic loss during the EUV-saturation phase of the young Sun.

Tian et al. (2009) applied a 1D multi-component hydrodynamic thermosphere-ionosphere model and a coupled electron transport-energy deposition model to Mars and found that for EUV fluxes >10 times that of today's Sun, CO_2 molecules dissociate efficiently resulting in less IR-cooling of CO_2 molecules in the thermosphere so that a CO_2 atmosphere was most likely also not stable on early Mars after the EUV-saturation phase ended. According to this study, the flux of the produced C and O atoms is $>10^{11}\ \mathrm{cm\,s^{-1}}$ before ~4 Gyr ago and was of the same order as the fluxes from volcanic outgassing (Tian et al. 2009).

Although this result seems logical, because the results of Tian et al. (2009) are model dependent and contain various uncertainties we estimate the possible growth of such a secondary CO_2 atmosphere from the outgassing rates shown in Fig. 5 or Grott et al. (2011), by considering that the outgassed CO_2 flux exceeded the thermal escape since 4.3, 4.2, 4.1 or 4 Gyr ago. Figure 9 shows the possible scenarios for a build up of a secondary CO_2 atmosphere and Fig. 10 shows secondary outgassed H_2O amounts in units of bar. Table 3 summarizes the accumulated outgassed CO_2 amount in units of bar after the outgassed flux becomes more efficient compared to the escape flux for the same oxygen fugacity the related iron-wustite buffer (IW) and f_p scenarios shown in Fig. 9. If we consider that the escape flux of CO_2 was less than that from the interior 4.3 Gyr ago (Fig. 9a) a secondary CO_2 atmosphere of ~0.7 bar could build up ~4 Gyr ago, which is about 100 times denser than the present atmosphere if one assumes a global melt channel. By considering mantle plumes only, the outgassing would be finished about 4 Gyr ago and a CO_2 atmosphere of ~0.5 bar could have been built up. If the escape flux could balance the volcanic outgassing for longer times, depending if one assumes a global melt channel or mantle plumes, only CO_2 atmospheres with lower upper densities of ~0.2–0.4 bar could build up ~2.5–4 Gyr ago. For cases with low oxygen fugacity, the secondary CO_2 atmosphere would only have a surface density between ~50–100 mbar.

Low CO_2 surface pressure values would also agree with a study by Zahnle et al. (2008), which is based on a photochemical CO_2 stability problem discussed by McElroy and Donahue (1972), that a martian CO_2 atmosphere much denser than several 100 mbar may be not

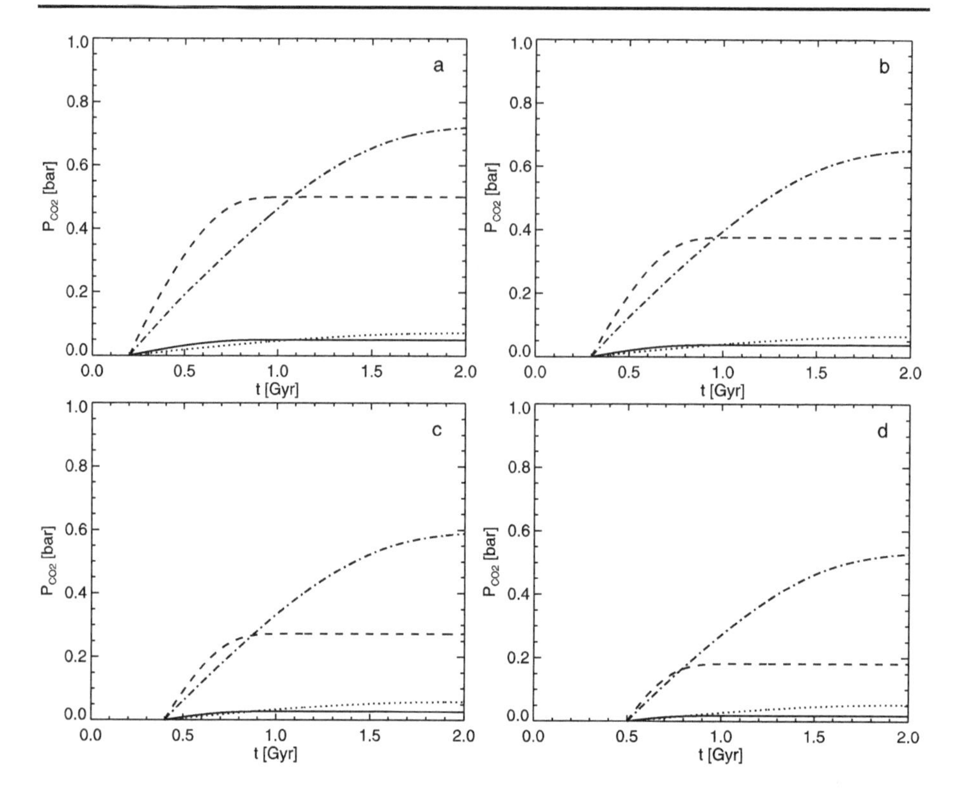

Fig. 9 CO_2 partial surface pressure as a function of time with the same assumptions as in Fig. 4 but for various onset times for the build up of a secondary CO_2 atmosphere after total loss of the earlier outgassed CO_2 content. *Dashed lines*: IW = 1, Surface fraction of the melt channel $f_P = 1$; *dashed-dotted lines*: IW = 1, $f_P = 0.01$; *solid lines*: IW = 0, $f_P = 1$; *dotted lines* IW = 0 and $f_P = 0.01$. The onset for atmospheric growth of a secondary CO_2 atmosphere is assumed in a: 4.3 Gyr (**a**), 4.2 Gyr (**b**), 4.1 Gyr (**c**), and 4 Gyr (**d**) ago

be stable for a long time because the CO_2 will be photochemically converted into CO over timescales of ~0.1–1 Gyr.

Chevrier et al. (2007) investigated the geochemical conditions which prevailed on the martian surface during the Noachian period by applying calculations of aqueous equilibria of phyllosilicates. These authors found that Fe^{3+}-rich phyllosilicates most likely precipitated under weakly acidic alkaline pH, which was a different environment compared to the following period which was dominated by strong acid weathering that led to the observed martian sulphate deposits. Chevrier et al. (2007) applied thermodynamic calculations which indicate that the oxidation state of the martian surface should have been also high during early periods, which supports our results of an early efficient escape of hydrogen.

However, equilibrium with carbonates implies that the precipitation of phyllosilicates occurs at low CO_2 partial pressure. Thus, from these considerations one would expect that the lower surface CO_2 pressure shown in Fig. 8 and Table 3 may have represented the martian atmosphere ~4 Gyr ago. If geochemical processes prevented the efficient formation of carbonates then a dense CO_2 atmosphere could not have been responsible for a long-term greenhouse effect which is necessary to enable liquid water to remain stable at the surface in the post-Noachian period. In such a case other greenhouse gases such as CH_4, SO_2, H_2S,

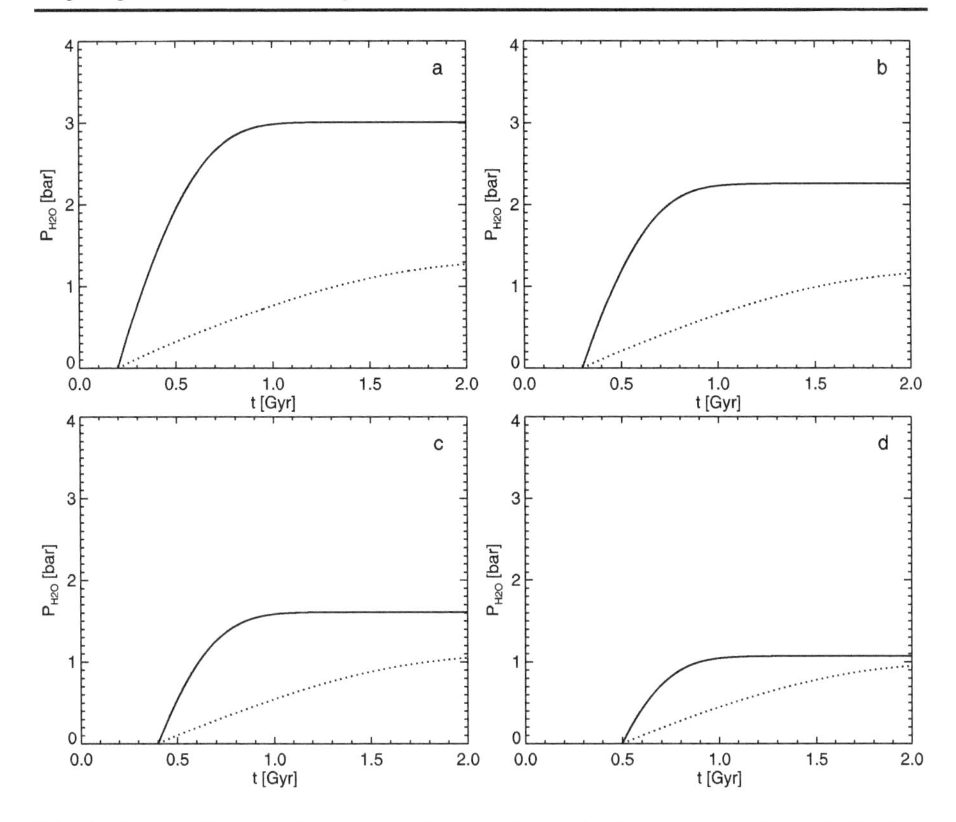

Fig. 10 Outgassed H_2O in units of bar as a function of time with the same assumptions as in Fig. 4 but for various onset times where the outgassing flux exceeded the escape flux after the total loss of the earlier outgassed water content. *Solid lines*: Surface fraction of the melt channel $f_p = 1$; *dotted lines*: $f_p = 0.01$. The bulk concentration of water in the mantle is assumed to be 100 ppm and the outgassing efficiency is assumed to be 0.4

Table 3 Secondary outgassed CO_2 partial surface pressure in units of bar as function of iron-wustite (IW) buffer and surface fraction of the melt channel (f_p) (see Sect. 3 and Grott et al. 2011) and time t after Mars' origin in Myr when the outgassing flux is assumed larger than the escape flux

IW and f_p *scenarios*	$t = 200$ Myr	$t = 300$ Myr	$t = 400$ Myr	$t = 500$ Myr
IW = 1; $f_p = 0.01$	~0.7 bar	~0.65 bar	~0.6 bar	~0.55 bar
IW = 1; $f_p = 1$	~0.5 bar	~0.37 bar	~0.25 bar	~0.18 bar
IW = 0; $f_p = 0.01$	~0.07 bar	~0.12 bar	~0.11 bar	~0.1 bar
IW = 0; $f_p = 1$	~0.05 bar	~0.085 bar	~0.05 bar	~0.03 bar

etc. (Kasting 1997) would be needed to solve the greenhouse-liquid water problem during the late Noachian.

Depending on the surface fraction of the melt channel f_p and the onset time of accumulation from Fig. 10 and Table 4 one can see that the outgassed amount of H_2O by volcanos would correspond to values ~1–3 bar, that is a ≈20–60 m EGL. One should also note that in addition to the secondary outgassed atmosphere significant amounts of water and carbon may have been brought later by comets. According to Morbidelli et al. (2000), Lunine et al.

Table 4 Secondary outgassed H_2O partial pressure in units of bar with similar conditions and assumptions as shown in Fig. 10

f_P *scenarios*	$t = 200$ Myr	$t = 300$ Myr	$t = 400$ Myr	$t = 500$ Myr
$f_P = 1$	~3 bar	~2.25 bar	~1.6 bar	~1.1 bar
$f_P = 0.01$	~1.3 bar	~1.2 bar	~1.1 bar	~1 bar

(2003) the equivalent of ~0.1 terrestrial ocean of H_2O, that is a ~300 m deep EGL of water, could have been provided to Earth by comets during the few 100 Myr following main accretion. The net budget of cometary impacts could hypothetically have also resulted in a net accretion of several bars or even tens of bars of H_2O (and several 100 mbar of CO_2) coming from infalling comets until the late Noachian or during the LHB as discussed in Sect. 6. A fraction of this impact delivered H_2O, if all was not lost due to the high thermal escape rate (e.g. Tian et al. 2009) could be stored in the crust (Lasue et al. 2012; Niles et al. 2012). Although it is not clear at the present how much CO_2 was in the martian atmosphere ~4 Gyr ago, the secondary outgassed and accumulated atmosphere was most likely denser compared to the 7 mbar of today.

6 Environmental Effects of the Late Heavy Bombardment

Although the previous sections have shown that due to the high EUV flux of the young Sun, atmospheric escape models do not favor a dense CO_2 atmosphere during the first Gyr, we now investigate possible effects of the late heavy bombardment (LHB) period. The ratios between critical mass vs. tangent mass n as discussed in Sect. 4 determines whether impacts cause atmospheric erosion or if they are rather a source of volatile. For investigating the upper limit of atmospheric erosion related to the LHB, we consider only the lowest limit of $n = 10$ and examine the number of impactors which are necessary to erode the martian atmosphere since the late Noachian so that we end up with ~7 mbar.

If the martian atmosphere can be eroded by impacts only, the numbers of impacts above the critical mass has to be higher compared to the number computed in the exponentially decaying impact flux model. We found from our calculations that depending on the initially assumed surface pressure of ~0.1–1 bar, one would need ~8000–15000 impactors with masses equal or larger than m_{crit} to erode the martian atmosphere to a surface pressure of ~7 mbars over the last 3.8 Gyr. However an exponentially decaying impactor flux model gives only ~86 for ~0.1 bar and ~30 for ~1 bar for the number of impacts above m_{crit} over this time period. These numbers are ~100–500 times smaller than the necessary number of large impacts. Therefore, by considering an exponentially decaying impact flux, it is unlikely that the martian atmosphere with a surface pressure $\geq$0.1 bar was eroded by impacts during the past 3.8 Gyr.

The LHB, on the other hand, can provide the number of large impactors which is necessary to remove the atmosphere from Mars. After a period which can most likely be characterized by a weak bombardment rate ~3.9 Gyr ago, the planets experienced the LHB. The LHB was a cataclysmic episode characterized by a high bombardment rate, during a time-span of ~50–300 Myr. The Nice model (e.g., Gomes et al. 2005) which simulates the orbital evolution of the Solar System with slow migration of the giant outer planets, followed by a chaotic phase of orbital evolution, yields an estimate of impactor mass distribution during this period. The impactor masses could be distributed as presented in Fig. 11 during the late heavy bombardment period (data provided by Morbidelli, private communication). By

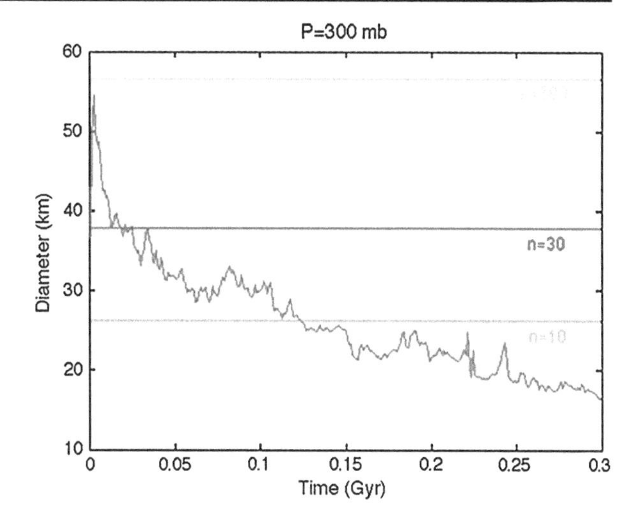

Fig. 11 Impactor diameters for the LHB for the best erosion case and for an initial pressure of ~300 mbar

investigating the best case for the erosion efficiency, we consider that the largest impactor provides the first impact.

The maximum diameter of the impactors hitting Mars as a function of time can be compared with the critical impactor diameter d_{crit} that can erode the atmosphere. For an assumed initial surface pressure of $\sim$300 mbar, the critical mass m_{crit} and the corresponding critical impactor diameter d_{crit} (with $\rho = 2000\ \mathrm{kg\,m^{-3}}$) for different values of $n = m_{\mathrm{crit}}/m_{\mathrm{tan}}$ can be calculated. By assuming $n = 10$, from these calculations one obtains an upper limit for the amount of atmosphere which can be eroded by impacts of $\sim$150 mbar over an intense bombardment period of $\sim$0.3 Gyr. Lower values of n can erode primordial atmospheres of $\sim$400 mbar ($n \approx 3$) and even $\sim$1 bar in the case of ($n \sim 1$). From these results one can see that impacts could have removed a major fraction of an accumulated secondary CO_2 atmosphere (see Fig. 6). The main problem with the impact studies remains to be the choice of parameter n. As discussed in Sect. 4 studies which assume values for $n \geq 30$ deliver volatiles to the martian surface. Under this consideration, small n, and hence atmospheric erosion due to impacts as discussed before, is questionable considering that recent hydrocode simulations suggest at least an order of magnitude larger value of n (Svetsov 2007; Pham et al. 2009, 2010, 2011). In such a case the LHB would have accumulated volatiles additively to the secondary outgassed atmosphere. By assuming that delivered CO_2 corresponds to $\sim$1 % of the impactors this accumulation could result in an amount of impact delivered CO_2 of $\sim$300 mbar. Thus, the H_2O which could have been brought to Mars by impacts (Levison et al. 2001) especially during the LHB-period, where the solar EUV flux and related thermal escape processes were much lower compared to their early values, could also be an important contribution to the planets present water inventory. The cometary bombardment, is largely unconstrained but can deliver up to or even more then $\sim$5 bar of H_2O, that corresponds to a $\approx$130 m deep GEL. These numbers should be considered as upper limits for the assumed total mass of the comets and asteroids (7×10^{22} g and 4×10^{22} g, respectively) which may have fallen to Mars during the LHB (data provided by Morbidelli et al. 2009, private communication).

Geomorphological and geological evidence shows that liquid water flowed on the martian surface, particularly in the Noachian period (Baker 2001; Squyres and Knoll 2005). In order to have liquid water stable on the martian surface, CO_2 surface pressures of several bar are necessary to obtain temperatures above freezing (Kasting 1991). If one considers scattering of infrared radiation from CO_2-ice clouds (Forget and Pierrehumbert 1997) or additional greenhouse gases such as CH_4, SO_2 and H_2S, which could have also been released by volcanism (Kasting 1997) this value can be achieved for ~0.5–1.0 bar. On the other hand aqueous solutions with lower melting points may have existed (e.g., Fairén 2010, and references therein; Möhlmann 2012) making it possible that Mars might have been "cold-and-wet" with average surface temperatures of ~245 K (Fairén 2010; Gaidos and Marion 2003). Furthermore, water released by large impacts during the LHB could also have liberated huge amounts of water so that transient wet and warm conditions on the surface (Segura et al. 2002; Toon et al. 2010) could have occurred.

However, if impacts delivered volatiles additionally to the secondary atmosphere during the LHB period, this portion should have been lost partly to space during the Hesperian and Amazonian by various nonthermal atmospheric escape processes and partly weathered out of the atmosphere into the surface and ice.

7 Escape and Surface Weathering of the Secondary Atmosphere Since the End of the Noachian

From Mars Express ASPERA-3 ion escape data, Barabash et al. (2007) estimated the fraction of CO_2^+ molecular ions lost to space since the end of the Noachian when the martian dynamo stopped to work equivalent to a surface pressure of about ~0.2–4 mbar. The present CO_2^+ escape rates are about two orders of magnitude lower compared to the O loss and are on the order of $\sim 8 \times 10^{22}$ s^{-1} (Barabash et al. 2007).

That direct escape of CO_2^+ ions from Mars was low is also in agreement with various MHD and hybrid model results which yield an integrated CO_2^+ ion loss (IL) since the end of the Noachian on the order of ~0.8–100 mbar (e.g., Ma et al. 2004; Modolo et al. 2005; Chassefière et al. 2007; Lammer et al. 2008; Manning et al. 2010). Moreover from a recent study of Ma and Nagy (2007) who calculated an escape rate of carbon of about 1.8 times larger at solar maximum than at solar minimum which is in good agreement with the dependency of the escape ion rate calculated in 2009 the estimated amount of CO_2 lost by ion loss since ~4 Gyr is most likely not in excess of ~1 mbar (Chassefière and Leblanc 2011a). Thus, from these studies we can consider that the realistic CO_2^+ molecular ion loss by pick up and outflow through the martian tail in the theoretical range of about 0.8–100 mbar given in Manning et al. (2010) should be considered closer to the lower values.

On the other hand one should mention that all the previous ion escape models did not use an accurately modeled neutral atmosphere and ionosphere which corresponds to higher EUV fluxes expected before 2.5 Gyr ago. In such a case one can expect that more carbon dioxide will be dissociated in the thermosphere so that it can be heated to higher temperatures. As shown by Tian et al. (2009) a hotter thermosphere leads to an expansion of the upper atmosphere and thus more extended coronae. In such a case the solar wind interaction area would be larger and one may expect higher ion loss rates too.

One should also note that solar wind induced forcing of Mars can also result in outflow and escape of ionospheric ions. ASPERA-3 observations indicate that the replenishment of cold ionospheric ions starts in the dayside at low altitudes at ~300–800 km, where ions move at a low velocity of ~5–10 $km\,s^{-1}$ in the direction of the external magnetosheath flow (Lundin 2011). The dominating energization and outflow process, applicable for the inner magnetosphere of Mars, leads to outflow at energies of ~5–20 eV. These energized "cool" ionospheric ions can be picked up, accelerated by the current sheet, by waves and parallel electric fields (Lundin 2011). The latter acceleration process can be observed above

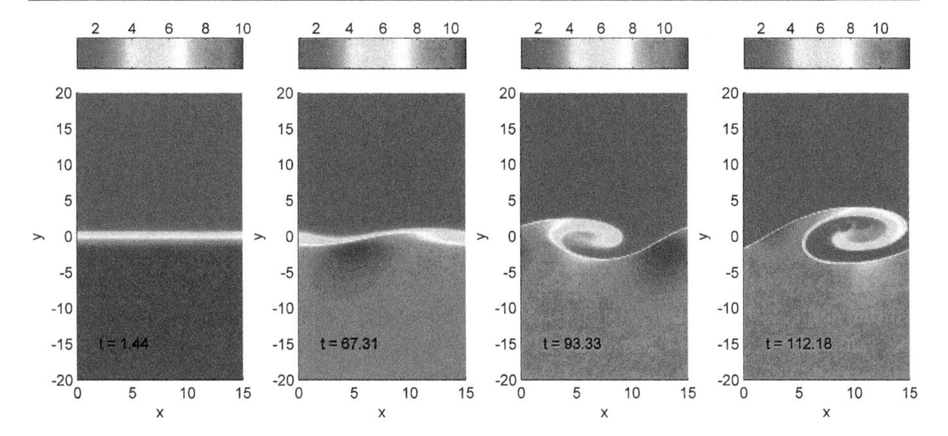

Fig. 12 Nonlinear evolution of the Kelvin-Helmholtz instability. The time series of the mass density is shown, from an MHD simulation with periodic boundary conditions in the x-direction. The mass density changes from the upper to the lower plasma layer and exhibits an increase of up to ten times (see the color code; *blue*: low density, *red*: high density). In the upper layer, the plasma flows from *left* to *right*. In the lower layer, the plasma is at rest. Initially small perturbations of the boundary layer separating the two plasma layers evolve into a KH vortex

martian crustal magnetic field regions. But even if we assume that cold ionospheric ions may enhance the ion escape for carbon bearing species the escape related to these processes most likely remains within the range given by Manning et al. (2010).

The Kelvin-Helmholtz (KH) plasma instability has also been regarded as a possible non-thermal atmospheric loss process around unmagnetized planets since Pioneer Venus Orbiter observed detached plasma structures, termed plasma clouds which contained ionospheric particles, downstream to the terminator in the magnetosheath of the planet (Brace et al. 1982; Wolff et al. 1980). Around planets, magnetopauses or ionopauses form boundaries with velocity shears, where the KH instability might be able to develop. On their way along the boundary from the subsolar point to the terminator, waves of initially small amplitudes grow and eventually form vortices in their nonlinear stage. When the vortex is able to detach, it carries ionospheric particles away and thus can contribute to the loss of ions (Brace et al. 1982).

Amerstorfer et al. (2010) and Möstl et al. (2011) performed recent numerical simulations of the KH instability with input parameters suitable for the boundary layers around unmagnetized planets. Figure 12 shows a time series of the normalized mass density at different times during one of their simulations. After the linear growth time of the instability, a regular-structured vortex has evolved in the nonlinear stage. For this simulation, the density of the lower plasma layer is only ten times the density of the upper layer—a larger density jump stabilizes the boundary layer. The results of Möstl et al. (2011) indicates that the martian ionopause should be stable with regard to the KH instability due to the stabilizing effect of the large mass density of the ionosphere. However, the induced magnetopause (Venus) or magnetic pile-up boundary (Mars) might be KH unstable during high solar activity. For this boundary, the atmospheric loss of planetary ions might not be as severe as if the ionopause was the unstable boundary. Thus this recent result indicates that the loss due to the KH instability is not as significant as previously thought (Penz et al. 2004). Furthermore, at the altitudes where one can expect that under extreme solar conditions plasma clouds may detach from the upper atmosphere, atomic oxygen ions should be the dominant species and CO_2^+ or CO^+ ions are most likely negligible constituents.

Atmospheric sputtering (SP) has been identified as an escape process of heavy atoms from planetary bodies with low gravity such as Mars (Luhmann and Kozyra 1991). Leblanc and Johnson (2002) studied sputter escape of CO_2 and CO from the martian atmosphere during the past 3.5 Gyr with a coupled test particle Monte Carlo molecular dynamic model which considers collisions between photochemically produced suprathermal atoms and background molecules for EUV fluxes which are 3 times and 6 times higher than that of today's Sun. These authors obtained an escape of CO_2 caused by sputtering since about 3.6–4 Gyr on the order of $\sim$50–60 mbar (see also Chassefière et al. 2007). More recently, it has been argued that the flux of pick-up ions reimpacting Mars' atmosphere follows a logarithmic slope with the EUV flux of $\sim$1.8 (Chassefière and Leblanc 2011a), much flatter than the value of $\sim$8 used in Chassefière et al. (2007), yielding a cumulated sputtering escape rate since $\sim$4 Gyr which gives a CO_2 loss $\leq$1 mbar.

One can see that the sputter loss was probably similar to that of ion erosion and for sure not efficient enough to cause the loss of hundreds of mbar of CO_2 could be lost. Further, we note that sputtering is a highly nonlinear process that depends on the EUV flux and the life time of the martian magnetic dynamo (Dehant et al. 2007). Furthermore, it was shown by Terada et al. (2009) that, due to the extreme solar wind atmosphere interaction caused by the young Sun before $\sim$4 Gyr ago, a stronger induced magnetic field in the upper atmosphere could have decreased sputtering during the transition period when the planet's intrinsic dynamo stopped working.

Besides ion escape and sputtering the loss of exothermal photochemically produced suprathermal atoms such as O, C, N and H could have been more effective compared to both escape processes discussed before. Exothermal processes such as dissociative recombination (DR) of O_2^+, N_2^+ or CO^+ ions produce neutral atoms in the ionosphere with higher kinetic energy compared to the background atmosphere (e.g., Ip 1988; Nagy et al. 1990; Kim et al. 1998; Lammer et al. 2000; Fox 2004; Fox and Hać 2009; Krestyanikova and Shematovich 2006; Chaufray et al. 2007; Valeille et al. 2010; Gröller et al. 2010, 2012). These newly created particles collide with the cooler background gas, lose energy by collisions, transfer energy so that a cold atom could become more energetic and finally a fraction of them reach the exobase level and if their energy is larger than the escape energy they are lost from the planet as neutrals. The production of these hot O and C atoms originating from DR of O_2^+ and CO^+ molecular ions is also strongly related to the solar EUV flux, and to an electron temperature dependent rate coefficient, where the total energy of these newly produced O(^{3}P, ^{1}D), O(^{3}P, ^{1}S) and C(^{3}P, ^{1}D), O(^{3}P, ^{1}D, ^{1}S) atoms is a sum of their released energies (ΔE) according to a DR reaction channel of kinetic and internal energy, the latter being stored in molecules as vibrational and rotational energy.

Excited C atoms can also be produced via photo-dissociation (PD) of CO molecules from $CO + h\nu \rightarrow C(^3P) + O(^3P) + \Delta E$, with ΔE obtained as the difference of the photon energy and the energy which is needed to dissociate the molecule and excite the newly produced atoms. Fox (2004) made a complete calculation of all possible photochemical channels for the production of carbon escaping particles and found that between 7.5×10^{23} C cm^{-2} s^{-1} to $\sim 4.5 \times 10^{24}$ C cm^{-2} s^{-1} may escape from solar minimum to maximum conditions with photo-dissociation of CO being the most efficient process.

If we apply a recently developed hot atom Monte Carlo model which selects the magnitude of the initial velocity of a newly produced hot particle randomly from the calculated velocity distribution, which considers collisions that are based on the energy dependent total and differential cross sections for elastic, inelastic and quenching collisions and numerous cascaded hot particles (Gröller et al. 2010, 2012) to present martian conditions, we obtain total escape rates for "hot" O atoms of $\sim 9 \times 10^{25}$ s^{-1} and for hot C atoms $\sim 2.7 \times 10^{25}$ s^{-1}

Table 5 Estimated min. and max. outgassed (VO: volcanic outgassing) CO_2 in units of bar ~4 Gyr ago and the expected min. and max. range of impact eroded (IE) or delivered (ID) atmosphere during the late heavy bombardment (LHB) period. CO_2 escape of various atmospheric loss processes (IL: ion loss; KH: Kelvin Helmholtz instability triggered ionospheric detached plasma clouds; SP: sputtering; DR: dissociative recombination; PD: photo dissociation) from observations and models integrated since that time

Sources and loss processes	CO_2 [bar]
VO ~4 Gyr ago, max.	~0.2–0.5 bar
VO ~4 Gyr ago, min.	~0.05 bar
ID ~ LHB	~0–0.3 bar
IE ~ LHB	~0–0.15 bar
IL since ~4 Gyr ago	~0.001–0.1 bar
KH since ~4 Gyr ago	~0.001 bar
SP since ~4 Gyr ago	≤0.001–0.05 bar
DR since ~4 Gyr ago	~0.001–0.1 bar
PD since ~4 Gyr ago	≤0.005–0.05 bar

during high solar activity conditions and $\sim 3 \times 10^{25}$ s^{-1} for "hot" O atoms and $\sim 3 \times 10^{24}$ s^{-1} for hot C atoms during low solar wind conditions. Our obtained hot O escape rates from present Mars are in good agreement with Chaufray et al. (2007) and Valeille et al. (2010) and about a factor of 5 to 3 lower compared to results from Fox and Hać (2009) for high and low solar activity, respectively.

If we apply our model to the thermosphere/ionosphere profiles modeled by Tian et al. (2009) during the earlier periods of martian history, preliminary estimates yield an integrated total CO_2 loss from today to 4 Gyr ago of ≤100 mbar. A smaller value of ≤10 mbar has been proposed by Chassefière and Leblanc (2011a).

Table 5 compares expected max. and min. outgassed and accumulated CO_2 atmosphere in units of bar at ~4 Gyr and the integrated min. and max. CO_2 escape of various nonthermal atmospheric loss processes from observations and models since that time. Table 5 shows a very large dispersion of one or two orders of magnitude yielding a range of total CO_2 loss from less than ~10 mbar up to ~100 mbar, which is definitely smaller than the expected upper values of an outgassed and accumulated secondary CO_2 atmosphere. However, a similar amount of CO_2 or even a higher one could be stored in the surface sinks which we discuss in the next section.

8 Surface Sinks of CO_2 and H_2O

Estimates of the total amount of CO_2 degassed into the martian atmosphere since accretion vary between ~5–12 bar, while as shown in Sect. 5 most of it was lost to space during the period of the young and active Sun. However, most of the strong EUV-powered atmospheric loss occurred most likely prior or during the early period of the Noachian and the Hellas impact (~4.0 Gyr). As discussed in the previous sections, the amount of CO_2 lost to space (accumulated from volcanic outgassing and/or was delivered by impacts during the LHB), is most likely not higher than ≈100 mbar. Therefore, if Mars had a denser CO_2 atmosphere ~3.5–4 Gyr ago, most of it should be hidden below the planet's surface.

8.1 Sequestration of CO_2 in Carbonates

Possible sinks for this CO_2 atmosphere include loss to space, adsorption on the regolith, deposits of CO_2-ice, and deposits of carbonate minerals. The maximum CO_2 adsorbed on the regolith has been estimated to be on the order of ~30–40 mbar and is likely to be less than that (Zent and Quinn 1995). Recent discoveries of buried CO_2-ice deposits in the polar regions indicate that modern CO_2-ice deposits could approach ~20–30 mbar equivalent CO_2 (Phillips et al. 2011). Combining these sinks with the amount of CO_2 likely lost to space since the past ~4 Gyr (see Table 5), it becomes clear that the majority of CO_2 that could be accounted for is most likely <150 mbar.

Therefore, the amount of CO_2 which is possibly stored as carbonate becomes the key to understanding the density of the ancient martian atmosphere at the end of the Noachian since it is the only sink that can accommodate a dense Noachian atmosphere. Early studies modeling a dense atmosphere on Mars predicted abundant carbonates in the martian crust (Pollack et al. 1987), but after a decade of intense exploration of Mars from orbit and on the surface, abundant carbonate deposits have not been discovered (e.g., Bibring et al. 2006). However, the detections of carbonates that have been made to date have revealed deposits that are either buried or widely dispersed making detection difficult (Ehlmann et al. 2008; Michalski and Niles 2010; Morris et al. 2010). Therefore, orbital detection of carbonates may not provide a complete view of the carbonate crustal reservoir on Mars which may be larger than currently expected. Nevertheless, information from martian meteorites, landers, rovers, and orbiters is now available to construct a fairly consistent story of the carbonate reservoir on Mars.

The most direct evidence for carbonates on Mars comes from martian meteorites, many of which contain carbonate minerals in trace abundances (<1 %) (Bridges et al. 2001). There are currently 40–50 known meteorites which are derived from 3–6 distinct sites on Mars (Eugster et al. 2002). These rocks are derived from energetic ejection events from the martian surface which likely destroyed all but the strongest igneous rocks providing a selection bias. Therefore it is clear that while martian meteorites are invaluable samples of the martian surface, they likely do not provide an adequate sample by which to judge the carbonate crustal reservoir of Mars. They do indicate that weathering and carbonate formation on Mars has been active at least at very low levels throughout the Noachian, Hesperian, and Amazonian periods (Gooding et al. 1988; Mittlefehldt 1994; Bridges et al. 2001; Niles et al. 2010). Localized deposits of more concentrated carbonates have also been found in some of the oldest Noachian terrains on Mars. These deposits, which have mostly been identified spectroscopically from orbit, are typically mixtures of carbonate minerals with other phases including serpentine, olivine, smectite clays, and pyroxene minerals (Ehlmann et al. 2008; Michalski and Niles 2010; Morris et al. 2010). They are also buried underneath younger volcanic or ejecta deposits indicating ages that likely date to the early to mid Noachian.

The carbonate abundance in these deposits ranges from ~10–30 % (Ehlmann et al. 2008; Michalski and Niles 2010; Morris et al. 2010). In Gusev crater, a carbonate deposit was identified by the Spirit rover using spectroscopic, chemical, and Mössbauer data. This deposit also consists of carbonate mixed with volcanoclastic materials, with carbonate abundances of ~16–34 %. Carbonates have also been detected in the dust and soils of Mars to be present at abundances between ~2–5 % (Bandfield et al. 2003). This is perhaps our best means for estimating the crustal reservoir of carbonate on Mars as the dust on Mars is globally mixed (McSween and Keil 2000) and may be representative of the average composition of the upper crust of the planet. However, the dust is very fine grained and might be highly susceptible

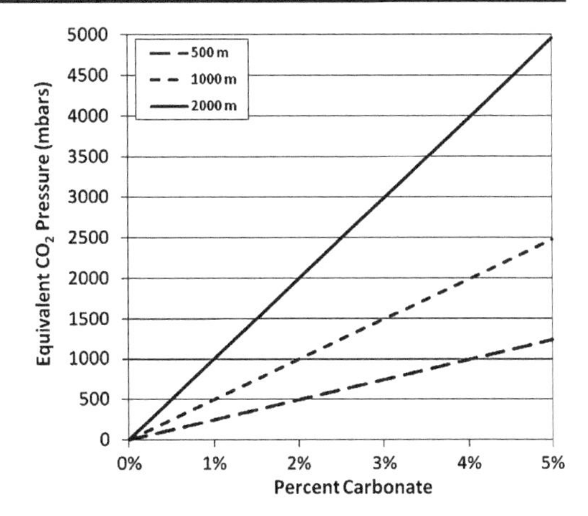

Fig. 13 Calculation of equivalent CO_2 atmospheric pressures based on carbonate reservoir size calculated from volume percentage of carbonate and depth of carbonate mineralization

to weathering under current martian conditions (Shaheen et al. 2010), and so might reflect an artificially elevated carbonate abundance due to atmospheric or near surface weathering. Nevertheless, the carbonate content of the dust can provide at least a hypothetical upper boundary for the carbonate content of the crust.

Figure 13 shows the equivalent atmospheric pressure for different crustal abundances of carbonate, assuming 5 % as an upper boundary in the crust and crustal thickness up to 2 km. As one can see, a maximum amount of CO_2 which could hypothetically be stored in the crust could be up to ~5 bar with the assumption that the dust is a representative sample of the top 2 km. Of course this estimate is for the total CO_2 reservoir in the martian crust, and only part of this or even a tiny fraction of a carbonate reservoir formed since the Noachian.

It is difficult to determine how much of this carbonate reservoir can be attributed to the CO_2 presence during the Noachian when many of the valley networks and phyllosilicates formed, but EUV-powered atmospheric escape was high. The assumption that not much carbonates may have formed after the early Noachian is in agreement with the nature of the carbonate deposits which have been discovered so far, as they all are dated at the early to mid-Noachian or earlier (Ehlmann et al. 2008; Michalski and Niles 2010; Morris et al. 2010).

Major outstanding issues remain in our understanding of the carbonate reservoir on Mars including:

- What was and is the total carbonate reservoir on Mars and how much carbonates are stored in the deep crust (>5 km)?
- How efficient were environmental processes, which acted against carbonate formation?
- How much carbonate has formed since the start of the Noachian and how is it distributed through time?
- How efficiently are carbonates recycled back into the atmosphere on Mars and through what mechanisms?

8.2 CH_4 as a Clue to an Active Long-Term Carbon Cycle

During the last decade, CH_4 has been detected in Mars atmosphere by different instruments from both Earth and spacecraft in orbit around Mars (Planetary Fourier Spectrometer (PFS) on Mars Express, Thermal Emission Spectrometer (TES) on Mars Global Surveyor (MGS)) at an average ~10–20 ppb level (Krasnopolsky et al. 2004; Formisano et al. 2004; Mumma et al. 2009; Fonti and Marzo 2010). One of the most striking characteristics of the observed CH_4 is its high temporal and spatial variability, implying a lifetime of ~200 days (Lefèvre and Forget 2002), much shorter than the currently admitted value of ~300 yr based on existing photochemical models (e.g. Krasnopolsky 2006). Such a large discrepancy, together with the small signal to noise ratio of spacecraft data, led some scientists to consider the detection of CH_4 as questionable. Concerning data acquired from Earth (Mumma et al. 2009), it has been argued that the coincidence between martian and telluric lines could have led to an erroneous retrieval of the Mars CH_4 abundance (Zahnle et al. 2011). Nevertheless, spacecraft data from MGS and Mars Express which now span ~6 martian years (MY24–MY29) and Earth-based measurements show a globally consistent and reproducible seasonal behaviour of the CH_4 mixing ratio with similar abundance levels and amplitudes of variation (see Fig. 1 in Chassefière and Leblanc 2011b; Mousis et al. 2012, this issue), reasonably suggesting that CH_4 is effectively present in the martian atmosphere. All existing observations show that CH_4 concentrations are spatially and temporally highly variable, and its lifetime has to be on the order of ~90 days. CH_4 could be oxidized in the superficial regolith layer through heterogeneous chemistry processes involving gaseous oxidants like H_2O_2 and/or direct oxidation by ferric iron at the surface of grains (Chassefière and Leblanc 2011a).

The calculated present CH_4 release flux, as deduced from present abundance measurements (~10 ppbv CH_4 in a seasonal average, Geminale et al. 2008, 2011; Fonti and Marzo 2010) and estimates of the CH_4 lifetime (200 days, Lefèvre and Forget 2002) is on the order of $\sim 1.0 \times 10^8$ cm^{-2} s^{-1}. Such a high flux is difficult to explain by an external (meteoritic) source and direct volcanism releasing only trace amounts of reduced carbon can be similarly ruled out as a major source of CH_4 (Atreya et al. 2006). According to these authors, hydrothermalism and/or biological activity may be at the origin of CH_4 on Mars. It has been suggested by Dohm et al. (2008) that Mars is still internally active, with a potential for continuing magmatic-driven activity (Hauber et al. 2011), including volcanism and hydrothermal activity, supporting a possible hydrothermal origin.

Hydrothermalism may result in the production of CH_4, either by reduction of CO_2 through water-rock interaction at low oxygen fugacity (Lyons et al. 2005), or through serpentinization followed by the conversion of H_2 in CH_4 (Oze and Sharma 2005). Whatever the origin of CH_4, assumed to take place in the crust below the water table, the fate of the produced CH_4 is to be transported upward by ascending hydrothermal fluids and to be stored in the cryosphere in the form of CH_4 clathrate hydrates (Chassefière and Leblanc 2011a; Mousis et al. 2012, this issue).

For how long has CH_4 been released at the present rate to the atmosphere? A few arguments in favour of a long-term and relatively continuous phenomenon have been proposed:

- If CH_4 oxidation is at the origin of the suspected present redox imbalance between the H and O escape fluxes, the release rate of methane averaged over the last 10^3 yr (photochemical lifetime of H_2) required to explain the presumably small O escape rate, is close to the present release rate (Chassefière and Leblanc 2011b). This suggests that methane could have been released at an average release rate similar to the present one for at least a few thousand years.
- As pointed out by Chassefière and Leblanc (2011b, 2011c), the quantity of carbon contained in the superficial layer of CO_2-ice covering the permanent south polar cap is comparable to the amount of CH_4 which would be delivered to the atmosphere, then converted to CO_2, over a time interval of 3 Myr, that is the time since the last obliquity transition period (Levrard et al. 2004). This small amount of CO_2, ~1 % of the atmospheric mass, would be at condensation equilibrium with larger amounts of CO_2-ice sequestered

in the martian south polar layered deposits, recently discovered by the Shallow Subsurface Radar (SHARAD) on Mars Reconnaissance Orbiter (MRO) (Phillips et al. 2010). The thin superficial CO_2-ice layer covering the south polar cap could therefore originate in the hydrothermal CH_4, as well as possibly some volcanic CO_2, released since the time of the transition. This suggests a CH_4 release rate similar to the present value over the last few million or ten million years.

- If, as may seem likely, CH_4 is released to the atmosphere from a CH_4 clathrate-rich cryosphere, the release rate of CH_4, produced sporadically at depth by hydrothermal activity, is considerably smoothed over a time scale of the order of the lifetime of CH_4 clathrate, that is the time for the cryosphere to fully sublimate to the atmosphere and possibly be recycled to the crust, which may have occurred in the past. This time is estimated to be or the order of $\sim 10^8$–10^9 yr (Mousis et al. 2012, this issue), suggesting that CH_4 could have been continuously released to the atmosphere over geological time scales.

By scaling the CH_4 release rate on the level of hydrothermal activity, assumed to be proportional to the lava extrusion rate as estimated from the geomorphological analysis of the surface (Greeley and Schneid 1991; Craddock and Greeley 2009), a cumulated amount of CO_2 resulting from CH_4 release since the Noachian of ~ 2 bar has been estimated (Chassefière and Leblanc 2011a). Assuming that CH_4 release is due to serpentinization, a released amount larger than ~ 0.4 bar is not consistent with the present D/H ratio (see Sect. 3.1) in the martian atmosphere (Chassefière and Leblanc 2011c). Up to ~ 0.4 bar of CO_2 could therefore have been released in the form of CH_4, then oxidized. It should be noted that CH_4 outgassing is not an additional source of carbon with respect to the CO_2 volcanic source described in the previous section. Either magmatic CO_2 is converted to CH_4 before being released to the atmosphere through fluid-rock interaction in deep hydrothermal fluids (Lyons et al. 2005), or it precipitates in crustal carbonates, with further hydrothermal decomposition to CO_2, reduced to CH_4 by the molecular hydrogen produced by serpentinization (Oze and Sharma 2005) and/or direct thermodynamical equilibration (Lyons et al. 2005). Subsurface hydrothermal activity could be responsible for both the release of CH_4 to the atmosphere, and the recycling of atmospheric CO_2 to the crust with further precipitation of carbonates, recycled later to the atmosphere under reduced form through hydrogeochemical processes (Chassefière and Leblanc 2011a).

Carbonate mineral deposits may occur in subsurface hydrothermal systems from liquid water, rich in dissolved CO_2. Provided relevant subsurface zones are not entirely sealed from the atmosphere, some CO_2 can be transferred from the atmosphere to subsurface water reservoirs and further precipitate in carbonates, as observed on Earth and expected to occur on Mars from geochemical modelling (Griffith and Shock 1995). Such a long term carbon cycle, with a progressive net removal of CO_2 from the atmosphere and subsequent carbonate deposition in the subsurface, could explain the disappearance of a possible Noachian CO_2 atmosphere built by volcanism and late impacts.

Because a carbon atom may have been cycled several times through the crust since its release from the mantle by volcanism, a cumulated CH_4 release rate up to ~ 0.4 bar should not be interpreted as the content of a subsurface isolated reservoir. It rather suggests that an efficient carbon cycle has been maintained by hydrothermal processes, probably until the early Amazonian and possibly the present epoch, with a substantial fraction of the volcanic outgassed carbon being cycled one or several times through crustal carbonates.

8.3 Storage Capability of H_2O in the Martian Crust

The present inventory of water on Mars is poorly constrained. The total water content of the two perennial polar caps corresponds to a EGL of H_2O of $\sim$16 m depth (Smith et al. 2001), and the ice deposits sequestered in the Dorsa Argentea Formation (DAF), near the south polar cap, could have represented $\sim$15 m in the past (Head and Pratt 2001). Nevertheless, only a fraction of the initial water could remain today in the DAF reservoir, corresponding to $\sim$5–7.5 m. Other reservoirs, expected to have been active during late Amazonian, could be present in tropical and mid-latitude regions (e.g., Watters et al. 2007; Holt et al. 2008; Head and Marchant 2009), but they probably represent only a minor contribution to the global reservoir. The total inventory of the known reservoir, including near-surface repositories that are distributed across middle to high latitudes, has been estimated to correspond to a 35 m thick EGL (Christensen 2006). The megaregolith capacity is large, with up to $\sim$500 m hypothetically trapped in the cryosphere, and hypothetically several additional hundreds of meters (up to $\sim$500 m) of ground water surviving at depth below the cryosphere (Clifford et al. 2010). It has been suggested that most of ground ice has been lost by sublimation at low latitudes, and that only small amounts of ground water would survive today (Grimm and Painter 2009), with therefore less water in the megaregolith. Carr (1986) suggested that a $\sim$500 m thick EGL of water has to be required to explain the formation of outflow channels and most of this H_2O could be trapped today as water ice, and possibly deep liquid water, in the subsurface.

Some of the water present on Mars may have reacted with minerals to form clay minerals and sulfates (Bibring et al. 2006; Mustard et al. 2008; Ehlmann et al. 2012, this issue). The presence of these hydrated minerals at the surface of Mars suggests that hydration processes have been active on Mars in the past. They may have been formed, either at the surface of Mars during the Noachian, when liquid water was flowing at the surface of the planet, or in the subsurface by aqueous alteration of subsurface rocks, and possibly by impacts able to provide subsurface water to the impacted material (Bibring et al. 2006). Existing geochemical model calculations show that hydrothermal hydration of martian crust is an efficient process (Griffith and Shock 1997). Calculations have been made for a temperature in the range from 150–250 °C. The results do not depend much on the oxygen fugacity, and are similar for a moderately oxidized crust and for a highly oxidized medium. The final conclusion of this study is that water storage via hydrous minerals can account for $\sim$5 wt.% of crustal rocks. The capacity of the upper 10 km of the crust in storing water in the form of hydrated minerals therefore hypothetically corresponds to a few hundreds meter depth EGL of H_2O, but the crustal content of hydrated minerals is basically unknown and may be much lower than this upper limit.

The effectiveness and amplitude of aqueous alteration processes in Mars' crust are basically unknown. SNC meteorites, originating in the martian crust, provide information on crustal geochemical processes. Formed of mantle material modified through interaction with crustal material during the upward migration of lava through the crust, they have been shown to record a wide range of oxidation conditions (Wadhwa 2001). Shergottites present a range of redox conditions from close to the IW-buffer up to that of the quartz-fayalite-magnetite buffer, which can be interpreted as the result of oxidation of the crust by a process such as aqueous alteration, through the oxidation of ferrous iron into ferric iron. Whereas Nakhlites and Chassignites are oxidized, the ancient ALH 84001 is relatively reduced. These results suggest that the silicates of SNC meteorites originate in a water-depleted martian mantle. These silicates would have been oxidized through assimilation of oxidized crustal material. In order to produce such an oxidation, a significant proportion of crustal material (10–30 %)

with a high Fe^{3+}/Fe ratio (~50 %) must have been mixed with mantle material (Herd et al. 2002). ALH84001, because it has crystallized at the end of the Noachian ~3.9 Gyr ago, could have not been mixed with oxidized crust material.

The occurrence in the crust of aqueous alteration processes, as proved by oxidation processes recorded in SNC meteorites, suggests that the conditions for an efficient hydration of crustal rocks may have been met at some places and times in the past. Such hydrated minerals have been found in Nakhlites. A particular hydration process occurring in Earth's crust is serpentinization, which generates H_2 from the reaction of water with ferrous iron derived from minerals, primarily olivine and pyroxene (McCollom and Back 2009). In the reaction, ferrous iron is oxidized by the water to ferric iron, which typically precipitates as magnetite, while hydrogen from water is reduced to H_2. Iron oxidation is accompanied by the storage of a large number of water molecules in serpentine, an hydrated mineral which has been recently observed by the Compact Reconnaissance Imaging Spectrometer for Mars (CRISM) on MRO in and around the Nili Fossae region (Ehlmann et al. 2009). Serpentinization occurring in crustal hydrothermal systems is a plausible process at the origin of the methane observed in the martian atmosphere (Oze and Sharma 2005). Based on an analysis of the present Mars' D/H ratio, it has been suggested that a water GEL of up to ~300–400 m depth could have been stored in crustal serpentine since the late Noachian due to hydrothermalism triggered by magmatic activity (Chassefière and Leblanc 2011a, 2011b, 2011c). Massive serpentinization of the southern crust could have been at the origin of both the crustal dichotomy and the strong remanent magnetic field of old southern terrains (Quesnel et al. 2009).

Although there is no direct observational evidence of active hydration processes in Mars' crust, aqueous alteration processes, e.g. serpentinization, are potentially able to store several hundreds of meters of H_2O in crustal hydrates. This amount is comparable to the estimated value of the water ice content of the cryosphere. Depending on the efficiencies of the various atmospheric escape processes and volatile delivery by impacts, several hundreds of meters of water could hypothetically be trapped in the subsurface in the form of H_2O-ice and/or hydrated minerals, and possibly liquid water below the cryosphere. All these reservoirs could exchange with each other, as well as with the atmosphere and polar caps (Lasue et al. 2012, this issue) at the occasion of magmatic and hydrothermal events, and could contain hypothetically up to ~1 km thick EGL of water.

9 Conclusions

The latest hypotheses on the formation, outgassing and evolution of the martian atmosphere from the early Noachian up to the present time as illustrated in Fig. 14 have been discussed. Depending on the captured nebula gas and/or outgassed amounts of volatiles, we show that due to the high EUV flux of the young Sun the planet's hydrogen-rich protoatmosphere was lost via hydrodynamic escape of atomic hydrogen which could have dragged heavier atoms such as C and O during the first tens or hundreds of Myr after Mars finished its accretion. The early Noachian impacts may have kept the protoatmosphere in vapor form and may not have much contributed to atmospheric growth because the delivered volatiles would have also escaped. After Mars lost its protoatmosphere the atmospheric escape rates were most likely balanced with a secondary outgassed atmosphere and delivered volatiles by impacts until the activity of the young Sun decreased so that the atmospheric sources could dominate over the losses. Depending on assumptions related to geochemical conditions such as the pH of the early martian environment, the existence of a global melt channel vs. melting

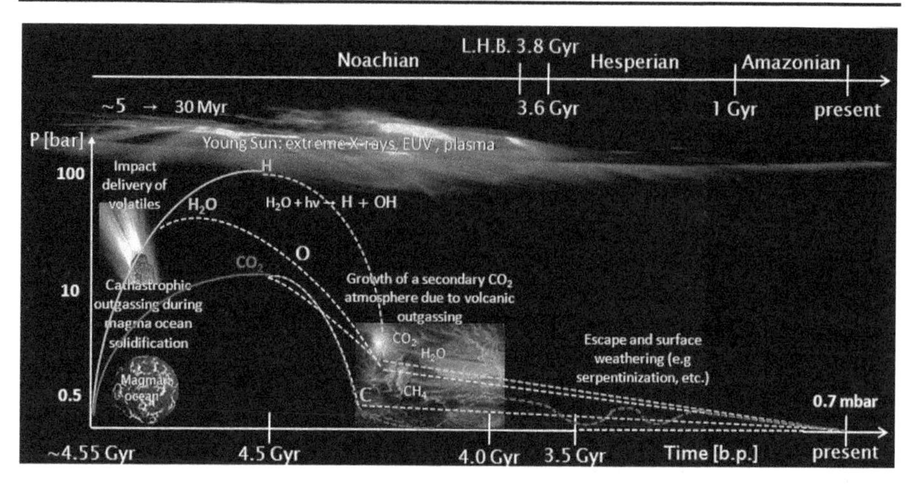

Fig. 14 Illustration of Mars' atmosphere evolution after the outgassing of volatiles and fast growth of a dense water vapour dominated H_2O/CO_2 atmosphere during and after the magma ocean solidification process. *Blue lines*: water and hydrogen; *yellow* lines: oxygen; *red lines* CO_2. *Solid lines*: Outgassed or delivered species; *dashed-lines*: escaping species. A complex interplay between the young Sun's EUV activity, impacts and atmospheric escape processes during the first ~500 Myr kept early Mars most of the time cool and dry. After the EUV flux of the young Sun decreased at ~4–4.3 Gyr ago, volcanic outgassing and impacts during the LHB could have resulted in the build-up of a denser and slightly warmer and wetter CO_2 atmosphere of $\leq$1 bar. If the secondary CO_2 atmosphere after the Noachian had a surface pressure which was larger than ~100 mbar, only a minor fraction escaped to space during the Hesperian and Amazonian and most of it should be stored so far undiscovered in the subsurface. CO_2 can be released from surface reservoirs during periods of climate change or impacts (non-linear *dashed red line* from the late Noachian until present) and may have modified the atmospheric surface pressure several times during the planet's history

in mantle plumes, variable oxygen fugacities, in combination with atmospheric escape and impact delivery during the late heavy bombardment Mars may have built up a secondary CO_2 atmosphere which was <1 bar and accumulated a water inventory equivalent to <10 bar at ~3.5–4 Gyr ago. By reviewing the latest observations and model studies on the escape of the martian CO_2 atmosphere during the Hesperian and Amazonian epochs we expect that the planet may have lost most likely less than ~150 mbar of CO_2. If a CO_2 atmosphere of several 100 mbar was indeed present at the end of the Noachian, most of it should have been removed to the crust by sequestration in carbonate rocks and partially recycled to the atmosphere under reduced and/or oxidized form. The water contained in a several 10 m or even 100 m deep EGL at the Noachian could have been trapped hypothetically in the crust in the form of H_2O-ice, hydrated minerals, and possibly liquid water under the cryosphere. On the other hand, if there are not enough hidden CO_2 deposits under the planet's surface, then other greenhouse gases are necessary for the explanation of standing bodies of liquid water on the planet's surface $\geq$3.5–4 Gyr ago.

Acknowledgements D. Breuer, E. Chassefière, M. Grott, H. Gröller, E. Hauber, H. Lammer, P. Odert and A. Morschhauser acknowledges support from the Helmholtz Alliance project "Planetary Evolution and Life". E. Chassefière acknowledges support from CNRS EPOV interdisciplinary program. H. Lammer acknowledge the support by the FWF NFN project S116 "Pathways to Habitability: From Disks to Active Stars, Planets and Life", and the related FWF NFN subproject, S116607-N16 "Particle/Radiative Interactions with Upper Atmospheres of Planetary Bodies Under Extreme Stellar Conditions". H. Gröller and H. Lammer acknowledges also support from the Austrian FWF project P24247-N16 "Modelling of non-thermal processes in early upper atomospheres exposed to extreme young Sun conditions" and support from the joined Russian-Austrian project under the RFBR grant 09-02-91002-215-ANF-a and the Austrian Science Fund (FWF) grant

I199-N16. P. Odert was supported via the FWF project grant P19446-N16 and the research by U. Möstl was funded by the FWF project grant P21051-N16. O. Karatekin thanks A. Morbidelli for the discussions related to impact studies and the LHB; O. Karatekin, V. Dehant and L.B.S. Pham acknowledges the support of Belgian PRODEX program managed by the ESA in collaboration with the BELSPO. O. Mousis acknowledges support from CNES. P. Niles acknowledges support from NASA Johnson Space Center and the Mars Fundamental Research Program. The authors also thank ISSI for hosting the conference and the Europlanet RI-FP7 project and its related Science Networking (Na2) working groups. Finally, the authors thank guest editor M. Toplis and two anonymous referees for their suggestions and recommendations which helped to improve the article.

References

Y. Abe, Thermal and chemical evolution of the terrestrial magma ocean. Phys. Earth Planet. Int. **100**, 27–39 (1997)

T.J. Ahrens, Impact erosion of terrestrial planetary atmospheres. Annu. Rev. Earth Planet. Sci. **21**, 525–555 (1993)

F. Albarède, J. Blichert-Toft, The split fate of the early Earth, Mars, Venus and Moon. C. R. Géosci. **339**, 917–927 (2007)

U.V. Amerstorfer, N.V. Erkaev, U. Taubenschuss, H.K. Biernat, Influence of a density increase on the evolution of the Kelvin-Helmholtz instability and vortices. Phys. Plasmas **17**, 072901 (2010). doi:10.1063/1.3453705

J.C. Andrews-Hanna, M.T. Zuber, W.B. Banerdt, The Borealis basin and the origin of the martian crustal dichotomy. Nature **453**, 1212–1215 (2008)

S.K. Atreya, R.P. Mahaffy, A.S. Wong, Methane and related trace species on Mars: origin, loss, implications for life, and habitability. Planet. Space Sci. **55**, 358–369 (2006)

V.R. Baker, Water and the martian landscape. Nature **412**, 228–236 (2001)

J.L. Bandfield, T.D. Gloch, P.R. Christensen, Spectroscopic identification of carbonate minerals in the martian dust. Science **301**, 1084–1086 (2003)

S. Barabash, A. Fedorov, R. Lundin, J.-A. Sauvaud, Martian atmospheric erosion rates. Science **315**, 501–503 (2007)

R.H. Becker, R.N. Clayton, E.M. Galimov, H. Lammer, B. Marty, R.O. Pepin, R. Weiler, Isotopic signatures in terrestrial planets. Space Sci. Rev. **106**, 377–410 (2003)

J.-P. Bibring, Y. Langevin, A. Gendrin, B. Gondet, F. Poulet, M. Berthé, A. Soufflot, R. Arvidson, N. Mangold, J. Mustard, P. Drossart, The OMEGA team, Mars surface diversity as revealed by the OMEGA/Mars Express observations. Science **307**, 1576–1581 (2005)

J.-P. Bibring, Y. Langevin, J.F. Mustard, F. Poulet, R. Arvidson, A. Gendrin, B. Gondet, N. Mangold, P. Pinet, F. Forget, The OMEGA team, Global mineralogical and aqueous Mars history derived from OMEGA/Mars Express data. Science **312**, 400–404 (2006)

L.E. Borg, D.S. Draper, A petrogenetic model for the origin and compositional variation of the Martian basaltic meteorites. Meteorit. Planet. Sci. **38**, 1713–1731 (2003)

L.H. Brace, R.F. Theis, W.R. Hoegy, Plasma clouds above the ionopause of Venus and their implications. Planet. Space Sci. **30**, 29–37 (1982)

R. Brasser, The formation of Mars: building blocks and accretion time scale. Space Sci. Rev. (2012, in press). doi:10.1007/s11214-012-9904-2

D. Breuer, T. Spohn, Viscosity of the Martian mantle and its initial temperature: constraints from crust formation history and the evolution of the magnetic field. Planet. Space Sci. **54**, 153–169 (2006)

J.C. Bridges, D.C. Catling, J.M. Saxton, T.D. Swindle, I.C. Lyon, M.M. Grady, Alteration assemblages in martian meteorites: implications for near-surface processes. Space Sci. Rev. **96**, 365–392 (2001)

R.M. Canup, E. Asphaug, Origin of the Moon in a giant impact near the end of the Earth's formation. Nature **412**, 708–712 (2001)

M.H. Carr, Mars—a water-rich planet? Icarus **68**, 187–216 (1986)

M.H. Carr, Retention of an atmosphere on early Mars. J. Geophys. Res. **1042**, 21897–21910 (1999)

E. Chassefière, Hydrodynamic escape of oxygen from primitive atmospheres: applications to the cases of Venus and Mars. Icarus **124**, 537–552 (1996)

E. Chassefière, F. Leblanc, Mars atmospheric escape and evolution, interaction with the solar wind. Planet. Space Sci. **52**, 1039–1058 (2004)

E. Chassefière, F. Leblanc, Constraining methane release due to serpentinization by the observed D/H ratio on Mars. Earth Planet. Sci. Lett. **310**, 262–271 (2011a)

E. Chassefière, F. Leblanc, Explaining the redox imbalance between the H and O escape fluxes at Mars by the oxidation of methane. Planet. Space Sci. **59**, 218–226 (2011b)
E. Chassefière, F. Leblanc, Constraining methane release due to serpentinization by the observed D/H ratio on Mars. Earth Planet. Sci. Lett. **310**, 262–271 (2011c)
E. Chassefière, F. Leblanc, B. Langlais, The combined effects of escape and magnetic field histories at Mars. Planet. Space Sci. **55**, 343–357 (2007)
J.Y. Chaufray, R. Modolo, F. Leblanc, G. Chanteur, R.E. Johnson, J.G. Luhmann, Mars solar wind interaction: formation of the Martian corona and atmospheric loss to space. J. Geophys. Res. **112**(E9), E09009 (2007)
V. Chevrier, F. Poulet, J.-P. Bibring, Early geochemical environment of Mars as determined from thermodynamics of phyllosilicates. Nature **448**, 60–63 (2007)
P. Christensen, Water at the poles and in permafrost regions of Mars. Elements **2**, 151–155 (2006)
M.W. Claire, J. Sheets, M. Cohen, I. Ribas, V.S. Meadows, D.C. Catling, The evolution of solar flux from 0.1 nm to 160 μm: quantitative estimates for planetary studies. Astrophys. J. **757**, 95 (2012), 12 pp.
S.M. Clifford, J. Lasue, E. Heggy, J. Boisson, P. McGovern, M.D. Max, Depth of the Martian cryosphere: revised estimates and implications for the existence and detection of subpermafrost groundwater. J. Geophys. Res. **115**, E07001 (2010). doi:10.1029/2009JE003462
R.A. Craddock, R. Greeley, Minimum estimates of the amount and timing of gases released into the martian atmosphere from volcanic eruptions. Icarus **204**, 512–526 (2009)
N. Dauphas, The dual origin of the terrestrial atmosphere. Icarus **165**, 326–339 (2003)
N. Dauphas, A. Pourmand, Hf-W-Th evidence for rapid growth of Mars and its status as a planetary embryo. Nature **473**, 489–493 (2011)
V. Dehant, H. Lammer, Y.N. Kulikov, J.-M. Grießmeier, D. Breuer, O. Verhoeven, O. Karatekin, T. van Hoolst, O. Korablev, P. Lognonne, Planetary magnetic dynamo effect on atmospheric protection of early Earth and Mars. Space Sci. Rev. **129**, 279–300 (2007)
J.M. Dohm, R.C. Anderson, N.G. Barlow, H. Miyamoto, A.G. Davies, G.J. Taylor, V.R. Baker, W.V. Boynton, J. Keller, K. Kerry, D. Janes, A.G. Fairén, D. Schulze-Makuch, M. Glamoclija, L. Marinangeli, G.G. Ori, R.G. Strom, J.-P. Williams, J.C. Ferris, J.A.P. Rodríguez, M.A. de Pablon, S. Karunatillake, Recent geological and hydrological activity on Mars: the Tharsis/Elysium corridor. Planet. Space Sci. **56**, 985–1013 (2008)
T.M. Donahue, Evolution of water reservoirs on mars from D/H ratios in the atmosphere and crust. Nature **374**, 432–434 (1995)
T.M. Donahue, Pre-global surveyor evidence for Martian ground water. Proc. Natl. Acad. Sci. **98**, 827–830 (2001)
T.M. Donahue, Accretion, loss, and fractionation of martian water. Icarus **167**, 225–227 (2004)
J.D. Dorren, M. Güdel, E.F. Guinan, X-ray emission from the Sun in its youth and old age. Astrophys. J. **448**, 431–436 (1995)
B.L. Ehlmann, J.F. Mustard, S.L. Murchie, F. Poulet, J.L. Bishop, A.J. Brown, W.M. Calvin, R.N. Clark, D.J. des Marais, R.E. Milliken, L.H. Roach, T.L. Roush, G.A. Swayze, J.J. Wray, Orbital identification of Carbonate-bearing rocks on Mars. Science **322**, 1828–1832 (2008)
B.L. Ehlmann, J.F. Mustard, G.A. Swayze, R.N. Clark, J.L. Bishop, F. Poulet, D.J. des Marais, L.H. Roach, R.E. Milliken, J.J. Wray, O. Barnouin-Jha, S.L. Murchie, Identification of hydrated silicate minerals on Mars using MRO-CRISM: Geologic context near Nili Fossae and implications for aqueous alteration. J. Geophys. Res. **114**, E00D08 (2009). doi:10.1029/2009JE003339
B.L. Ehlmann et al., Geochemical consequences of widespread clay formation in Mars' Ancient Crust. Space Sci. Rev. (2012, this issue). doi:10.1007/s11214-012-9930-0
L.T. Elkins-Tanton, Linked magma ocean solidification and atmospheric growth for Earth and Mars. Earth Planet. Sci. Lett. **271**, 181–191 (2008)
O. Eugster, H. Busemann, S. Lorenzetti, D. Terrebilini, Ejection ages from krypton-81-krypton-83 dating and pre-atmospheric sizes of martian meteorites. Meteorit. Planet. Sci. **37**, 1345–1360 (2002)
A.G. Fairén, A cold and wet Mars. Icarus **208**, 165–175 (2010)
J. Filiberto, A.H. Treiman, Martian magmas contained abundant chlorine, but little water. Geology **37**, 1087–1090 (2009)
S. Fonti, G.A. Marzo, Mapping the methane on Mars. Astron. Astrophys. **512**, A51 (2010)
F. Forget, R.T. Pierrehumbert, Warming early Mars with carbon dioxide clouds that scatter infrared radiation. Science **278**, 1273–1276 (1997)
V. Formisano, S. Atreya, T. Encrenaz, N. Ignatiev, M. Giuranna, Detection of methane in the atmosphere of Mars. Science **306**, 1758–1761 (2004)
J.L. Fox, CO_2^+ dissociative recombination: a source of thermal and nonthermal C on Mars. J. Geophys. Res. **109**, A08306 (2004)

J.L. Fox, A.B. Hać, Photochemical escape of oxygen from Mars: a comparison of the exobase approximation to a Monte Carlo method. Icarus **204**, 527–544 (2009)
A.A. Fraeman, J. Korenaga, The influence of mantle melting on the evolution of Mars. Icarus **210**, 43–57 (2010)
E. Gaidos, G. Marion, Geological and geochemical legacy of a cold early Mars. J. Geophys. Res. **108**(E6), 5055 (2003), pp. 9-1
A. Geminale, V. Formisano, M. Giuranna, Methane in Martian atmosphere: average spatial, diurnal, and seasonal behaviour. Planet. Space Sci. **56**, 1194–1203 (2008)
A. Geminale, V. Formisano, G. Sindoni, Mapping methane in Martian atmosphere with PFS-MEX data. Planet. Space Sci. **59**, 137–148 (2011)
T.M. Gerlach, E.J. Graeber, Volatile budget of Kilauea volcano. Nature **313**, 273–277 (1985)
R. Gomes, H.F. Levison, K. Tsiganis, A. Morbidelli, Origin of the cataclysmic Late Heavy Bombardment period of the terrestrial planets. Nature **435**, 466–469 (2005)
J.L. Gooding, S.J. Wentworth, M.E. Zolensky, Calcium-carbonate and sulfate of possible extraterrestrial origin in the Eeta-79001 meteorite. Geochim. Cosmochim. Acta **52**, 909–915 (1988)
R. Greeley, B.B. Schneid, Magma generation on Mars—amounts, rates and comparisons with Earth, Moon, and Venus. Science **254**, 996–998 (1991)
L.P. Greenland, Composition of gases from the 1984 eruption of Mauna Loa volcano, in *Volcanism in Hawaii*. U.S. Geol. Surv. Prof. Pap. 1350, vol. 1 (U.S. Gov. Printing Office, Washington, 1987a), pp. 781–790
L.P. Greenland, Hawaiian eruptive gases, in *Volcanism in Hawaii*. U.S. Geol. Surv. Prof. Pap. 1350, vol. 1 (U.S. Gov. Printing Office, Washington, 1987b), pp. 781–790
L.L. Griffith, E.L. Shock, A geochemical model for the formation of hydrothermal carbonates on Mars. Nature **377**, 406–408 (1995)
L.L. Griffith, E.L. Shock, Hydrothermal hydration of Martian crust: illustration via geochemical model calculations. J. Geophys. Res. **102**, 9135–9143 (1997)
R.E. Grimm, S.L. Painter, On the secular evolution of groundwater on Mars. Geophys. Res. Lett. **36**, L24803 (2009). doi:10.1029/2009GL041018
H. Gröller, V.I. Shematovich, H.I.M. Lichtenegger, H. Lammer, M. Pfleger, Yu.N. Kulikov, W. Macher, U.V. Amerstorfer, H.K. Biernat, Venus' atomic hot oxygen environment. J. Geophys. Res. **115**, E12017 (2010)
H. Gröller, H. Lammer, H.I.M. Lichtenegger, M. Pfleger, O. Dutuit, V.I. Shematovich, Yu.N. Kulikov, H.K. Biernat, Hot oxygen atoms in the Venus nightside exosphere. Geophys. Res. Lett. **39**, L03202 (2012). doi:10.1029/2011GL050421
M. Grott, A. Morschhauser, D. Breuer, E. Hauber, Volcanic outgassing of CO_2 and H_2O on mars. Earth Planet. Sci. Lett. **308**, 391–400 (2011)
M. Grott, D.D. Baratoux, E.E. Hauber, V.V. Sautter, J.J. Mustard, O.O. Gasnault, S.S. Ruff, S.-I. Karato, V.V. Debaille, M.M. Knapmeyer, F.F. Sohl, T.T. Van Hoolst, D.D. Breuer, A.A. Morschhauser, M.J. Toplis, Long-term evolution of the crust-mantle system. Space Sci. Rev. (2012, accepted). doi:10.1007/s11214-012-9948-3
M. Güdel, E.F. Guinan, R. Mewe, J.S. Kaastra, S.L. Skinner, A determination of the coronal emission measure distribution in the young solar analog EK draconis from ASCA/EUVE spectra. Astrophys. J. **479**, 416–426 (1997)
A.N. Halliday, The origin of the earliest history of the Earth. Treatise Geochem. **1**, 509–557 (2003)
K. Hamano, Y. Abe, Pressure dependence of atmospheric loss by impact-induced vapor expansion, in *The 37th Lunar and Planetary Science Conference* (2006), abs. 1562
P. Hartogh, D.C. Lis, D. Bockelée-Morvan, M. de Val-Borro, N. Biver, M. Küppers, M. Emprechtinger, E.A. Bergin, J. Crovisier, M. Rengel, R. Moreno, S. Szutowicz, G.A. Blake, Ocean-like water in the Jupiter-family comet 103P/Hartley 2. Nature **478**, 218–220 (2011)
E. Hauber, P. Brozž, F. Jagert, P. Jodlowski, T. Platz, Very recent and wide-spread basaltic volcanism on Mars. Geophys. Res. Lett. **38**, L10201 (2011)
S.A. Hauck II, R.J. Phillips, Thermal and crustal evolution of Mars. J. Geophys. Res. **107**, 5052 (2002), 19 pp.
C. Hayashi, K. Nakazawa, H. Mizuno, Earth's melting due to the blanketing effect of the primordial dense atmosphere. Earth Planet. Sci. Lett. **43**, 22–28 (1979)
J.W. Head, D.R. Marchant, Inventory of ice-related deposits on mars: evidence for burial and long-term sequestration of ice in non-polar regions and implications for the water budget and climate evolution, in *The 40th Lunar and Planetary Science Conference* (2009), abs. 1356
J.W. Head, S. Pratt, Extensive Hesperian-aged south polar ice sheet on mars: evidence for massive melting and retreat, and lateral flow and ponding of meltwater. J. Geophys. Res. **106**, 12275–12300 (2001)
C.D.K. Herd, L.E. Borg, J.H. Jones, J.J. Papike, Oxygen fugacity and geochemical variations in the martian basalts: implications for martian basalt petrogenesis and the oxidation state of the upper mantle of mars. Geochim. Cosmochim. Acta **66**, 2025–2036 (2002)

M.M. Hirschmann, A.C. Withers, Ventilation of CO_2 from a reduced mantle and consequences for the early Martian greenhouse. Earth Planet. Sci. Lett. **270**, 147–155 (2008)
J.R. Holloway, Graphite melt equilibria during mantle melting: constraints on CO_2 in MORB magmas and the carbon content of the mantle. Chem. Geol. **147**, 89–97 (1998)
J.R. Holloway, V. Pan, G. Gudmundsson, High-pressure fluid-absent melting experiments in the presence of graphite: oxygen fugacity, ferric/ferrous ratio and dissolved CO_2. Eur. J. Mineral. **4**, 105–114 (1992)
J.W. Holt, A. Safaeinili, J.J. Plaut, J.W. Head III, R.J. Phillips, R. Seu Roberto, S.D. Kempf, P. Choudhary, D.A. Young, N.E. Putzig, D. Biccari, Y. Gim, Radar sounding evidence for buried glaciers in the southern mid-latitudes of Mars. Science **322**, 1235–1238 (2008)
J. Horner, O. Mousis, J.-M. Petit, B.-W. Jones, Differences between the impact regimes of the terrestrial planets: implications for primordial D: H ratios. Planet. Space Sci. **57**, 1338–1345 (2009)
D.M. Hunten, Atmospheric evolution of the terrestrial planets. Science **259**, 915–920 (1993)
D.M. Hunten, R.O. Pepin, J.C.G. Walker, Mass fractionation in hydrodynamic escape. Icarus **69**, 532–549 (1987)
W.-H. Ip, On a hot oxygen corona of Mars. Icarus **76**, 135–145 (1988)
N. Iro, D. Gautier, F. Hersant, D. Bockelée-Morvan, J.-I. Lunine, An interpretation of the nitrogen deficiency in comets. Icarus **161**, 511–532 (2003)
B.A. Ivanov, Mars/Moon cratering rate ratio estimates. Space Sci. Rev. **96**, 87–104 (2001)
B.A. Ivanov, V.V. Shuvalov, N.A. Artemieva, Meteoritic bombardment in the Noachian time: influence on geological and atmospheric evolution, in *Proc. 35th Microsymposium of the Brown University and the Vernadsky Institute* (2002)
B.M. Jakosky, R.O. Pepin, R.E. Johnson, J.L. Fox, Mars atmospheric loss and isotopic fractionation by solar-wind-induced sputtering and photochemical escape. Icarus **111**, 271–288 (1994)
M.C. Johnson, M.J. Rutherford, P.C. Hess, Chassigny petrogenesis: Melt compositions, intensive parameters and water contents of Martian magmas. Geochim. Cosmochim. Acta **55**, 349–366 (1991)
J.F. Kasting, CO_2 condensation and the climate of early Mars. Icarus **94**, 1–13 (1991)
J.F. Kasting, The early Mars climate question heats up. Science **278**, 1245 (1997)
J.F. Kasting, J.B. Pollack, Loss of water from Venus I. Hydrodynamic escape of hydrogen. Icarus **53**, 479–508 (1983)
R.F. Katz, M. Spielman, C.H. Langmuir, A new parameterization of hydrous mantle melting. Geochem. Geophys. Geosyst. **4**, 1073 (2003)
J. Kim, A.F. Nagy, J.L. Fox, T.E. Cravens, Solar cycle variability of hot oxygen atoms at Mars. J. Geophys. Res. **103**, 29339–29342 (1998)
T.T. Koskinen, M.J. Harris, R.V. Yelle, P. Lavvas, The escape of heavy atoms from the ionosphere of HD 209458b. I. A photochemical-dynamical model of the thermosphere. Icarus (2012, accepted). arXiv:1210.1536
V. Krasnopolsky, Note: on the deuterium abundance on Mars and some related problems. Icarus **148**, 597–602 (2000)
V.A. Krasnopolsky, Some problems related to the origin of methane on Mars. Icarus **180**, 359–367 (2006)
V.A. Krasnopolsky, P.D. Feldman, Detection of molecular hydrogen in the atmosphere of Mars. Science **294**, 1914–1917 (2001)
V.A. Krasnopolsky, G.L. Bjoraker, M.J. Mumma, D.F. Jennings, High resolution spectroscopy of Mars at 3.7 and 8 μm. J. Geopys. Res. **102**, 6525–6534 (1997)
V.A. Krasnopolsky, M.J. Mumma, G.R. Gladstone, Detection of atomic Deuterium in the upper atmosphere of Mars. Science **280**, 1576–1580 (1998)
V.A. Krasnopolsky, J.-P. Maillard, T.C. Owen, Detection of methane in the Martian atmosphere: evidence for life? Icarus **172**, 537–547 (2004)
M.A. Krestyanikova, V.I. Shematovich, Stochastic models of hot planetary and satellite coronas: a hot oxygen corona of Mars. Sol. Syst. Res. **40**, 384–392 (2006)
H. Lammer, W. Stumptner, S.J. Bauer, Upper limits for the Martian exospheric number density during the planet B/Nozomi mission. Planet. Space Sci. **48**, 1473–1478 (2000)
H. Lammer, C. Kolb, T. Penz, U.V. Amerstorfer, H.K. Biernat, B. Bodiselitsch, Estimation of the past and present martian water-ice reservoirs by isotopic constraints on exchange between the atmosphere and the surface. Int. J. Astrobiol. **2**(3), 195–202 (2003)
H. Lammer, J.F. Kasting, E. Chassefière, R.E. Johnson, Y.N. Kulikov, F. Tian, Atmospheric escape and evolution of terrestrial planets and satellites. Space Sci. Rev. **139**, 399–436 (2008)
H. Lammer, P. Odert, M. Leitzinger, M.L. Khodachenko, M. Panchenko, Yu.N. Kulikov, T.L. Zhang, H.I.M. Lichtenegger, N.V. Erkaev, G. Wuchterl, G. Micela, T. Penz, H.K. Biernat, J. Weingrill, M. Steller, H. Ottacher, J. Hasiba, A. Hanslmeier, Determining the mass loss limit for close-in exoplanets: what can we learn from transit observations? Astron. Astrophys. **506**, 399–410 (2009)

H. Lammer, K.G. Kislyakova, P. Odert, M. Leitzinger, R. Schwarz, E. Pilat-Lohinger, Yu.N. Kulikov, M.L. Khodachenko, M. Güdel, A. Hanslmeier, Pathways to Earth-like atmospheres: extreme ultraviolet (EUV)-powered escape of hydrogen-rich protoatmospheres. Orig. Life Evol. Biosph. **41**, 503–522 (2012)

J. Lasue, N. Mangold, E. Hauber, S. Clifford, W. Feldman, O. Gasnault, C. Grima, S. Maurice, O. Mousis, Quantifying the Martian hydrosphere: current evidence, time evolution and implications for the habitability of the planet. Space Sci. Rev. (2012, accepted). doi:10.1007/s11214-012-9946-5

F. Leblanc, R.E. Johnson, Role of molecular species in pick up ion sputtering of the Martian atmosphere. J. Geophys. Res. **107**, 1–6 (2002)

F. Lefèvre, F. Forget, Observed variations of methane on Mars unexplained by known atmospheric chemistry and physics. Nature **460**, 720–723 (2002)

L.A. Leshin, S. Epstein, E.M. Stolper, Hydrogen isotope geochemistry of SNC meteorites. Geochim. Cosmochim. Acta **60**, 2635–2650 (1996)

H.F. Levison, L. Dones, C.R. Chapman, A.S. Stern, M.J. Duncan, K. Zahnle, Could the lunar "late heavy bombardment" have been triggered by the formation of Uranus and Neptune. Icarus **151**, 286–306 (2001).

B. Levrard, F. Forget, F. Montmessin, J. Laskar, Ice-rich deposits formed at high latitude on Mars by sublimation of unstable equatorial ice during low obliquity. Nature **431**, 1072–1075 (2004)

J.G. Luhmann, J.U. Kozyra, Dayside pickup oxygen ion precipitation at Venus and Mars: spatial distributions, energy deposition and consequences. J. Geophys. Res. **96**, 5457–5467 (1991)

J.G. Luhmann, R.E. Johnson, M.H.G. Zhang, Evolutionary impact of sputtering of the Martian atmosphere by o(+) pickup ions. Geophys. Res. Lett. **19**, 2151–2154 (1992)

R. Lundin, Ion acceleration and outflow from Mars and Venus: an overview. Space Sci. Rev. **162**, 309–334 (2011). doi:10.1007/s11214-011-9811-y

R. Lundin, H. Lammer, I. Ribas, Planetary magnetic fields and solar forcing: implications for atmospheric evolution. Space Sci. Rev. **129**, 245–278 (2007)

J.I. Lunine, J. Chambers, A. Morbidelli, L.A. Leshin, The origin of water on Mars. Icarus **165**, 1–8 (2003)

J.I. Lunine, D.P. O'Brien, S.N. Raymond, A. Morbidelli, T. Qinn, A.L. Graps, Dynamical models of terrestrial planet formation. Adv. Sci. Lett. **4**, 325–338 (2011)

J. Lyons, C. Manning, F. Nimmo, Formation of methane on Mars by fluid-rock interaction in the crust. Geophys. Res. Lett. **32**, L13201.1–L13201.4 (2005)

Y.-J. Ma, A.F. Nagy, Ion escape fluxes from Mars. Geophys. Res. Lett. **34**, L08201 (2007)

Y. Ma, A.F. Nagy, I.V. Sokolov, K.C. Hansen, Three-dimensional, multispecies, high spatial resolution MHD studies of the solar wind interaction with Mars. J. Geophys. Res. **109**(A7), A07211 (2004)

C.V. Manning, C.P. McKay, K.J. Zahnle, Thick and thin models of the evolution of carbon dioxide on Mars. Icarus **180**, 38–59 (2006)

C.V. Manning, Y. Ma, D.A. Brain, C.P. McKay, K.J. Zahnle, Parametric analysis of modeled ion escape from Mars. Icarus **212**, 131–137 (2010)

B. Marty, A. Meibom, Noble gas signature of the late heavy bombardment in the Earth's atmosphere. Earth Discuss. **2**, 99–113 (2007)

T. Matsui, Y. Abe, Impact-induced atmospheres and oceans on Earth and Venus. Nature **322**, 526–528 (1986)

T. McCollom, W. Back, Thermodynamic constraints on hydrogen generation during serpentinization of ultramafic rocks. Geochim. Cosmochim. Acta **73**, 856–875 (2009)

M.B. McElroy, T.M. Donahue, Stability of the Martian atmosphere. Science **177**, 986–988 (1972)

H.Y. McSween, P.P. Harvey, Outgassed water on Mars: constraints from melt inclusions in SNC meteorites. Science **259**, 1890–1892 (1993)

H.Y. McSween, K. Keil, Mixing relationships in the Martian regolith and the composition of globally homogeneous dust. Geochim. Cosmochim. Acta **64**, 2155–2166 (2000)

H.Y. McSween, T.L. Grove, R.C. Lentz, J.C. Dann, A.H. Holzheid, L.R. Riciputi, J.G. Ryan, Geochemical evidence for magmatic water within Mars from pyroxenes in the Shergotty meteorite. Nature **409**, 487–490 (2001)

H.J. Melosh, A.M. Vickery, Impact erosion of the primordial atmosphere of Mars. Nature **338**, 487–489 (1989)

J.R. Michalski, P.B. Niles, Deep crustal carbonate rocks exposed by meteor impact on Mars. Nat. Geosci. **3**, 751–755 (2010)

D.W. Mittlefehldt, ALH84001, a cumulate orthopyroxenite member of the Martian meteorite clan. Meteoritics **29**, 214–221 (1994)

H. Mizuno, K. Nakazawa, C. Hayashi, Dissolution of the primordial rare gases into the molten Earth' material. Earth Planet. Sci. Lett. **50**, 202–210 (1980)

R. Modolo, G.M. Chanteur, E. Dubinin, A.P. Matthews, Influence of the solar EUV flux on the Martian plasma environment. Ann. Geophys. **23**, 1–12 (2005)

D. Möhlmann, Widen the belt of habitability! Orig. Life Evol. Biosph. **42**, 93–100 (2012)
A. Morbidelli, J. Chambers, J.I. Lunine, J.M. Petit, F. Robert, G.B. Valsecchi, K. Cyr, Source regions and timescales for the delivery of water to Earth. Meteorit. Planet. Sci. **35**, 1309–1320 (2000)
A. Morbidelli, W.F. Bottke, D. Nesvorný, H.F. Levison, Asteroids were born big. Icarus **204**, 558–573 (2009)
R.V. Morris, S.W. Ruff, R. Gellert, D.W. Ming, R.E. Arvidson, B.C. Clark, D.C. Golden, K. Siebach, G. Klingelhöfer, C. Schröder, I. Fleischer, A.S. Yen, S.W. Squyres, Identification of carbonate-rich outcrops on Mars by the spirit rover. Science **329**, 421–424 (2010)
A. Morschhauser, M. Grott, D. Breuer, Crustal recycling, mantle dehydration, and the thermal evolution of Mars. Icarus **212**, 541–558 (2011)
U.V. Möstl, N.V. Erkaev, M. Zellinger, H. Lammer, H. Gröller, H.K. Biernat, D. Korovinskiy, The Kelvin-Helmholtz instability at Venus: what is the unstable boundary? Icarus **216**, 476–484 (2011)
O. Mousis, J.I. Lunine, J.-M. Petit, S. Picaud, B. Schmitt, D. Marquer, J. Horner, C. Thomas, Impact regimes and post-formation sequestration processes: implications for the origin of heavy noble gases in terrestrial planets. Astrophys. J. **714**, 1418–1423 (2010)
O. Mousis, J.I. Lunine, E. Chassefière, F. Montmessin, A. Lakhlifi, S. Picaud, J.M. Petit, D. Cordier, Mars cryosphere: a potential reservoir for heavy noble gases? Icarus **218**, 80–87 (2012)
M.J. Mumma, G.L. Villanueva, R.E. Novak, T. Hewagama, B.P. Bonev, M.A. DiSanti, A.M. Mandell, D.M. Smith, Strong release of methane on Mars in northern summer 2003. Science **323**, 1041–1045 (2009)
J.F. Mustard, S.L. Murchie, S.M. Pelkey, B.L. Ehlmann, R.E. Milliken, J.A. Grant, J.-P. Bibring, F. Poulet, J. Bishop, E. Noe Dobrea, L. Roach, F. Seelos, R.E. Arvidson, S.R. Green, H. Hash, D. Humm, E. Malaret, J.A. McGovern, K. Seelos, T. Clancy, R. Clark, D.D. Marais, N. Izenberg, A. Knudson, Y. Langevin, T. Martin, P. McGuire, R. Morris, M. Robinson, T. Roush, M. Smith, G. Swayze, H. Taylor, T. Titus, M. Wolff, Hydrated silicate minerals on Mars observed by the Mars reconnaissance orbiter CRISM instrument. Nature **454**, 305–309 (2008)
A.F. Nagy, J. Kim, T.E. Cravens, Hot hydrogen and oxygen atoms in the upper atmospheres of Venus and Mars. Ann. Geophys. **8**, 251–256 (1990)
G. Neukum, D.U. Wise, Mars: a standard crater curve and possible new time scale. Science **194**, 1381–1387 (1976)
G. Neukum, B.A. Ivanov, W.K. Hartmann, Cratering records in the inner Solar System in relation to the Lunar reference system. Space Sci. Rev. **96**, 55–86 (2001)
G. Newkirk Jr., Solar variability on time scales of 10^5 years to $10^{9.6}$ years. Geochim. Cosmochim. Acta, Suppl. **13**, 293–301 (1980)
P.B. Niles, W.V. Boynton, J.H. Hoffman, D.W. Ming, D. Hamara, Stable isotope measurements of martian atmospheric CO_2 at the Phoenix landing site. Science **329**, 1334–1337 (2010)
P.B. Niles, D.C. Catling, G. Berger, E. Chassefière, B.L. Ehlmann, J.R. Michalski, R. Morris, S.W. Ruff, B. Sutter, Geochemistry of carbonates on Mars: implications for climate history and nature of aqueous environments. Space Sci. Rev. (2012). doi:10.1007/s11214-012-9940-y
M.D. Norman, The composition and thickness of the crust of Mars estimated from REE and nd isotopic compositions of Martian meteorites. Meteorit. Planet. Sci. **34**, 439–449 (1999)
L.E. Nyquist, D.D. Bogard, C.-Y. Shih, A. Greshake, D. Stöffler, O. Eugster, Ages and geologic histories of martian meteorites. Space Sci. Rev. **96**, 105–164 (2001)
D. Olsson-Steel, Collisions in the solar system. IV: cometary impacts upon the planets. Mon. Not. R. Astron. Soc. **227**, 501–524 (1987)
C. O'Neill, A. Lenardic, A.M. Jellinek, W.S. Kiefer, Melt propagation and volcanism in mantle convection simulations, with applications for Martian volcanic and atmospheric evolution. J. Geophys. Res. **112**, E07003 (2007)
T. Owen, A. Bar-Nun, Comets, impacts and atmospheres. Icarus **116**, 215–226 (1995)
J.E. Owen, A.P. Jackson, Planetary evaporation by UV & X-ray radiation: basic hydrodynamics. Mont. Not. R. Astron. Soc. **425**, 2931–2947 (2012)
T. Owen, A. Bar-Nun, I. Kleinfeld, Possible cometary origin of heavy noble gases in the atmospheres of Venus, Earth, and Mars. Nature **358**, 43–46 (1992)
C. Oze, M. Sharma, Have olivine, will gas: serpentinization and the abiogenic production of methane on Mars. Geophys. Res. Lett. **32**, L10203 (2005)
T. Penz, N.V. Erkaev, H.K. Biernat, H. Lammer, U.V. Amerstorfer, H. Gunell, E. Kallio, S. Barabash, S. Orsini, A. Milillo, W. Baumjohann, Ion loss on Mars caused by the Kelvin-Helmholtz instability. Planet. Space Sci. **52**, 1157–1167 (2004)
R.O. Pepin, On the origin and early evolution of terrestrial planet atmospheres and meteoritic volatiles. Icarus **92**, 2–79 (1991)
R.O Pepin, Origin of noble gases in the terrestrial planets. Ann. Rev. Earth Planet. Sci. **20**, 389–430 (1992)

R.O. Pepin, Evolution of Earth's noble gases: consequences of assuming hydrodynamic loss driven by giant impact. Icarus **126**, 148–156 (1997)
R.O. Pepin, Atmospheres on the terrestrial planets: clues to origin and evolution. Earth Planet. Sci. Lett. **252**, 1–14 (2006)
L.B.S. Pham, Ö. Karatekin, V. Dehant, Effect of meteorite impacts on the atmospheric evolution of Mars. Astrobiology **9**, 45–54 (2009)
L.B.S. Pham, Ö. Karatekin, V. Dehant, Effect of an meteorites and asteroids bombardments on the atmospheric evolution of Mars. EPSC Proc. **5**, EPSC2010-127 (2010), 2 pp.
L.B.S. Pham, Ö. Karatekin, V. Dehant, Effects of impacts on the atmospheric evolution: comparison between Mars, Earth and Venus. Planet. Space Sci. **59**, 1087–1092 (2011)
R.J. Phillips, M.T. Zuber, S.C. Solomon, M.P. Golombek, B.M. Jakosky, W.B. Banerdt, D.E. Smith, R.M.E. Williams, B.M. Hynek, O. Aharonson, S.A. Hauck II, Ancient geodynamics and global-scale hydrology on Mars. Science **291**, 2587–2591 (2001)
R.J. Phillips, B.J. Davis, S. Byrne, B.A. Campbell, L.M. Carter, R.M. Haberle, J.W. Holt, M.A. Kahre, D.C. Nunes, J.J. Plaut, N.E. Putzig, I.B. Smith, S.E. Smrekar, K.L. Tanaka, T.N. Titus, SHARAD finds voluminous CO_2 ice sequestered in the martian South Polar layered deposits. AGU, Fall Meeting, abs. P34A-01.2010 (2010)
R.J. Phillips, B.J. Davis, K.L. Tanaka, S.M. Byrne, T. Michael, N.E. Putzig, R.M. Haberle, M.A. Kahre, A. Melinda, B.A. Campbell, L.M. Carter, I.B. Smith, J.W. Holt, S.E. Smrekar, D.C. Nunes, J.J. Plaut, A.F. Egan, T.N. Titus, R. Seu, Massive CO_2 ice deposits sequestered in the South Polar layered deposits of Mars. Science **332**, 838–841 (2011)
E. Pierazzo, G. Collins, A brief introduction to hydrocode modeling of impact cratering, in *Submarine Craters and Ejecta-Crater Correlation*, ed. by P. Claeys, D. Henning (Springer, New York, 2003), pp. 323–340
R. Pierrehumbert, E. Gaidos, Hydrogen greenhouse planets beyond the habitable zone. Astrophys. J. **734**, L13 (2011)
J.B. Pollack, J.F. Kasting, S.M. Richardson, K. Poliakoff, The case for a wet, warm climate on early Mars. Icarus **71**, 203–224 (1987)
Y. Quesnel, C. Sotin, B. Langlais, S. Costin, M. Mandea, M. Gottschalk, J. Dyment, Serpentinization of the martian crust during Noachian. Earth Planet. Sci. Lett. **277**, 184–193 (2009)
R.R. Rafikov, Atmospheres of protoplanetary cores: critical mass for nucleated instability. Astrophys. J. **648**, 666–682 (2006)
I. Ribas, E.F. Guinan, M. Güdel, M. Audard, Evolution of the solar activity over time and effects on planetary atmospheres. I. High-energy irradiances (1–1700 Å). Astrophys. J. **622**, 680–694 (2005)
T.L. Segura, O.B. Toon, A. Colaprete, K. Zahnle, Environmental effects of large impacts on Mars. Science **298**, 1977–1980 (2002)
R. Shaheen, A. Abramian, J. Horn, G. Dominguez, R. Sullivan, M.H. Thiemens, Detection of oxygen isotopic anomaly in terrestrial atmospheric carbonates and its implications to Mars. Proc. Natl. Acad. Sci. **107**, 20213–20218 (2010)
C.K. Shearer, G. McKay, J.J. Papike, J.M. Karner, Valence state partitioning of vanadium between olivine liquid: estimates of the oxygen fugacity of Y980459 and application to other olivine phyric Martian basalts. Am. Mineral. **91**, 1657–1663 (2006)
V.V. Shuvalov, N.A. Artemieva, Atmospheric erosion and radiation impulse induced by impacts, in *Proc. International Conference on Catastrophic Events and Mass Extinctions: Impacts and Beyond* (2001), abs. 3060
A. Skumanich, J.A. Eddy, Aspects of long-term variability in Sun and stars, in *Solar Phenomena in Stars and Stellar Systems* (Reidel, Dordrecht, 1981), pp. 349–397
D.E. Smith, M.T. Zuber, H.V. Frey, J.B. Garvin, J.W. Head, D.O. Muhleman, G.H. Pettengill, R.J. Phillips, S.C. Solomon, H.J. Zwally, W.B. Banerdt, T.C. Duxbury, M.P. Golombek, F.G. Lemoine, G.A. Neumann, D.D. Rowlands, O. Aharonson, P.G. Ford, A.B. Ivanov, C.L. Johnson, P.J. McGovern, J.B. Abshire, R.S. Afzal, X. Sun, Mars orbiter laser altimeter: experiment summary after the first year of global mapping of Mars. J. Geophys. Res. **106**, 23689–23722 (2001)
S.W. Squyres, A.H. Knoll, Sedimentary rocks at MeridianiPlanum: origin, diagenesis, and implications for life on Mars. Earth Planet. Sci. Lett. **240**, 1–10 (2005)
B.D. Stanley, M.M. Hirschmann, A.C. Withers, CO_2 solubility in Martian basalts and Martian atmospheric evolution. Geochim. Cosmochim. Acta **75**, 5987–6003 (2011)
V.V. Svetsov, Atmospheric erosion and replenishment induced by impacts of cosmic bodies upon the Earth and Mars. Sol. Syst. Res. **41**, 28–41 (2007)
N. Terada, Yu.N. Kulikov, H. Lammer, H.I.M. Lichtenegger, T. Tanaka, H. Shinagawa, T.-L. Zhang, Atmosphere and water loss form early Mars under extreme solar wind and extreme ultraviolet conditions. Astrobiology **9**, 55–70 (2009)

F. Tian, O.B. Toon, A.A. Pavlov, H. De Sterck, A hydrogen-rich early Earth atmosphere. Science **308**, 1014–1017 (2005)

F. Tian, J.F. Kasting, H. Liu, R.G. Roble, Hydrodynamic planetary thermosphere model: 1. The response of the Earth's thermosphere to extreme solar EUV conditions and the significance of adiabatic cooling. J. Geophys. Res. **113**(E5), E05008 (2008)

F. Tian, J.F. Kasting, S.C. Solomon, Thermal escape of carbon from the early Martian atmosphere. Geophys. Res. Lett. **36**(2), L02205 (2009)

O.B. Toon, T. Segura, K. Zahnle, The formation of martian river valleys by impacts. Annu. Rev. Earth Planet. Sci. **38**, 303–322 (2010)

G. Turner, S.F. Knott, R.D. Ash, J.D. Gilmour, Ar-Ar chronology of the Martian meteorite ALH 84001: evidence for the timing of the early bombardment of mars. Geochim. Cosmochim. Acta **61**, 3835–3850 (1997)

A. Valeille, M.R. Combi, V. Tenishev, S.W. Bougher, A.F. Nagy, A study of suprathermal oxygen atoms in Mars upper thermosphere and exosphere over the range of limiting conditions. Icarus **206**, 18–27 (2010)

A.M. Vickery, Impacts and atmospheric erosion on the early Earth, in *Proc. International Workshop on Meteorite Impact on the Early Earth* (1990), pp. 51–52

M. Wadhwa, Redox state of Mars: upper mantle and crust from Eu anomalies in shergottite pyroxenes. Science **291**, 1527–1530 (2001)

K.J. Walsh, A. Morbidelli, S.N. Raymond, D.P. O'Brien, A.M. Mandell, A low mass for Mars from Jupiter's early gas-driven migration. Nature **475**, 206–209 (2011)

H. Wänke, G. Dreibus, Chemistry and accretion history of Mars. Philos. Trans. R. Soc. Lond. **349**, 285–293 (1994)

P.H. Warren, G.W. Kallmeyen, Siderophile trace elements in ALH84001, other SNC meteorites and eucrites: evidence of heterogeneity, possibly time-linked, in the mantle of Mars. Meteorit. Planet. Sci. **31**, 97–105 (1996)

A.J. Watson, T.M. Donahue, J.C.G. Walker, The dynamics of a rapidly escaping atmosphere: applications to the evolution of Earth and Venus. Icarus **48**, 150–166 (1981)

L.L. Watson, I.D. Hutcheon, S. Epstein, E.M. Stolper, Water on Mars: clues from deuterium/hydrogen and water contents of hydrous phases in SNC meteorites. Science **265**, 86–90 (1994)

T.R. Watters, B. Campbell, L. Carter, C.J. Leuschen, J.J. Plaut, G. Picardi, R. Orosei, A. Safaeinili, S.M. Clifford, W.M. Farrell, M. William, A.B. Ivanov, R.J. Phillips, E.R. Stofan, Radar sounding of the Medusae Fossae formation Mars: equatorial ice or dry, low-density deposits? Science **318**, 1125–1128 (2007)

G.W. Wetherill, Accumulation of terrestrial planets and implications concerning lunar origin, in *Origin of the Moon*, ed. by W.K. Hartmann, R.J. Phillips, G.J. Taylor (Arizona Press, Tucson, 1986), pp. 519–550

R.S. Wolff, B.E. Goldstein, C.M. Yeates, The onset and development of Kelvin-Helmholtz instability at the Venus ionopause. J. Geophys. Res. **85**, 7697–7707 (1980)

R. Wordsworth, Transient conditions for biogenesis on low-mass exoplanets with escaping hydrogen atmospheres. Icarus **219**, 267–273 (2012)

Y.L. Yung, J.S. Wen, J.P. Pinto, M. Allen, K.K. Pierce, S. Paulson, HDO in the Martian atmosphere: implications for the abundance of crustal water. Icarus **76**, 146–159 (1988)

K.J. Zahnle, Xenological constraints on the impact erosion of the early Martian atmosphere. J. Geophys. Res. **98**, 10899–10913 (1993)

K.J. Zahnle, J.F. Kasting, Mass fractionation during transonic escape and implications for loss of water from Mars and Venus. Icarus **68**, 462–480 (1986)

K.J. Zahnle, J.C.G. Walker, The evolution of solar ultraviolet luminosity. Rev. Geophys. **20**, 280–292 (1982)

K.J. Zahnle, J.F. Kasting, J.B. Pollack, Evolution of a steam atmosphere during Earth's accreation. Icarus **74**, 62–97 (1988)

K.J. Zahnle, J.B. Pollack, D. Grinspoon, Impact-generated atmospheres over Titan, Ganymede and Callisto. Icarus **95**, 1–23 (1992)

K.J. Zahnle, R.M. Haberle, D.C. Catling, J.F. Kasting, Photochemical instability of the ancient Martian atmosphere. J. Geophys. Res. **113**(E11), E11004 (2008)

K.J. Zahnle, R.S. Freedman, D.C. Catling, Is there methane on Mars? Icarus **212**, 493–503 (2011)

A.P. Zent, R.C. Quinn, Simultaneous adsorption of CO_2 and H_2O under Mars-like conditions and application to the evolution of the martian climate. J. Geophys. Res. **100**, 5341–5349 (1995)

Space Sci Rev (2013) 174:155–212
DOI 10.1007/s11214-012-9946-5

Quantitative Assessments of the Martian Hydrosphere

Jeremie Lasue · Nicolas Mangold · Ernst Hauber · Steve Clifford · William Feldman · Olivier Gasnault · Cyril Grima · Sylvestre Maurice · Olivier Mousis

Received: 13 December 2011 / Accepted: 20 October 2012 / Published online: 30 November 2012

Abstract In this paper, we review current estimates of the global water inventory of Mars, potential loss mechanisms, the thermophysical characteristics of the different reservoirs that water may be currently stored in, and assess how the planet's hydrosphere and cryosphere evolved with time. First, we summarize the water inventory quantified from geological analyses of surface features related to both liquid water erosion, and ice-related landscapes. They indicate that, throughout most of Martian geologic history (and possibly continuing through to the present day), water was present to substantial depths, with a total inventory ranging

Chapter in ISSI book "Quantifying the Martian geochemical reservoirs".

J. Lasue (✉) · O. Gasnault · S. Maurice · O. Mousis
UPS-OMP, IRAP, Université de Toulouse, Toulouse, France
e-mail: jeremie.lasue@irap.omp.eu

J. Lasue · O. Gasnault · S. Maurice · O. Mousis
CNRS, IRAP, 9 Av. colonel Roche, BP 44346, 31028 Toulouse cedex 4, France

N. Mangold
Laboratoire Planétologie et Géodynamique de Nantes, CNRS and University of Nantes, 2, rue de la Houssinière, 44322 Nantes, France

E. Hauber
DLR-Institut für Planetenforschung, Rutherfordstrasse 2, 12489 Berlin-Adlershof, Germany

S. Clifford
Lunar and Planetary Institute, 3600 Bay Area Boulevard, Houston, TX 77058, USA

W. Feldman
Planetary Science Institute, 1700 East Fort Lowell, Tucson, AZ 85719, USA

C. Grima
Institute for Geophysics, Jackson School of Geosciences, University of Texas at Austin, Austin, TX 78758, USA

O. Mousis
Institut UTINAM, CNRS/INSU, UMR 6213, Observatoire des Sciences de l'Univers THETA, Université de Franche-Comté, Besançon, France

from several 100 to as much as 1000 m Global Equivalent Layer (GEL). We then review the most recent estimates of water content based on subsurface detection by orbital and landed instruments, including deep penetrating radars such as SHARAD and MARSIS. We show that the total amount of water measured so far is about 30 m GEL, although a far larger amount of water may be stored below the sounding depths of currently operational instruments. Finally, a global picture of the current state of the subsurface water reservoirs and their evolution is discussed.

Keywords Planetary Sciences · Mars · Water · Hydrosphere · Cryosphere

1 Introduction

Geologic evidence from spacecraft investigations to detect water and its alteration products on Mars has transformed the general view of Mars as a desert-like planet with only a thin external veneer of water, to the current understanding that water was (and may continue to be) pervasive beneath the surface, and present to substantial depths. This water has played a major role in the geologic evolution of the surface and may have provided suitable conditions for the evolution and persistence of life (e.g. Carr 1986, SSES 2003, CASSEP 2008). Understanding the inventory of water on Mars and how it has evolved with time is a major crosscutting theme for the exploration of the planet (CASEM 2007, CPSDS 2011). In the following work, we review current estimates of the past and present inventory of water on Mars, including potential loss mechanisms and reservoirs, based on the latest discoveries, and then assess how each of these various factors may have influenced the evolution of the planet's hydrosphere.

While the initial inventory of water on Mars remains elusive, arguments based on solar system dynamics and meteorite compositions indicate that if the bulk of the water in terrestrial planets was delivered by water-rich embryos that condensed far from the Sun, Mars may have possessed an initial inventory, expressed as a Global Equivalent Layer (GEL), up to several kms deep (e.g. 600 to 2700 m GEL, Lunine et al. 2003). This estimate could be more than an order of magnitude higher if water-rich accreting bodies formed closer to the early Sun in a cold solar nebula (Drake and Righter 2002; Solomon et al. 2005; Raymond et al. 2006).

Early solar activity is expected to have resulted in the loss of volatiles from the early Martian atmosphere. Processes of atmospheric escape include ions and neutral species fluxes (Tian et al. 2009, Lammer 2012, this issue). Estimates of atmospheric water loss during the first 150 My after the formation of the planet, based on ion escape, are $\sim$10 to 70 m GEL (e.g. Terada et al. 2009) but several times more could be expected to be lost from neutral species escape (e.g. Fox 2004, Fox and Hác 2009; Chassefière 1996). The total loss estimates are still under investigation, but all escape processes were very strong during the first 500 My of the planet's history. For the lifetime of the planet, estimates of water atmospheric loss range from 10 m to 90 m GEL after 3.5 Gy (see e.g. Jakosky and Phillips 2001; Valeille et al. 2010; Lammer 2012, and references therein). Such losses could be balanced by similar volcanic outgassing values from the planet's mantle amounting to between $\sim$20 and 60 m of GEL (Grott et al. 2011) and direct injection of volatiles to the planetary inventory by the impacts of H_2O-rich bodies, such as comets or C-type asteroids. The loss rate corresponding to solar wind interactions measured by ASPERA-3 on-board Mars Express is lower than these previous estimates, and would lead to the escape of a few cm of water over the past 3.5 Gy, indicating that major reservoirs of water and carbon dioxide are still to be discovered in the subsurface of Mars (Barabash et al. 2007). However, the fractionation of the deuterium over hydrogen ratio due to the Martian atmosphere erosion (the D/H ratio of Mars is five times higher than the terrestrial value) is difficult to explain with the current loss rate which would indicate an enhanced loss rate of hydrogen from early Mars (Donahue 2004). It could be explained by a significant reduction of the water reservoir exchangeable with the atmosphere, estimated depending on the model, to be between 3 and 60 m (Lammer et al. 2003). Another possible explanation is that some of the observed D/H fractionation is due to crust oxidation, with further release of H to the atmosphere, then to space through Jeans escape (e.g. Chassefière and Leblanc 2011).

Studies of the total extent and drainage density of valley networks on Mars (Carr and Chuang 1997; Carr 1996; Luo and Stepinski 2009), and the observed widespread occurrence of phyllosilicates in Noachian age terrains by the orbital spectrometers OMEGA (Poulet et al. 2005; Bibring et al. 2006; Loizeau et al. 2008; Mangold et al. 2007) and CRISM (Mustard et al. 2008; Murchie et al. 2009), strongly support the conclusion that early Mars was water-rich. High-resolution images reveal the presence of mineral-filled fractures (Okubo and McEwen 2007) and deep occurrence of phyllosilicate minerals in km-scale vertical exposures (Mustard et al. 2008; Murchie et al. 2009; Ehlmann et al. 2011), indicating aqueous activity that has spanned geologic time. The processes and conditions that formed these minerals indicate the potential occurrence of aqueous alterations at a variety of depths, along with evidence for low- to moderate-temperature hydrothermal circulation of vapor and liquid water in the deep subsurface of Mars that may continue to the present day (Clifford 1993; Bibring et al. 2006).

Carr (1986, 1996) has estimated that, at the time of peak outflow channel activity which ranged from the Late Hesperian to Early Amazonian ($\sim$3.4 to $\sim$2 Gy, e.g. Tanaka 1986; Hartmann and Neukum 2001), Mars possessed a planetary inventory of water equivalent to about 0.5 to 1 km GEL. As this peak activity post-dates the period when the most efficient water loss processes (hydrodynamic escape and atmospheric erosion by large impacts) were active (during the Pre- and Early-Noachian, >4.2 Gy ago), the majority of this water is expected to have survived, in some form, to the present day.

Any water that survived is expected to be stored in several distinct subsurface reservoirs: (1) as ground ice, located within the near-surface region of perennially frozen ground known as the cryosphere, (which mirrors the first-order variation in topography, extending down to km depths), and (2) as groundwater, saturating the lowermost porous regions of the crust, where radiogenic heating increases lithospheric temperatures above the freezing point (see Fig. 1, Fanale 1976; Rossbacher and Judson 1981; Kuzmin 1983; Clifford 1993; Carr 1996; Clifford et al. 2010). Because the distribution of ground ice is determined by the thermal structure of the crust, while the distribution of groundwater is governed by gravity and crustal porosity, the vertical distances separating these two reservoirs may range from zero, beneath the northern plains, to many kilometers, beneath the southern highlands resulting in an unsaturated subpermafrost vadose zone (Clifford 1993; Clifford et al. 2010). (3) Large amounts of water may also be stored in hydrated minerals such as phyllosilicates (e.g. Bibring et al. 2006), but due to the necessarily incomplete knowledge of the global abundance of these minerals it is impossible to quantify these reservoirs. A particularly large reservoir would be provided by large-scale crustal serpentinization as hypothesized by Quesnel et al. (2009), and Chassefière and Leblanc (2011) have estimated that up to 400 m GEL of water may have been stored in serpentine since the Noachian. However, so far there is no firm evidence for such large amounts of serpentine on Mars, and only a few localized outcrops have been identified (Ehlmann et al. 2010).

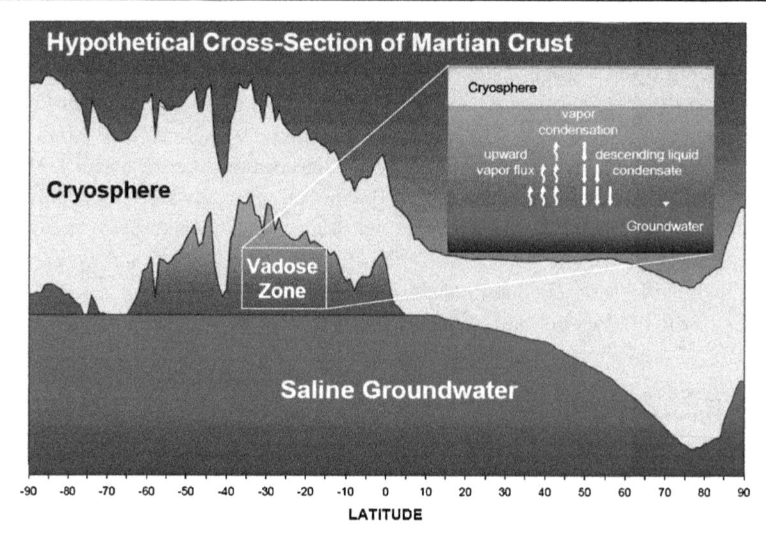

Fig. 1 Hypothetical meridian cross section of the present-day Martian crust (along 203°E), illustrating the potential relationship between surface topography, ground ice, and groundwater. Surface elevations are from the MOLA Mission Experiment Gridded Data Record with a 2° smoothing function (Smith et al. 2003). At those locations where the base of the cryosphere is in contact with the water table, the presence of dissolved salts may reduce the thickness of frozen ground by depressing the freezing point of the groundwater. Where the cryosphere and groundwater are not in direct contact, low-temperature hydrothermal convection may result in basal melting temperatures closer to 273 K (Clifford 1993). Figure adapted from Clifford and Parker (2001)

Evidence that Mars remains water-rich is provided by the amount: (1) visible in the polar layered deposits, (2) inferred to be present as ice and hydrated minerals in the shallow (top 50 cm) subsurface, direct detections of which include abundant near surface hydrogen at mid- to high-latitudes by GRS on board Mars Odyssey (Boynton et al. 2002; Feldman et al. 2004), the observation of anomalously low near-surface dielectric values (suggestive of the presence of a large volumetric component of ice within the top 100 m of the surface) by the MARSIS orbital radar sounder on Mars Express (Mouginot et al. 2010, 2012), and of mid-latitude glacial deposits (Holt et al. 2008), (3) landforms including the widespread occurrence of apparently young glacial and periglacial features (gullies, polygonal terrains, lobate debris aprons, etc.) throughout mid- to high-latitudes in both hemispheres (e.g. Squyres and Carr 1986; Squyres 1989; Malin and Edgett 2001; Mangold et al. 2004; Levy et al. 2010), and (4) direct observations of ice by the Phoenix lander (Smith et al. 2009; Mehta et al. 2011), and the direct exposure of ice (in disequilibrium with current surface temperatures and atmospheric relative humidity) by recent impacts at mid-latitudes (Byrne et al. 2009). Together with the detection of ice, the recent discovery of perchlorates at the Phoenix landing site (Hecht et al. 2009), has renewed interest in the potential existence of liquid water in the form of low freezing point brines and the low-temperature aqueous alteration of the Martian crust (e.g. Fairén et al. 2009).

Given the evidence for abundant water on Mars there is a critical need to understand the quantity and state of water in the crust and how it has varied with time. To this end, we will review the geomorphic and geophysical evidence for ground-ice and groundwater as well as the spectroscopic evidence for aqueous activity—assessing the age and amount of water required by each. A review of the last decade of measurements will also give us a lower limit to the quantity of water still present in the subsurface of Mars. Finally, we will consider where the surviving water reservoirs are most likely located and their implications for present day habitability.

2 Geomorphological Analysis of the Hydrosphere and Cryosphere of Mars

2.1 The Terrestrial Cryosphere as a Reference for Mars: Processes and Landscapes

2.1.1 Definition

The terrestrial cryosphere is defined as "the subsystem of the Earth characterized by the presence of snow, ice, and permafrost" (Slaymaker and Kelly 2007, p. 1). In applying this concept to Mars, these same conditions should be met. Snow and ice have been observed on Mars by various instruments from orbit and in-situ, on the surface. Permafrost is defined as "soil, bedrock, or other earth materials, whose temperature remains below 0 °C for at least 2 consecutive years" (e.g. van Everdingen 1998). Note that this definition is based on the thermal state of the crust, not on material properties. Thus, Mars can be considered to be a permafrost planet. Therefore, the concept of a cryosphere may be readily applied to Mars, for which the terrestrial cryosphere serves as the basic reference.

2.1.2 Areal Extent and Ice Volume

With the exception of high-altitude environments in mid- and low-latitudes (e.g. the Qinghai-Xizang (Tibet) Plateau, China, highest regions of the Alps), the terrestrial cryosphere is limited to the Arctic and Antarctica (Table 1). The most obvious manifestations of cryospheric surface features are probably huge ice caps (e.g. Greenland, Antarctica) or alpine glaciers such as in the Himalayas or Alaska, but the various components of the cryosphere also include snow, ice, and frozen ground (Fig. 2). The most important component in terms of spatial coverage is permafrost, which underlies roughly 25 % of the Earth's land surface, mainly in northern Asia, Russia and North America (Brown et al. 1997). The total amount of ice stored in the various components of the Earth's cryosphere is about 28×10^6 km^3 (Table 2). Most of this ice forms huge ice sheets in Antarctica and Greenland,

Table 1 Summary characteristics of Arctic and Antarctic ecosystems (shortened from French 2011)

	Arctic		Antarctic
	Low Arctic	High Arctic	Continent
Climate	Very cold winters, temperate summers, low precipitation 3.5–5 months >0 °C	Very cold winters, cold summers, very low precipitation 2–3 months >0 °C	Extremely cold, short summers, very low precipitation, strong winds ~1 month >0 °C
Snow-free period	~3–4 months	~1–1.5 months	~1–2 months
Permafrost	Continuous: temperature is ~−3 to −4 °C at 10–30 m depth	Continuous: temperature is ~−10 to −14 °C at 10–30 m depth	Continuous: temperature is ~−8 to −18 °C at 10–30 m depth
Active-layer depth	~30–50 cm in silt/clay	~30–50 cm in silt/clay	30–50 cm in gravel and ablation till
	~2–5 m in sand	~70–120 cm in sand	~1–2 m in bedrock

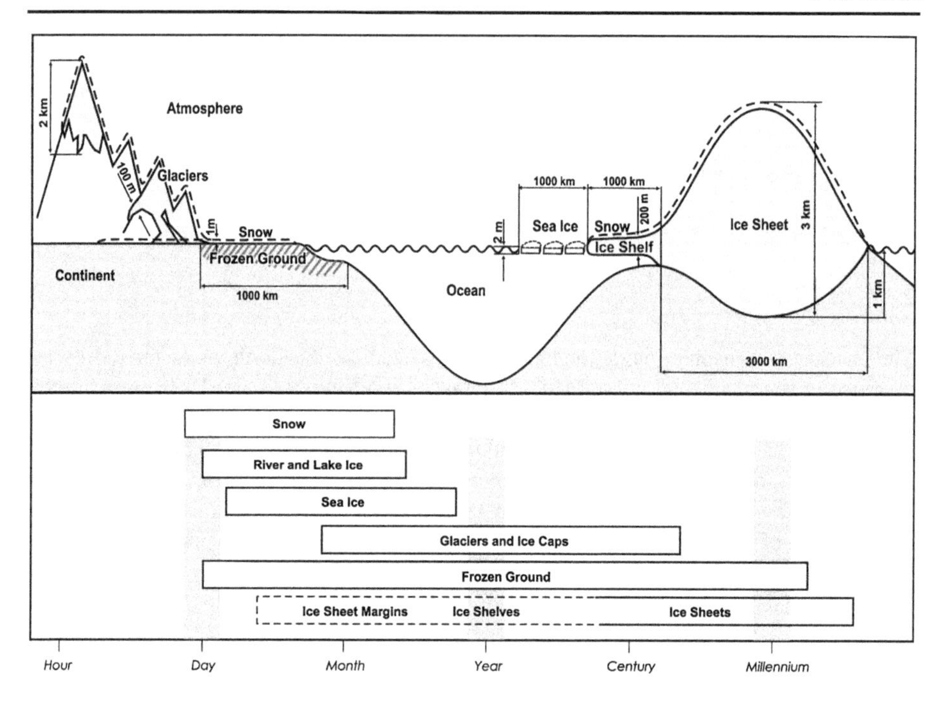

Fig. 2 Components of the cryosphere and their time scales (from Lemke et al. 2007, Source: The Intergovernmental Panel on Climate Change, 2007 Fourth Assessment Report, The Physical Science Basis, Chap. 4, pp. 337–383, http://www.ipcc.ch)

Table 2 Components of the cryosphere: areas and volumes of ice (from Lemke et al. 2007)

Cryosphere component	Area [10^6 km^2]	Ice volume [10^6 km^3]
Snow on land (NH)	1.9–45.2	0.0005–0.005
Sea ice	19–27	0.019–0.025
Glaciers and ice caps	0.51–0.54	0.05–0.13
Ice shelves	1.5	0.7
Ice sheets	14	27.6
Greenland	1.7	2.9
Antarctica	12.3	24.7
Seasonally frozen ground (NH)	5.9–48.1	0.006–0.065
Ground ice (NH)[a]	22.8	0.0056–0.015
Ground ice (NH)[b]	22.8	0.011–0.037

Note: NH = Northern Hemisphere

[a]Based on IPA (International Permafrost Association) map categories

[b]Based on assumptions used by Zhang et al. (2008a)

Source: The Intergovernmental Panel on Climate Change (2007). Fourth Assessment Report, The Physical Science Basis, Chap. 4, pp. 337–383 (http://www.ipcc.ch)

while ground ice in permafrost regions accounts only for a tiny fraction of the total volume (5.6 to 36.5×10^3 km^3, depending on the assumptions used in the calculations), equivalent to a terrestrial GEL of just a few centimetres, while, in comparison, the Earth's oceans amount to about 3700 m (Charette and Smith 2010). The thickness of permafrost is controlled by many factors, including climatic conditions and the geothermal gradient. Permafrost thicknesses can reach several hundred meters in parts of northern Canada, but reach a maximum thickness of more than 1.4 km in the north Central Siberian Plateau (Yershov 1989).

2.1.3 Geomorphic Processes in the Cryosphere

The geomorphology of the cryosphere is shaped by glacial and periglacial processes, which can create micro- and macro-scale landforms. A comprehensive view of glacial and periglacial processes and landforms is beyond the scope of this paper, but can be found in Benn and Evans (1998) and French (1988, 2007). Here we briefly review those processes and landforms that are most relevant to Mars.

Glacial Processes Glaciers and icecaps are dynamic systems which transport ice and other materials from accumulation zones to ablation zones. Glacial processes, therefore, modify the landscape by erosion and deposition. If a glacier is warm-based, basal melting will occur and the glacier erodes the underlying substrate as it slides downward. On the other hand, cold-based glaciers cause relatively minor erosion (Atkins et al. 2002), as they are frozen to the ground and their deformation is mainly by shear within the ice. In reality, this two-fold classification is overly simplistic, as some glaciers are polythermal, i.e. they are cold-based in one region and warm-based in another (Lorrain and Fitzsimons 2011). The thermal state of the base of a glacier determines what geomorphic processes are associated with glaciation (Fig. 3).

Periglacial Processes Permafrost is defined by its thermal state (Fig. 4a and b) and is only affected by two processes: freezing and thawing, (also described as aggradation and degradation, e.g. Dobiński 2011). Nevertheless, there are various geomorphic processes associated with non-glaciated permafrost terrain, which are commonly called periglacial processes. It is important to note, however, that periglacial processes are not confined to regions of permanently frozen ground, but can also operate in seasonally frozen ground. In fact, one possible definition of periglacial climates includes all areas where the Mean Annual Air Temperature (MAAT) is below +3 °C (French 2007, p. 31).

Weathering is an important process in cold-climates (both glacial and periglacial). An important, though perhaps commonly overestimated, component is physical weathering, e.g. by frost action and freeze-thaw cycles (frost shattering or frost wedging as a result of thermal fatigue of materials). Chemical weathering, despite low temperatures, is also known to be a significant contributor to cold-climate weathering (e.g. Dixon et al. 2008), which can be enhanced by the presence of water-soluble, freezing-point depressing salts (Marion 1995). Ground ice is a volumetrically significant component of permafrost (Mackay 1972). Permafrost degradation results from the melting or sublimation of ground ice, potentially leading to the subsidence of the originally ice-rich terrain (thermokarst) or the development of thaw lakes from the pooling of meltwater.

As thaw lakes drain and freeze, any surviving liquid water beneath the lake sediments (taliks, Fig. 4c) can freeze by permafrost aggradation, creating an ice lens which uplifts the overlying sediments as the ice expands. This resulting mound, which can range from as

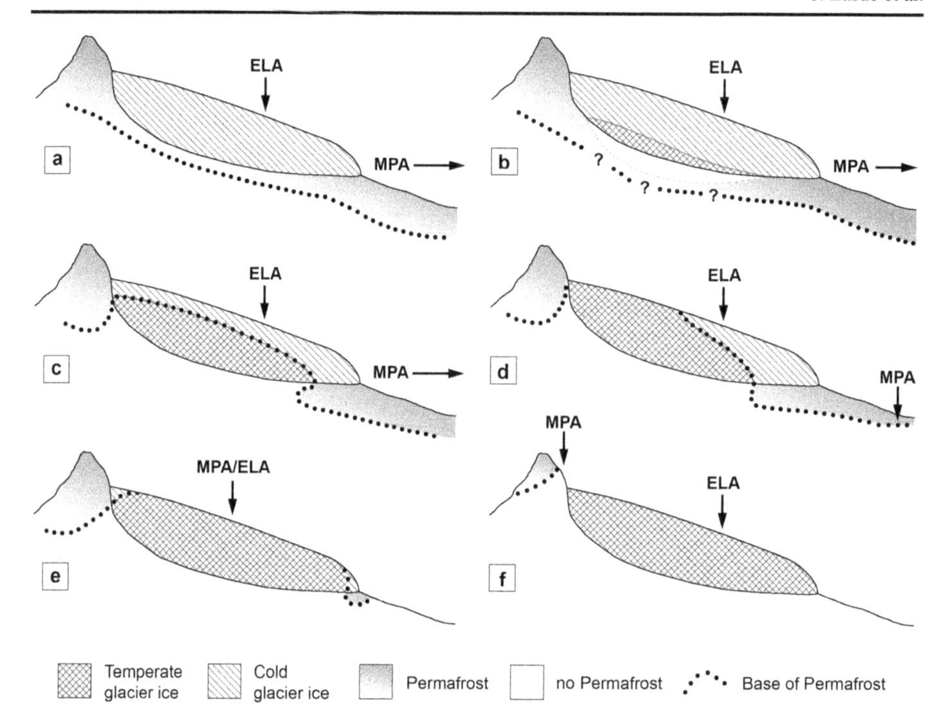

Fig. 3 Schematic illustration of thermal regimes in glaciers related to permafrost in mountain areas (redrawn from Etzelmüller and Hagen 2005). (**a**) Entire cold glacier in a permafrost environment. Mountain Permafrost Altitude (MPA) $\ll$ Equilibrium Line Altitude (ELA). (**b**) Mostly cold-based glacier with temperate layers along the glacier-ground interface. MPA $\ll$ ELA. (**c**), (**d**) Polythermal glacier with temperate glacier ice along most of the bottom and in the accumulation area. MPA < ELA. (**e**) Mostly warm-based glacier with some cold patches, either locally in the frontal areas or in dry, cold accumulation areas. MPA $\sim$ ELA or MPA > ELA. For entirely temperate glaciers, MPA $\gg$ ELA. (**f**) If the ELA is located well below the MPA, temperate glaciers dominate

little as several meters in breadth and height to as much as 500 m in diameter and 50 m in height, is called a hydrostatic or closed-system pingo. Pingos can also result in thermally heterogeneous environments where taliks are present between expanses of discontinuous permafrost. When such a talik experiences freezing within the seasonally active layer, liquid water can be drawn from beneath the permafrost and ascend to the base of the downward propagating freezing front—where it can once again form an ice lens that lifts any overlying material, forming what is known as a hydraulic or open-system pingo.

Seasonal freezing can result in large and rapid drops in temperature, producing steep thermal gradients and thermal stresses that can result in the contraction and cracking of cohesive ground. These cracks may then be filled by the infiltration of meltwater or eolian sediments, which, over many seasonal cycles, can result in the formation of ice- or sand-wedge polygons.

The near-surface region of rock and soil, above the permafrost, that undergoes seasonal freezing and thawing is called the active layer (Fig. 4a). Active layer processes such as frost heave (due to the volumetric expansion by $\sim$10 % of freezing water), ice segregation, mass wasting (e.g. frost creep, solifluction), thaw consolidation and cryoturbation (frost churning of soils) produce a variety of landform assemblages (French 1988).

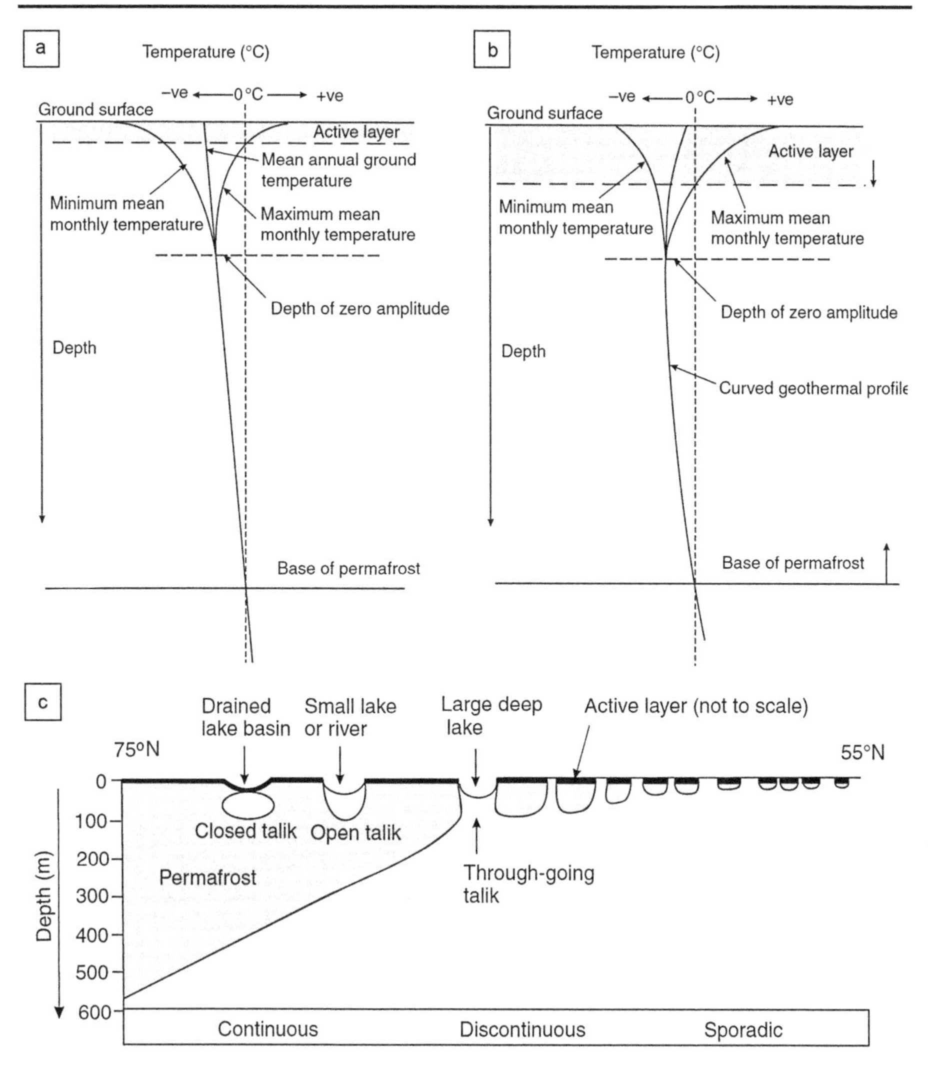

Fig. 4 Basic permafrost characteristics. (**a**) and (**b**) Thermal profiles in permafrost (from Harris 2004). (**a**) In equilibrium, positive temperatures induce thawing while negative ones induce freezing of the active layer, (**b**) during thermal adjustment to surface warming (permafrost degradation). (**c**) Idealized sketch of permafrost characteristics along a north–south transect (from Harris 2004, after Lewkowicz 1989)

Landscapes of the Cryosphere The terrestrial cryosphere exhibits a variety of geomorphic features that are morphologically analogous to landforms on Mars. Their unambiguous identification on Mars, however, can be problematic. It is well known that different processes can produce very similar results (equifinality, e.g. Slaymaker 2004), and our ability to distinguish between the formative processes on Mars is further hindered by the limited data availability and the ubiquitous dust cover of the Martian surface. A promising approach to mitigate this dilemma is the analysis of landscapes, not landforms. The identification of an assemblage of landforms, which is typical for a specific landscape, increases the confidence in an interpretation that might be questionable if the surface feature were investigated in isolation. Suitable terrestrial landscape analogues exist for different aspects of Martian cryospheric

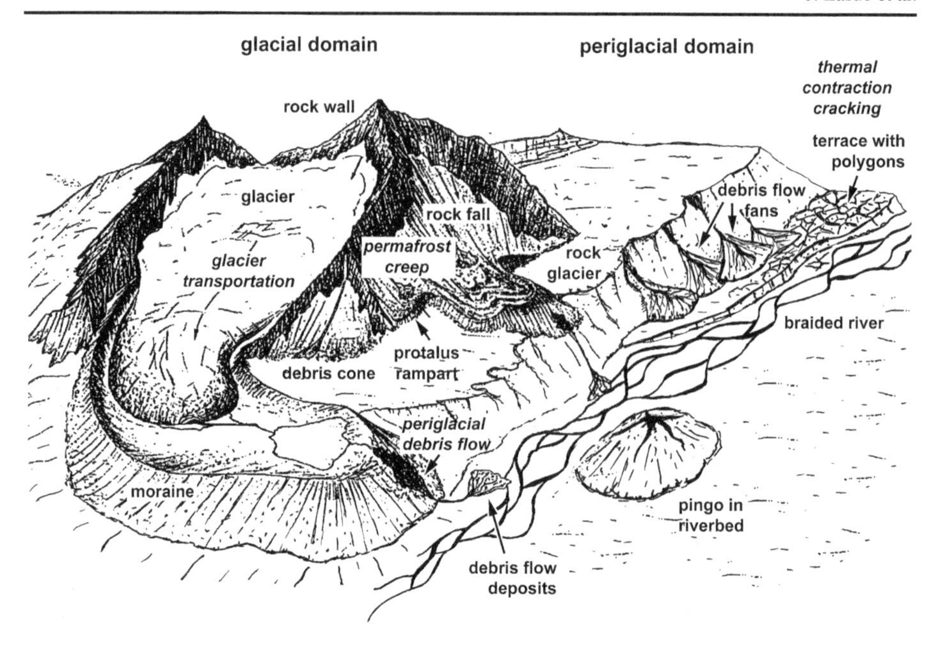

Fig. 5 Idealized assemblage of landscape elements in high arctic alpine permafrost terrain. This qualitative sketch is not meant to represent a real situation anywhere, but to illustrate the spatial arrangement of the landforms that is observed in such environment (e.g. on Svalbard). Morphologically similar landforms have been observed in Martian mid-latitude craters, often in comparably close spatial proximity (from Hauber et al. 2011a, originally modified from Haeberli 1985, Fig. 1)

studies. For example, landscapes of lowland permafrost such as in northern Siberia (Ulrich et al. 2010) or on the Tuktoyaktuk peninsula (Alaska, Soare et al. 2011) display thermal contraction polygons, thermokarst phenomena and hydrostatic pingos that have morphological analogues in the Martian mid-latitude plains (e.g. Utopia Planitia). On the other hand, alpine or mountain permafrost terrain (e.g. in the High Arctic of Svalbard) exhibits a characteristic landform assemblage (e.g. rock glaciers and other frost creep features; patterned ground, open-system pingos; Fig. 5) that shares many morphological similarities with Martian mid-latitude regions with significant topographic relief (Hauber et al. 2011b). Although the above mentioned examples are very useful as process-analogues, the climates of Siberia, Alaska or Svalbard are, despite their arctic character, much warmer and wetter than the environmental conditions on present-day Mars. The hyper-arid cold deserts of continental Antarctica (e.g. Bockheim and Hall 2002) have long been considered to be the best terrestrial climate-analogue to Mars, providing insights into the geomorphology and biology of extremely cold and dry environments (e.g. Anderson et al. 1972; Marchant and Head 2007; Doran et al. 2010). A particularly interesting feature of continental Antarctica is "dry permafrost" (e.g. Bockheim and Tarnocai 1998), which tends to blur the distinction between the active layer and the permafrost and might be representative for current conditions at many locations on Mars.

2.2 Geomorphological Evidence for Current and Recent Water Ice on Mars

Pioneering studies using Viking imagery interpreted several landforms as due to processes related to ice. More recent high resolution imagery has enabled scientists to better understand these landforms, confirming the presence of ice for many of them, contradicting it for others. The summary below groups these landforms by type of processes involved, as far as these are understood at the time of this review.

2.2.1 Polar Caps

The polar terrains are among the youngest surfaces to be found on Mars. In the north, their extreme youth is supported by high-resolution images of the surface that fail to reveal any craters with diameters >300 m within the $\sim 10^6$ km^2 covered by the deposits (Herkenhoff et al. 1997). This observation argues for an age of the exposed surface that is less than $\sim 10^5$ years—a result that may reflect either the deposition of new material or the continuous reworking of older material at the pole (Toon et al. 1980; Herkenhoff and Plaut 2000). However, there is evidence that the age of the geologically distinct north polar basal unit, which underlies the north polar layered deposits, may be much older—perhaps as much as 1 Gy (Tanaka et al. 2008).

Based on a detailed examination of the MOLA altimetry data and MOC high-resolution images of the south polar region, Koutnik et al. (2002) identified ~ 36 craters with diameters >800 m, as well as an additional 12 whose identity has yet to be confirmed. These results indicate a surface age of 30–100 My, depending on the recent Martian cratering flux (Herkenhoff and Plaut 2000; Koutnik et al. 2002; Kolb and Tanaka 2006). These results imply a minimum difference in surface age between the north and south polar layered deposits of 2–3 orders of magnitude.

The recent surfacing (deposition or erosion) rates for the polar layered deposits that are implied by the observed crater densities are ~ 5–10 m/My in the south and ~ 1200 m/My in the north (Herkenhoff and Plaut 2000; Koutnik et al. 2002). These results are at odds with the concept that the processes that form the layers are hemispherically symmetric, and suggest that the ~ 6 km elevation difference between the polar regions, combined with Martian orbit/axial variations that are thought to produce global climate variations on $\sim 10^5$–10^8 years timescales (Laskar et al. 2002; Levrard et al. 2007) have resulted in major differences in polar deposition and erosion between the poles.

Geomorphological assessments, and direct radar measurements have shown that the polar caps are composed of nearly pure water ice. They are covered with a seasonal veneer of CO_2 and include at least one massive deposit of CO_2 ice embedded near the central portion of the South Polar Layered Deposit (SPLD). However, this CO_2 still comprises only a small fraction of the total ice forming the cap (Phillips et al. 2011). From both altimetric and radar sounding data, the total volume of the North Polar Layered Deposit (NPLD) has been estimated to be between 1.2 and 1.7×10^6 km^3 (from 8.3 to 11.7 m GEL, Zuber et al. 1998; Selvans et al. 2010). The volume of the SPDL has also been estimated at between 1.6 to 2.3×10^6 km^3 (from 11 to 15.9 m GEL) based on both stereographic Viking imaging (Schenk and Moore 2000) and observations by the MARSIS deep-sounding radar (Plaut et al. 2007).

2.2.2 Ice Deformation Related Landforms

Mid-Latitude Glaciers As summarized by Squyres and Carr (1986), Viking-era investigations identified a wide range of geomorphic evidence indicative of the presence of ice at mid-latitudes. Features called lobate debris aprons (LDA), lineated valley fill and concentric crater fills were observed in orbital images and attributed to the viscous flow of ice (Carr and Schaber 1977; Lucchitta 1981; Squyres 1978, 1989; Jankowsky and Squyres 1992; Carr 1996; Mangold 2003a; Kargel 2004; Head et al. 2005). In the northern hemisphere, lobate debris aprons are most evident at the foot of 1–2 km high scarps in the

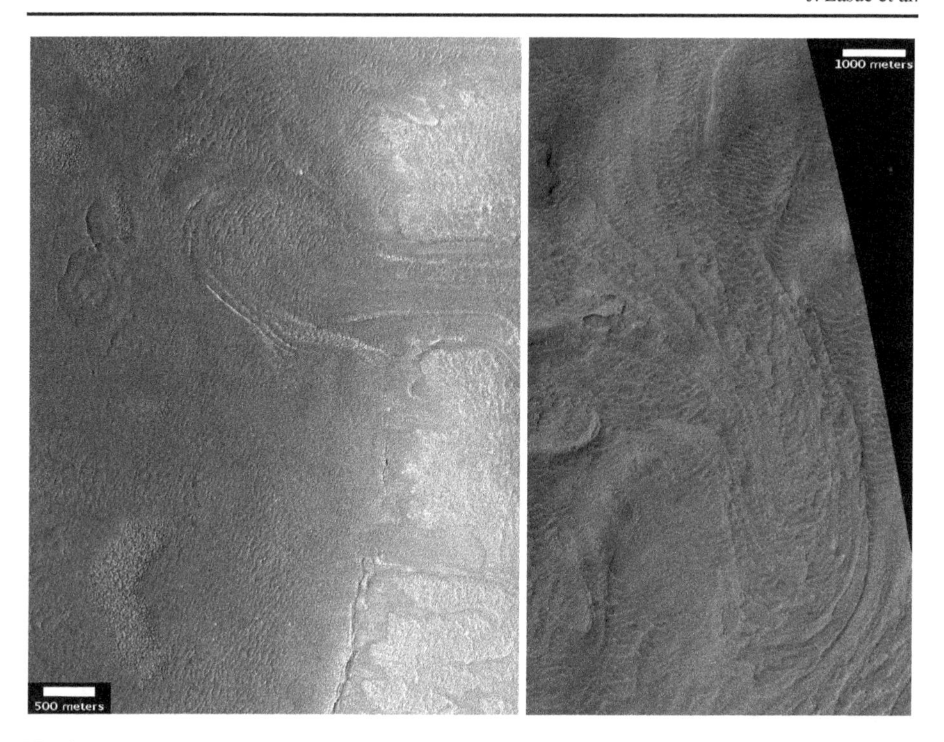

Fig. 6 Lobate Debris Apron located in the Deuteronilus Mensae region along the highland/lowland boundary, in the Northern hemisphere of Mars. The mesa top and the Lobate Debris Apron extending from its base appear covered with ice-rich mantling materials thought to have been deposited around 10 Ma ago during a period of high obliquity. Image taken with the HiRISE instrument on-board MRO (43.6°N, 28.6°E, ESP_016959_2240, NASA/JPL/University of Arizona). Lobate Debris Apron located East of Hellas Planitia. Image taken with the HiRISE instrument on-board MRO (41.7°S, 108.6°E, PSP_005696_1380, NASA/JPL/University of Arizona)

regions of Deuteronilus and Protonilus Mensae while in the southern hemisphere, they are most abundant in the region east of Hellas Planitia. Lineated valley fills and concentric crater fills occur within these same geographic regions and exhibit similar evidence of rheologic deformation—the former typically occurring within valleys and the latter in crater floors. The systematic presence of deformational features (lineations, flow lines, etc.) in these thick landforms (100 m to 1 km) has led scientists to propose that the viscous flow of ice was involved in their formation. Other, more recent, datasets—such as laser altimetry and radar sounding profiles—have revealed that the LDA possess parabolic profiles consistent with the shape and basal stress of terrestrial glaciers, which strongly supports an origin by viscous deformation (Mangold and Allemand 2001). Dielectric values for the LDAs, deduced from radar sounding data, indicate that they contain a high volumetric content of ice (80 %, Holt et al. 2008; Plaut et al. 2009). Although the potential origin of this ice has been debated for a long time (i.e., whether it was derived from the atmosphere or subsurface), recent studies suggest that the predominant process of formation was by atmospheric precipitation as snow (Head et al. 2005; Forget et al. 2006), which means that the LDA can be correctly described as mid-latitude glaciers (Fig. 6).

Lobate shapes interpreted as residual moraines from tropical glaciers can be observed at the foot of the Tharsis volcanoes (e.g. Lucchitta 1981; Head et al. 2005). The material

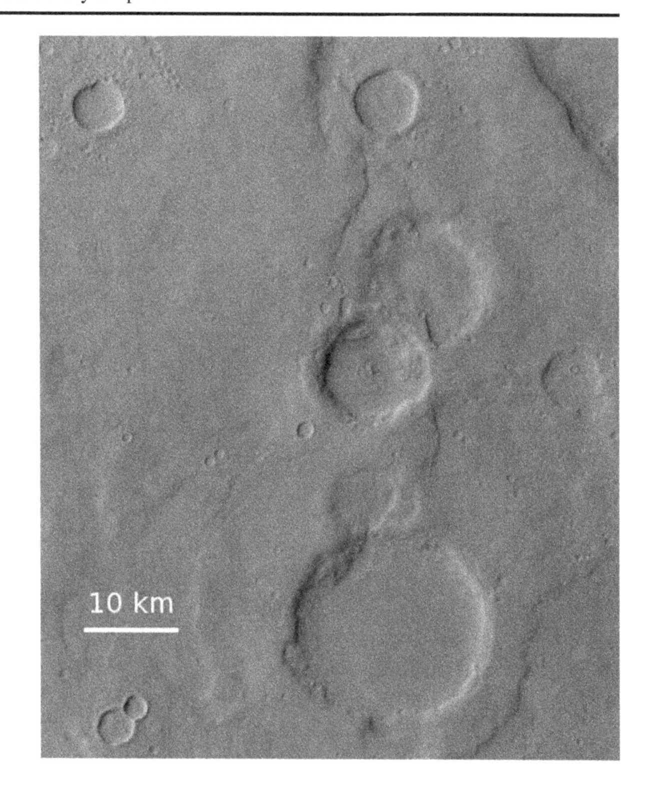

Fig. 7 Example of craters in softened terrains (33.4°N, 313.2°W, adapted from HRSC image H2963_0000_ND3, Neukum et al. 2004; Jauman et al. 2007)

left behind is similar to the sublimation tills observed in Antarctica (Marchant et al. 2002). These areas are consistent with the regions of snowfall predicted by Global Circulation Models (GCM) at times of high obliquity (Forget et al. 2006). Other remnants of lobate aprons exist in regions located at about 20° North and South latitude that likely correspond to similar processes of ice accumulation, but that are now completely desiccated (Hauber et al. 2008). Overall, little or no evidence of surface ice is found near the equator compared to mid-latitude regions where ice is massively present close to the surface, usually protected from sublimation by a dusty lag.

Deformation of Ice in the Permafrost While LDAs appear predominantly related to atmospheric processes, the presence of subsurface ice in the cryosphere and its rheologic deformation under the influence of gravity has also been suggested as the explanation behind the origin of the muted or 'softened' morphology of mid-latitude terrains—especially craters (Squyres 1989). The latter correspond to low-relief (low depth-to-diameter ratio) craters with rounded rims (Fig. 7). The softened appearance of these craters has been interpreted as being due to either deformation of the ice-rich crater walls and floor (Squyres 1989; Jankowsky and Squyres 1992) or infilling by eolian/ice deposits (Zimbelman et al. 1989). Topographic profiles (using photoclinometry) have been analyzed to estimate the depth under which this creep occurred, giving a rough estimate of 1 km (Jankowsky and Squyres 1992).

Similarly, the lobate form of the ejecta seen around impact craters has been linked to melt-water interactions during deposition (Wohletz and Sheridan 1983; Mouginis-Mark 1987). Surveys of the morphologies of the craters have shown that ejecta morphologies can be linked to impacts into material with varying proportions of volatiles, following latitude-depth relationships for the presence of ice (Barlow and Bradley 1990). Furthermore, comparison with fluid flow experiments shows that the viscosity variations with depth are compatible with the estimate of a 1 km thick volatile rich layer located above a less porous material (Baratoux et al. 2002). The presence of ice at such depth still needs to be demonstrated as it has not yet been detected by instruments such as radar.

Recent results obtained by automatic crater detection based on the Mars Orbiter Laser Altimeter (MOLA, Smith et al. 2001) topographic data show that the depth over diameter ratio of craters indicates a global latitudinal variation coherent with the presence of a global cryosphere at least 1 km deep at 40° and higher latitudes (Stepinski and Urbach 2009; Stepinski 2010).

Moreover, when deformation takes place in craters, the poleward-facing slopes have a colder surface temperature than the equatorward-facing slopes, a situation that is reversed during times of high obliquity. Overall, a rheologic signature of this thermal asymmetry should be visible in the NS topography of impact craters. Parsons and Nimmo (2009) used MOLA data to test this expectation, and did not observe it, concluding that, if ice is present, its volumetric contribution must be less than the amount required to initiate creep. However, a similar (but better resolution) study, based on Digital Elevation Models (DEM) generated from images taken by the High-Resolution Stereo Camera on Mars Express, did find a topographic asymmetry that can be attributed to the deformation of ice-rich terrain (Conway et al. 2011). This observation provides support for the presence of a significant amount of ice in the subsurface down to a depth of at least 100 m, at mid-latitudes, where craters displaying this asymmetry are found.

Experimental tests were done to determine the minimum amount of water ice required for ice-rich sediments to flow. The threshold for creep to initiate depends on the size and types of particles. Using Martian permafrost analogs with different particle sizes, it has been found that a minimum water ice content of ~28 % is necessary for ground ice to creep (Mangold et al. 2002). This threshold value has been confirmed recently with similar deformation tests (Durham et al. 2009). Jankowsky and Squyres (1993) determined that softened terrain covers $\sim 3 \times 10^7$ km^2 of the Martian surface, while the morphology of the terrains suggests a viscous deformation over at least a 1 km deep permafrost. Assuming a mean ice content of 25 %, they calculated a minimum amount of 55 m of GEL necessary to produce these features.

However, terrain softening is generally not evident throughout most of the Northern plains because there is little topographic relief to relax. This does not mean that ice is not present, as shown by the distribution of polygonal landforms and gullies. Assuming that the available pores are saturated with ice poleward of 30° latitude at a global scale, this ice could amount to up to 125 m GEL distributed over a 1 km thick layer (Jankowsky and Squyres 1993). If porosity is saturated by ice from top to bottom, the estimation is substantially increased because ice content would vary from 40 % by volume at the surface to 28 % at 1 km depth. The calculations then give ~80 m of GEL, taking into account regions where terrain softening is observed, and about 200 m taking into account all regions poleward of 35° (Mangold et al. 2002). These estimations are calculated only from ice-deformation related evidence, but much more ice could be stored below the 1 km deep and 25 % ice content boundaries as will be discussed in Sect. 4.

Ice Thermal Contraction Related Landforms The occurrence of polygons on Mars, similar to those found on Earth in periglacial regions has been the subject of debates for three decades. Large polygonal systems were identified on Viking images of the Northern plains (Pechmann 1980; Lucchitta 1983). These polygons, 2–10 km across with bounding trough

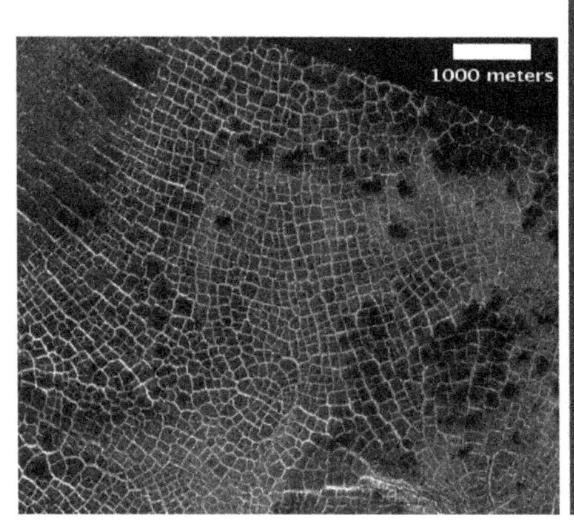

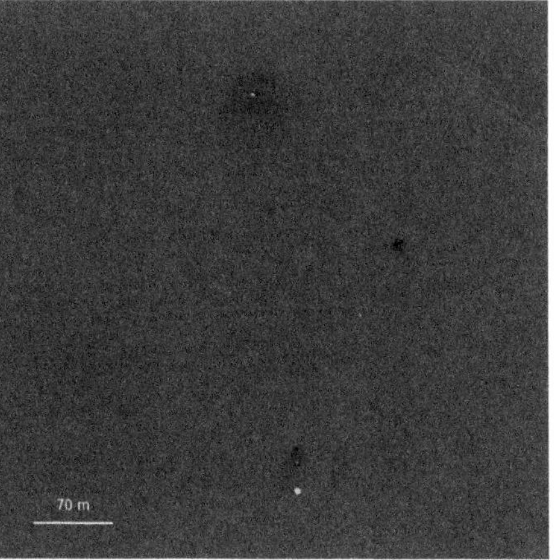

Fig. 8 Example of polygonal terrains at different scales. (**a**) Polygonal patterned grounds on the floor of a crater, smaller polygons can be distinguished within the larger polygonal structure (size about 200 m). (69.8°N, 64.9°E, adapted from HiRise image ESP_016641_2500). (**b**) Polygonal patterned grounds with a size of the order of 10 meters (Mellon et al. 2009b), this stunning picture represents the landing site of Phoenix lander where the parachute, heat shield impact and lander with deployed solar arrays are visible. (68.2°N, 125.7°W, adapted from HiRise image PSP_008591_2485)

widths of 200 to 800 m, are too large to be analogues of terrestrial periglacial polygons and more likely originate from tectonic stress, possibly related to the loading or unloading associated with a past ocean, see Sect. 2.3.2 (Hiesinger and Head 2000). High resolution images by the Mars Observer Camera (MOC) have revealed much smaller polygons, similar in size (ranging from several to several 10's of m) to those associated with terrestrial patterned ground (Baker 2001; Malin and Edgett 2001; Masson et al. 2001; Kuzmin et al. 2002; Kuzmin and Zabalueva 2003; Yoshikawa 2002, 2003; Leverington 2003; Mangold et al. 2003, 2004; Mangold 2005). Polygons found in the Utopia and Elysium regions are typically several hundreds of meters in size and could originate from the thermal cracking of Late Amazonian sediments or by thawing of ice-rich ground (Seibert and Kargel 2001). At high latitudes, polygonal shapes are ubiquitous in both hemispheres (Mangold et al. 2004; Mangold 2005; Levy et al. 2009a, 2009b). Recent HiRISE data have shown that much smaller polygons exist (<20 m) in regions of lower latitudes, roughly down to 35° of latitude in both hemispheres (Levy et al. 2009a). Figure 8 illustrates some examples of Martian polygonal terrains.

Martian polygonal landforms display bounding cracks that are similar to those found in terrestrial periglacial environments, which result from the thermal contraction of the ground in response to the propagation of the cold thermal wave during winter (e.g. French 1996). Models show that the same effect can occur under current Martian conditions (Mellon 1997), where thermal contraction due to seasonal temperature variations can produce polygons ranging in size from 15 to 300 m. Therefore, water ice must be close enough to the surface for thermal contraction to occur in response to seasonal temperature variations (i.e., roughly 1–2 meters), but to explain the width of polygons (>100–200 m) ice must extend down to a depth equivalent to $\sim 1/3$ the size of the individual polygons—or to a minimum of several tens of meters. Lastly, high-latitude polygons are sparsely cratered indicating that they are very recent, if not currently active (Mangold 2005).

The geographic location of polygonal terrains on Mars is known to be latitudinally constrained (e.g. Levy et al. 2009a; Saraiva et al. 2009) and to follow the subsurface water

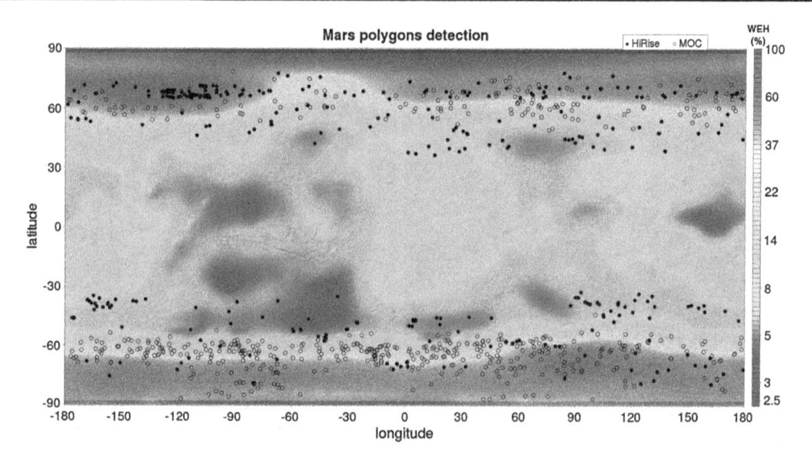

Fig. 9 Superposition map of MOLA shaded Mars altimetric data, Mars Odyssey Neutron Spectrometer (MONS) Hydrogen equivalent water content of the near subsurface and the position of polygons detected amongst HiRise (*black circles*) and MOC (*open circles*) data (adapted from Smith et al. 2001; Maurice et al. 2011; Levy et al. 2009a, 2009b; Mangold et al. 2004)

content detected by Mars Odyssey Gamma-Ray and Neutron Spectrometers (e.g. Mangold et al. 2004). Polygonal terrains typically cluster in regions where the Water Equivalent Hydrogen content (WEH) of the subsurface is 15 wt.% or higher as shown on Fig. 9. The number and size of polygonal terrains existing poleward of 30° latitude is another indication of the depth of ice rich soil on Mars.

2.2.3 Ice Sublimation Related Landforms

Sublimation on Glacial and Periglacial Landforms High-resolution Viking images (i.e., ~30–40 m/pixel) revealed the presence of pits and closed depressions on mid-latitude lobate debris aprons that Squyres (1989) tentatively interpreted as being due to sublimation of glacial ice. Better resolution images later confirmed this interpretation (Mangold 2003a). At high resolution, mid-latitude glaciers exhibit a variety of dissected morphologies that are consistent with the progressive degradation of the debris apron by the sublimation of ice (Malin and Edgett 2001; Mangold 2003a; Crown et al. 2005; Kress and Head 2008; Kress et al. 2010; Mangold 2011). Of particular interest is the presence of fractures indicative of glacial flow, which can have both a longitudinal and curved orthogonal pattern. These fractures form a preferential zone for ice sublimation. When ice grains sublime from a fracture, nearby solid grains become unstable and may be ejected (Mangold 2003a; Levy et al. 2009a). In such a removal process, ice grains close to the fracture are more rapidly exposed to the atmosphere and can sublime more rapidly than on a flat surface. This effect likely explains why the surfaces of landforms related to ice deformation display such a variety of morphologies. With this loss of ice, a residual lag forms that may limit further sublimation.

Polygonal landforms are also affected by sublimation. Polygons display differently evolved shapes ranging from (1) fresh polygons at latitudes >70° where cracks are thin, and polygonal structure well defined, to (2) slightly degraded polygons at latitudes 60–70°, where a degradation of the crack network is visible but polygons are still well identifiable, and (3) strongly degraded polygons at latitudes roughly below 55–60° where polygonal structure is modified and more difficult to identify (Mangold 2005). The progressive degradation of polygon morphology from the highest to the lower latitudes may be explained by climatic variations in the latitudinal stability of ground ice and the role of fractures in amplifying the effect of sublimation, which is higher at low latitudes.

Sublimation of a Mid-Latitude Mantle Based on an analysis of Mariner 9 orbital imagery, Soderblom et al. (1973) noted that, poleward of 30° latitude in both hemispheres, the Martian surface is covered by a mantle of debris that has undergone differential erosion—which has left it well preserved in some areas and with extensive pits, knobs, and remnant mesas in others. They inferred that these mantles consisted of eolian debris, derived from the polar layered deposits, that was redistributed to lower latitudes as the result of climatically varying insolation and wind. This mantle was more fully resolved and characterized by MOC (Mustard et al. 2001). The thickness of this dissected unit, often referred to as Latitude-Dependent Mantle (LDM) appears to range from meters to 10s of meters. Fresh impact craters of any size are very rare suggesting ages less than 1 My. Detailed mapping of the mantles shows that they are approximately symmetrical, concentrated in two latitude bands: 30° to 70°N and 25° to 65°S (Mustard et al. 2001). These observations have reinforced the belief that these eolian mantles are derived from the PLD at times of high obliquity, when both ice and dust are removed from the polar deposits and atmospherically redistributed to lower latitudes.

The eroded character of many areas within the LDM suggests a volume change or differential removal of material. The preference for erosional morphologies to occur on slopes that experience high solar insolation over at least some part of the obliquity cycle, supports the belief that volatiles constitute a significant volume fraction of the deposits—having been emplaced within the mantling dust and that subsequent ice sublimation created the observed erosional landforms (Malin and Edgett 2001; Mustard et al. 2001; Head et al. 2003a; Schon et al. 2009; Mangold 2011).

Recently, new impacts 10–100 m in diameter have exposed near-surface ice in the Martian mid-latitudes (Byrne et al. 2009). This ice is observed to slowly fade over timescales of months. Water-ice stability models, based on this discovery, indicate that such ice can persist in the subsurface at 45° latitude if it is buried beneath a dry layer ~15–50 cm thick (Dundas and Byrne 2010), confirming interpretations made for pitted landforms at these latitudes. One possible explanation is that the shallow subsurface is made of atmospheric ice deposited with a low dust content forming an excess ice layer, which is different from regolith-supported pore ice.

Utopia Planitia and Malea/Peneus Planum (south of Hellas Planitia) are mid-latitude regions where ice sublimation landforms are typically found (Morgenstern et al. 2007; Lefort et al. 2009, 2010; Zanetti et al. 2010). For example, scalloped terrains, shown in Fig. 10, are found in these two regions and are rarely observed elsewhere. Scalloped depressions are rimless, shallow and ovoid in form, ranging from circular to elongate with a morphology that is independent of altitude. The typical features of a scalloped depression are a 15 to 30° steep poleward-facing scarp and opposite a gentle equatorward-facing slope, sometimes almost flat, with a slope of typically 2° (Lefort et al. 2009). It is hypothesized that scalloped terrain formation is initiated by slight hummocks or depressions that have relatively high near-surface temperatures on their equatorward-facing slopes, which leads in turn to enhanced sublimation of ground ice on these slopes (Lefort et al. 2009, 2010). Over time this process deepens and extends the scallop, mainly by erosion of the equatorward-facing slope. Alternatively, it has been proposed that the poleward-facing slopes degrade in response to the higher insolation that occurs at times of high obliquity (Séjourné et al. 2011).

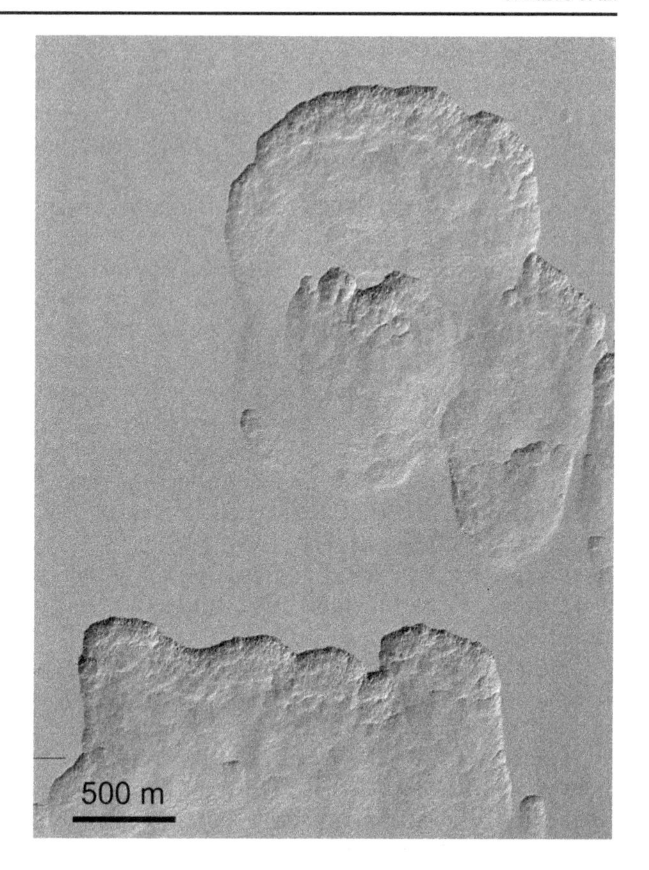

Fig. 10 Scalloped terrains near Amphititres Patera (Southern Hemisphere). HiRISE image PSP_005698_1225 (57° S, 51.3° E). *Light* from the top left. *Arcuate depressions* indicating sublimation of a volatile-rich material present in the first tens of meters

The distribution of these landforms in Utopia Planitia and Malea Planum suggest that the thickness of the ice-rich mantle is both greater and has undergone more recent degradation than other areas at the same latitude (Lefort et al. 2009, 2010; Zanetti et al. 2010; Séjourné et al. 2011).

There is extensive evidence that impact crater morphology has also been influenced by the effects of sublimation. For example, the ejecta from craters formed after the emplacement of the LDM may armor this volatile-rich unit, protecting the buried ground ice from sublimation. This effect is spectacular in the case of pedestal craters that are observed in the northern regions of the Utopia and Chryse Planitia, which have large excess ejecta volumes when compared to the cavity volume (Meresse et al. 2006). Such craters sometimes exhibit pits at the location of the ejecta boundary, suggesting that enhanced sublimation on the scarps of the pedestals has contributed to the shaping of these landforms (Kadish et al. 2009). Thus, the combination of sublimation and wind deflation appears responsible for the apparent excess ejecta volumes of pedestal craters, resulting in craters with pedestal heights of between 40 and 50 meters, near the SPLD (Kadish et al. 2010; Kadish and Head 2011).

Overall, the distribution and thickness of the LDM, combined with the volumetric contribution of embedded ice inferred from presence of pits and other sublimation-related landforms, suggests an associated reservoir of water ice equivalent to a few meters GEL.

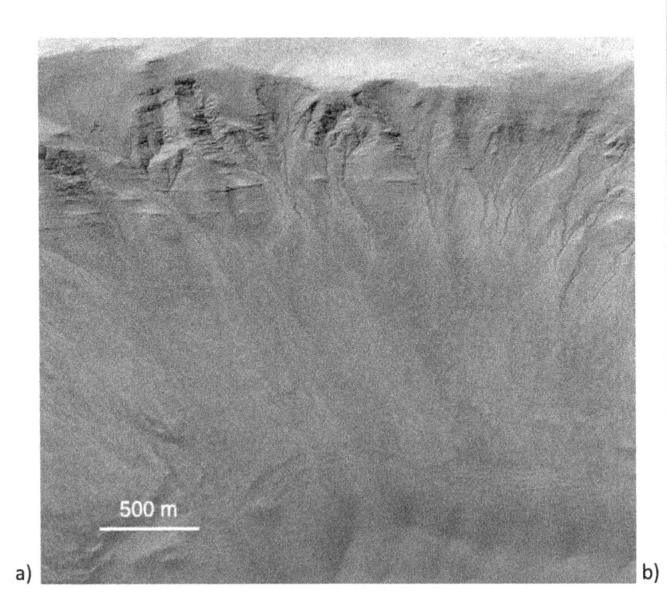

Fig. 11 (**a**) Examples of gully landforms located on the rim of a crater in Newton crater located at 41.8° S, 158.0° W. (MGS MOC Release No. MOC2-317 NASA/JPL/MSSS). (**b**) Close up of a gully showing clearly the alcove, channel and fan parts of the landform (gully located on the internal part of the wall of eastern Newton crater 40.1° S, 155.3° W) (adapted from Head et al. 2008; Goldspiel and Squyres 2011)

2.2.4 Ice Melting-Related Landforms

Recent Gullies Under current climatic conditions, liquid water is unstable at the Martian surface (Ingersoll 1970; Haberle et al. 2001). However, transient liquid water (i.e., water derived from the melting of surface or subsurface ice, that then undergoes rapid evaporation or is preserved under a thickening cover of ice, Goldspiel and Squyres 2011) has been proposed as a possible explanation for several landforms. Of particular interest are the geologically recent Martian gullies, discovered by the Mars Orbiter Camera (MOC) (Malin and Edgett 2000a). These features exhibit characteristic morphologies with an alcove (a physiographic depression located high on a steep slope), an incised and typically sinuous channel leading down from the alcove, and, finally, an apron of deposited material, shown in Fig. 11. The similarity to terrestrial features formed by flowing water or water-rich slurries led Malin and Edgett (2000a) to suggest that the Martian gullies, too, were formed by the action of water, seeping from shallow underground aquifers.

Gullies form only on steep slopes, primarily on impact crater walls, but also on the banks of fluvial channels, the sides of sand dunes and polar pits. Observations that 10–20 % of gullies occur on isolated knobs, hills and the central peaks of craters, are difficult to reconcile with formation from groundwater sources, as any aquifer would be isolated and small. The same conclusion is drawn from observations of gullies starting near the top of impact crater walls (Costard et al. 2002; Balme et al. 2006). They are also fairly common in pitted terrain in the far south. This has led further studies to suggest processes related to surficial activity including snowmelt (Costard et al. 2002; Christensen 2003), the action of CO_2-based debris flow (Hoffman 2000, Musselwhite et al. 2001), granular avalanches or mass wasting of CO_2 frost (Ishii and Sasaki 2004; Hansen et al. 2011), fluidization of a regolith bed by CO_2 frost sublimation (Cedillo-Flores et al. 2011) or dry mass wasting (Treiman 2001)—although the failure of dry materials, such as eolian sediments, is unlikely to produce the sinuous shapes that are characteristic of many gullies (Mangold et al. 2010).

Martian gullies are principally found at mid-latitudes, between 30° and 50° in both hemispheres (Costard et al. 2002; Heldmann and Mellon 2004; Kneissl et al. 2010; Malin et al. 2010), but are also common at latitudes >70°, near the South Pole (Balme et al. 2006). Gullies are preferentially found on poleward-facing slopes, especially between 30°S and 40°S, but examples can be found with all orientations. The range of gully morphology and occurrence at different latitudes, altitudes, and orientations, suggests that they have not formed by a single process.

Indeed, farther south, gullies show little preference for orientation. This latitudinal dependence of orientation and abundance suggests both slope distribution and climate are key controls on gully formation. These observations reinforce the role of insolation and atmospheric conditions on gully formation. Thus, gully formation depends on (1) the presence of steep slopes (Reiss et al. 2009), (2) the stability of water and/or CO_2 at the surface and near-surface, and (3) changes in insolation and/or atmospheric pressure, humidity and temperature due to variations in obliquity (Balme et al. 2006). The most recent work based on gully variations tracked over time strongly indicates that both CO_2 (Dienega et al. 2010) and water (Reiss et al. 2010; Schon and Head 2011) generate gullies, the activity of which depends on seasonal and climatic variations.

Like the polygonal terrains, the geographic distribution of gullies is known to be latitudinally constrained, suggesting a relationship to the presence of shallow sources of water (e.g. Heldmann and Mellon 2004; Balme et al. 2006; Levy et al. 2009a, 2009b; Kneissl et al. 2010). Gully locations also correlate with the subsurface hydrogen content detected by Mars Odyssey Gamma-Ray and Neutron Spectrometers. They typically cluster in high-slope regions where the Water Equivalent Hydrogen content (WEH) of the subsurface is 5 wt.% or higher as shown on Fig. 12. It has been observed that CO_2 frost deposition does not occur at mid-latitudes where it should theoretically be possible. This absence of CO_2 frost deposition can be explained if higher thermal inertia near-surface water ice is present (Vincendon et al. 2010). Such local ice is far below the resolution of the neutron data but might still be sufficient to serve as a source of water for the formation of the mid-latitude gullies.

Comparing the latitudinal distribution of both gullies and polygons (as shown in Fig. 13) indicates a clear transition in their respective distributions at approximately 60° latitude, with gullies being more abundant below this latitude and polygons more abundant above. This may reflect the latitudinal transition from rougher terrains, where gullies form easily, to smoother terrain where polygons are more prominent (Kreslavsky and Head 2000). It could also be linked to the latitudinally-varying depth of subsurface ice. However, a high number of gullies are detected on hillslopes at latitudes as high as 70° South, where temperatures are too low for liquid water to exist under current obliquity conditions. Such gullies could

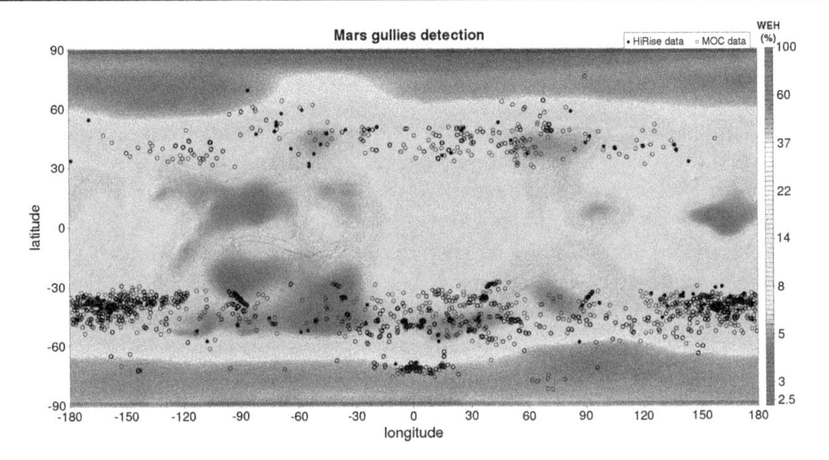

Fig. 12 Superposition map of MOLA shaded Mars altimetric data, MONS Hydrogen equivalent water content of the near subsurface and the position of gullies features detected amongst HiRise (*black circles*) and MOC (*open circles*) data (adapted from Smith et al. 2001; Maurice et al. 2011; Levy et al. 2009a, 2009b; Mangold et al. 2004; Malin et al. 2010)

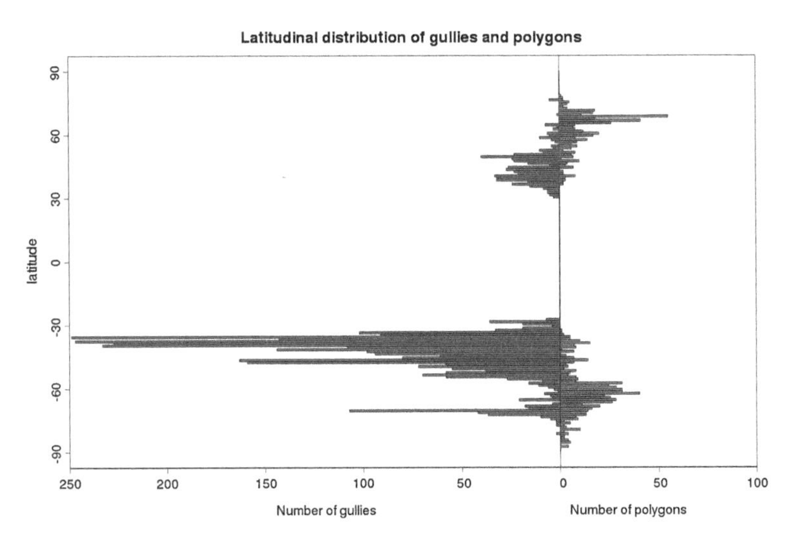

Fig. 13 Latitudinal distribution of gullies (*blue* on the left) and polygons (*red* on the right) detected by HiRise and MOC with a 1° step

be formed by the fluidization of eolian sediment by the sublimation of CO_2 and, thus, may be independent of the presence of subsurface water (e.g. Cedillo-Flores et al. 2011).

Freeze-Thaw Cycle Related Landforms Sorted stone circles and stripes are features found in terrestrial periglacial environments that are formed by repeated cycles of seasonal freeze–thaw. The scale of these features is typically on the order of one to several meters—making them difficult to identify even at HiRISE resolution. Parallel sorted stone stripes on hillslopes were tentatively identified in MOC images (Mangold 2005), and convincing examples have been imaged by HiRISE (Johnsson et al. 2010), although the occurrence of this landform does not appear to be widespread. Sorted stone circles and polygonal ground are rarely found close to the equator, although they are present in high resolution images of the Cerberus plains and Marte Vallis channel (Page 2007; Balme et al. 2009).

Solifluction lobes are formed by the deformation of ice-rich soil that undergoes periodic freeze-thaw on local slopes. Solifluction lobes result from the creep of surface material in response to the reduced frictional strength of the soil due to the presence of seasonal melt water in the soil pores. Solifluction is limited to the thickness of the active layer, which, on Earth, is typically on the order of 1 m. High resolution images of the Martian surface have revealed the presence of lobes very similar to terrestrial solifluction lobes (Mangold 2005; Gallagher et al. 2011; Johnsson et al. 2012), with lengths of a few hundred meters, potential thickness of a few meters, and slopes of 5° to 15°.

The possibility of pingo detection on Mars is of particular interest because of the associated implications for liquid water, since they are formed from ground upheaval due to the freezing of injected liquid water, as discussed in Sect. 2.1.3. Pingos develop surface fractures due to extension of the frozen ground over the ice core, which help to identify mounds from orbit. While the putative identification of these landforms was limited by the resolution at MOC scale (Mangold 2003b), HiRISE images provide convincing evidence for local pingos (Dundas and McEwen 2010). These features are confined to the Martian mid-latitudes, in the bands where gullies are also most common.

Evidence of purely dry erosional processes, such as dunes and slope failure, are globally distributed. However, the distribution of landforms potentially indicative of periglacial processes, such as slope stripes, pingos and solifluction lobes, are found at mid- to high-latitudes where they may have been formed by freeze-thaw processes associated with past episodes of high obliquity (i.e., 35° or more, Mangold 2005). Such past episodes of freeze–thaw cycles may have been enhanced in specific environments such as crater interiors or local slopes, explaining their scarcity anywhere else. Liquid water generated by such processes is fundamental to the understanding of climatic variations and has strong exobiological implications. Nevertheless, the water content represented by these landforms is not a significant contributor to the planetary inventory when considered at the global scale. Current views on their formation suggest that they originate from the melting of surface and near-surface ice, over a maximum depth of just a few meters.

2.3 Geomorphological Evidence for Water in the Past

While many ice-related landforms are very recent, i.e. less than 10 My, other landforms are related to more ancient periods. The following discussion describes the types of evidence and related water content inferred, going from the oldest to the youngest water-related features of the planet.

2.3.1 Valley Networks

Since the Mariner 9 mission in 1972, networks of small valleys have been observed, dissecting the planet's most heavily cratered terrain found predominantly in the southern hemisphere (Sharp and Malin 1975; Carr and Clow 1981; Carr 1996). The valleys are arranged in a branching pattern similar to that of fluvial valley networks on Earth. The largest valleys have widths of a few kilometers and a depth of hundreds of meters (Williams and Phillips 2001). Inferred discharge rates have been estimated where inner channels are visible in these valleys at few hundreds to as much as 5000 $m^3\ s^{-1}$ (Irwin et al. 2005a), values consistent with fluvial activity on Earth and far below the estimated discharge rates of the outflow channels discussed in Sect. 2.3.3. The valley networks are found almost exclusively in Noachian age (>3.7 Gy) terrain appearing to have largely formed near the end of the Noachian or just at the beginning of the Hesperian (Fassett and Head 2008). Several dendritic fluvial valley networks date from Late Hesperian regions, suggesting that liquid water was still present at the surface, at least episodically, during this period (e.g. Mangold et al. 2004; Bouley et al. 2010). Due to the small number of absolute ages available for Hesperian-aged fluvial features, it is not possible to place quantitative constraints on this episodicity. Nevertheless, the size of the valleys and amplitude of this activity appear to decline progressively with time (Ansan et al. 2008).

Because of their resemblance to terrestrial fluvial valley networks, they have been interpreted as having formed mainly by surface runoff associated with rainfall (Carr 1981; Craddock and Howard 2002; Mangold et al. 2004; Howard et al. 2005; Irwin et al. 2005b; Ansan and Mangold 2006). However, the relatively low drainage density and the low degree of valley organization suggest precipitation levels comparable to those of the most hyper-arid regions of Earth (e.g. Carr and Chuang 1997; Luo and Stepinski 2009; Stepinski and Luo 2010). Alternatively, it has been suggested that groundwater sapping, triggered by geothermal or hydrothermal heating, was predominant (e.g. Sharp and Malin 1975; Pieri 1980; Goldspiel and Squyres 2000), which is supported by the occurrence of the very youngest (Hesperian age) valley networks with regional volcanic and tectonic context (Gulick and Baker 1990).

The amount of water required to form the valley networks is still poorly constrained. Based on the associated amount of eroded material, it likely required a volume of water equal to at least several hundred meters GEL but a more accurate estimate would require an understanding of the sediment load and the extent to which the water involved may have been recycled by precipitation, runoff, and evaporation.

2.3.2 Standing Bodies of Water

Past Oceans The existence of the valley networks suggest that early Mars supported a complex hydrological system. This belief is also supported by the geologic evidence and arguments which suggest that the planet once hosted a northern ocean (and possibly large seas) covering up to one-third of the planet's surface (Parker et al. 1987, 1989, 1993; Baker and Milton 1991; Head et al. 1999; Clifford and Parker 2001).

The possibility that a large, and progressively ice-covered, ocean once occupied the northern plains of Mars is based largely on the work of Parker et al. (1987, 1989, 1993), who identified evidence in high-resolution Viking Orbiter images of a series of nested levels—which they interpreted as shorelines—located along the highland/lowland boundary. The highest and oldest of these was called the 'Arabian Level' and is believed to date back to the Late Noachian (Clifford and Parker 2001).

In the much higher resolution (~0.2–20 m/pixel) MOC, HiRise and HRSC images, the Arabian Level exhibits evidence of terracing (potentially indicative of wave-cut erosion); however, the lower, younger levels/'shorelines' do not. The interior plains encompassed by these lower levels include vast expanses of cold-climate landforms, such as polygonal ground and pingos, a relationship that is consistent with either an initially warm, but progressively cooling, marine environment—or initial conditions that were cold from the outset. In either case, the flow-front-like morphologies associated with the lower levels may have resulted from ice-shoving due to short-lived transgressive events caused by later episodes of outflow channel activity around the northern plains (Parker and Calef 2012).

The combination of high-resolution orbiter images with MOLA gridded topography has enabled the compilation of regional and global maps of the proposed shorelines. Apparent discrepancies between the absolute elevation of one of these proposed shorelines with the perimeter of an equipotential surface was cited as a potential serious weakness of the paleo-ocean hypothesis (Carr and Head 2003). However, improved shoreline maps, based on the recent influx of new, higher-resolution images, combined with recognition of the potential impact of polar wander on the post-ocean deformation of shorelines (Perron et al. 2007) have helped resolve much of this disagreement.

The existence of an early ocean is also supported by consideration of the hydraulic implications of the elevated source regions of the Late Hesperian outflow channels, extrapolated backward in time. Current models suggest that the most likely source of the water that carved the channels was the discharge of subpermafrost groundwater from the Martian southern highlands, where the thickness of frozen ground during the Late Hesperian is thought to have been large enough to confine an elevated reservoir of groundwater in disequilibrium with the global topography. However, while such confinement may have been possible given the generally cooler geothermal conditions that existed in the Late Hesperian, the mean global heat flow is expected to have been several times higher in the Late Noachian—resulting in a cryosphere that was too thin to support an elevated groundwater table. Thus, the water that was stored as ground ice and groundwater in the Late Hesperian would have been available to flood any low-lying topography—including the northern plains as well as the larger impact basins, such as Hellas and Argyre. As the global climate and planet's internal heat flow cooled with time, these bodies of water would have frozen throughout. At low- to mid-latitudes, the exposure of this ice would have led to its sublimation and cold-trapping at higher latitudes—as well as its possible re-introduction into the subsurface by the basal melting of the more extensive early polar layered deposits (Clifford 1987; Clifford and Parker 2001). By these processes, the planet's declining heat flow would have led to the development of a cryosphere that was thick enough to hydraulically confine an elevated reservoir of groundwater—creating the conditions necessary for the later development of the outflow channels.

The ocean hypothesis has been challenged by some on the basis of the paucity of well-defined shoreline features in the latest high-resolution orbiter imagery (Malin and Edgett 1999; Ghatan and Zimbelman 2006) and by the relatively small amounts of carbonate and hydrated minerals detected from orbit (e.g. Ehlmann et al. 2008). However, Parker and Calef (2012) have conducted an extensive review of the latest high-resolution imagery and topographic data and find compelling support for the ocean hypothesis. As for the lower than expected quantities of aqueous minerals, Fairén et al. (2004, 2011); Fairén (2010), Ghatan and Zimbelman (2006) and Chevrier et al. (2007), have noted that the formation of these minerals would have been inhibited, given the expected high salinity, low temperature, and acidic nature of an early Martian ocean.

While the presence of an early ocean remains uncertain, a variety of geomorphological evidence, ranging from the global distribution of deltas and alluvial fans along the dichotomy boundary (Di Achille et al. 2010) and valley networks in the southern highlands (Luo and Stepinski 2009) remain supportive of the presence of an early ocean in the northern plains.

Estimates of the total volume of water required to form a northern ocean vary from $\sim$160 m GEL (Carr and Head 2003) up to 700 m GEL (Clifford and Parker 2001; Fairén et al. 2003), consistent with the planetary inventory of water necessary to carve the valley networks and outflow channels.

Frozen Seas As discussed by Clifford and Parker (2001), one of the greatest agents of early topographic and geomorphic change on Mars were large impacts (Schultz et al. 1982;

Schultz and Frey 1990). Of the readily identifiable impact basins, the three largest—Hellas (∼2300 km in diameter), Argyre (∼1500 km), and Isidis (∼1100 km)—all date back to the Noachian (Tanaka et al. 1992). Another Hellas-size (∼3200 km) impact buried under Late Hesperian material was discovered in Utopia centered at approximately 45°N, 248°W (McGill 1989; Frey et al. 1999; Thomson and Head 2001). Because of their size and the limited confinement capability of the early cryosphere, it is likely that all four of these basins—as well as many others whose record has been erased or only partly preserved—hosted interior ice-covered lakes or seas sometime during the planet's first half billion years (Abramov and Kring 2005).

In Utopia and Isidis, resurfacing by Late Hesperian plains material has eliminated any interpretable record of a previous lacustrine environment. Despite this fact, consideration of the initial depth of these basins, and their location in the northern plains, argues for an initially flooded condition. Various lines of geomorphic evidence indicate that this state may have persisted through, or episodically reoccurred during, the Late Hesperian and Early Amazonian—a probable consequence of the concurrent activity of the circum-Chryse and Elysium outflow channels (Lucchitta et al. 1986; McGill 1986; Grizzaffi and Schultz 1989; Scott et al. 1992; Parker et al. 1993; Chapman 1994; Costard and Kargel 1995; Hiesinger and Head 2000). Repetitive flooding of the basins, alternating with periods of eolian deposition and volcanism, would have led to extensive resurfacing of the basins which could have preserved thick deposits of massive ice (Clifford and Parker 2001).

In light of the potential for significant flooding during the Noachian, and the geomorphic and topographic evidence that massive ice deposits may still survive beneath the northern plains, it seems reasonable to expect that some frozen remnant of this early water might also survive in the low-lying interiors of Hellas and Argyre. However, orbital radar investigations have yet to provide any evidence for the presence of massive ice deposits in these locations. This result may reflect either the absence of such deposits or the masking of their geomorphic and topographic expression by later processes and events. Nevertheless, Moore and Wilhelms (2001) have interpreted several landforms as being the results of ice deformation on the floor of Hellas. Ansan et al. (2011) have shown that the huge 2 km thick deltaic deposits in Terby crater, a crater located at the northern margin of Hellas, could have been related to a deep lake in Terby and a related sea covering the whole of Hellas basin on most of the northern rim, for which shorelines were suggested by Wilson et al. (2007).

Based on the crater-diameter scaling relationships of Grieve and Cintala (1992), Clifford (1993) has estimated that the formation of a $\sim 1.5 \times 10^3$ km diameter impact basin will produce $\sim 10^7$ km^3 of impact melt—the majority of which is expected to have pooled in the basin's interior. In addition, the intense fracturing of the underlying crust is likely to have focused considerable post-impact igneous activity in these locations (Schultz and Glicken 1979; Wichman and Schultz 1989; Tanaka and Leonard 1995). As a result, the lithology of the basin floors, as well as much of the surrounding terrain, may well be dominated by a combination of coherent igneous rock and re-welded breccias. Given the size and expected longevity of the thermal disturbances that resulted from these impacts, and the inferred water-rich nature of the surrounding crust, it appears likely that the impacts triggered an extended period of hydrothermal activity that may have eventually led to the formation of an interior lake or sea (Abramov and Kring 2005; Schwenzer and Kring 2009). This conclusion is consistent with the presence of interior layered deposits (ranging from hundreds of meters to as much as several kilometers thick and varying widely in apparent age) in both basins (Moore and Edgett 1993; Tanaka and Leonard 1995; Parker 1996; Tanaka 2000; Tanaka et al. 2002). Then, the basin lakes eventually froze, and the exposed ice sublimed away.

Past Lakes Despite the modest spatial resolution of Viking Orbiter images, some mid-size impact craters (10s to 100s km in diameter) were found to contain alluvial fans, deltas, sedimentary terraces, and evidence of paleoshorelines suggesting the craters once hosted interior lakes (e.g. de Hon 1992, Newsom et al. 1996, Cabrol and Grin 1999, 2001, 2003). Martian paleolake interpretations have been locally confirmed by post-Viking data in several craters, such as Holden, Eberswalde, Gale and possibly Gusev crater (e.g. Malin and Edgett 2000b; Ori et al. 2000; Fassett and Head 2005; Pondrelli et al. 2005, 2008; Mangold and Ansan 2006; Grant et al. 2008; Hauber et al. 2009; Dehouck et al. 2010; Morris et al. 2012; Carter and Poulet 2012). However, the frequency of such craters and the duration of the sedimentary activity related to their filling are still a matter of debate. The evidence for sub-aqueous deposition is often uncertain, especially in the case of the partial preservation of a depositional fan. Only a detailed geometric analysis for sediment structure (as well as the context of deposition) can demonstrate an aqueous origin, as shown in Holden crater (Grant et al. 2008), Eberswalde crater (Pondrelli et al. 2008), Terby crater (Ansan et al. 2011), or Jezero crater (Fassett and Head 2005). Other depressions out of well-defined craters have also been identified as lacustrine deposits, such as lakes at the source of Ma'adim Vallis, in the southern mid-latitudes, which left behind shorelines and degraded sediments still visible to this day (Irwin et al. 2002).

The amount of water required for this lacustrine activity is dependent on the time necessary to form the crater's deposits. Impact related hydrothermal lakes (Newsom et al. 1996) may have had only a short duration. Several paleolakes could have formed in a reduced time period such as in Eberswalde or Jezero crater where the paleolake could have been episodic, on the scale of several hundreds to several 10,000 years (Jerolmack et al. 2004; Fassett and Head 2005). In addition, recent studies show that some lakes could have formed as late as during the Hesperian (as for Eberswalde, Mangold et al. 2012; Irwin 2011, or in Claritas, Mangold and Ansan 2006) while others could have formed over longer durations in much more sustained climate (i.e., Irwin et al. 2002). A full understanding of the implications of paleolakes on the total water reservoir is currently missing, especially given the lack of well-defined age and duration for the lacustrine activity.

2.3.3 Outflow Channels

Evidence that Mars has experienced numerous episodes of catastrophic flooding is found at many locations across the Martian surface, although principally concentrated along the dichotomy boundary. This evidence consists of broad scoured depressions, 10s of km wide, up to a km deep, and 100s to 1000s of km long, with streamlined islands along their beds. Typically these outflow channels begin abruptly, from a fracture or a region of collapsed and jumbled chaotic terrain (Fig. 14) suggesting they were formed by the catastrophic discharge of a large reservoir of subpermafrost groundwater. This interpretation is supported by their geomorphic similarity to the Channeled Scablands in the western United States, which were formed by the sudden and massive release of water from the ice-dammed Lake Missoula (e.g. Baker and Milton 1979; Baker 2001). Discharge rates of the Martian events are estimated to have reached $10^7\ m^3\ s^{-1}$, more than 100 times greater than the present discharge of the Amazon river on Earth (e.g. Burr et al. 2002a, 2002b; Manga 2004; Williams and Malin 2004). With such a discharge, the outflow channels may have formed in as little as several hours to several days (Manga 2004; Andrews-Hanna and Phillips 2007), although they may have persisted for months or years if the discharge rates were smaller.

A reservoir of deep subpermafrost groundwater, confined beneath a thick layer of frozen ground, could conceivably have been released by any event or process capable of disrupting the hydraulic barrier (e.g., by a large impact, earthquake or local igneous intrusion).

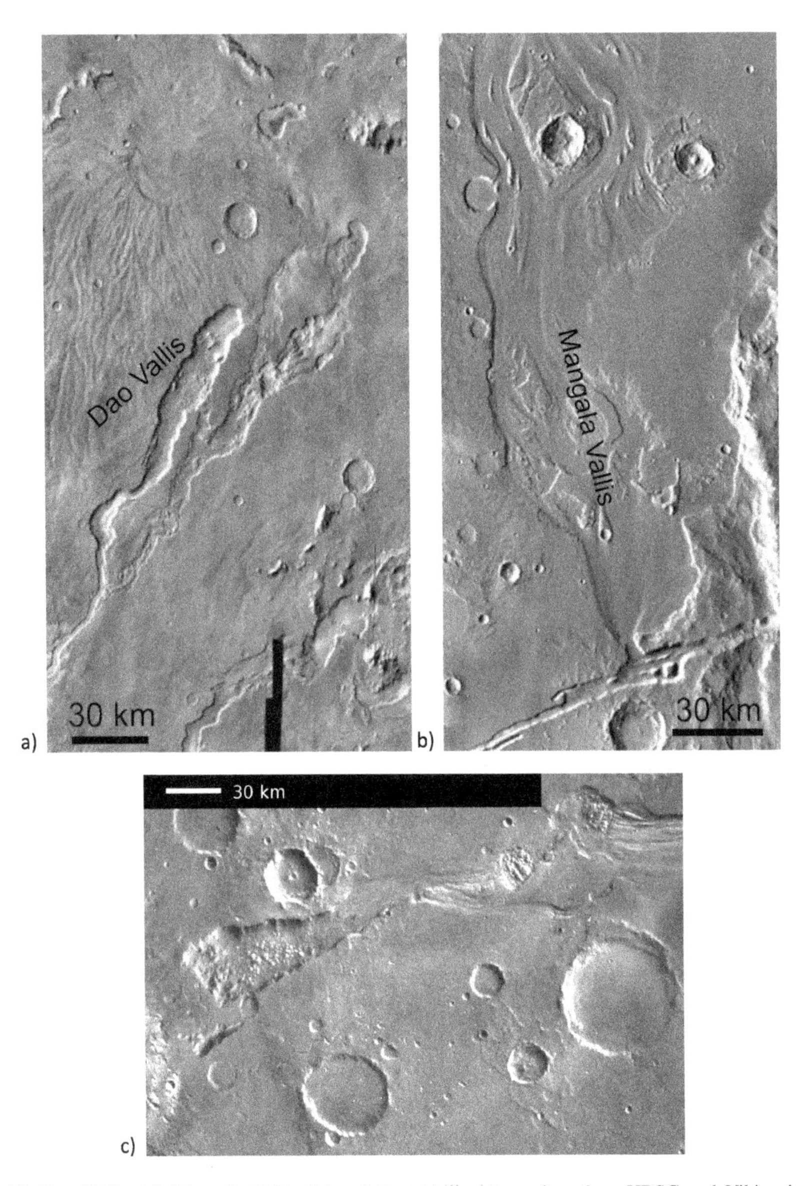

Fig. 14 Dao Vallis (**a**), Mangala Vallis (**b**) and Ravi Vallis (**c**) as viewed on HRSC and Viking images illustrate different types of outflow channels. Channels are defined as outflows when the valley looks like a wide braided channel with a unique source different from dendritic patterns of valley networks. Dao Vallis flows to the SW with a valley head in amphitheater resulting likely from the interaction between subsurface volatiles and the nearby volcano Hadriarca Patera visible north of Dao head. Mangala Vallis flows to the north with a valley head inside a hundreds of km long fracture. This valley head suggests subsurface volatiles were collected at depth and released along this fracture. Ravi Vallis originates from a chaotic region called Aromatum Chaos and extends roughly 200 kms to the East. Its flooding process is generally regarded as involving the release under pressure of a fluid (probably water) from the subsurface of Aromatum chaos. Outflow channels are unique landforms indicating the presence of deep aquifers of water ice filled cryosphere

Table 3 Brief summary of outflow channels age formation and, minimum water volumes associated. The global estimates take into account the likely extent of the aquifer necessary to provide the water to carve these features, leading to an order of magnitude higher GEL. (see text for details. References: 1. Irwin et al. 2002, 2. Irwin et al. 2004, 3. Leask et al. 2006, 4. Harrison and Grimm 2008, 5. Andrews-Hanna and Phillips 2007, 6. Warner et al. 2009, 7. Carr 1986, 8. Carr 1996, 9. Ghatan et al. 2005, 10. Hanna and Phillips 2006, 11. Burr et al. 2002a, 2002b, 12. Neukum et al. 2010)

Outflow channel	Range of ages	Minimum estimated water volume (km^3)	Minimum estimated GEL range (m)	References
Ma'adim	1–3.5 Gy	3.5×10^4–5.6×10^5	0.2–3.8	1, 2, 12
Ravi	3.5 Gy	1.1×10^4–6.5×10^4	0.1–0.5	3
Kasei	1.5–3 Gy	5.5×10^5	3.8	4, 12
Ares	2.5–3.6 Gy	3×10^4–2.2×10^5	0.2–1.5	5, 6
Total for Circum-Chryse formation	2.5–3.6 Gy	1.5×10^6–5×10^6	10.4–34.6	4, 7, 8
Mangala	0.5–3 Gy	2×10^4–4×10^4	0.1–0.3	9, 10, 12
Athabasca	2 My–30 My	2×10^2–1.4×10^4	1.4×10^{-3}–0.1	10, 11
Global estimates	2 My–3.6 Gy	7×10^7–1.4×10^8	5×10^2–10^3	7, 8

Potential examples of this include Mangala Valles (Southwest of the Tharsis region, see Fig. 14) and Athabasca Valles (southeast of Elysium region) which have channel characteristics similar to those of many other outflow channels, but whose source areas consist of a single fracture, probably related to volcano-tectonic activity (Burr et al. 2002a; Head et al. 2003b) or tectonic processes (Hanna and Phillips 2006).

Although an origin by the catastrophic discharge of groundwater appears most consistent with the preponderance of geologic evidence regarding the origin of the outflow channels (Carr 1979; Baker et al. 1992; Rodriguez et al. 2005; Coleman 2005; Meresse et al. 2008), in reality, they may have formed by a variety of processes—including the melting of ground ice by massive igneous intrusions (Head and Wilson 2007), erosion by glaciers (Lucchitta 1982) or, less probably, by the discharge of liquid CO_2 (Hoffman 2000).

Some outflow channels, such as those found in the Xanthe-Margaritifer region, east of Valles Marineris, date back to the Hesperian period (>3 Gy ago). However, others, such as those associated with Cerberus Fossae and the Athabasca Valles, southeast of Elysium, are much more recent—dating back just a few million years (Berman and Hartmann 2002; Burr et al. 2002a, 2002b; Neukum et al. 2010). Carr (1986, 1996) suggests that the peak in outflow channel activity occurred during the Late Hesperian to Early Amazonian (~3.4 to ~2 Gy, e.g. Tanaka 1986; Hartmann and Neukum 2001; Tanaka et al. 2005), although activity has continued to persist episodically to very recent times (Neukum et al. 2010).

Table 3 presents some estimates of the age and water volumes required to erode outflow channel formations on Mars. It is to be noted that the water volume estimates given here are minimum estimates using very conservative sediment loads of 40 % (Komar 1980; Carr 1986) and considering the lowest number, duration and intensity of flood events possible. Using more realistic values for the sediment loads (few percents) and considering the fact that the aquifers feeding the flood are unlikely to have been drained and were part of a deep underground reservoir of water, estimates for the values given in Table 3 would be an order of magnitude larger (Carr 1986). For example, the minimal amount of water necessary to carve the Circum-Chryse outflow channels amounts to 10.4–34.6 m GEL in a very

Table 4 Total planetary water content estimated from geomorphological evidence (see text for details and references)

Geological feature	Formation age	Estimated inventory of H_2O (m GEL)
Northern Ocean and Hellas Basin sea	Episodically from ~200 My–4.5 Gy (with maximum areal extent during the Noachian)	~160–1000
Valley networks	Mid-Noachian to Early Hesperian	~300–1000
Outflow channels	2 My–3.6 Gy (mostly Hesperian)	~500–1000
South Polar Layered Deposit (SPLD)	10 My–100 My	11–16
North Polar Layered Deposit (NPLD)	4 My–1 Gy	8–12
Lobate Debris Aprons	Late Amazonian	~1
Softened terrains permafrost	Late Amazonian	50–200
Ice-sublimation mantle	<1 My	few meters
Total amount of water necessary for SPDL, NPDL and young, cold climate surface features	–	80–240

conservative approach, but it likely required several 100s m GEL and thousands of different flood events to be formed (Carr 1986, 1996; Harrison and Grimm 2008). Because of the many uncertainties involved, it is difficult to accurately estimate the volume of water necessary to carve all the channels, but a plausible range is between 0.5 and 1 km GEL (Carr 1986, 1996; Carr and Head 2003). This estimate could be much higher if the sediment load was not as conservative as mentioned above. Because the time of formation of the outflow channels ranges from Hesperian to late-Amazonian terrains, which post-dates the period of the most efficient water loss mechanisms (hydrodynamic escape and atmospheric erosion by large impacts), the bulk of the water necessary to form the channels should still survive to the present day.

2.4 Summary of Water Content from Geomorphological Evidence

A summary of the formation age and quantity of water (in GEL) deduced to be necessary to form these geomorphological features is given in Table 4. While the upper limit of the quantity of water is not that constrained, some young features (Hesperian or later) indicate the potential for at least several hundreds of meters of GEL to still be present underground. In the next section, we will describe the challenges and successes of the instrumental detection of current subsurface water on Mars.

3 Detection of Subsurface Water on Mars

In this section, we detail the direct and indirect detection of subsurface water on Mars and assess its inferred inventory.

3.1 Direct and Indirect Detection of Global Shallow-Subsurface Ice

Following its 25 May 2008 landing, the Phoenix mission investigated the Northern arctic plains environment for about 5 months (location: 68.2°N, 125.7°W). The analysis of local patterned ground by the robotic arm (Fig. 8) uncovered a shallow ice table under a nearby polygon at depths of 5 cm near its center and of 18 cm near its edge (Smith et al. 2009). The presence of abundant ground ice beneath a relatively thin layer of desiccated regolith is consistent with vapor-diffusive equilibrium with the mean annual atmospheric water content under present-day climatic conditions. The emplacement of this ice can then be explained by the condensation of atmospheric water vapor in the regolith pores when diurnal and seasonal temperature variations cause local soil temperatures to fall below the H_2O frost point. The volume fraction of ice can be further increased by seasonal and obliquity-induced variations in soil temperature that result in the melting and refreezing of ground ice, a process similar to that responsible for frost heave on Earth, which may be aided by the presence of potent freezing-point depressing salts in the regolith. In this way, the resulting meltwater migrates from the warmer to colder regions of the regolith where it may result in the formation of ice lenses, needle ice, and similar structures (Mellon et al. 2009a). The presence of calcium carbonates and other aqueous minerals and salts (3 to 5 wt.%) could be explained by atmospheric processes such as chemical reaction of atmospheric CO_2 with thin water films on the surface of soil particulates (Boynton et al. 2009).

Indications of the presence of ground ice at greater depths, at mid- and higher latitudes, were obtained by recent high-resolution observations of fresh impact craters, 10 to 100 m in diameter, which exposed bright interior deposits, interpreted to be water ice, that faded and eventually disappeared over timescales of several months (Byrne et al. 2009). Water-ice stability models show that the ice excavated is buried beneath a dry layer of 15 to 50 cm thickness (Dundas and Byrne 2010), confirming interpretations made for pitted landforms at these latitudes. Recent identification of a buried deposit of 50 wt.% Water Equivalent Hydrogen (WEH) throughout a large portion of Arcadia Planitia that contained these craters also supports these models (Feldman et al. 2011). These results suggest that the shallow subsurface is made of atmospheric ice deposited with a low dust content at times of high obliquity (Soderblom et al. 1973) forming an excess ice layer, at least partially, different from regolith-supported pore ice. The detection of such features at five different locations and surrounding terrain within a large portion of Arcadia Planitia hints at a global layer of shallow subsurface ice that could be present down to 30°N.

It was later confirmed that shallow water ice should be more pervasive by indirect detection from the behavior of seasonal CO_2 frost deposits. CO_2 frost deposits can be detected by the OMEGA spectro-imager which showed seasonal deposition and sublimation of the atmospheric CO_2. Deposition models indicate that the poleward slopes of craters and dunes should be covered down to about 25°S, while observations show no presence of CO_2 on the surface northward of 35°S. This discrepancy can be explained should the subsurface of Mars present a thermal inertia closer to that of water ice rather than typical rocks and regoliths. This indicates that water ice is present within one meter of the surface in spotty areas down to about 25°S with depths of at least 2 to 3 meters (Vincendon et al. 2010). Global interpolation of this subsurface water ice amounts to about 0.5 to 1 m GEL.

3.2 Mars Odyssey Gamma-Ray and Neutron Spectrometers

The Mars Odyssey Gamma-Ray Spectrometer (GRS) and Neutron Spectrometer (MONS) comprise several detectors that are sensitive to the presence of hydrogen in the soil. These detectors can detect both the gamma photons emitted by hydrogen atoms at 2223 keV and the epithermal neutrons (0.3 eV–700 keV) that are liberated by collisions between cosmic rays and elemental nuclei, whose velocities are rapidly moderated by the presence of hydrogen. Because these observations are conducted from orbit on rapidly moving spacecrafts and at high-energy, the spatial resolution is limited to about 500 km (full-width half maximum). As both gamma rays and thermal neutrons are strongly attenuated by the regolith, this technique is useful to detect hydrogen mass fractions integrated over a sampling depth of several tens of centimeters. Although it is the hydrogen atom that is detected, H is conventionally expressed as H_2O or water-equivalent hydrogen (WEH) in the literature.

Two quantities are measured that allow to constrain two models for the distribution of H in the subsurface. In the first case, the surface is assumed to have a uniform H-distribution over a depth of several tens of centimeters, in the second case, the same surface is assumed to possess a two-layer structure with a dryer top sheet (as shown by recent in situ detections, see Sect. 3.1). The results based on these two models do not differ much in terms of spatial distribution of the near-surface hydrogen reservoirs on Mars, but the two-layer model does predict a greater hydrogen content at depth versus the average value obtained with a uniform model (Maurice et al. 2011). In a layered model, the soil in contact with the atmosphere within 60° of the equator is estimated to be as dry as 2 wt.% H_2O (Feldman et al. 2011).

Figure 9 shows the distribution of water equivalent hydrogen at depth in the two-layer model with a logarithmic scale. The dynamic from the equator to the poles is very important, from 2 wt.% H_2O to nearly pure ice. There is evidence, close to the surface, of very high hydrogen abundances at the poles, although obscured by the residual polar cap of CO_2 in the South. The high near-surface hydrogen abundances detected from orbit were later confirmed by in-situ investigations during the Phoenix mission (e.g. Mellon et al. 2009a). Near the North Pole, a local minimum can be seen in the region of Olympia Undae sand dunes (81°N, −179°E) (Feldman et al. 2007, 2008a, 2008b). At the equator, at latitudes less than 45°, the data reveal two large areas of enhanced hydrogen content (up to about 13 wt.% H_2O in the two-layer model). These areas are located in the region of Sabaea-Arabia and Amazonis-Tharsis, which are mostly dominated by a layer of unconsolidated fines at the scale of the gamma-ray spectrometer (Newsom et al. 2007) and show a distinctive composition (high in H and Cl, low in Si and Fe) (Gasnault et al. 2010). The burial depth of the hydrogen in these regions seems to vary by a factor of 10 from a minimum of 4 g/cm^2 to a maximum of 40 g/cm^2 (Diez et al. 2008; Maurice et al. 2011). At intermediate latitudes (45–75°), especially around 55° (North and South), the boundary between the equatorial regions depleted in hydrogen and the polar regions enriched in hydrogen is relatively sharp, with a strong negative correlation between the hydrogen abundance and the depth of that hydrogen (Maurice et al. 2011). Figures 9 and 12 also show the strong correlation between this boundary and the locations of near-surface periglacial features such as gullies and polygonal terrains. As the two-layer model cannot realistically be applied near the polar regions of the planet, the single layer model where the hydrogen is assumed to be distributed uniformly with depth gives a maximum abundance of 80 wt.% H_2O (with error bars between 2 and 4.5 wt.% Maurice et al. 2011). At first order this is consistent with the gamma-ray (Boynton et al. 2008).

From the MONS data, it is possible to estimate the amount of water present in the near subsurface of Mars (Feldman et al. 2008c). The WEH map directly provides a quantitative measure of the distribution of water molecules in the form of water ice that fills regolith pore volumes or H_2O/OH that is physically or chemically attached to a variety of minerals near the surface. Though the method does not differentiate between the two types of water, the consistency of the model with the physical characteristics of the soil has been used to determine conditions under which water ice would be in excess of pore volume (Feldman et al. 2011). The regions showing WEH values larger than the threshold value of $\sim$26 % include the latitudes poleward of 60°, and some zones around the mid-latitudes. It has been demonstrated that excess ice value is enhanced in a region of Arcadia Planitia within which

HiRISE has observed fresh icy craters (Byrne et al. 2009; Feldman et al. 2011). Using the more accurate two-layer model for calculating the volume of water in the shallow subsurface with an upper layer containing 2 wt.% H_2O, and assuming that the penetration depth of the neutrons reaches about 50 cm, the total integrated volume of H_2O over the surface of the planet is equivalent to an 11 cm GEL. This value is quite small, but it should be reiterated that the detected amount corresponds to the H_2O fraction of the soil over a thickness of 50 cm. If this fraction remains constant over a larger depth, as is most certainly the case, as suggested by geomorphological evidence correlated with enhanced WEH zones (see Sect. 2), then the total water amount could reach the 10s to 100s of m GEL inferred from geomorphological evidence, but the upper limit cannot be constrained from this measurement.

3.3 Direct Detection by Radar Sounders

The most quantitative measurements made to estimate the water inventory of the planet are made by radar instruments due to their penetration depths and coverage. Here we describe the instruments sent to Mars and the current state of knowledge of the quantity of water ice detected.

3.3.1 The Martian Radar Sounders

Radar (RAdio Detection And Ranging) sounders on Earth, also known as Ground Penetrating Radar (GPR), are commonly employed from surface and airborne platforms to investigate the lithology, structure and distribution of water in the subsurface, down to depths ranging from 1 to 10 wavelengths, depending on the electromagnetic properties of the local rock and soil (Olhoeft 1996). However, because soil moisture can cause the rapid attenuation of radar, and because precipitation on the Earth is widespread, no attempt has been made to conduct radar sounding investigations from spacecraft in Earth orbit. However, because the environmental conditions on other planetary bodies currently preclude rainfall, radar sounding is considered a potentially powerful tool in their investigation. This led to the Apollo Lunar Sounder Radar System (ALSE), the first orbital radar designed for subsurface investigations of an extra-terrestrial body (Porcello et al. 1974). Surprisingly, despite the development of unmanned space exploration, it was not until the late 1990s that radar instruments were selected to fly on-board space missions to Mars. The first of these was the Mars Advanced Radar for Subsurface and Ionospheric Sounding (MARSIS), which flew on ESA's Mars Express spacecraft (Picardi et al. 2005; Jordan et al. 2009). It was closely followed by the SHallow RADar (SHARAD) on the Mars Reconnaissance Orbiter (MRO) (Seu et al. 2007). Both radar instruments were provided by the Italian Space Agency (ASI) and successfully retrieved the first deep underground sounding from the red planet.

While the depth investigated by most remote sensing techniques, such as orbital imaging and spectroscopy, is typically on the order of a few microns, radar sounders can probe the subsurface to depths of $\sim$4–5 km in nearly pure H_2O ice and to a maximum depth of $\sim$100–200 m in lithic environments.

MARSIS and SHARAD are High Frequency (HF), nadir-looking sounding radars with dipole antennas transmitting successive electromagnetic 10-W pulses along the spacecraft trajectory, having a Pulse Repetition Frequency (PRF) of hundreds of Hz (Table 5). The interactions of radar waves with planetary surfaces, and their propagation into the subsurface, are largely determined by the dielectric properties of the materials they encounter. When a radar pulse reaches the boundary between two materials with differing dielectric properties (such as the atmosphere and ground), a portion of the incident energy (an echo) is

Table 5 MARSIS and SHARAD specifications. Vertical resolution and penetration depth depend on the radar wavelength and permittivity of the sounded material, while horizontal resolution is linked to the radar's antenna design, speed of the spacecraft, pulse repetition frequency, and surface roughness. From Jordan et al. (2009) and Seu et al. (2007)

	Frequencies	PRF	Resolution (along × across-track)	Maximum penetration depth in H_2O ice	Resolution (vertical)
MARSIS	1.8, 3, 4, 5 $\pm$ 0.5 MHz	127 Hz	5–10 × 15–30 km	<5 km	75–150 m
SHARAD	20 $\pm$ 5 MHz	700 Hz	0.3–1 × 3–6 km	<1.5 km	10–24 m

reflected backward, while the remainder continues to propagate into the subsurface, where it may suffer additional losses due to scatter by imbedded objects and absorption by the host material itself. Aside from the effect of the interface roughness that may weaken the received signal, the higher the dielectric contrast, the brighter the echo. Since liquid water has a relative dielectric constant (ε) of $\sim$80, much higher than common geological materials—ice to rock dielectric constants differ by factors from 3 to 12 (Ulaby et al. 1986; Campbell and Ulrichs 1969)—radar sounders are thus particularly suitable for the detection of aquifers. The reflection of the transmitted pulse continues until the signal is absorbed in the medium. A signal is considered undetectable when it reaches $\sim$3 dB above the galactic noise (Seu et al. 2007).

Attenuation mainly depends on the intensity and number of reflections, scattering due to heterogeneities within the medium, and dielectric absorption. Once again, ice has a very low loss tangent ($<1 \times 10^{-3}$) compared to other geological material (1×10^{-3} to 100×10^{-3}), making radar sounders excellent instruments to deeply probe the cryosphere (Ulaby et al. 1986; Campbell and Ulrichs 1969). Echoes from the same initial pulse are recorded in a single frame with various delays (the short-time dimension). The well-known radargram visualization proceeds from the concatenation of these frames, given a cross-section of the soil with a time vertical-axis. The velocity of radar waves in matter is proportional to its dielectric constant, implying that the depth, d, reached can be derived by simple inversion to give the relation $d = \frac{tc}{2\sqrt{\varepsilon}}$, where t is the time delay and c the velocity of light in vacuum.

The performances of the two Martian radar sounders are complementary (Table 5). MARSIS can be set to several frequencies from 1.8 to 5 MHz that are sensitive to the ionosphere crossing (Safaeinili et al. 2003) and allows a theoretical penetration depth up to 5 km, designed for the search of deep aquifers, while SHARAD is set to 20 MHz and cannot probe beyond 1.5 km depth. The vertical resolution (δ_v) is achieved by the "range processing" that aims to compress the µs echoes and to increase their signal-to-noise ratio. The efficiency of this process depends on the bandwidth B, so that $\delta_v = \frac{cK}{2B\sqrt{\varepsilon}}$ where K is an apodization factor. Thus, SHARAD achieves a vertical resolution $\sim$7 times higher than MARSIS, allowing finer structural and morphologic studies of geological objects. The horizontal resolution (δ_h) is a function of the surface roughness. When the latter is smooth and specular, integration of the signal by the receiver maintains the coherent information from a small circular footprint (Fresnel zone) directly at nadir, while peripheral areas are out of phase because of a longer propagation path. The horizontal resolution is determined by $\delta_h = 2\sqrt{\frac{h\lambda}{2}}$ where h is the altitude and λ the wavelength. On the contrary, for a rough surface, the signal is no longer coherent anywhere, and the horizontal resolution is then limited by the pulse length so that $\delta_h = 2\sqrt{\frac{hc}{B}}$. From there, the along-track resolution is commonly improved by applying a

Doppler processing (Wiley 1965). Once again, SHARAD has better capabilities for probing targets at regional and sub-regional scales with horizontal resolution around 1 km, whereas the ~10 km MARSIS footprint is clearly designed for covering wider areas.

3.3.2 Identified Reservoirs and Their Water Content

Polar Layered Deposits Because of the low permittivity of H_2O ice (~3), MARSIS and SHARAD achieved their deepest penetration of the Martian subsurface in the PLD. Radar pulses penetrate deep into the ice before being backscattered by the basal interface (bedrock). While MARSIS records this echo for both structures, SHARAD only detects the North Polar Layered Deposit (NPLD) bedrock, even for depths twice as large as some regions of the South Polar Layered Deposit (SPLD). The reason for this is still not understood but may be the result of attenuation, a rough basal interface, or a low dielectric contrast between the basal ice and underlying bedrock, or some combination of all three.

Plaut et al. (2007) were the first to use the basal detection to make a global map of the SPLD thickness with MARSIS. From the delay between surface and bedrock echoes, the authors derived the ice thickness assuming a dielectric constant of pure water ice ($\varepsilon = 3$) and obtained a maximum thickness of 3.7 ± 0.4 km and a volume of $1.6 \pm 0.2 \times 10^6$ km^3 corresponding to a water GEL of 11 ± 1.6 m. Signal losses through the ice indicate low values $2 \times 10^{-3} < \tan\delta < 5 \times 10^{-3}$. The true value of the dielectric constant is an important issue as it could modify the inferred volume of the ice. Indeed, it is sensitive to the porosity/impurity ratios, temperature of the medium, and radar wavelength (Kofman et al. 2010). For the latter parameter, MARSIS and SHARAD wavelengths (HF frequencies) are commonly considered as nearly similar at first order for simplicity. Using SHARAD, Grima et al. (2009) measured ε for a representative part of the NPLD using a reconstruction of the bedrock, by interpolation of the surrounding topography, to independently infer the ice thickness. Coupled to basal-echoes delays, they retrieved the average permittivity of a vertical column of ice down to the bedrock with a total average value $\varepsilon = 3.10$ ($\sigma = 0.12$) and $\tan\delta < 0.0026$ ($\sigma = 0.0005$) where σ is a 68 % confidence interval, consistent with an impurity rate lower than 5 % in volume. By analogy these values are sometimes used for the SPLD as well, confirming the nearly pure water ice assumption made by Plaut et al. (2007) and the derived volume for the SPLD. However, the only real constraint on the SPLD dielectric constant comes from Zhang et al. (2008b) with values in the range $3 < \varepsilon < 5$, obtained by MARSIS data inversion using Bayesian inference, or even slightly more if the bedrock is wet. Hence an accurate measurement of the SPLD bulk ice remains to be done.

At a global scale, contrary to Planum Australe, Planum Boreum is made of two geological units: (i) the NPLD that partly covered (ii) the basal unit (BU, shown in Fig. 15a). Several radar studies have derived volume estimates of these units (Putzig et al. 2009; Grima et al. 2009; Selvans et al. 2010). All estimates converge to values of $\sim 0.8 \times 10^6$ km^3 for the NPLD, that is a ~5.5 m GEL of pure water, and $0.45 \pm 0.1 \times 10^6$ km^3 for the BU. The impurity rate of the latter is not well known but is believed to be higher than the NPLD since its albedo is darker (Byrne and Murray 2002; Fishbaugh and Head 2005). If we arbitrarily vary this rate from 0 % to 50 %, the corresponding global layer of water decreases from 3.1 ± 0.1 m to 1.6 ± 0.1 m respectively. To summarize, according to current knowledge of ice dielectric properties and available radar resolutions, both polar deposits should encompass together a GEL of water from 16.5 m for the lower bound to a maximum of 21.2 m. These estimated values appear to be smaller than the ones derived from geomorphological evidence, as a consequence of the fact that radars have measured a well defined basal boundary, though the values obtained from radar measurements and from geomorphological assessments are consistent with one another.

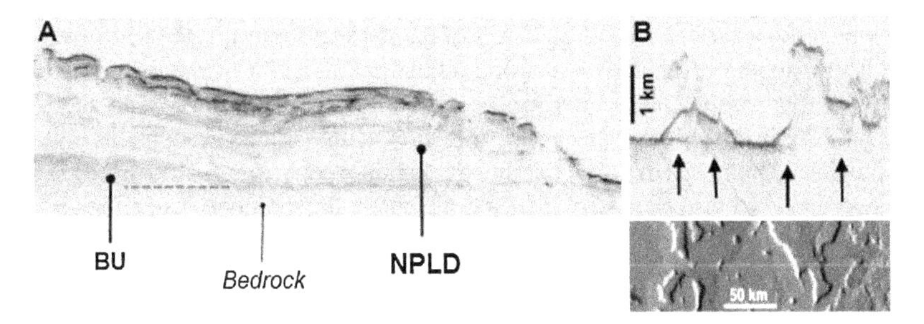

Fig. 15 Migrated SHARAD radargrams ($\varepsilon = 3$) with the same vertical and horizontal axis. Vertical axis is exaggerated. (**A**) Observation 491501 across half of the NPLD's diameter. A part of the underlying BU is seen. The BU/bedrock interface is not detected by SHARAD but symbolized by a *dotted line*. (**B**) Observation 719502, in Deuterolinus Mensae, along with shaded topography (light comes from the left), adapted from Plaut et al. (2009). LDAs at cliffs' bases are obvious thanks to the detected basal interfaces (*black arrows*)

Mid-Latitude Lobate Debris Aprons Since the tilt of Mars' rotation axis could have reached as much as 60° during the last 20 million years (Laskar et al. 2002, 2004), it is believed that the sublimation of polar ice at times of high obliquity resulted in the redistribution of ice and dust to mid-latitudes, where it set the stage for the development of episodic 'ice ages' (e.g. Head et al. 2003a; Forget et al. 2006). Around the northern dichotomy and the margins of Argyre and Hellas Planitia (30° to 60° latitude), there are hundreds of candidates for remnant glaciers called "lobate debris aprons" (see Sect. 2.2.2, Carr and Schaber 1977; Squyres 1979). LDAs occur at cliff bases as aprons extending tens of km outward and exhibiting evidences of flow with gently sloping surfaces from 300–800 m above the surrounding valley-floors to convex upward margins (Fig. 15b). There has been a long debate on whether LDAs were derived from the deformation of the local ice-rich terrain, similar to rock glaciers, or were the result of massive deposits of ice that were later covered by rock debris and regolith (e.g. Squyres 1978; Lucchitta 1984; Colaprete and Jakosky 1998). A survey of SHARAD radargrams enables the identification of the basal interface of LDAs in many regions of both hemispheres (Holt et al. 2008; Plaut et al. 2009). The vertical migration of this reflector to fit with the valley-floors agrees with a dielectric constant from 3 to 3.2 while signal attenuation is equivalent to $3 \times 10^{-3} < \tan\delta < 4 \times 10^{-3}$, consistent with nearly-pure ice (impurities <10 %). The similarities of these measurements for regions distant by more than 5,000 km (nominally Deuterolinus Mensae and Eastern rim of Hellas impact) argue for a common composition of the LDAs at planetary scale. The equivalent water content from all the LDAs has not been measured yet, but the one corresponding to Hellas region is estimated to be ~0.2 m (Holt et al. 2008) so that it is not expected to be much greater than a meter GEL for the entire planet.

Permafrost In Sect. 2, we reviewed the geomorphological evidence that the Martian subsurface is likely ice-rich down to significant depth on a global scale. Since the depth of the cryosphere may range from ~5 to 20 km, increasing with latitudes (Sect. 4), the formation of ice-rich permafrost has likely been a major sink for ground water (Clifford et al. 2010).

Based on an analysis of MARSIS radar sounding data, Mouginot et al. (2010) constructed a global map of surface reflectivity (i.e. the power of the initial surface echo), from which the dielectric properties of the surface can be derived. The surface reflectivity includes contributions from both surface roughness and crustal permittivity (averaged over a depth of one wavelength, which for MARSIS represents the first 60 to 80 m). To remove the roughness component, the authors simulated its contribution using the MOLA topography (Smith et al.

2001), which has a surface resolution that is of the same order as the MARSIS radar wavelength. The mapping result shows a strong latitudinal variation of the dielectric constant down to 50°–60° latitude, especially for the northern hemisphere. Moreover, anomalously low near-subsurface dielectric values were measured within the northern plains region interior to previously proposed martian shorelines (Mouginot et al. 2012). The striking similarity of behavior between the reflectivity map and the potential 'shorelines' identified by Parker et al. (1993) and Clifford and Parker (2001) suggests that the low dielectric values are either the product of an icy relic of a former ocean or high-porosity sediments that were deposited in the northern plains from the formation of the outflow channels. The frozen relic of an ancient ocean may have been preserved to the present day after its burial by eolian sediments and volcanic ashes through the process of thermal vapor diffusion (Clifford 1993) with redistribution of surviving ice from greater depth. This led the authors to interpret this low permittivity as the result of a water ice-regolith mixing. Using a mixing formula, they estimated the volume of water in this near-surface permafrost as being $\sim 10^6$ km^3, which is approximately 7 m of GEL. This estimate could be much higher as only the uppermost $\sim$100 m were probed using this method.

3.3.3 Ambiguous and Unconfirmed Reservoirs

Medusa Fossae The Medussa Fossae Formation (MFF) is a set of massive, eroded, and geologically young appearing deposits occurring along the equator between 140° and 280°E longitude, straddling the dichotomy boundary (Harrison et al. 2010). The total volume of the formation is believed to be similar to the SPLD. The MFF have undergone extensive wind-erosion with slopes smoothly decreasing towards the lower plains over 400–800 km. It has not been possible yet to determine if the MFF is an aeolian deposit, a mound of volcaniclastic material, or a paleo-deposit of ice (Mandt et al. 2008 and references therein). Since the basal interface of the MFF is detected by both MARSIS and SHARAD, the same migration method used for the LDAs has been applied to obtain a dielectric constant around 3 and a permittivity range of $2 \times 10^{-3} < \tan\delta < 6 \times 10^{-3}$ (Watters et al. 2007; Carter et al. 2009). These results are consistent with either an ice-rich or a dry and low-density deposit but, contrarily to the LDAs, there is no evidence that these deposits have undergone flow, which would support a composition dominated by ice. However, a recent indication of buried excess WEH has been found in neutron thermal and epithermal counting-rate data (Feldman et al. 2011), in support of the ice-related interpretation of the radar data.

Gully Sources As discussed in Sect. 2.2.4, gullies are thin carved channels that could reach hundred meters long at steep slopes on mid-latitude regions for both hemispheres. Seasonal activity is attested by high-resolution images, showing morphological changes and albedo variations within a Martian year (Malin et al. 2006, Dundas et al. 2010). It has been suggested that gully origin could include processes as different as seasonal liquid discharge or dry flows of sand. To check the first hypothesis, Nunes et al. (2010) did a partial survey of a widespread population of gullies using SHARAD to search for bright underground interfaces. No indication of gully sources such as aquifers or ground-ice reservoirs were found. Even if only a restricted number of gullies sites were observed, it is possible that SHARAD resolution is insufficient to detect such features, especially if liquid water is limited to shallow transient melting, unrelated to aquifers.

Polar Basal Water An early study of the potential thermal conditions at the base of the PLD suggested that, under the thickest portions of the deposits (then estimated to be on the order of ∼4–6 km) basal temperatures may have been high enough that, with the assistance of potent freezing point depressing salts, large expanses of the basal ice might be at the melting point (Clifford 1987). Although there is geomorphic evidence that suggests that basal melting may have occurred beneath the south polar layered deposits during the Hesperian, the acquisition of improved topographic data from MOLA made it clear that the present polar deposits are both too thin to undergo widespread basal melting, although the possibility of melting associated with local geothermal anomalies still exists (Clifford and Parker 2001).

However, if melting does presently occur beneath the polar deposits, the MARSIS radar is ideally suited to detect it, due to the high dielectric contrast between liquid water and most other common geological materials. Indeed, Plaut et al. (2007) reported evidence of an unusual basal reflector in the MARSIS data within the 310°–0°E longitude sector of the SPLD. The reflection is brighter than the corresponding surface echo. The authors emphasized that this region is one of the coldest on the planet, further lowering the attenuation of radar by reducing the amount of supercooled liquid water in the polar ice. This, in turn, could explain the bright basal reflection, which appears to be a far more plausible explanation than meltwater. Other regional searches for evidence of polar basal melting have also been conducted. Farrell et al. (2008) also did a MARSIS survey of Chasma Australe, a reentrant of the SPLD, without any indication of basal melting. Grima et al. (2009) reached the same conclusion with SHARAD measurements of Gemina Lingula, one-fourth of the NPLD area.

3.3.4 Limitations on the Geophysical Detection of Subsurface H_2O

The distribution of water and ice in the subsurface of Mars may be highly heterogeneous in response to local variations in diffusive, thermal and hydraulic properties, differences in mean annual temperature, geothermal heat flow, and the presence of potent freezing-point depressing salts. As a result, local geophysical investigations are likely to yield results that are unrepresentative of the planet. Orbital investigations (such as GRS, MARSIS and SHARAD) have the ability to conduct a global reconnaissance. Unfortunately, a number of factors have limited the efficacy of these investigations to shallow depths (<100 m), making any deeper H_2O essentially undetectable.

Radar sounders' capabilities are limited by (i) their own specifications such as horizontal resolution, making most of the sub-kilometer objects difficult to detect, and (ii) orbital coverage (complete for MARSIS, about 50 % so far for SHARAD). Sounding radars are also affected by their height h above the surface, which reduces the strength of the returned signal by $\sim 1/h^4$, and by the shape of the spacecraft's orbit—which, given the highly eccentric orbit of Mars Express influences the strength of the returned signal.

Also, the optimal conditions for subpermafrost groundwater detection, such as a sharp transitional boundary, a well saturated crustal porosity (>10 %) and a low surface roughness, characterize about 20 % of the planet's surface (Picardi et al. 2003). From the observation that effective penetration depth seems to be lower than that predicted over most of the planet, possibly down to ∼100 m in dry volcanic terrains (Heggy et al. 2006), Farrell et al. (2009) show that a crustal material with a plausible high conductivity would explain the lack of detection of the presumed subpermafrost water-table. In addition another crucial limiting factor is scattering of the signal. This process occurs when the cells of a random network of dielectric discontinuities are of the order of a radar wavelength. The power is then weakened due to destructive interferences and scattered in all directions of space, greatly weakening the signal in the receiver direction. It has been shown that scattering due to surface roughness is a dominant process over Mars (Grima et al. 2012) while it can even

Table 6 Total water content on Mars directly measured by orbiter instruments. Near-surface permafrost correspond to the first 60–80 m depth, extended from PLDs margins to latitudes of 50°–60° (see text for details)

Geological feature	Detected GEL	Instrument
Global near-surface water content	>11 cm	MONS
	1 m	HiRISE, OMEGA
SPLD	11 ± 1.6 m	MARSIS
NPLD	~5.5 m	MARSIS
Near-surface permafrost	>7 m	MARSIS
Basal unit	1.6 m to 3.1 m	SHARAD
Lobate Debris Aprons	1 m	SHARAD
Total	24.5 m to 29.3 m	–

cancel out the surface echo in some regions (Mouginot et al. 2010). Volume scattering is also likely to occur underground due to subsurface heterogeneities. Considering that the emitted wave crosses the same units twice along its propagation path before reaching the emitter-receiver, scattering can limit the penetration depth of the signal. Moreover, the known "long trail" of energy associated to scattering within an observation frame can hide some weak subsurface echoes, making some near-surface reflectors inaccessible (Farrell et al. 2009; Pommerol et al. 2010).

3.4 Summary

Table 6 summarizes the current water reservoirs positively detected by orbiting instruments in the Martian subsurface. When compared with the geomorphological subsurface water estimates presented in Table 4, the water detected amounts to about the quantity of water from the most recent and most readily visible features (Polar Layered Deposits, Basal Units, Lobate Debris Aprons). The water linked to softened terrain morphologies and outflow channel erosions that could amount to several hundreds of meters of GEL is conspicuously missing from this compilation of measurements. As argued in the previous section, while some water reservoirs may be inaccessible to radar detection, other reasons may prevent a strong signal from being detected. Foremost is the possibility that all the water inventory of the planet is trapped in the subsurface cryosphere, leading to very deep, tenuous, or nonexistent aquifers. Moreover, continuous water content variations between the cryosphere and potential aquifers, e.g. by circulation within the vadose zone, would also dampen the radar signal and prevent a clear positive detection (Clifford et al. 2010). Further work is on-going to accurately interpret the radar measurements, for which a global coverage of the planet is still missing (in the case of SHARAD), and better constrain the water content of the Martian subsurface.

4 Characteristics of the Cryosphere and Evolution with Time

Based on a realistic range of globally-averaged subsurface properties, Clifford et al. (2010) investigated the plausible extent of the present-day Martian cryosphere and considered its evolution with time. In this section, we review this analysis and consider its implications for the exchange of water with other volatile reservoirs (atmosphere, polar caps, and the formation of hydrated minerals) and the consequences for subsurface habitability.

4.1 Current State of the Martian Cryosphere

4.1.1 Martian Subsurface Properties and Potential Water Content

Today, subsurface water on Mars may exist in two thermally distinct reservoirs: as ground ice, within the cryosphere, and as liquid groundwater at greater depth, where the geothermal heat flux elevates the lithospheric temperature above the freezing point (Fanale 1976; Rossbacher and Judson 1981; Kuzmin 1983; Clifford 1993).

The Martian cryosphere encompasses a region of the crust extending from the surface down to a local depth, z, given as a first approximation by the solution to the one-dimensional heat conduction equation: $z = k_{(T)}\frac{(T_{mp}-T_{ms})}{Q_g}$, where $k_{(T)}$ is the temperature-dependent thermal conductivity of the crust, T_{ms} is the mean annual surface temperature (which ranges from about 218 K at the equator to about 154 K at the poles), T_{mp} is the melting temperature of ice at the base of the cryosphere for which values range from 273 K for pure water ice or lower values if freezing-point depressing salts are present, and Q_g is the geothermal heat flux (Clifford 1993; Clifford and Parker 2001; Clifford et al. 2010).

Thermal Conductivity The basalt thermal conductivity under Martian conditions shows an increase when temperatures get lower (Clauser and Huenges 1995; Lee and Deming 1998). Water ice presents a similar behavior at equivalent temperatures (Clifford et al. 2010) and can be assumed to follow the same expression which varies from a lower bound of 2.26 W m^{-1} K^{-1} at 273 K, to an upper bound of 3.64 W m^{-1} K^{-1} at 154 K (Hobbs 1974).

In its pure state, the thermal conductivity of gas hydrate is ~0.5 W m^{-1} K^{-1} (Davidson 1983; Sloan 1997), or approximately one-fifth that of water ice, a value that recent laboratory measurements have demonstrated remains fairly constant over a broad range of subfreezing temperatures (Krivchikov et al. 2005). Its presence may thus significantly reduce the effective thermal conductivity of the subsurface, especially at shallow depths, where the porosity of the crust is expected to remain relatively high.

Finally, at latitudes below 40°, where water ice is diffusively unstable with respect to the relative humidity of the atmosphere, the regolith may be desiccated to depths ranging from several meters to as much as ~1 km, depending on the diffusive and thermal properties of the subsurface (Clifford and Hillel 1983; Fanale et al. 1986; Mellon and Jakosky 1995). This desiccated region could have an effective thermal conductivity as low as 0.05 W m^{-1} K^{-1}— or even lower, given a regolith of fine-grained sediments or volcanic ash (Presley and Christensen 1997).

To calculate the composite thermal conductivity of a multi-component regolith (with varying porosity and lithic and volatile composition), we use a geometric mean mixing rule which represents a good approximation of the arithmetic mean of the non-parametric rules (Maxwell 1892; Robertson and Peck 1974, Horai 1991), over the range of Martian temperatures and compositions considered here. Moreover, Sass et al. (1971) has shown good agreement between laboratory measurements and theoretical values calculated with a geometric mean mixing rule for a variety of materials. Thus, we use this approach to calculate the thermal conductivity of porous basalt filled with varying amounts of water ice, gas hydrates, and air.

Possible Presence of Groundwater Freezing Point Depression Salts The freezing point of Martian groundwater can be depressed by the presence of dissolved salts. The evolution of Martian groundwater into a highly mineralized brine is an expected consequence of three processes: (1) the leaching of crustal rocks beneath the groundwater table, (2) the increased concentration of dissolved minerals in groundwater as more of its inventory is cold trapped by the growth of the cryosphere over geologic time, and (3) the influx of minerals leached from the vadose zone by the low-temperature hydrothermal convection of vapor between the water table and the base of the cryosphere (Clifford 1991, 1993).

The presence of dissolved salts may significantly reduce the depth of frozen ground. NaCl brines have a freezing point of 252 K at their eutectic and other chloride compounds ($CaCl_2$, $MgCl_2$, etc.) could lower the freezing point to as little as ~210 K (Brass 1980; Clark and van Hart 1981; Clifford 1993; Knauth and Burt 2002) though sulfate-rich brines can strongly limit the amplitude of this effect. Evidence that Martian soils, at high latitude, contain up to 1 wt.% of perchlorate was provided by the Phoenix Lander investigations (Hecht et al. 2009). The discovery of perchlorate is significant because it is a highly potent freezing-point depressing salt, with its $Mg(ClO_4)_2$ form having a eutectic temperature of ~203 K (Clifford et al. 2010).

As discussed by Clifford (1991, 1993) and shown in Fig. 1, the presence of a geothermal temperature gradient in the vadose zone between the local groundwater table and the base of the cryosphere will lead to the development of a low-temperature hydrothermal circulation system of ascending water vapor and descending liquid condensate within the crust. Given a geothermal gradient of ~15 $\mathrm{K\,km^{-1}}$ and reasonable estimates of crustal porosity and pore size, the equivalent of 1 km GEL could be cycled through the vadose zone by this process every ~10^6–10^7 years. An important mineralogical consequence of this activity would have been the depletion of any easily dissolved substances from the vadose zone and their concentration, potentially up to the point of saturation and precipitation, in the underlying groundwater. As a consequence, where the cryosphere and saline groundwater are not in direct contact, low-temperature hydrothermal convection is likely to have depleted the intervening crust of any potent freezing-point depressing salts, resulting in a basal temperature near ~273 K, thus, maximizing the local depth of frozen ground. Conversely, where the cryosphere and saline groundwater are in contact, eutectic concentrations of perchlorate or other dissolved salts may dramatically reduce—and potentially eliminate—the local thickness of frozen ground (Clifford et al. 2010).

Heat Flow Early estimates of the present-day mean global heat flow of Mars varied between 15–45 $\mathrm{mW\,m^{-2}}$ (see discussion in Clifford 1993), with a nominal value of ~30 $\mathrm{mW\,m^{-2}}$ based on the assumption that Mars possesses a chondritic composition (Fanale 1976). However, more recent rheologic estimates of the elastic lithosphere thickness suggest a significantly lower range of values from about 8 to 25 $\mathrm{mW\,m^{-2}}$ (Solomon and Head 1990; McGovern et al. 2004; Phillips et al. 2008). A lower value of geothermal heat flow (~13 to 24 $\mathrm{mW\,m^{-2}}$) is also consistent with the absence of a present day magnetic dynamo on Mars and models of the mantle convection (Nimmo and Stevenson 2000; Li and Kiefer 2007). A much lower average value of 6.4 $\mathrm{mW\,m^{-2}}$ (with local variations from 1 to 13 $\mathrm{mW\,m^{-2}}$) has even been inferred from Mars Odyssey GRS instrument measurements (Hahn et al. 2011). If this lower estimate of global heat flow is correct, it would increase all previous estimates of the cryosphere thickness.

Given, on the one hand, that continental heat flow on Earth varies by ±50 % about the global mean (Pollack et al. 1993) and, on the second hand, the evidence for a similar level of geologic diversity on Mars (e.g. Nimmo and Tanaka 2005; Solomon et al. 2005; Watters et al. 2007), it is reasonable to expect that the value of Martian heat flow will exhibit a comparable range of local variability (Clifford 1993; Clifford et al. 2010). Moreover, significant spatial variations in heat transport and therefore cryosphere thickness, can arise from low-temperature hydrothermal convection within a subpermafrost aquifer (Travis et al. 2003; Travis and Feldman 2009).

The available pore volume of the Martian subsurface, considered in the context of the total planetary inventory of water, has important implications for understanding the hydrologic and mineralogic evolution of the crust. For example, the volume of H_2O that is potentially stored within the cryosphere can be estimated by calculating the pore volume of the crust between the Martian surface and the base of the cryosphere, based on the porosity vs. depth relationship of Clifford (1993) that assumes a surface porosity, $\Phi(0)$, and an exponential decay constant of 2.82 km obtained by gravitationally scaling the porosity of the lunar crust (inferred from an analysis of the Apollo seismic data, Binder and Lange 1980) to the appropriate value for Mars.

Using a low value of $\Phi(0) = 20\ \%$ for the surface porosity value of the Moon, consistent with measured values for lunar breccias, gives a total pore volume capacity of about 600 m GEL. If a more realistic near-surface porosity value of $\Phi(0) = 35\ \%$ is used (consistent with in situ measurements and geomorphology, see Sect. 2.2.2), the volume available increases to 1 km GEL. As discussed in Sect. 2, this is more than enough to segregate most of the water necessary to erode the geomorphology features seen at the surface of the planet.

4.1.2 Current State of the Cryosphere

Cryosphere Consisting of Water Ice Previous best estimates of the thickness of the present day cryosphere suggested that it varied from an average depth of ~2.5 km at the equator to ~6.5 km at the poles, noting the potential for significant (±50 %) local variations, due to the likely heterogeneity of crustal heat flow and thermal properties (Fanale 1976; Clifford 1993). However, more recent calculations indicate that the current depth of the Martian cryosphere may be up to 2 to 3 times greater, a consequence of the consideration of the temperature-dependent thermal conductivity of ice, rock and recently revised estimates of the planet's present-day geothermal heat flow as discussed above.

Clifford et al. (2010) presented the extent and thermal structure of the Martian cryosphere as the result of a one-dimensional finite difference thermal model (Clifford and Bartels 1986) enhanced to include the temperature-dependent thermal conductivity of ice and rock, the potential presence of hydrates, the exponential decline in crustal porosity with depth, different values of lithospheric heat flow, and long-term (~10^5–10^7 yr) astronomically-induced variations in insolation. These depths were calculated based on average surface temperatures derived from the 20 My nominal insolation history of Laskar et al. (2004), which represents a more relevant (and dynamic) boundary condition than the present-day mean annual surface temperatures. No attempt was made to simulate the dynamic redistribution of surface ice in response to climate change. Rather, fixed perennial polar caps, encompassing the area poleward of 80° latitude, were assumed. Diurnal, seasonal, and mean annual surface temperatures were then calculated at appropriate intervals to maximize convergence at the desired temporal resolution. Interestingly, although these astronomically driven changes in insolation have led to significant variations in mean annual surface temperature over the last 20 My, the mean latitudinal surface temperature over this period is generally within a few degrees of the current annual mean. The last 20 My surface temperature variations are shown in Fig. 16 at latitudes of 0°, 30°, 60°, and 90°, showing the zone around 60° where obliquity influence is minimal as shown by Schorghofer (2008).

Figure 17, from Clifford et al. (2010), presents the depth of a water-ice cryosphere, for three groundwater freezing temperatures (273 K for pure water, 252 K for a eutectic solution of NaCl, and 203 K for a eutectic solution of $Mg(ClO_4)_2$), two values of mean global

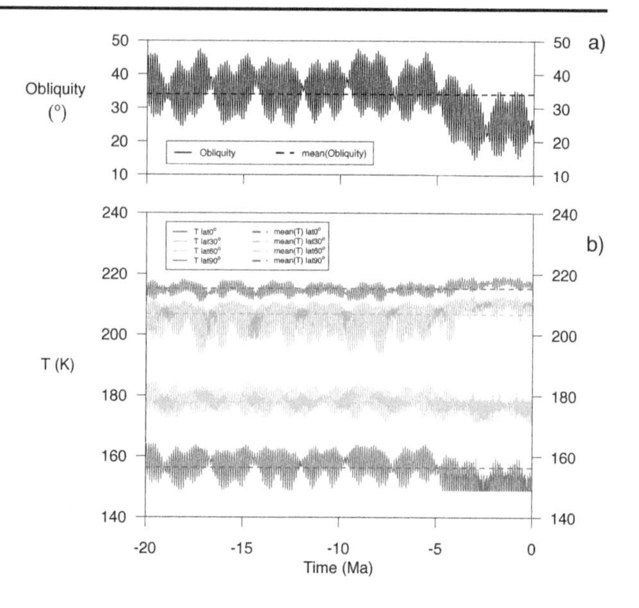

Fig. 16 Orbital parameters induced variations in mean surface temperature at four Mars latitudes over the past 20 Ma. (**a**) The 20 Ma nominal obliquity history of Mars (Laskar et al. 2004). (**b**) Corresponding variation in mean annual surface temperature at 0°, 30°, 60° and 90° latitude. Note that the corresponding 20 Ma mean surface temperatures (*horizontal dashed lines*) are 216 K for 0°, 208 K for 30°, 178 K for 60°, and 157 K for 90°

heat flow (15 m Wm^{-2} and 30 mW m^{-2}) and two values of desiccated equatorial regolith thermal conductivity (0.1 Wm^{-1} K^{-1} and 1.0 Wm^{-1} K^{-1}). The plots show the depth of the hypothetical cryosphere over the 0° to 90° latitude range (x-axis).

Comparing Figs. 17a and 17b illustrates the impact ($\Delta z \sim 4$ km) that a fine-grained, desiccated, low-conductivity regolith layer has on reducing the maximum depth of the cryosphere near the equator. This effect vanishes by mid-latitudes, as the depth of regolith desiccation declines. Assuming a basal temperature of 273 K, and the previous best estimate of lithospheric heat flow ($Q_g = 30$ mW m^{-2}), the depth of the equatorial cryosphere ranges from 0.5 to 4.5 km (from the 'low-k' to the 'high-k' model). This depth increases to 10.5 km under the poles. When Q_g is reduced to our present best estimate of 15 mW m^{-2}, these estimates increase by a factor of two. Significant reductions in depth occur when the base of the cryosphere is in contact with a eutectic brine of NaCl (freezing temperature of 252 K). Moreover, if sufficient perchlorate is present in the crust to lower the freezing point of groundwater to 203 K, then the cryosphere will be absent at all latitudes equatorward of 35° where it is in direct contact with the perchlorate brine—although it will still exist at higher latitudes, reaching a maximum thickness of between 4.5 km (30 mW m^{-2}) and 9 km (15 mW m^{-2}) at the poles.

Possible Presence of Clathrates The current potential for the biotic or abiotic production of methane at depth (which is contingent on the presence of liquid water) raises the possibility that the chief volatile component of the cryosphere is gas hydrate, rather than water ice (Fisk and Giovannoni 1999; Max and Clifford 2000).

It is also possible that if groundwater no longer survives on Mars, substantial amounts may still survive in the form of gas hydrate, produced during earlier epochs when groundwater was abundant (Prieto-Ballesteros et al. 2006).

Mousis et al. (2012a) have compared the stability domain of CO_2-dominated clathrate against several mean surface temperature profiles of ancient Mars expressed as a function of CO_2 atmospheric pressure. They concluded that, despite potentially significant greenhouse effects, it is perfectly possible that large amounts of clathrate deposits have formed at favorable Martian latitudes. Assuming a CO_2-dominated Martian paleoatmosphere, whose sur-

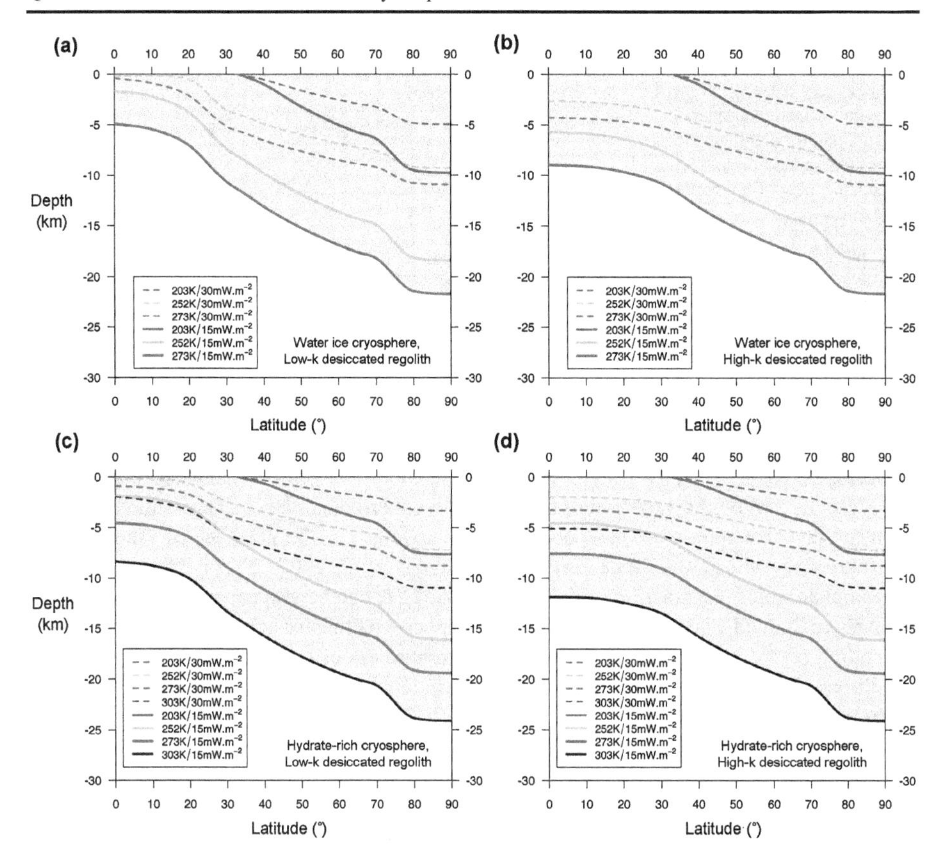

Fig. 17 Latitudinal variation in the depth of a water-ice cryosphere for three different groundwater freezing temperatures (273, 252, and 203 K), two values of assumed lithospheric heat flow (15 $\mathrm{mW\,m^{-2}}$ and 30 $\mathrm{mW\,m^{-2}}$), and two models of the thermal conductivity of the desiccated equatorial regolith for a (**a**) "low-k" model (0.1 $\mathrm{Wm^{-1}\,K^{-1}}$) and (**b**) a "high-k" model (1.0 $\mathrm{Wm^{-1}\,K^{-1}}$). (**c**) and (**d**) Subsurface extent of a gas hydrate cryosphere for the same variables as in Figs. 4a and 4b but assuming a basal temperature of 303 K, which approximates the base of the gas hydrate stability zone for the pressure conditions expected at depths of $\geq$10 km on Mars. See text and Clifford et al. (2010) for additional details

face pressure ranges from its current value up to 3 bars, and containing both CH_4 (whether derived from biogenic or abiogenic processes) and SO_2 (derived from volcanism) as minor species, these deposits would have trapped the minor species in differing proportions. Investigations of clathrate formation with varying initial CH_4 atmospheric mixing ratios and a SO_2 mixing ratio of 10^{-6} (approximately the long-term average, Johnson et al. 2008) allow to determine the mole fractions of guest molecules trapped in Martian clathrates (Mousis et al. 2012b, this issue). The relative abundance of a guest species in a clathrate is defined as the ratio of the average number of guest molecules to the average total number of molecules incorporated. Calculations of the occupancy fractions of the guest species for a given type of cage and for a given type of clathrate are determined from the Langmuir constants, related to the strength of the interaction between each guest species and each type of cage (van der Waals and Platteeuw 1959).

Figure 18 was calculated using a set of intermolecular parameters fitted from recent experimental and simulations work for CO_2–H_2O, CH_4–H_2O and SO_2–H_2O interactions

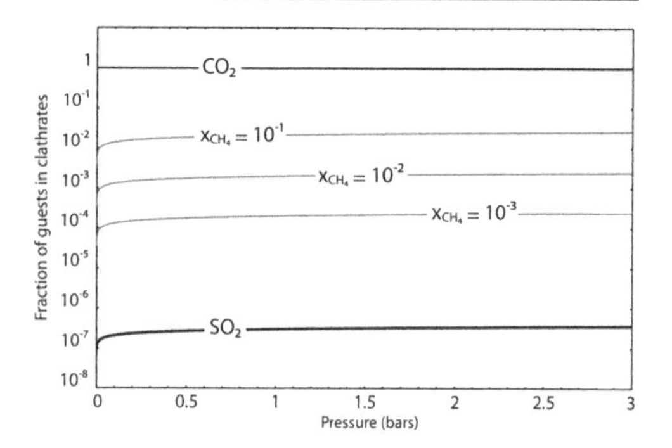

Fig. 18 Mole fractions of CO_2, SO_2 and CH_4 encaged in clathrates calculated as a function of the atmospheric surface pressure and for different values of the atmospheric mole fraction of methane x_{CH_4}. Four cases are considered: $x_{CH_4} = 0, 10^{-3}$, 10^{-2} and 10^{-1}. The *curves* describing the mole fractions of CO_2 and SO_2 as a function of x_{CH_4} appear superimposed

(Sloan and Koh 2008; Mousis et al. 2010, 2012a). It shows the mole fractions of species encaged in clathrates as a function of the surface pressure for four cases of CH_4 partial pressure (0, 10^{-3}, 10^{-2}, 10^{-1}). Interestingly, the mole fractions of CO_2 and CH_4 are almost not affected by the variation of the CH_4 atmospheric abundance. This behavior results from the fact that minor species poorly interact together in the presence of the dominant gas (here CO_2) during clathrate formation (Thomas et al. 2007, 2008). The calculations show that an important fraction of the atmospheric CH_4 can be sequestered into the Martian cryosphere. Indeed, about 4 to 40 % of the atmospheric methane can be trapped in clathrates, depending on the adopted surface pressure and regardless of its initial abundance. In these conditions, if large amounts of CO_2 are presently stored in the form of clathrates into the Martian cryosphere, then important masses of CH_4 should be equally present. Moreover, a maximum of 50 % of the initial abundance of atmospheric SO_2 is found to be trapped in CO_2-dominated clathrates, also depending on the considered surface pressure. The calculated fraction of sequestrated SO_2 is probably a minimum value since recent experiments (Beeskow-Strauch et al. 2011) show that the trapping efficiency of this compound should be even higher in these clathrates. In any case, given the expected very low atmospheric abundance of SO_2 (about 10^{-6} based on the long-term average atmospheric SO_2 value from Johnson et al. 2008), the full trapping of this compound in clathrates should not affect their ability to act as an efficient sink of CH_4. This confirms that another sink mechanism, most likely photochemistry (Johnson et al. 2008), must be invoked in order to explain the complete extinction of this molecule in the Martian atmosphere.

If the cryosphere is mostly constituted of clathrates, this has the potential to lower the effective thermal conductivity of the crust in proportion to its relative volumetric contribution. Figures 17c and 17d represent the gas hydrate equivalents of the water-ice cryospheres, where we have assumed all the available pore space to be completely saturated with clathrate hydrates and a surface porosity of 0.35. While the lower effective thermal conductivity of hydrates means shallower isotherm depths for a saturated cryosphere, under a confining pressure of $\sim$100 MPa, equivalent to the lithostatic pressure at a depth of $\sim$10 km on Mars (assuming a crustal density of 2.5×10^3 kg m^{-3}), hydrate is stable to temperatures in excess of $\sim$303 K (Max and Clifford 2000). Thus, a gas hydrate cryosphere extends to even greater depths than one composed of water ice alone (Figs. 17c and 17d).

Consequences for the Global Inventory of Water Clifford and Parker (2001) calculated that a planetary inventory of 500 m GEL could have been cold trapped into the thickening cryosphere by the end of the Late Hesperian ($\sim$3 Gy). With the nearly twofold increase in the potential maximum thickness of the cryosphere suggested here the prospects for long-term survival of subpermafrost groundwater are in greater doubt. Table 7 gives storage capacity of the Martian cryosphere based on the isotherm profiles shown in Fig. 17. In the

Table 7 Occupied cryosphere pore volume expressed in meters of GEL based on Fig. 17. Low-k ($0.1\ \mathrm{W\,m^{-1}\,K^{-1}}$) and high-k ($1.0\ \mathrm{W\,m^{-1}\,K^{-1}}$) models of the desiccated equatorial regolith are shown. $\Phi(0)$ correspond to the surface porosity used to infer the porosity decrease with depths (see Clifford et al. 2010)

Basal melting/Groundwater freezing T (K)	$Q_g = 15\ \mathrm{mW\,m^{-2}}$ High-k equatorial regolith		$Q_g = 30\ \mathrm{mW\,m^{-2}}$ Low-k equatorial regolith	
	$\Phi(0) = 0.2$	$\Phi(0) = 0.35$	$\Phi(0) = 0.2$	$\Phi(0) = 0.35$
273	585	1025	435	760
252	565	990	340	595
203	205	360	135	235

preceding section, we determined that for $\Phi(0) = 0.2$ and 0.35, the total pore volume of the crust is between ~600 m GEL and ~1000 m GEL respectively. Therefore, even assuming a high ($30\ \mathrm{mW\,m^{-2}}$) heat flow and a eutectic solution of NaCl, the present-day occurrence of groundwater on Mars appears highly unlikely, given the smaller (<500 m GEL) estimates for the total planetary inventory given in Table 4. But even under these limited conditions, some subpermafrost groundwater may still survive in places where it is perchlorate-rich, although such occurrences are likely to be restricted to isolated pockets rather than any system of regional or global extent.

If the planetary inventory of H_2O is closer to the upper estimate of ~1 km GEL from Table 4, then subpermafrost groundwater may still persist beneath much of the surface, but will generally reside at depths ≥5 km (Fig. 17), or roughly twice as deep as previously thought (Clifford 1993; Clifford and Parker 2001). However, natural variations in crustal thermal conductivity, heat flow, and groundwater composition, may permit isolated occurrences of groundwater to exist at significantly shallower depths (Clifford et al. 2010).

4.2 Mars Geological History and the Evolution of the Cryosphere

4.2.1 Climate Variations of Mars

Because the obliquity and orbital parameters of Mars vary with time, so too does the mean annual insolation and temperature at any given latitude (Clifford et al. 2010). The obliquity of Mars (currently 25.2°) varies with a period of $\sim 1.2 \times 10^5$ yrs, having maximum-to-minimum fluctuations of as much as 20–30 degrees. The amplitude of this oscillation is also modulated with a period of 1.3×10^6 years (Ward 1992) with extreme values ranging from 0° to over 60° (Touma and Wisdom 1993; Laskar et al. 2004). The system becomes chaotic on timescales >20 My. The mean obliquity is $\mu_{\mathrm{obl}} = 34.6°$ with a standard deviation of $\sigma_{\mathrm{obl}} = 9.96°$, while the mean eccentricity is $\mu_{\mathrm{ecc}} = 0.07$ with a standard deviation of $\sigma_{\mathrm{ecc}} = 0.026$ (Laskar et al. 2004).

For a given obliquity and eccentricity, the mean annual insolation, $S(l)$, at a latitude l can be estimated from the formula of Ward (1974):

$$S(l) = \left[\frac{S_0}{2\pi\sqrt{1-e^2}}\right] \cdot \int_0^{2\pi} \left[1 - \left(\sin(l)\cos(i) - \cos(l)\sin(i)\sin(\varphi)\right)^2\right]^{1/2} d\varphi \qquad (1)$$

where i is the obliquity, e the eccentricity, S_0 the solar constant at the orbit's semi-major axis and φ is the diurnal variation of insolation. From this mean insolation, the corresponding mean annual surface temperature can then be calculated based on the assumption of radiative equilibrium:

$$T_m(l) \approx \left[1.05 \times \frac{(1-A)S(l)}{\varepsilon\sigma}\right]^{1/4} \tag{2}$$

where $T_m(l)$ is the mean annual temperature at a latitude l, A is the albedo, ε is the emissivity, σ is the Stefan-Boltzmann constant, and the numerical coefficient '1.05' reflects the 5 % increase in mean annual radiation from the atmosphere, relative to the mean annual insolation at the surface as measured by the Viking landers (Clifford et al. 2010).

Mean annual temperatures that are calculated using this approach are generally accurate at latitudes below ~45° and at latitudes above the perimeter of the perennial polar caps (~80°–85°). However, at those latitudes that lie in between, Eq. (2) yields mean annual temperatures that consistently exceed observed Martian values. This is because, above ~45°, part of the mean annual insolation is reflected back into space due to the formation of the seasonal polar caps. Therefore the diurnal and seasonal condensation and sublimation of atmospheric CO_2 has been taken into account (Clifford et al. 2010).

4.2.2 Evolution of the Cryosphere

Various lines of geomorphic and mineralogical evidence suggest that early Mars was at least intermittently warm and wet. The improved preservation state of post-Noachian terrains suggests a geologically rapid transition to less erosive conditions, similar to those which exist on Mars today (Carr 1999). With the transition to a colder climate, a freezing front developed in the planet's crust, creating a growing cold-trap for both atmospheric and subsurface H_2O. The downward propagation of this freezing front, in response to the planet's long-term decline in geothermal heat flux, had two important consequences for the nature of the hydrologic cycle and the state and distribution of subsurface water (Clifford 1993; Clifford et al. 2010).

First, condensation of ice within the near-surface regolith would have prevented communication between groundwater and the atmosphere, and resulted in a ground ice distribution governed by the thermal structure of the crust mirroring the first order variations in surface topography (Clifford 1993). The elimination of atmospheric recharge would have led to the decay of any precipitation-driven influence on the shape of the global water table in ~10^6–10^8 years. In the absence of major seismic or thermal disturbances, this would have resulted in an aquifer in effective hydrostatic equilibrium saturating the lowermost porous regions of the crust. The vertical distance between the base of the cryosphere and the groundwater table may have varied considerably, creating an intervening unsaturated zone whose thickness was maximized in regions of high elevation and minimized (or absent) at low elevation (see Fig. 1, Clifford 1993).

Second, as the cryosphere deepened with time, it would have created a growing sink for the planet's inventory of groundwater. Where the cryosphere and water table were in direct contact, the groundwater would have frozen to the base of the cryosphere as the freezing front propagated downward with time. However, in many locations throughout the highlands, the vertical distances separating the base of the cryosphere from the water table may have been up to several kilometers or more. Under such conditions, the depletion of groundwater is expected to have occurred by the thermally induced diffusion of vapor from warmer depths to the colder pores behind the advancing freezing-front (Clifford 1991, 1993).

We can estimate the thickness of the cryosphere at the time of this transition based on the assumptions of Sect. 4.1 and assuming (1) a mean value of Late-Noachian geothermal

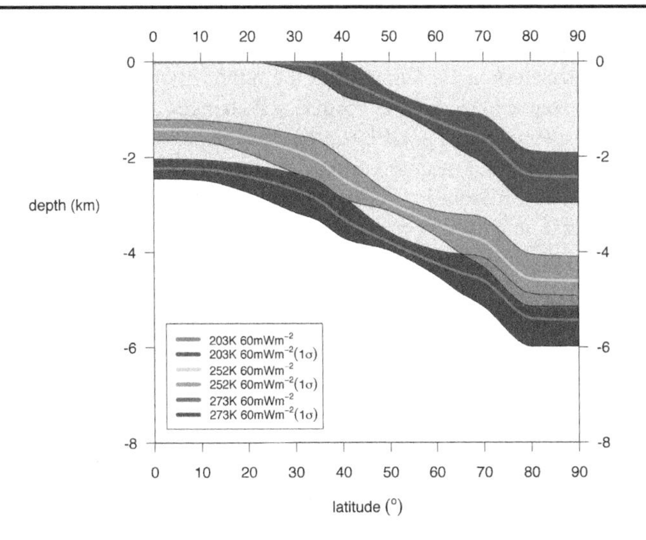

Fig. 19 Latitudinal variation in the depth of a water-ice cryosphere for three different groundwater freezing temperatures (273, 252, and 203 K), a geothermal heat flow of 60 mW m^{-2}, and obliquity and orbital eccentricity characteristics given by the probability distribution functions of Laskar et al. (2004). *Solid lines* represent cryosphere depths calculated from the average values of obliquity and orbital eccentricity, while the *shaded region* around each correspond to the maximum and minimum depths calculated based on the 1σ high and low extremes of obliquity and orbital eccentricity

heat flux of about 60 mW m^{-2} (based on the topographic relief of Noachian-age topography and its rheologic implications for the thickness of the elastic lithosphere at the time, McGovern et al. 2004; Solomon et al. 2005) and (2) that the variations in Martian obliquity and orbital eccentricity (Sect. 4.2.1) were comparable to those of today (an assumption whose validity is dependent on the developmental state of Tharsis at that time, Ward 1979). Figure 19 illustrates the variation in the depth of the cryosphere as a function of latitude based on the average and extreme obliquity and orbital eccentricity values of Laskar et al. (2004), combined with an assumed Noachian geothermal heat flux of 60 mW m^{-2}, and three groundwater freezing temperatures (273 K, 252 K and 203 K, see Sect. 4.1 for details).

The resulting calculated maximum depth of the cryosphere (based on the assumption of a pure water freezing point) varies from ~2.3 km near the equator up to ~5.6 km near the poles, where mean annual temperatures are significantly colder due to the combination of lower insolation and the higher albedo of the seasonal and perennial polar ice deposits (which, for the purposes of these calculations, are assumed to have had an extent comparable to that of today's). For a freezing point appropriate to a saturated solution of magnesium perchlorate (i.e., 203 K), the extent of the cryosphere is much reduced, reaching a maximum thickness at the poles of ~2.5 km and progressively diminishing at lower latitudes until it completely disappears at latitudes of less than 30°–35°, where mean annual temperatures exceed the assumed perchlorate-depressed freezing point of groundwater (203 K). Based on calculations similar to Sect. 4.1, we find that a Noachian cryosphere could have cold-trapped the equivalent of ~300 m (±100 m) GEL of H_2O. This suggests that, by the end of the Noachian, the minimum surviving inventory of subpermafrost groundwater was still on the order of a couple of hundred meters (GEL), providing enough water for the later erosion of the Martian outflow channels.

From Fig. 19, we also observe that the latitudes from 50° to 60° are those which are least affected by obliquity-driven variations in surface temperature (Schorghofer 2008; Clifford et al. 2010). However, near the equator the variations in the extent of the cryosphere reaches as much as 500 m based on a 1σ variation of the obliquity, while at the poles, changes in cryospheric depth can reach as much as 1 km. Such changes are equivalent to the obliquity driven transport of several tens of m GEL (or the equivalent volume of both polar caps) from low-latitudes to high-latitudes, a quasiperiodic exchange that continues to the present day (e.g. Forget et al. 2006). In addition to the changes in cryospheric depth, extensive thawing of the near-surface may also occur at mid- to high-latitudes at times of high obliquity, a condition that will be enhanced by the presence of freezing-point depressing salts.

The development of an ice-rich cryosphere several kms deep is consistent with a large initial inventory of water and the subsequent dynamic exchange of water between the polar and non-polar reservoirs by atmospheric transport, surface runoff and groundwater flow, the latter two driven by polar basal melting (Clifford 1987, 1993; Clifford and Parker 2001; Head and Pratt 2001).

However, should the surviving planetary inventory of water be much less than a 500 m GEL (either because of a small initial inventory, or because significant amounts of water have been lost by exospheric escape or the formation of hydrated minerals), then much, if not all, of the surviving inventory of water may have been cold-trapped into the cryosphere—precluding the potential for groundwater transport, driven by polar basal melting, to complement insolation-driven, long-term atmospheric exchange (Clifford 1993).

Grimm and Painter (2009) considered one possible scenario of the hydrologic evolution of Mars, with an initially small inventory of water. Their 2D longitudinal model, which assumed that lateral heterogeneities inhibited horizontal groundwater flow, predicted a progressive drying out of the crust at low-latitudes (<30°) in less than 100 My, although subpermafrost groundwater survived at higher latitudes for several hundred million years longer. However, it appears difficult to reconcile this vision of Mars with the fact that outflow channel activity (which has occurred primarily at equatorial latitudes) peaked around 3.4–2 Gy, with some activity occurring as recently as several million years ago (see discussion in Sect. 2.3.3 and Table 3). These observations reinforce the conclusion that Mars has remained water-rich throughout most of its geologic history—and may continue to be so through the present day.

4.3 Implications for Habitability on Mars

Present thinking suggests that, for an environment to be habitable, liquid water must be available in a temperature range suitable for life (Bennett and Shostak 2007). As we have seen in the previous discussion, there are several Martian environments that would satisfy this requirement, which we discuss here, as well as their relevance to the further exploration of the planet.

Impact Induced and Volcanic Hydrothermal Systems Impact driven or volcanic processes include: temperature rise, and shock-induced fracturing, and partial melting of the crust and the Martian cryosphere. Under today's Martian climatic conditions, liquid water on Mars' surface would boil in most places, because the vapor pressure of water at 274 K (6.57 mbar) is above the mean atmospheric pressure on the surface of Mars (6.36 mbar). However, in a cold climate a standing body of water could freeze on its surface while being kept liquid at depth by residual heat (Newsom 2010). The prolonged existence of liquid water and the resulting production and accumulation of sediments will further enhance the diversity of potential habitats in the crater.

At the same time, fracturing linked to these phenomena creates new surfaces but also opens pathways and enhances the permeability of the target. For biology, the fractures might provide important shelter from adverse surface conditions such as UV irradiation or desiccation (e.g. Cockel et al. 2005). The fractured crust will promote the circulation of warm water to the surface, where it may accumulate in crater interiors, forming long-lived lakes. In this way, the presence of fractures and the occurrence of hydrothermal circulation will significantly affect both the extent of the cryosphere and the available pathways for the circulation of water from deep aquifers and other liquid water reservoirs, permitting the migration of both nutrients and potential organisms from one habitat to the other. Studies have shown that medium-sized impact craters (10s of km in diameter) are sufficient to puncture the cryosphere, fracture the crust, form crater lakes, and sustain the hydrothermal circulation of water between liquid water reservoirs and potential habitats (Newsom 1980; Abramov and Kring 2005; Schwenzer et al. 2012).

Near-Surface Thawing Cycles We have discussed that at low-latitudes, local thawing can happen even today depending on the availability and concentration of freezing-point depressing salts, such as calcium chloride and magnesium perchlorates in the near-subsurface, consistent with the inferred origin of the seasonally recurrent dark slope streaks observed by McEwen et al. (2011). On obliquity variation timescales, such melting could potentially create transient habitable environments in the shallow subsurface down to depths as great as several 10s of meters. In such instances, other environmental stresses (e.g., desiccation, oxidation and radiation exposure) become the primary deterrent to organism survival (Johnson et al. 2011). Any hypothetical Martian ecosystem may have evolved physiological traits similar to terrestrial cold and desiccation-tolerant organisms that allow sporadic metabolism during periods of increased water activity.

Aquifer Habitability Finally, deep aquifers below the cryosphere may have provided a hydraulic connection between various subpermafrost habitats (Clifford 1993). If Mars were ever inhabited, these hydraulic connections would likely have provided a means for biota to be transported from one habitable environment to another. An analogous system is fracture networks within or under permafrost in the terrestrial arctic. These systems harbor sulfate-reducing microorganisms and other anaerobic taxa that can grow within the cold, saline conditions of the permafrost (Onstott et al. 2009). Analogous conditions may exist within the Martian deep-subsurface where impact-generated fractures may have allowed both microorganisms and nutrients to migrate from one habitat to another—even ones arising from recent impacts and their associated hydrothermal environments, if habitats on Mars were inhabited and life existed on that planet (Schwenzer et al. 2012).

Out of these three potential habitats, the first two are the most easily accessible and will be probed further by the next Martian missions, such as Mars Science Laboratory that has successfully landed on Mars in August 2012 and is currently exploring the complex hydrous system of Gale crater. By their global occurrence, aquifers could give the highest probability of past or present life detection, however, the likely depth of subpermafrost aquifers makes them improbable targets for the near-term exploration of Mars.

5 Conclusions

In this work, we have reviewed the global inventory of water on Mars based on the latest theoretical studies, surface analyses and measurements. The geomorphological evidence suggests that, at the peak of outflow channel activity (∼2–3.4 Gy), Mars possessed a total water inventory ranging from several 100 m up to 1 km GEL. On the other hand, current subsurface investigations, including deep penetrating radar measurements by SHARAD and MARSIS account for a global total of about 30 m GEL, far short of that required to explain all the inferred water-related geological features of the planet. This discrepancy could be explained by the inability of present radar investigations to sound to the depth of subpermafrost groundwater, whether due to its great depth or the highly attenuating nature of the intervening crust (Clifford et al. 2010). Another possibility is that most of the planetary inventory of water necessary to explain Hesperian and Amazonian aqueous landforms has now been bound up in the formation of hydrated minerals. A true understanding of the current state of the Martian subsurface must await global investigations by a variety of seismic and electromagnetic geophysical techniques, aided by numerical simulations. This will, in turn, provide valuable insights into whether the Martian deep subsurface still retains the potential for harboring life.

Acknowledgements The authors thank the editor Mike Toplis, and 3 referees for careful comments on this manuscript. The authors also thank the staff of the International Space Science Institute (ISSI) in Bern, Switzerland for hospitality and assistance. Additional thanks to the HiRISE, MOC, MOLA, TES, THEMIS, SHARAD and MARSIS teams for providing excellent data to the scientific community. This work was partly supported by Los Alamos LDRD program. This is SC LPI contribution number 1704.

References

O. Abramov, D.A. Kring, J. Geophys. Res. (2005). doi:10.1029/2005JE002453
D.M. Anderson, L.W. Gatto, F.C. Ugolini, Antarct. J. U.S. **7**, 114 (1972)
J.C. Andrews-Hanna, R.J. Phillips, J. Geophys. Res. (2007). doi:10.1029/2006JE002881
V. Ansan, N. Mangold, Planet. Space Sci. **54**, 219 (2006)
V. Ansan, N. Mangold, P. Masson, E. Gailhardis, G. Neukum, J. Geophys. Res. (2008). doi:10.1029/2007JE002986
V. Ansan et al., Icarus **211**, 273 (2011)
C.B. Atkins, P.J. Barrett, S.R. Hicock, Geology **30**, 659 (2002)
V.R. Baker, Nature **412**, 228 (2001)
V.R. Baker, D.J. Milton, Icarus **23**, 27 (1979)
V.R. Baker, D.J. Milton, Nature **352**, 589 (1991)
V.R. Baker, M.H. Carr, V.C. Gulick, C.R. Williams, M.S. Marley, in *Mars*, ed. by H.H. Kieffer, B.M. Jakosky, C.W. Snyder, M. Matthews (University Arizona Press, Tucson, 1992), pp. 493–522
M.R. Balme, N. Mangold, D. Baratoux, F. Costard, M. Gosselin, P. Masson, P. Pinet, G. Neukum, J. Geophys. Res. (2006). doi:10.1029/2005JE002607
M.R. Balme, C.J. Gallagher, D.P. Page, J.B. Murray, J.P. Muller, Icarus **200**, 30 (2009)
S. Barabash, A. Fedorov, R. Lundin, J.A. Sauvaud, Science **315**, 501 (2007)
D. Baratoux, C. Delacourt, P. Allemand, Geophys. Res. Lett. (2002). doi:10.1029/2001GL013779
N.G. Barlow, T.L. Bradley, Icarus **87**, 156 (1990)
B. Beeskow-Strauch, J.M. Schicks, E. Spangenberg, J. Erzinger, Chemistry (2011). doi:10.1002/chem.201003262
D.I. Benn, D.J.A. Evans, *Glaciers and Glaciation* (Edward Arnold, London, 1998)
J. Bennett, S. Shostak, *Life in the Universe* (Pearson Education, San Francisco, 2007)
D.C. Berman, W.K. Hartmann, Icarus **159**, 1 (2002)
J.P. Bibring et al., Science **312**, 400 (2006)
A.B. Binder, M.A. Lange, J. Geophys. Res. **85**, 3194 (1980)
J.G. Bockheim, K.J. Hall, South Afr. J. Sci. **98**, 82 (2002)
J.G. Bockheim, C. Tarnocai, in *Proc. Int. Conf. Permafrost* 7 (1998), p. 57
S. Bouley, R.A. Craddock, N. Mangold, V. Ansan, Icarus **207**, 686 (2010)
W.V. Boynton et al., Science **297**, 81 (2002)
W.V. Boynton, G.J. Taylor, S. Karunatillake, R.C. Reedy, J.M. Keller, in *The Martian Surface: Composition, Mineralogy, and Physical Properties*, ed. by J. Bell (Cambridge University Press, New York, 2008), pp. 105–124
W.V. Boynton et al., Science **325**, 61 (2009)
G.W. Brass, Icarus **42**, 20 (1980)

J. Brown, O.J. Ferrians, J.A. Heginbottom, E.S. Melnikov, Circum-arctic map of permafrost and ground ice conditions: International Permafrost Association, ed., scale 1:10,000,000, U.S. Geol. Survey, Circum-Pacific Map Series, Map CP-45 (1997)
D.M. Burr, A.S. McEwen, S.E.H. Sakimoto, Geophys. Res. Lett. (2002a). doi:10.1029/2001GL0013345
D.M. Burr, J.A. Grier, A.S. McEwen, L.P. Keszthelyi, Icarus **159**, 53 (2002b)
S. Byrne, B.C. Murray, J. Geophys. Res. (2002). doi:10.1029/2001JE001615
S. Byrne et al., Science **325**, 1674 (2009)
N.A. Cabrol, E.A. Grin, Icarus **142**, 160 (1999)
N.A. Cabrol, E.A. Grin, Icarus **149**, 291 (2001)
N.A. Cabrol, E.A. Grin, Glob. Planet. Change **35**, 199 (2003)
M.J. Campbell, J. Ulrichs, J. Geophys. Res. (1969). doi:10.1029/JB074i025p05867
M.H. Carr, J. Geophys. Res. **84**(B6), 2995 (1979)
M.H. Carr, in *The Geology of Terrestrial Planets*, ed. by M.H. Carr (1981), pp. 207–263. NASA Spec. Pub. SP-469, Washington
M.H. Carr, Icarus **68**, 187 (1986)
M.H. Carr, *Water on Mars* (Oxford University Press, New York, 1996)
M.H. Carr, J. Geophys. Res. **104**, 21897 (1999)
M.H. Carr, F.C. Chuang, J. Geophys. Res. **102**(E4), 9145 (1997)
M.H. Carr, G.D. Clow, Icarus **48**, 91 (1981)
M.H. Carr, J.W. Head III, J. Geophys. Res. (2003). doi:10.1029/2002JE001963
M.H. Carr, G.G. Schaber, J. Geophys. Res. **82**, 4039 (1977)
L.M. Carter, F. Poulet, Icarus **219**, 250 (2012)
L.M. Carter et al., Icarus **199**, 295 (2009)
Y. Cedillo-Flores, A.H. Treiman, J. Lasue, S.M. Clifford, Geophys. Res. Lett. (2011). doi:10.1029/2011GL049403
M.G. Chapman, Icarus **109**, 393 (1994)
M.A. Charette, W.H.F. Smith, Oceanography **23**, 112 (2010)
E. Chassefière, Icarus **124**, 537 (1996)
E. Chassefière, F. Leblanc, Earth Planet. Sci. Lett. **310**, 262 (2011)
V. Chevrier, F. Poulet, J.P. Bibring, Nature **448**, 60 (2007)
P.R. Christensen, Nature **422**, 45 (2003)
B.C. Clark, D.C. van Hart, Icarus **45**, 370 (1981)
C. Clauser, E. Huenges, in *Rock Physics and Phase Relations: A Handbook of Physical Constants*, ed. by T.J. Ahrens. AGU Ref. Shelf, vol. 3 (AGU, Washington, 1995), pp. 105–126
S.M. Clifford, J. Geophys. Res. **92**, 9135 (1987)
S.M. Clifford, Geophys. Res. Lett. **18**, 2055 (1991)
S.M. Clifford, J. Geophys. Res. **98**, 10973 (1993)
S.M. Clifford, C.J. Bartels, Lunar Planet. Sci. **XVII**, 142 (1986)
S.M. Clifford, D. Hillel, J. Geophys. Res. **88**, 2456 (1983)
S.M. Clifford, T.J. Parker, Icarus (2001). doi:10.1006/icar.2001.6671
S.M. Clifford, J. Lasue, E. Heggy, J. Boisson, P. McGovern, M.D. Max, J. Geophys. Res. (2010). doi:10.1029/2009JE003462
C.S. Cockel, P. Lee, P. Broady, D.S.S. Lim, G.R. Osinski, J. Parnell, C. Koeberl, L. Pesonen, J. Salminen, Meteorit. Planet. Sci. **40**, 1901 (2005)
A. Colaprete, B.M. Jakosky, J. Geophys. Res. **103**, 5897 (1998)
N.M. Coleman, J. Geophys. Res. (2005). doi:10.1029/2005JE002419
Committee on an Astrobiology Strategy for the Exploration of Mars, *An Astrobiology Strategy for the Exploration of Mars* (National Academies Press, Washington, 2007). http://www.nap.edu/catalog.php?record_id=11937. Accessed 19 July 2011
Committee on Assessing the Solar System Exploration Program, *Grading NASA's Solar System Exploration Program: a Midterm Review*, (National Academies Press, Washington, 2008). http://www.nap.edu/catalog.php?record_id=12070. Accessed 19 July 2011
Committee on the Planetary Science Decadal Survey, *Vision and Voyages for Planetary Science in the Decade 2013–2022* (National Academies Press, Washington, 2011), http://www.nap.edu/catalog.php?record_id=13117. Accessed 19 July 2011
S.J. Conway, N. Mangold, V. Ansan, Lunar Planet. Sci. **XLII**, 2174 (2011)
F.M. Costard, J.S. Kargel, Icarus **114**, 93 (1995)
F.M. Costard, F. Forget, N. Mangold, J.P. Peulvast, Science **295**, 110 (2002)
R.A. Craddock, A.D. Howard, J. Geophys. Res. (2002). doi:10.1029/2001JE001505
D.A. Crown, L.F. Bleamaster III, S.C. Mest, J. Geophys. Res. (2005). doi:10.1029/2005JE002496

D. Davidson, in *Natural Hydrates: Properties, Occurrence and Recovery*, ed. by J.L. Cox, (Butterworth, Woburn, 1983), pp. 1–16
R.A. de Hon, Earth Moon Planets **56**, 95 (1992)
E. Dehouck, N. Mangold, S. Le Mouélic, V. Ansan, F. Poulet, Planet. Space Sci. (2010). doi:10.1016/j.pss.2010.02.005
G. Di Achille, B.M. Hynek, N.T. Bridges, C.M. Dundas, A.S. McEwen, Geology (2010). doi:10.1130/G31287.1
S. Dienega, S. Byrne, N.T. Bridges, C.M. Dundas, A.S. McEwen, Geology (2010). doi:10.1130/G31287.1
B. Diez, W.C. Feldman, S. Maurice, O. Gasnault, T.H. Prettyman, M.T. Mellon, O. Aharonson, N. Schorghofer, Icarus (2008). doi:10.1016/j.icarus.2008.02.006
J.C. Dixon, C.E. Thorn, R.G. Darmody, Z. Geomorphol. **52**, 27 (2008)
W. Dobiński, Earth-Sci. Rev. **108**, 158 (2011)
T.M. Donahue, Icarus **167**, 225 (2004)
P.T. Doran, W.B. Lyons, D.M. McKnight (eds.), *Life in Antarctic Deserts and Other Cold Dry Environments* (Cambridge University Press, Cambridge, 2010)
M.J. Drake, K. Righter, Nature **416**, 39 (2002)
C.M. Dundas, S. Byrne, Icarus **206**, 716 (2010)
C.M. Dundas, A.S. McEwen, Icarus **205**, 244 (2010)
C.M. Dundas, A.S. McEwen, S. Diniega, S. Byrne, S. Martinez-Alonso, Geophys. Res. Lett. (2010). doi:10.1029/2009GL041351
W.B. Durham, A.V. Pathare, L.A. Stern, H.J. Lenferink, Geophys. Res. Lett. (2009). doi:10.1029/2009GL040392
B.L. Ehlmann et al., Nature **322**, 1828 (2008)
B.L. Ehlmann, J.F. Mustard, S.L. Murchie, Geophys. Res. Lett. **37**, L06201 (2010). doi:10.1029/2010GL042596
B.L. Ehlmann, J.F. Mustard, S.L. Murchie, J.P. Bibring, A. Meunier, A.A. Fraeman, Y. Langevin, Nature **479**, 53 (2011)
B. Etzelmüller, J.O. Hagen, in *Cryospheric Systems Glaciers and Permafrost*, ed. by C. Harris, J.B. Murton. Special Publications, vol. 242 (Geological Society, London, 2005), pp. 11–27
A.G. Fairén, Icarus **208**, 165 (2010)
A.G. Fairén, J.M. Dohm, V.R. Baker, M.A. de Pablo, J. Ruiz, J.C. Ferris, R.C. Anderson, Icarus **165**, 53 (2003)
A.G. Fairén, D. Fernandez-Remolar, J.M. Dohm, V.R. Baker, R. Amils, Nature **431**, 423 (2004)
A.G. Fairén, A.F. Davila, L. Gago-Duport, R. Amils, C.P. McKay, Nature **459**, 401 (2009)
A.G. Fairén, A.F. Davila, L. Gago-Duport, J.D. Haqq-Misra, C. Gil, C.P. McKay, J.F. Kasting, Nat. Geosci. **4**, 667 (2011)
F.P. Fanale, Icarus **28**, 179 (1976)
F.P. Fanale, J.R. Salvail, A.P. Zent, S.E. Postawko, Icarus **67**, 1 (1986)
W.M. Farrell et al., J. Geophys. Res. (2008). doi:10.1029/2007JE002974
W.M. Farrell, J.J. Plaut, S.A. Cummer, D.A. Gurnett, G. Picardi, T.R. Watters, A. Safaeinili, Geophys. Res. Lett. (2009). doi:10.1029/2009GL038945
C.I. Fassett, J.W. Head III, Geophys. Res. Lett. (2005). doi:10.1029/2005GL023456
C.I. Fassett, J.W. Head III, Icarus **195**, 61 (2008)
W.C. Feldman et al., J. Geophys. Res. (2004). doi:10.1029/2003JE002160
W.C. Feldman, M.T. Mellon, O. Gasnault, B. Diez, R.C. Elphic, J.J. Hagerty, D.J. Lawrence, S. Maurice, T.H. Prettyman, Geophys. Res. Lett. (2007). doi:10.1029/2006GL028936
W.C. Feldman, M.C. Bourke, R.C. Elphic, S. Maurice, J. Bandfield, T.H. Prettyman, B. Diez, D.J. Lawrence, Icarus (2008a). doi:10.1016/j.icarus.2007.08.044
W.C. Feldman, J.L. Bandfield, B. Diez, R.C. Elphic, S. Maurice, S.M. Nelli, J. Geophys. Res. (2008b). doi:10.1029/2007JE003020
W.C. Feldman, M.T. Mellon, O. Gasnault, S. Maurice, T.H. Prettyman, in *The Martian Surface: Composition, Mineralogy, and Physical Properties*, ed. by J. Bell (Cambridge University Press, New York, 2008c), pp. 125–148
W.C. Feldman, A. Pathare, S. Maurice, T.H. Prettyman, D.J. Lawrence, R. Milliken, B.J. Travis, J. Geophys. Res. (2011). doi:10.1029/2011JE003806
K.E. Fishbaugh, J.W. Head III, Icarus **174**, 444 (2005)
M.R. Fisk, S.J. Giovannoni, J. Geophys. Res. **104**, 11805 (1999)
F. Forget, R.M. Haberle, F. Montmessin, B. Levrard, J.W. Head, Science **311**, 368 (2006)
J.L. Fox, Geophys. Res. Lett. **109**, A11310 (2004)
J.L. Fox, A.B. Hác, Icarus **204**, 527 (2009)

H.M. French, in *Advances in Periglacial Geomorphology*, ed. by M.J. Clark (Wiley, Chichester, 1988), pp. 151–198
H.M. French, in *Canada's Cold Environments*, ed. by H.M. French, O. Slaymaker (McGill-Queen's University Press, Montreal, 1996), pp. 143–170
H.M. French, *The Periglacial Environment*, 3rd edn. (Wiley, Chichester, 2007)
H.M. French, Periglacial, in *Encyclopedia of Snow, Ice and Glaciers*, ed. by V.P. Singh, P. Singh, U.K. Haritashya (Springer, Dordrecht, 2011), pp. 827–841
H. Frey, S. Sakimoto, J. Roark, Lunar Planet. Sci. **XXX**, 1500 (1999)
C. Gallagher, M.R. Balme, S.J. Conway, P.M. Grindrod, Icarus **211**, 458 (2011)
O. Gasnault, G.J. Taylor, S. Karunatillake, J. Dohm, H. Newsom, O. Forni, P. Pinet, W.V. Boynton, Icarus (2010). doi:10.1016/j.icarus.2009.11.010
G.J. Ghatan, J.R. Zimbelman, Icarus **185**, 171 (2006)
G.J. Ghatan, J.W. Head III, L. Wilson, Earth Moon Planets **96**, 1 (2005)
J.M. Goldspiel, S.W. Squyres, Icarus **148**, 176 (2000)
J.M. Goldspiel, S.W. Squyres, Icarus **211**, 238 (2011)
J.A. Grant et al., Geology **36**, 195 (2008)
R.A.F. Grieve, M.J. Cintala, Meteoritics **27**, 526 (1992)
C. Grima, W. Kofman, J. Mouginot, R.J. Phillips, A. Hérique, D. Biccari, R. Seu, M. Cutigni, Geophys. Res. Lett. (2009). doi:10.1029/2008GL036326
C. Grima, W. Kofman, A. Hérique, R. Orosei, R. Seu, Icarus **220**, 84 (2012)
R.E. Grimm, L. Painter, Geophys. Res. Lett. (2009). doi:10.1029/2009GL041018
P. Grizzaffi, P.H. Schultz, Icarus **77**, 358 (1989)
M. Grott, A. Morschhauser, D. Breuer, E. Hauber, Earth Planet. Sci. Lett. **308**, 391 (2011)
V.C. Gulick, V.R. Baker, J. Geophys. Res. **95**, 14325 (1990)
R.M. Haberle, C.P. McKay, J. Schaeffer, N.A. Cabrol, E.A. Grin, A.P. Zent, R. Quinn, J. Geophys. Res. **106**(E10), 23317 (2001)
W. Haeberli, Nor. Geogr. Tidsskr. **50**, 3 (1985)
B.C. Hahn, S.M. McLennan, E.C. Klein, Geophys. Res. Lett. (2011). doi:10.1029/2011GL047435
J.C. Hanna, R.J. Phillips, J. Geophys. Res. (2006). doi:10.1029/2005JE002546
C.J. Hansen et al., Science **331**, 575 (2011)
C. Harris, Permafrost, in *Encyclopedia of Geomorphology*, ed. by A.S. Goudie (Routledge, London, 2004), pp. 777–779
K.P. Harrison, R.E. Grimm, J. Geophys. Res. (2008). doi:10.1029/2007JE002951
S.K. Harrison, M.R. Balme, A. Hagermann, J.B. Murray, J.P. Muller, Icarus **209**, 405 (2010)
W. Hartmann, G. Neukum, Space Sci. Rev. **96**, 165 (2001)
E. Hauber, S. van Gasselt, M.G. Chapman, G. Neukum, J. Geophys. Res. (2008). doi:10.1029/2007JE002897
E. Hauber et al., Planet. Space Sci. **57**, 944 (2009)
E. Hauber et al., in *Martian Geomorphology*, ed. by M.R. Balme, A.S. Bargery, C.J. Gallagher, S. Gupta. Special Publications, vol. 356 (Geological Society of London, London, 2011a), pp. 111–131,
E. Hauber, et al., in *Analogs for Planetary Exploration*, ed. by W.B. Garry, J.E. Bleacher. Special Paper, vol. 483 (Geol. Soc. Am., Boulder, 2011b), pp. 177–201. doi:10.1130/2011.2483(12)
J.W. Head III, S. Pratt, J. Geophys. Res. (2001). doi:10.1029/2000JE001359
J.W. Head III, L. Wilson, Ann. Glac. **45**, 1 (2007)
J.W. Head III, H. Hiesinger, M.A. Ivanov, M.A. Kreslavsky, S. Pratt, B.J. Thomson, Science **286**, 2134 (1999)
J.W. Head III, J.F. Mustard, M.A. Kreslavsky, R.E. Milliken, D.R. Marchant, Nature **426**, 797 (2003a)
J.W. Head III, L. Wilson, K.L. Mitchell, Geophys. Res. Lett. (2003b). doi:10.1029/2003GL017135
J.W. Head III et al., Nature **434**, 346 (2005)
J.W. Head III, D.R. Marchant, M.A. Kreslavsky, Proc. Natl. Acad. Sci. USA **105**, 13258 (2008)
M.H. Hecht et al., Science **325**, 64 (2009)
E. Heggy, S.M. Clifford, R.E. Grimm, C.L. Dinwiddie, D.Y. Wyrick, B.E. Hill, J. Geophys. Res. (2006). doi:10.1029/2005JE002589
J.L. Heldmann, M.T. Mellon, Icarus **168**, 285 (2004)
K.E. Herkenhoff, J.J. Plaut, Icarus **144**(2), 243 (2000)
K.E. Herkenhoff, J.J. Plaut, S.A. Nowicki, Lunar Planet. Sci. **XXVIII**, 1575 (1997)
H. Hiesinger, J.W. Head, J. Geophys. Res. **105**(E5), 11999 (2000)
P.V. Hobbs, *Ice Physics* (Clarendon, Oxford, 1974)
N. Hoffman, Icarus **146**, 326 (2000)
J.W. Holt, S.D. Kempf, P. Choudhary, D.A. Young, D.E. Putzig, D. Biccari, Y. Gim, Science **322**, 1235 (2008)
K. Horai, J. Geophys. Res. **93**(B3), 4125 (1991)

A.D. Howard, J.M. Moore, R.P. Irwin III, J. Geophys. Res. (2005). doi:10.1029/2005JE002459
A.P. Ingersoll, Science **168**, 972 (1970)
R.P. Irwin III, Lunar Planet. Sci. **LXII**, 2748 (2011)
R.P. Irwin III, T.A. Maxwell, R.A. Craddock, A.D. Howard, D.W. Leverington, Science **296**, 2209 (2002)
R.P. Irwin III, A.D. Howard, T.A. Maxwell, J. Geophys. Res. (2004). doi:10.1029/2004JE002287
R.P. Irwin III, R.A. Craddock, A.D. Howard, Geology **33**, 489 (2005a)
R.P. Irwin III, A.D. Howard, R.A. Craddock, J.M. Moore, J. Geophys. Res. (2005b). doi:10.1029/2005JE002460
T. Ishii, S. Sasaki, Lunar Planet. Sci. **XXXV**, 1556 (2004)
B.M. Jakosky, R.G. Phillips, Nature **412**, 237 (2001)
D.G. Jankowsky, S.W. Squyres, Icarus **100**, 26 (1992)
D.G. Jankowsky, S.W. Squyres, Icarus **106**, 365 (1993)
R. Jauman et al., Planet. Space Sci. **55**, 928 (2007)
D.J. Jerolmack, D. Mohrig, M.T. Zuber, S. Byrne, Geophys. Res. Lett. (2004). doi:10.1029/2004GL021326
S.S. Johnson, M.A. Mischna, T.L.J. Grove, M.T. Zuber, J. Geophys. Res. (2008). doi:10.1029/2007JE002962
A.P. Johnson et al., Icarus **211**, 1162 (2011)
A. Johnsson et al., Lunar Planet. Sci. **XLI**, 2492 (2010)
A. Johnsson et al., Icarus **218**, 489 (2012)
R.L. Jordan et al., Planet. Space Sci. (2009). doi:10.1016/j.pss.2009.09.016
S.J. Kadish, J.W. Head, Icarus **213**, 443 (2011)
S.J. Kadish, N.G. Barlow, J.W. Head, J. Geophys. Res. (2009). doi:10.1029/2008JE003318
S.J. Kadish, J.W. Head, N.G. Barlow, Icarus **210**, 92 (2010)
J.S. Kargel, *Mars: A Warmer, Wetter Planet* (Springer, Chichester, 2004)
L.P. Knauth, D.M. Burt, Icarus **158**, 267 (2002)
T. Kneissl, D. Reiss, S. van Gasselt, G. Neukum, Earth Planet. Sci. Lett. **294**, 357 (2010)
W. Kofman, R. Orosei, E. Pettinelli, Space Sci. Rev. **153**, 249 (2010)
E.J. Kolb, K.L. Tanaka, Mars (2006). doi:10.1555/mars.2006.0001
P.D. Komar, Icarus **42**, 317 (1980)
M. Koutnik, S. Byrne, B. Murray, J. Geophys. Res. (2002). doi:10.1029/2001JE001805
M.A. Kreslavsky, J.W. Head III, J. Geophys. Res. **105**(E11), 26695 (2000)
A. Kress, J.W. Head III, Geophys. Res. Lett. (2008). doi:10.1029/2008GL035501
A. Kress, J.W. Head III, A. Safaeinili, J. Holt, J. Plaut, L. Posiolova, R. Philips, R. Seu (SHARAD Team), Lunar Planet. Sci. **XLI**, 1166 (2010)
A.I. Krivchikov, B.Y. Gorodilov, O.A. Korolyuk, V.G. Manzhelii, H. Conrac, W. Press, J. Low Temp. Phys. **139**, 693 (2005)
R.O. Kuzmin, *Cryolithosphere of Mars* (Nauka, Moscow, 1983)
R.O. Kuzmin, E.V. Zabalueva, Lunar Planet. Sci. **XXXIV**, 1912 (2003)
R.O. Kuzmin, E.D. Ershow, I.A. Komarow, A.H. Kozlov, V.S. Isaev, Lunar Planet. Sci. **XXXIII**, 2030 (2002)
H. Lammer, Space Sci. Rev. (2012, this issue). doi:10.1007/s11214-012-9943-8
H. Lammer, C. Kolb, T. Penz, U.V. Amerstorfer, H.K. Biernat, B. Bodiselitsch, Int. J. Astrobio. **2**(3), 195 (2003)
J. Laskar, B. Levrard, J.F. Mustard, Nature (2002). doi:10.1038/nature01066
J. Laskar, A.C.M. Correia, M. Gastineau, F. Joutel, B. Levrard, P. Robutel, Icarus **170**, 343 (2004)
H.J. Leask, L. Wilson, K.L. Mitchell, J. Geophys. Res. (2006). doi:10.1029/2005JE002550
Y. Lee, D. Deming, J. Geophys. Res. **103**, 2447 (1998)
A. Lefort, P.S. Russel, N. Thomas, A.S. McEwen, C.M. Dundas, R.L. Kirk, J. Geophys. Res. (2009). doi:10.1029/2008JE003264
A. Lefort, P.S. Russel, N. Thomas, Icarus **205**, 259 (2010)
P. Lemke et al., in *Climate Change 2007: The Physical Science Basis. Contribution of Working Group I to the Fourth Assessment Report of the Intergovernmental Panel on Climate Change*, ed. by S. Solomon, D. Qin, M. Manning, Z. Chen, M. Marquis, K.B. Averyt, M. Tignor, H.L. Miller (Cambridge University Press, Cambridge, 2007), pp. 337–384
D.W. Leverington, in *Proc. Mars Polar Sci. Conf. 3* (2003), 8013
B. Levrard, F. Forget, F. Montmessin, J. Laskar, J. Geophys. Res. (2007). doi:10.1029/2006JE002772
J.S. Levy, J.W. Head, D.R. Marchant, J. Geophys. Res. (2009a). doi:10.1029/2008JE003273
J.S. Levy, J.W. Head, D.R. Marchant, Icarus (2009b). doi:10.1016/j.icarus.2009.02.018
J.S. Levy, D.R. Marchant, J.W. Head, Icarus (2010). doi:10.1016/j.icarus.2009.09.005
A.G. Lewkowicz, Periglacial systems, in *Fundamentals of Physical Geography (Canadian Edition)*, ed. by D. Briggs, P. Smithson, T. Ball (Copp Clark Pitman, Toronto, 1989), pp. 363–397
Q. Li, W.S. Kiefer, Geophys. Res. Lett. (2007). doi:10.1029/2007GL030544
D. Loizeau et al., J. Geophys. Res. (2008). doi:10.1029/2007JE002974

R.D. Lorrain, S.J. Fitzsimons, Cold-based glaciers, in *Encyclopedia of Snow, Ice and Glaciers*, ed. by V.P. Singh, P. Singh, U.K. Haritashya (Springer, Dordrecht, 2011), pp. 157–161
B.K. Lucchitta, Icarus **45**(2), 264 (1981)
B.K. Lucchitta, J. Geophys. Res. **87**(B12), 9951–9973 (1982)
B.K. Lucchitta, in *Proc. Int. Conf. on Permafrost 4* (1983), p. 744
B.K. Lucchitta, J. Geophys. Res. **89**, B409 (1984)
B.K. Lucchitta, H.M. Ferguson, C. Summers, J. Geophys. Res. **91**, E166 (1986)
J.I. Lunine, J. Chambers, A. Morbidelli, L.A. Leshin, Icarus **165**, 1 (2003)
W. Luo, T. Stepinski, J. Geophys. Res. (2009). doi:10.1029/2009JE003357
J.R. Mackay, Ann. Assoc. Am. Geogr. **62**, 1 (1972)
M.C. Malin, K.S. Edgett, Geophys. Res. Lett. **26**, 3049 (1999)
M.C. Malin, K.S. Edgett, Science **288**, 2330 (2000a)
M.C. Malin, K.S. Edgett, Science **290**, 1927 (2000b)
M.C. Malin, K.S. Edgett, J. Geophys. Res. **106**(E10), 23429 (2001)
M.C. Malin, K.S. Edgett, L.V. Posiolova, S.M. McColley, E.Z.N. Dobrea, Science **314**, 1573 (2006)
M.C. Malin, K.S. Edgett, B.A. Cantor, M.A. Caplinger, G.E. Danielson, E.H. Jensen, M.A. Ravine, J.L. Sandoval, K.D. Supulver, Mars (2010). doi:10.1555/mars.2010.0001
K.E. Mandt, S.L. de Silva, J.R. Zimbelmann, D.A. Crown, J. Geophys. Res. (2008). doi:10.1029/2008JE003076
M. Manga, Geophys. Res. Lett. (2004). doi:10.1029/2003GL018958
N. Mangold, J. Geophys. Res. (2003a). doi:10.1029/2002JE001885
N. Mangold, in *Proc. Int. Permafrost Conf. 8* (2003b), p. 723
N. Mangold, Icarus **174**, 336 (2005)
N. Mangold, Geomorphology (2011). doi:10.1016/j.geomorph.2010.11.009
N. Mangold, P. Allemand, Geophys. Res. Lett. **28**(3), 407 (2001)
N. Mangold, V. Ansan, Icarus **180**, 75 (2006)
N. Mangold, P. Allemand, P. Duval, Y. Geraud, P. Thomas, Planet. Space Sci. **50**, 385 (2002)
N. Mangold, S. Maurice, W.C. Feldman, F. Costard, F. Forget, in *Proc. Mars Polar Sci. Conf. 3* (2003), 8043
N. Mangold, S. Maurice, W.C. Feldman, F. Costard, F. Forget, J. Geophys. Res. (2004). doi:10.1029/2004JE002235
N. Mangold, F. Poulet, J.F. Mustard, J.P. Bibring, B. Gondet, Y. Langevin, V. Ansan, Ph. Masson, C. Fassett, J.W. Head III, H. Hoffmann, G. Neukum, J. Geophys. Res. (2007). doi:10.1029/2006JE002835
N. Mangold, A. Mangeney, V. Migeon, V. Ansan, A. Lucas, D. Baratoux, F. Bouchut, J. Geophys. Res. (2010). doi:10.1029/2009JE003540
N. Mangold, E. Kite, M. Kleinhans, H. Newsom, V. Ansan, E. Hauber, E. Kraal, C. Quantin-Nataf, K. Tanaka, Icarus **220**, 530 (2012)
D.R. Marchant, J.W. Head III, Icarus **192**, 187 (2007)
G.M. Marion, CRREL Special Report 95 (1995)
P. Masson, M.H. Carr, F. Costard, R. Greeley, E. Hauber, R. Jaumann, Space Sci. Rev. **96**, 333 (2001)
S. Maurice, W. Feldman, B. Diez, O. Gasnault, D.J. Lawrence, A. Pathare, T. Prettyman, J. Geophys. Res. (2011). doi:10.1029/2011JE003810
M.D. Max, S.M. Clifford, J. Geophys. Res. **105**, 4165 (2000)
J.C. Maxwell, *A Treatise on Electricity and Magnetism*, 3rd edn., vol. 1 (Clarendon, Oxford, 1892)
A.S. McEwen, L. Ojha, C.M. Dundas, S.S. Mattson, S. Byrne, J.J. Wray, S.C. Cull, S.L. Murchie, N. Thomas, V.C. Gulick, Science (2011). doi:10.1126/science.1204816
G.E. McGill, Geophys. Res. Lett. **13**, 705 (1986)
G.E. McGill, J. Geophys. Res. **94**, 2753 (1989)
P.J. McGovern et al., J. Geophys. Res. (2004). doi:10.1029/2004JE002286
M. Mehta et al., Icarus **211**, 172 (2011)
M.T. Mellon, J. Geophys. Res. **102**(E11), 25617 (1997)
M.T. Mellon, B.M. Jakosky, J. Geophys. Res. **100**(E6), 11781 (1995)
M.T. Mellon et al., J. Geophys. Res. (2009a). doi:10.1029/2009JE003417
M.T. Mellon et al., J. Geophys. Res. (2009b). doi:10.1029/2009JE003418
S. Meresse, F. Costard, N. Mangold, D. Baratoux, J.M. Boyce, Meteorit. Planet. Sci. (2006). doi:10.1111/j.1945-5100.2006.tb00442.x
S. Meresse, F. Costard, N. Mangold, P. Masson, G. Neukum (HSRC Co-I Team), Icarus (2008). doi:10.1116/j.icarus.2007.10.023
J.M. Moore, K.S. Edgett, Geophys. Res. Lett. **20**, 1599 (1993)
J.M. Moore, D.E. Wilhelms, Icarus **154**, 258 (2001)
A. Morgenstern, E. Hauber, D. Reiss, S. van Gasselt, G. Grosse, L. Schirrmeister, J. Geophys. Res. (2007). doi:10.1029/2006JE002869

R.V. Morris et al., Science (2012). doi:10.1126/science.1189667
P.J. Mouginis-Mark, Icarus **71**, 268 (1987)
J. Mouginot, A. Pommerol, W. Kofman, P. Beck, B. Schmitt, A. Hérique, C. Grima, A. Safaeinili, J.J. Plaut, Icarus (2010). doi:10.1016/j.icarus.2010.07.003
J. Mouginot, A. Pommerol, P. Beck, W. Kofman, S. Clifford, Geophys. Res. Lett. (2012). doi:10.1029/2011GL050286
O. Mousis, J.I. Lunine, S. Picaud, D. Cordier, Faraday Discuss. **147**, 509 (2010)
O. Mousis, J.I. Lunine, E. Chassefière, F. Montmessin, A. Lakhlifi, S. Picaud, J.M. Petit, D. Cordier, Icarus **218**, 80 (2012a)
O. Mousis et al., Space Sci. Rev. (2012b, this issue). doi:10.1007/s11214-012-9942-9
S.L. Murchie et al., J. Geophys. Res. (2009). doi:10.1029/2009JE003342
D.S. Musselwhite, T.D. Swindle, J.I. Lunine, Geophys. Res. Lett. **28**, 1283 (2001)
J.F. Mustard, C.D. Cooper, M.K. Rifkin, Nature **412**, 411 (2001)
J.F. Mustard et al., Nature (2008). doi:10.1038/nature07097
G. Neukum, R. Jaumann (the HRSC Co-Investigators and Experiment Team), *HRSC: The High Resolution Stereo Camera of Mars Express in Mars Express: The Scientific Payload* (ESA, Noordwijk, 2004)
G. Neukum, A.T. Basilevsky, T. Kneissl, M.G. Chapman, S. van Gasselt, G. Michael, R. Jaumann, H. Hoffmann, J.K. Lanz, Earth Planet. Sci. Lett. **294**, 204–222 (2010)
H.E. Newsom, Icarus **44**, 207–216 (1980)
H.E. Newsom, in *Lakes on Mars*, ed. by N. Cabrol, E. Grin (Elsevier, Amsterdam, 2010), pp. 91–110
H.E. Newsom, G.E. Brittelle, C.A. Hibbitts, L.J. Crossey, A.M. Kudo, J. Geophys. Res. (1996). doi:10.1029/96JE01139
H.E. Newsom et al., J. Geophys. Res. (2007). doi:10.1029/2006JE002680
F. Nimmo, D.J. Stevenson, J. Geophys. Res. **105**, 11969 (2000)
F. Nimmo, K. Tanaka, Annu. Rev. Earth Planet. Sci. **33**, 133 (2005)
D.C. Nunes, S.E. Smrekar, A. Safaeinili, J.W. Holt, R.J. Phillips, R. Seu, B.A. Campbell, J. Geophys. Res. (2010). doi:10.1029/2009JE003509
C.H. Okubo, A.S. McEwen, Science (2007). doi:10.1126/science.1136855
G.R. Olhoeft, in *Int. Conf. Ground Penetrating Radar 6* (1996), p. 1
T.C. Onstott et al., Microb. Ecol. **58**, 786 (2009)
G.G. Ori, L. Marinangeli, A. Baliva, J. Geophys. Res. **105**(E7), 17629 (2000)
D.P. Page, Icarus **189**, 83 (2007)
T.J. Parker, Lunar Planet. Sci. **XXVII**, 1003 (1996)
T.J. Parker, F.J. Calef, in *3rd Conference on Early Mars* (2012), 7085
T.J. Parker, D.M. Schneeberger, D.C. Pieri, R.S. Saunders, Symposium on Mars: evolution of its climate and atmosphere. LPI Tech. Rept. 87-01, pp. 96–98, 1987
T.J. Parker, R.S. Saunders, D.M. Schneeberger, Icarus **82**, 111 (1989)
T.J. Parker, D.S. Gorsline, R.S. Saunders, D.C. Pieri, D.M. Schneeberger, J. Geophys. Res. **98**, 11061 (1993)
R.A. Parsons, F. Nimmo, J. Geophys. Res. (2009). doi:10.1029/2007JE003006
J.C. Pechmann, Icarus **42**, 185 (1980)
J.T. Perron, J.X. Mitrovica, M. Manga, I. Matsuyama, M.A. Richards, Nature **447**, 840–843 (2007)
R.J. Phillips et al., Science (2008). doi:10.1126/science.1157546
R.J. Phillips et al., Science (2011). doi:10.1126/science.1203091
G. Picardi et al., in *Proc. Int. Conf. Radar* (2003). doi:10.1109/RADAR.2003.1278795
G. Picardi et al., Science (2005). doi:10.1126/science.1122165
D.C. Pieri, Science **210**, 895 (1980)
J.J. Plaut et al., Science (2007). doi:10.1126/science.1139672
J.J. Plaut, A. Safaeinili, J.W. Holt, R.J. Philips, J.W. Head, R. Seu, N.E. Putzig, A. Frigeri, Geophys. Res. Lett. (2009). doi:10.1029/2008GL036379
H.N. Pollack, S.J. Hurter, J.R. Johnston, Rev. Geophys. **31**, 3 (1993)
A. Pommerol, W. Kofman, J. Audouard, C. Grima, P. Beck, J. Mouginot, A. Hérique, A. Kumamoto, T. Kobayashi, T. Ono, Geophys. Res. Lett. (2010). doi:10.1029/2009GL041681
M. Pondrelli, A. Baliva, S. Di Lorenzo, L. Marinangeli, A.P. Rossi, J. Geophys. Res. (2005). doi:10.1029/2004JE002335
M. Pondrelli, A.P. Rossi, L. Marinangeli, E. Hauber, K. Gwinner, A. Baliva, S. Di Lorenzo, Icarus **197**, 429 (2008)
L.J. Porcello, R.L. Jordan, J.S. Zelenka, G.F. Adams, R.J. Phillips, W.E. Brown Jr., S.H. Ward, P.L. Jackson, Proc. IEEE **62**(6), 769 (1974)
F. Poulet, J.P. Bibring, J.F. Mustard, A. Gendrin, N. Mangold, Y. Langevin, R.E. Arvidson, B. Gondet, C. Gomez, Nature **438**, 623 (2005)
M.A. Presley, P.R. Christensen, J. Geophys. Res. **102**, 9221 (1997)

O. Prieto-Ballesteros, J.S. Kargel, A.G. Fairén, D.C. Fernández-Remolar, J.M. Dohm, R. Amils, Geology **34**, 149 (2006)
N.E. Putzig, R.J. Phillips, B.A. Campbell, J.W. Holt, J.J. Plaut, L.M. Carter, A.F. Egan, F. Bernardini, A. Safaeinili, R. Seu, Icarus **204**, 443 (2009)
Y. Quesnel, C. Sotin, B. Langlais et al., Earth Planet. Sci. Lett. **277**, 184 (2009)
S.N. Raymond, T. Quinn, J.I. Lunine, Icarus **183**, 265 (2006)
D. Reiss, H. Hiesinger, E. Hauber, K. Gwinner, Planet. Space Sci. **57**, 958 (2009)
D. Reiss, G. Erkeling, K.E. Bauch, H. Hiesinger, Geophys. Res. Lett. **37**, L06203 (2010). doi:10.1029/2009GL042192
E.C. Robertson, D.L. Peck, J. Geophys. Res. **79**, 4875 (1974)
J.A.P. Rodriguez, S. Sasaki, R.O. Kuzmin, J.M. Dohm, K.L. Tanaka, H. Miyamoto, K. Kurita, G. Komatsu, A.G. Fairén, J.C. Ferris, Icarus **175**, 36 (2005)
L.A. Rossbacher, S. Judson, Icarus **45**, 39 (1981)
A. Safaeinili, W. Kofman, J.F. Nouvel, A. Hérique, R.L. Jordan, Planet. Space Sci. (2003). doi:10.1016/S0032-0633(03)00048-5
J. Saraiva, L. Bandeira, J. Antunes, P. Pina, T. Barata, Finisterra **XLIV**(87), 71 (2009)
J.H. Sass, A.H. Lachenbruch, R.J. Munroe, J. Geophys. Res. **76**, 3391 (1971)
P.M. Schenk, J.M. Moore, J. Geophys. Res. **105**, 24529 (2000)
S.C. Schon, J.W. Head, Icarus **213**, 428 (2011)
S.C. Schon, J.W. Head, R.E. Milliken, Geophys. Res. Lett. (2009). doi:10.1029/2009GL038554
N. Schorghofer, Geophys. Res. Lett. (2008). doi:10.1029/2008GL034954
R.A. Schultz, H.V. Frey, J. Geophys. Res. **95**, 14175 (1990)
P.H. Schultz, H. Glicken, J. Geophys. Res. **84**, 8033 (1979)
P.H. Schultz, R.A. Schultz, J. Rogers, J. Geophys. Res. **87**, 9803 (1982)
S.P. Schwenzer, D.A. Kring, Geology **37**, 1091 (2009)
S.P. Schwenzer et al., Earth Planet. Sci. Lett. **335–336**, 9 (2012)
D.H. Scott, M.G. Chapman, J.W. Rice Jr., J.M. Dohm, Lunar Planet. Sci. **XXII**, 53 (1992)
N.M. Seibert, J.S. Kargel, Geophys. Res. Lett. **28**(5), 899 (2001)
A. Séjourné, F. Costard, J. Gargani, R.J. Soare, A. Fedorov, C. Marmo, Planet. Space Sci. **59**(5–6), 412 (2011)
M.M. Selvans, J.J. Plaut, O. Aharonson, A. Safaeinili, J. Geophys. Res. (2010). doi:10.1029/2009JE003537
R. Seu et al., J. Geophys. Res. (2007). doi:10.1029/2006JE002745
R.P. Sharp, M.C. Malin, Geol. Soc. Am. Bull. **86**, 593 (1975)
O. Slaymaker, in *Encyclopedia of Geomorphology*, ed. by A.S. Goudie (Routledge, London, 2004), pp. 321–323
O. Slaymaker, R.E.J. Kelly, *The Cryosphere and Global Environmental Change* (Blackwell Sci., Oxford, 2007)
E.D. Sloan Jr., *Clathrate Hydrates of Natural Gases* (Marcel Decker, New York, 1997)
E.D. Sloan Jr., C.A. Koh, *Clathrate Hydrates of Natural Gases*, 3rd edn. (CRC Press, Boca Raton, 2008)
D.E. Smith et al., J. Geophys. Res. **106**, 23689 (2001)
D.E. Smith, G. Neuman, R.E. Arvidson, E.A. Guinness, S. Slavney, Mars Global Surveyor Laser Altimeter Mission Experiment Gridded Data Record, in *NASA Planet. Data Syst.*, Greenbelt, MD (2003)
P.H. Smith et al., Science **325**, 58 (2009)
R.J. Soare, A. Séjourné, G. Pearce, F. Costard, G.R. Osinski, in *Analogs for Planetary Exploration*, ed. by W.B. Garry, J.E. Bleacher. Special Paper, vol. 483. (Geol. Soc. Am, Boulder, 2011), pp. 203–218. doi:10.1130/2011.2483(13)
L.A. Soderblom, T.J. Kreidler, H. Masursky, J. Geophys. Res. **78**, 4117 (1973)
Solar System Exploration Survey, *New Frontiers in Solar System Exploration* (National Academies Press, Washington, 2003), http://www.nap.edu/catalog.php?record_id=10898. Accessed 19 July 2011
S.C. Solomon, J.W. Head III, J. Geophys. Res. **95**, 11073 (1990)
S.C. Solomon et al., Science **307**, 1214 (2005)
S.W. Squyres, Icarus **34**, 600 (1978)
S.W. Squyres, J. Geophys. Res. **84**, 8087 (1979)
S.W. Squyres, Icarus **79**, 229 (1989)
S.W. Squyres, M.H. Carr, Science **231**, 249 (1986)
T.F. Stepinski, Lunar Planet. Sci. **LXI**, 1845 (2010)
T.F. Stepinski, W. Luo, Lunar Planet. Sci. **LXI**, 1350 (2010)
T.F. Stepinski, E.R. Urbach, Lunar Planet. Sci. **LX**, 1117 (2009)
K.L. Tanaka, J. Geophys. Res. **91**(B13), E139 (1986)
K.L. Tanaka, Icarus **144**, 254 (2000)
K.L. Tanaka, G.J. Leonard, J. Geophys. Res. **100**, 5407 (1995)

K.L. Tanaka, D.H. Scott, V.C. Gulick, R. Greeley, in *Mars*, ed. by H.H. Kieffer, B.M. Jakosky, C.W. Snyder, M. Matthews (University Arizona Press, Tucson, 1992), pp. 345–382
K.L. Tanaka, J.S. Kargel, D.J. McKinnon, T.M. Hare, N. Hoffman, Geophys. Res. Lett. **29**, 1195 (2002)
K.L. Tanaka, J.A. Skinner, T.M. Hare, *Geologic Map of the Northern Plains of Mars* (2005). U.S. Geol. Surv., Sci, Inv. Map 2888
K.L. Tanaka, J.A.P. Rodriguez, J.A. Skinner Jr., M.C. Bourke, C.M. Fortezzo, K.E. Herkenhoff, E.J. Kolb, C.H. Okubo, Icarus (2008). doi:10.1016/j.icarus.2008.01.021
N. Terada, Yu.N. Kulikov, H. Lammer, H.I.M. Lichtenegger, T. Tanaka, H. Shinagawa, T. Zhang, Astrobiology **9**, 55 (2009)
C. Thomas, O. Mousis, V. Ballenegger, S. Picaud, Astron. Astrophys. **474**, L17 (2007)
C. Thomas, S. Picaud, O. Mousis, V. Ballenegger, Planet. Space Sci. **56**, 1607 (2008)
B.J. Thomson, J.W. Head III, J. Geophys. Res. **106**, 1 (2001)
F. Tian, J.F. Kasting, S.C. Solomon, Geophys. Res. Lett. **2**, L02205 (2009)
O.B. Toon, J.B. Pollack, W. Ward, J.A. Burns, K. Bilski, Icarus **44**, 552 (1980)
J. Touma, J. Wisdom, Science **259**, 1294 (1993)
B.J. Travis, W.C. Feldman, Lunar Planet. Sci. **XL**, 1315 (2009)
B.J. Travis, N.D. Rosenberg, J.N. Cuzzi, J. Geophys. Res. **108**, 8040 (2003)
A.H. Treiman, J. Geophys. Res. (2001). doi:10.1029/2002JE001900
F.T. Ulaby, R.K. Moore, A.K. Fung, *Microwave Remote Sensing: Active and Passive, Vol. III: Volume Scattering and Emission Theory, Advanced Systems and Applications* (Artech House, Dedham, 1986)
M. Ulrich, A. Morgenstern, F. Günther, D. Reiss, K.E. Bauch, E. Hauber, S. Rössler, L. Schirrmeister, J. Geophys. Res. (2010). doi:10.1029/2010JE003640
A. Valeille, M.R. Combi, V. Tenishev, S.W. Bougher, A.F. Nagy, Icarus (2010). doi:10.1016/j.icarus.2008.08.018
J.H. van der Waals, J.C. Platteeuw, in *Advances in Chemical Physics*, vol. 2, ed. by I. Prigogine (Interscience, New York, 1959), pp. 1–57
R.O. van Everdingen (ed.), *Multi-Language Glossary of Permafrost and Related Ground-Ice Terms. International Permafrost Association* (Arctic Institute of North America, University of Calgary, Calgary, 1998), 268 pp.
M. Vincendon, J. Mustard, F. Forget, M. Kreslavsky, A. Spiga, S. Murchie, J.P. Bibring, Geophys. Res. Lett. (2010). doi:10.1029/2009GL041426
W.R. Ward, J. Geophys. Res. **79**, 3375 (1974)
W.R. Ward, J. Geophys. Res. **84**, 237 (1979)
W.R. Ward, in *Mars*, ed. by H.H. Kieffer, B.M. Jakosky, C.W. Snyder, M. Matthews (University Arizona Press, Tucson, 1992), pp. 298–320
N. Warner, G. Sanjeev, J.P. Muller, J.R. Kim, S.Y. Lin, Earth Planet. Sci. Lett. **288**, 58 (2009)
T.R. Watters et al., Science (2007). doi:10.1126/science.1148112
R.W. Wichman, P.H. Schultz, J. Geophys. Res. **94**, 17333 (1989)
C.A. Wiley, Pulsed Doppler radar methods and apparatus. U.S. Patent 3196436-86, 1965
R.M.E. Williams, M.C. Malin, J. Geophys. Res. (2004). doi:10.1029/2003JE002178
R.M.E. Williams, R.J. Phillips, J. Geophys. Res. (2001). doi:10.1029/2000JE001409
S.A. Wilson, A.D. Howard, J.M. Moore, J.A. Grant, J. Geophys. Res. (2007). doi:10.1029/2006JE002830
K.H. Wohletz, M.F. Sheridan, Icarus **56**, 15 (1983)
E.D. Yershov, *Geocryology of the USSR: Eastern Siberia and the Far East* (Nedra, Moscow, 1989). Orig. in Russian
K. Yoshikawa, Lunar Planet. Sci. **XXXIII**, 1159 (2002)
K. Yoshikawa, Geophys. Res. Lett. (2003). doi:10.1029/2003GL017165
M. Zanetti, H. Hiesinger, D. Reiss, E. Hauber, G. Neukum, Icarus **206**, 691 (2010)
T. Zhang, R.G. Barry, K. Knowles, J.A. Hegginbottom, J. Brown, Polar Geogr. **31**, 47 (2008a)
Z. Zhang, T. Hagfors, E. Nielsen, G. Picardi, A. Mesdea, J.J. Plaut, J. Geophys. Res. (2008b). doi:10.1029/2007JE002941
J.R. Zimbelman, S.M. Clifford, S.H. Williams, Lunar Planet. Sci. **XIX**, 397 (1989)
M.T. Zuber et al., Science **282**, 2053 (1998)

Space Sci Rev (2013) 174:213–250
DOI 10.1007/s11214-012-9942-9

Volatile Trapping in Martian Clathrates

Olivier Mousis · Eric Chassefière · Jérémie Lasue · Vincent Chevrier · Megan E. Elwood Madden · Azzedine Lakhlifi · Jonathan I. Lunine · Franck Montmessin · Sylvain Picaud · Frédéric Schmidt · Timothy D. Swindle

Received: 24 November 2011 / Accepted: 4 October 2012 / Published online: 25 October 2012

O. Mousis (✉) · A. Lakhlifi · S. Picaud
Institut UTINAM, CNRS/INSU, UMR 6213, Université de Franche-Comté, 25030, Besançon Cedex, France
e-mail: olivier.mousis@obs-besancon.fr

O. Mousis
UPS-OMP, CNRS-INSU, IRAP, Université de Toulouse, 14 Avenue Edouard Belin, 31400 Toulouse, France

E. Chassefière · F. Schmidt
Laboratoire IDES, UMR 8148, Univ. Paris-Sud, 91405 Orsay, France

E. Chassefière · F. Schmidt
CNRS, 91405 Orsay, France

J. Lasue
UPS-OMP, IRAP, Université de Toulouse, Toulouse, France

J. Lasue
CNRS, IRAP, 9 Av. colonel Roche, BP 44346, 31028 Toulouse cedex 4, France

V. Chevrier
W.M. Keck Laboratory for Space Simulation, Arkansas Center for Space and Planetary Sciences, University of Arkansas, Fayetteville, AR 72701, USA

M.E. Elwood Madden
School of Geology and Geophysics, University of Oklahoma, Norman, OK 73072, USA

J.I. Lunine
Center for Radiophysics and Space Research, Cornell University, Ithaca, NY, USA

F. Montmessin
LATMOS, CNRS/IPSL/UVSQ, Guyancourt, France

T.D. Swindle
Lunar and Planetary Laboratory, University of Arizona, Tucson, AZ, USA

Abstract Thermodynamic conditions suggest that clathrates might exist on Mars. Despite observations which show that the dominant condensed phases on the surface of Mars are solid carbon dioxide and water ice, clathrates have been repeatedly proposed to play an important role in the distribution and total inventory of the planet's volatiles. Here we review the potential consequences of the presence of clathrates on Mars. We investigate how clathrates could be a potential source for the claimed existence of atmospheric methane. In this context, plausible clathrate formation processes, either in the close subsurface or at the base of the cryosphere, are reviewed. Mechanisms that would allow for methane release into the atmosphere from an existing clathrate layer are addressed as well. We also discuss the proposed relationship between clathrate formation/dissociation cycles and how potential seasonal variations influence the atmospheric abundances of argon, krypton and xenon. Moreover, we examine several Martian geomorphologic features that could have been generated by the dissociation of extended subsurface clathrate layers. Finally we investigate the future in situ measurements, as well as the theoretical and experimental improvements that will be needed to better understand the influence of clathrates on the evolution of Mars and its atmosphere.

Keywords Mars · Clathrates · Polar caps · Cryosphere · Atmosphere

1 Introduction

Clathrate hydrates (hereafter clathrates) are crystalline solids which look like ice and form when water molecules constitute a cage-like structure around small "guest molecules". Clathrates have been extensively observed on Earth and are considered today to be one of the most important reservoirs of fossil energy (Sloan 1998; Sloan and Koh 2008). The most common guest molecules in terrestrial clathrates are methane (the most abundant), ethane, propane, butane, nitrogen, carbon dioxide and hydrogen sulfide. Water crystallizes in the cubic system in clathrates, rather than in the hexagonal structure of normal ice. Several different clathrate structures are known, the two most common ones being "structure I" and "structure II". In structure I, the unit cell is formed of 46 water molecules and can incorporate up to 8 guest molecules. In structure II, the unit cell consists of 136 water molecules and can incorporate at most 24 guest molecules.

Thermodynamic conditions prevailing on Mars are appropriate for clathrates to form in the deep cryosphere (depth down to a few kilometers) and near subsurface (depth down to a few meters). Yet many observations show that the dominant condensed phases on Mars are solid CO_2 ice and water ice (Kieffer et al. 1992; Longhi 2006). Meanwhile, the thermodynamic stability field of clathrates is such that conditions on the Martian surface or close subsurface are tantalizingly close to the stability boundary (Carr 1996). It is then difficult to assess whether clathrates are abundant, nonexistent, or something in between. Figure 1 illustrates the stability conditions of CO_2-dominated clathrate and CO_2 ice on modern Mars. The figure shows that the equilibrium curves of these two solid phases are very close under Martian atmospheric conditions. Assuming isolated porosity in the soil, CO_2-dominated clathrate is stable at the surface (and also at any depth) at $\sim$150 K, i.e. the coldest temperature reached during winter in the south pole region. Considering a higher surface temperature of 180 K in the same region, the clathrate dissociates at the surface and needs to be buried deeper than $\sim$1.1 m in the subsurface to remain stable. If one assumes an open porosity in the Martian soil at 180 K, the stability conditions of CO_2-dominated clathrate are never even

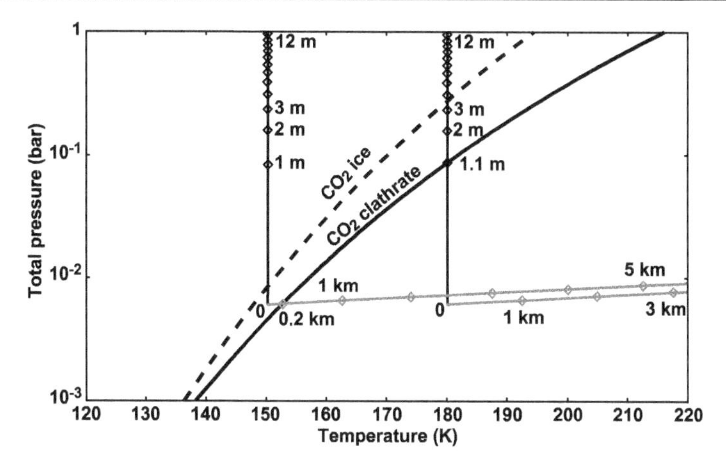

Fig. 1 Equilibrium pressure of CO_2-dominated clathrate (*thick black curve*), CO_2 ice (*dashed thick black curve*), lithostatic pressure (*black vertical lines*) and atmospheric hydrostatic pressure in a porous subsurface (*grey horizontal lines*) as a function of temperature. At given temperature, the clathrate is stable at pressures equal to or higher than its equilibrium pressure. Stability conditions of clathrate on Mars result from the intersection between its equilibrium curve and the lithostatic (resp. hydrostatic) pressure profile in the case of permafrost with isolated (resp. open) porosity. The two considered surface temperature are 150 K (i.e. the coldest temperature reached during winter in the south pole region) and 180 K. Depths to the base of the permafrost are several kilometers

met. At 150 K, Fig. 1 shows that a clathrate layer could form and remain stable in the first 200 meters depth of the subsurface. However, a surface temperature of ~150 K corresponds only to the stability region of CO_2 ice. This implies that CO_2 clathrate could be stable only in a very restricted region, which corresponds to the permanent CO_2 polar cap. It could be also confined within a very shallow layer beneath seasonal CO_2 ice. It is important to note that clathrate formation in the Martian subsurface requires the presence of available water ice. Hence, the existence of clathrate at the surface of the south pole region seems unlikely, even at the lowest temperature since observations of the residual CO_2 ice cap have revealed scant evidence of the H_2O (Kieffer et al. 2000) that is integral to clathrate formation. In any case, the figure shows that the cycle of clathrate formation/dissociation is susceptible to being affected not only by the nature of the soil (composition, porosity, ...), but also by climate change and seasonal variations.[1]

Clathrates on Mars have been repeatedly invoked to play an important role in the distribution and total inventory of that planet's volatiles. In particular, clathrates have been proposed to form a potential reservoir hidden in the cryosphere which could store an atmospheric equivalent up to ~0.3 bar of CO_2 (Mousis et al. 2010). Clathrate reservoirs present in the close subsurface would allow the sporadic release of methane into the atmosphere through dissociation or ablation (Chastain and Chevrier 2007; Elwood Madden et al. 2007, 2011; Chassefière 2009; Thomas et al. 2009; Chassefière and Leblanc 2011). The presence of large amounts of clathrates in the cryosphere has also been advocated to explain the two–orders–of–magnitude lower abundances of krypton and xenon (in mass/mass of the planet) in the Martian atmosphere compared to those sampled on Earth and Venus (Mousis et al.

[1]Changes in obliquity over long timescales are probably too strong (Mischna et al. 2003; Forget et al. 2006; Montmessin 2006) to allow the preservation of any information concerning the clathrate formation/dissociation cycles that took place at these epochs.

2010, 2012). This scenario implies that the abundances of krypton and xenon measured in the Martian atmosphere are not representative of its global noble gas budget since a large part of these volatiles may be sequestrated in CO_2-dominated clathrate. Interestingly, clathrate dissociation has also been proposed to be involved in the formation of several geomorphic features on Mars. Chaotic terrain is attributed primarily to localized collapse of the cratered upland owing to removal of subsurface material including clathrates (Baker and Milton 1974; Rodriguez et al. 2005; Baker 2009). The catastrophic dissociation of carbon dioxide clathrate has also been suggested to produce catastrophic floods on Mars, explaining the large water discharge and the missing volume of the chaotic terrains (Milton 1974; Hoffman 2000).

In this paper, we review the effects of proposed clathrates on Mars. This work aims to provide indirect observational tests that might be verified by future in situ measurements, thus enabling direct investigation of clathrates in the volatile exchange at the surface/atmosphere in both past and present climates. In Sect. 2, we discuss models of clathrate sources for modern atmospheric methane. We also review possible mechanisms of clathrate formation in the near subsurface and at the base of the cryosphere. The plausible mechanisms for methane release into the atmosphere from a clathrate layer are also discussed in this section. Section 3 is dedicated to the presentation of the thermodynamic statistical model used to calculate the composition of clathrates formed under Martian conditions. We then discuss the proposed links between clathrate formation/dissociation cycles and a potential seasonal variability of noble gas abundances. The influence of clathrates on the Martian geomorphology is discussed in Sect. 4. Section 5 discusses future work, in particular key in situ measurements that may be used to assess the presence of clathrates on Mars. We also detail future theoretical and experimental work needed to better understand the potential influence of clathrates on the evolution of the planet and its atmosphere. Section 6 is devoted to conclusions.

2 Clathrates and the Origin of the Martian Methane

The presence of methane in the Martian atmosphere at the $\sim$10–20 ppb level (Krasnopolsky et al. 2004; Formisano et al. 2004; Mumma et al. 2009; Fonti and Marzo 2010) suggests that CH_4 clathrates could be present in the subsurface of Mars. The most intriguing feature of the observed CH_4 is its high temporal and spatial variability, with a general seasonal trend showing more CH_4 in the atmosphere during northern spring and summer. A more detailed discussion of CH_4 observations may be found in the paper by Lammer et al. (this issue). From calculations using a General Circulation Model, the patchy structure of the atmospheric CH_4 field can be reproduced only with a short CH_4 lifetime of 200 days (Lefèvre and Forget 2009), three orders of magnitude shorter than its photochemical lifetime (300 yr). One then needs to invoke the presence of a significant sink for methane on Mars. Since homogeneous atmospheric photochemistry is not consistent with such a short lifetime, this sink is now expected to be the heterogeneous oxidation of CH_4 in the near subsurface. Hydrogen peroxide adsorbed in the regolith is not likely a significant sink for methane (Gough et al. 2011). Several other possible mechanisms have been extensively discussed by Chassefière and Leblanc (2011).

The origin of CH_4 still remains unknown. The spatial and temporal variability of the observed CH_4, requiring a sink able to remove CH_4 from the atmosphere in less than one Martian year, led some scientists involved in the photochemical modelling of the Martian atmosphere to consider the detection of CH_4 as questionable. Concerning data acquired from Earth in the 3025–3040 cm^{-1} wavenumber range (Mumma et al. 2009), where two

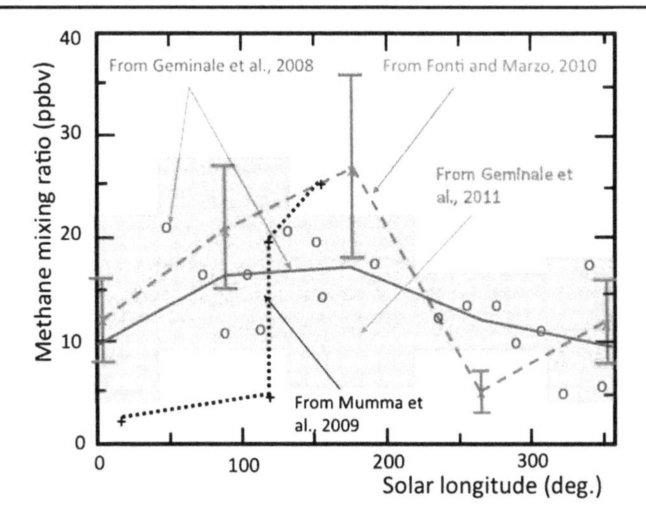

Fig. 2 Seasonal evolution of the methane mixing ratio. *Red crosses* and *red dashed line* correspond to TES data (Fonti and Marzo 2010). *Blue circles* correspond to PFS data (Geminale et al. 2008). The *blue solid line* connects the 4 points obtained by averaging 2008 PFS data in 4 bins, namely Ls $= 0° \pm 45°$, $90° \pm 45°$, $180° \pm 45°$ and $270° \pm 45°$, represented by the blue boxes. The *green area* delineates the more complete set of PFS data recently published (Geminale et al. 2011), with the exception of 7 extreme points (4 in the 30–40 ppb range, 3 at 0 ppb). The average mixing ratios of methane measured from Hawai'i (Mumma et al. 2009) are also plotted (+ *signs* and *black dotted line*)

CH_4 absorption lines are used simultaneously to retrieve the CH_4 abundance, it has been argued that the coincidence between Martian and telluric lines could have led to an erroneous retrieval of the Mars CH_4 abundance (Zahnle et al. 2011). Nevertheless, spacecraft data (PFS on Mars Express at 3020 cm^{-1}, TES on MGS in the thermal infrared at 1306 cm^{-1}) which now span ∼6 Martian years (MY24-MY29) and Earth-based measurements (CSHELL/IRTF in Hawaii) show a globally consistent and reproducible seasonal behavior of the CH_4 mixing ratio with similar abundance levels and amplitudes of variation (Fig. 2). CH_4 mixing ratios obtained by space measurements are reported with typical error bars of ±2–3 ppb, that is ±10–20 % in a relative amount, for PFS-Mars Express (Geminale et al. 2011) and ±20–30 % in a relative amount for TES-MGC (Fonti and Marzo 2010). The existing observations by three different instruments from both Earth and spacecraft in orbit about Mars, working at different wavelengths in the solar and thermal infrared, are globally so consistent with each other that the detection of CH_4 is hardly questionable. On the other hand, accepting that CH_4 has been detected, a highly variable CH_4, and therefore a short CH_4 lifetime smaller than one Martian year, is an unavoidable consequence of measurements.

A large methane release flux seems to exclude an external (meteoritic) source, as well as a direct magmatic origin (Atreya et al. 2007). More plausibly, CH_4 could be produced, either by hydrothermal alteration of basaltic crust (Lyons et al. 2005) or by serpentinization of ultramafic rocks producing H_2 and reducing crustal carbon into CH_4 through the reaction of dissolved CO_2 with H_2, observed in hydrothermal systems at mid-ocean ridges (Oze and Sharma 2005; Atreya et al. 2007). On Earth, a significant proportion of the CH_4 produced on mid-ocean ridges and in geothermal systems is expected to be biogenic (Emmanuel and Ague 2007), and some of the CH_4 released from Mars's hydrothermal systems could similarly be of biological origin. In this context, Martian clathrates have been proposed to form an intermediate storage reservoir in the subsurface that regularly releases methane into the atmosphere. Here we first discuss the trapping mechanism that may have enabled the formation and long term stability of CH_4 clathrate at the surface and in the close subsurface of Mars. According to this hypothesis, methane would have formed clathrates close to the surface in the past, at a time when the methane partial pressure in the atmosphere was sufficiently high, and these clathrates, possibly buried under thin layers of regolith would have survived until now. We then discuss the possibility that clathrates could have been formed, and possibly continue to be formed, at depth in the crust at the base of the cryosphere due to the production of gaseous methane in deep hydrothermal systems. We also provide an estimate of their lifetime in this environment. After having addressed the existing links between mechanisms of methane release into the atmosphere and the seasonal cycle, we finally discuss the possibility that this release may be due to the presence of metastable atmospheric particles of methane clathrates.

2.1 Clathrate Formation in the Shallow Subsurface

This section is mostly concerned with the feasibility of forming methane clathrates on Mars and conserving them to the present in order to account for the possible release of methane in the atmosphere. Most publications on the topic of clathrate formation using thermodynamic models (Chastain and Chevrier 2007; Thomas et al. 2008) agree on the pressure and temperature conditions required for their formation on the Martian surface. Low temperatures, down to $\sim$150 K, are required and these conditions can be reached in the polar regions (see Introduction). The main obstacle is usually the methane partial pressure required to induce trapping in the clathrate. Three main models have been used to determine the fraction of trapped methane as a function of the environmental conditions. Chastain and Chevrier (2007) determined the clathration conditions of a CO_2/CH_4 mixture from various pressure and temperature values, by using a program developed for industrial purposes (CSMHYD model; Sloan 1998). Alternatively, Thomas et al. (2009) used a dedicated thermodynamic statistical model including most of the minor and trace compounds in the Martian atmosphere. The same kind of model, constrained by existing observations, was used by Herri et al. (2011) and Herri and Chassefière (2012). In these three cases, all compositions were determined on the clathrate's equilibrium curve. The main results of these models show that clathrates form at lower pressure and higher temperature with increasing CO_2 abundance in the atmosphere (which can be considered as the total pressure most of the time).

But the effect of the temperature is much more important than the ratio between the different gases. For example, forming a clathrate at a typical early Mars pressure of $\sim$1 bar (Pollack et al. 1987) requires temperatures below 220 K, while the gas binary composition changes the pressure by only a few tenths of a bar. However, the fraction of clathrated methane significantly grows as the atmospheric mole fraction of methane increases, and more moderately with increasing temperature. Thomas et al. (2009) show that at 7 mbar and at the corresponding equilibrium temperature, the abundance ratio of methane in the clathrate to that in the atmosphere evolves from 0.166 to 0.551 when the molar abundance of methane in the gas phase evolves from 1×10^{-4} to 0.9 in the gas phase. However, since methane is only a trace gas in the atmosphere, all models predict that clathrate in equilibrium with the modern atmosphere would have a negligible concentration of methane, with a mixing ratio in the clathrate phase four to six times lower than in the gaseous phase (Herri et al. 2011; Thomas et al. 2009). The formation of methane clathrates in present-day conditions being prevented by the very low partial pressure of methane, their potential presence points to an early formation under different atmospheric conditions, when the partial pressure of methane was higher. Alternatively, the very small amounts of methane detected in the atmosphere, in the ppb range (Formisano et al. 2004;

Mumma et al. 2009) could also indicate more recent formation of clathrates with trace amounts of methane. Such higher pressure of methane could have helped maintain relatively high temperatures (Kasting 1997) to allow stability of liquid water necessary for the formation of the abundant Noachian phyllosilicate deposits (Chevrier et al. 2007), although recent studies have questioned the efficacy of methane in warming up the early Mars atmosphere (F. Tian; personal communication). This last scenario would favor trapping of methane in clathrates, since both thermodynamic models favor clathrate formation in cold martian climates.

Although geochemical (phyllosilicates) and geomorphological (valley networks) observations point to extensive activity of liquid water on early Mars, most atmospheric evolution models are not able to raise the temperature high enough to generate pure liquid water, at least with just CO_2 as a greenhouse gas (Halevy and Schrag 2009). Moreover, recent extrapolation of CO_2 escape from the atmosphere suggests that the overall pressure remained quite low since the Noachian (Barabash et al. 2007). Therefore, recent thermodynamic models of the martian surface mineralogy and geochemistry suggest that the majority of the alteration of the surface occurred in cold (below 273 K) conditions (Fairén et al. 2009) favoring the stability of water ice and clathrates as well.

Considering the possible release of methane in present-day conditions, these ancient clathrates must have remained at least partially stable several billion years. Burial remains a favored mechanism, especially if the geothermal gradient dropped quickly after their formation. Models of burial stability suggest that methane clathrates could have remained stable between several meters (polar regions) and several hundred meters deep in the subsurface, mostly depending on the nature of the subsurface (Chastain and Chevrier 2007). Thin layers of regolith covering thick compact basaltic layers present the best scenario since they allow quick heat transfer between the subsurface and the surface, and therefore lower temperatures in the deep subsurface. The problem with this model is that formation of clathrates under hot basaltic layers seems unlikely. Alternatively, a thick layer of ice-cemented soil could provide the necessary thermal properties while still being compatible with the formation processes of clathrates. The recent literature on radar sounding of the subsurface seems to indicate deep extensive deposits akin to the polar deposits (Phillips et al. 2011) some of which may include clathrates. In any case these observations suggest that clathrates could be present within extensive deposits at least in the polar regions. Other recent observations indicate the presence of mid-latitude glaciers (Plaut et al. 2009) as well as shallow ice in recent impact ejecta (Byrne et al. 2009). These environments could be ideal for the formation and preservation of clathrates and also suggest that the stability zone of clathrates on Mars could extend beyond the polar regions.

Along with the depth of the clathrate and its stability, another vexing issue is the release mechanism of the methane back to the surface. Indeed, the geothermal gradient is important for their stability but is by definition constant through time in the subsurface, and its variability is mostly due to the structure of the subsurface. Therefore, unless we are seeing the very end stage of clathrate destabilization, this process should be constant in time. That is, the flux should not really vary with time, unless some unknown process alters it very rapidly after release in the atmosphere. Atmospheric chemistry models (Lefèvre and Forget 2009) and experiments simulating the Martian soil (Gough et al. 2011) have failed to suggest a plausible mechanism. Therefore, it seems that some subsurface processes must be invoked to explain the methane release. For example some form of seismic activity could open cracks connecting the surface (diffusive region) to the subsurface where the clathrates are stocked, allowing a sudden release of methane.

2.2 Clathrate Formation and Lifetime at the Base of the Cryosphere

2.2.1 Clathrate Formation

Depending on the quantity of water available in the subsurface of the planet, the transition between the cryosphere of Mars and deep aquifers can occur as direct contact in low altitude areas, such as the northern plains, or via an unsaturated zone where water vapor circulates between deep aquifers and the base of the cryosphere (Clifford et al. 2010; Lasue et al. this issue). Assuming that CH_4 is produced at depth, either hydrothermally or biologically, this can lead to two different modes of clathrate formation at the base of the cryosphere: physical transport by convecting hydrothermal fluid, or gas phase transport through the unsaturated zone.

Most recent estimates of heat flux on Mars based either on lithostatic rheology calculations (McGovern et al. 2004) or inferred from radiogenic elements measured by gamma ray spectrometry on-board Mars Odyssey (Hahn et al. 2011) suggest low values for the present-day average crustal heat flow, between 6.4 $mW\,m^{-2}$ and 30 $mW\,m^{-2}$. These values are also consistent with the current volcanism rate of the planet (Li and Kiefer 2007) and the absence of an active core dynamo (Breuer et al. 2010). Such values approximately translate into a crustal thermal gradient between 3 and 12.5 $K\,km^{-1}$, assuming an average thermal conductivity of 2.4 $W\,m^{-1}\,K^{-1}$ consistent with ice-cemented soil. Such values are given as rough indications of those expected and will vary with types of material and pressure and temperature conditions. Using a representative value of 15 $mW\,m^{-2}$ for the crustal heat flow, thermal simulations of the cryosphere of Mars indicate that the depth of the base of a clathrate-rich cryosphere could range from 3 to 9 km below the equator and from 11 to 24 km below the poles (Clifford et al. 2010). It should be emphasized here that local heat flow differences, for example due to a magmatic event, are expected to modify the thermal structure of the subsurface (see e.g. Lyons et al. 2005), and may lead to convection which thins the permafrost layer (see e.g. Gulick 1998).

If the planetary inventory of water is low, or in high altitude terrains located around the equator, the base of the cryosphere will not be in contact with deep aquifers. However, thermal gradients in the crust will lead to vapor circulation in the unsaturated zone between the cryosphere and potential aquifers at depth which will replenish the cryospheric water content from below over short geological timescales (Clifford 1993). Methane gas would also circulate and get trapped at the base of the cryosphere (Max and Clifford 2000), forming gas or liquid methane deposits similar to what is characteristic of oceanic hydrate systems on Earth. The methane would react readily with water to form clathrates. The local pressure will be determined by the pore connectivity, resulting in variable pressures in the subsurface. Together with potential heterogeneities in the crustal lithology, permeability and thermal gradient, it will give rise to significant differences in the fractional occupancies and the type of guest sites occupied by the methane clathrate. If the amount of water in the subsurface of Mars is sufficient to allow a water-ice interface between the cryosphere and deep aquifers, the CH_4 may be transported upward by convecting hydrothermal fluids (Chassefière and Leblanc 2011). It has been shown that hydrothermal convection may easily occur in the Martian regolith, even under the sole effect of the background geothermal heating, and that the warm convecting fluid may locally thin the permafrost ice layer down to 1 km or less (Travis et al. 2003). CH_4-rich fluid encounters several phase changes of both H_2O and CH_4 as it moves upward through the crust, and crosses into the clathrate stability field. Typical trajectories of the fluid are represented in Fig. 3. Two trajectories are shown depending on the thermal gradient of the upper crust ($\sim$30 $K\,km^{-1}$ with cryosphere base depth at around

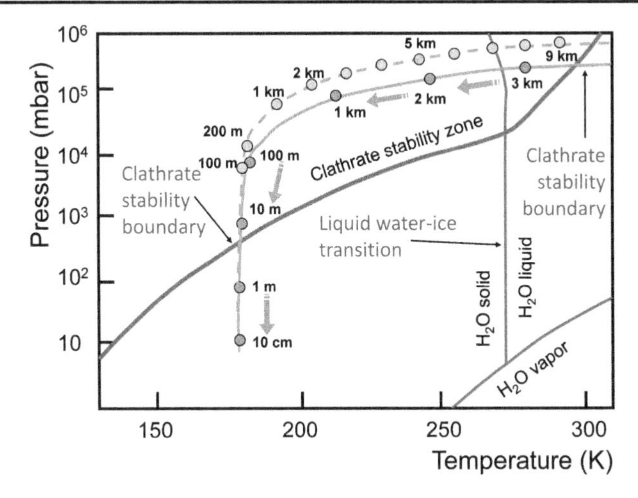

Fig. 3 Phase diagram of methane clathrate and water, and typical trajectory of the CH_4-rich fluid rising from a deep source up to the surface and, ultimately, to the atmosphere (modified from Chastain and Chevrier 2007). The top geothermal pressure-temperature profile (*dashed blue line*) uses a conductivity of 2.4 $W\,m^{-1}\,K^{-1}$ and a density of 2018 $kg\,m^{-3}$, consistent with an ice-cemented soil. The bottom geothermal profile (*solid blue line*) uses a conductivity of 0.9 $W\,m^{-1}\,K^{-1}$ and a density of 2338 $kg\,m^{-3}$, consistent with a dry sandstone. Properties are assumed constant with depth, with a surface temperature of 180 K and a geothermal heat flux of 30 $mW\,m^{-2}$ (see Mellon and Phillips 2001)

3 km for the solid line and $\sim$10 $K\,km^{-1}$ with cryosphere depth at around 9 km depth for the dashed line).

The trajectory of the fluid (blue lines), travelling from the right to the left (decreasing temperature) and from the top to the bottom (decreasing pressure), crosses several boundaries. Figure 4 presents a schematic view of the state of CH_4 over the whole column from 10 km depth up to the surface assuming a thermal gradient of 30 $K\,km^{-1}$. If the thermal gradient is 10 $K\,km^{-1}$, the liquid water ice transition is at 7 km depth, and the clathrate stability boundary at 10 km depth. At large depth, CH_4 is dissolved in the fluid. In existing models (e.g. Lyons et al. 2005), CH_4 is exsolved above 9 km depth, forming bubbles. When reaching the clathrate stability boundary at 3.5 km depth (Fig. 4), clathrates form at the interface between the gas inside the bubbles and ambient water under the form of a thin film surrounding the bubble. This phenomenon is observed in oceanic CH_4 plumes forming above deep-sea CH_4 sources (Sauter et al. 2006). At 2.8 km depth, the fluid encounters a second boundary, the water liquid-ice interface. In the presence of magmatic intrusions increasing the geothermal gradient, this depth can be significantly reduced. Indeed, hydrothermal fluids have been shown to be able to melt within a few thousand years through a 1–2 km thick permafrost layer (Gulick 1998). During CH_4-rich water freezing, the methane contained in clathrate-coated bubbles (gas and solid clathrate) is expected to be fully converted into clathrates and to fill the pore spaces. At this point, the clathrate solidifies and fills the pore space available, so there would not be any more convection (Elwood Madden et al. 2009). Clathrate can further expand upward over time through diffusion.

Another important gas that will be present at depth in significant quantities is CO_2. Through potential volcanic outgassing, and because it is unlikely that subsurface unsaturated zones are entirely sealed from the current atmosphere of the planet, we can expect CO_2 to be a major constituent of subsurface gases. The CO_2 clathrate equilibrium diagram is very close to that of CH_4 clathrates, except with somewhat larger energies required for dissociation of

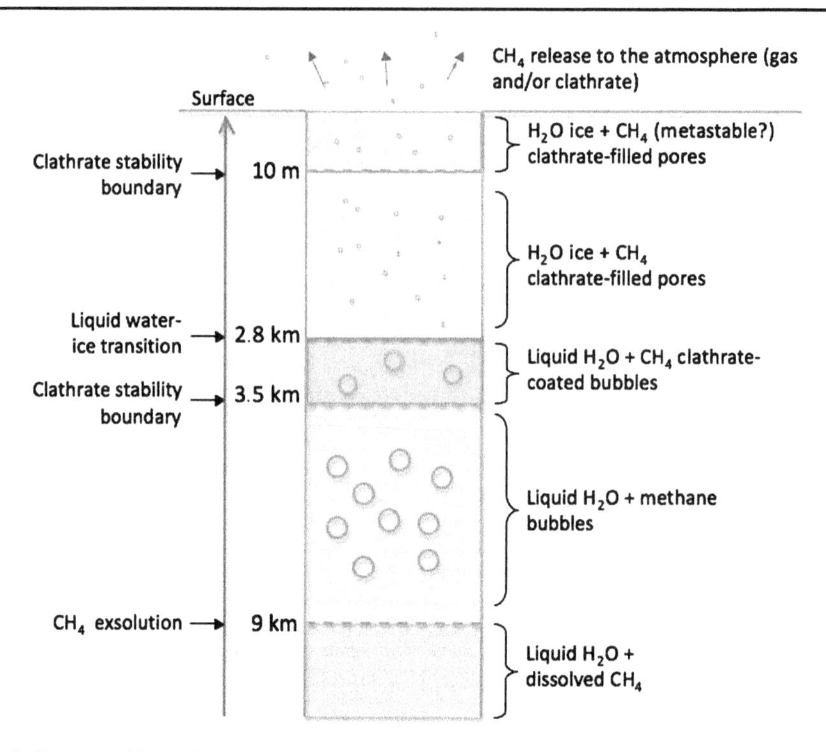

Fig. 4 Phase transition and physical state of water and methane in a rising flow of methane-rich water from a deep source up to the surface at Mars, using the bottom geothermal pressure-temperature profile in Fig. 3 (adapted from Chassefière and Leblanc 2011)

the clathrate (e.g. Longhi 2006). This implies that some fraction of the clathrate contained in the cryosphere should also trap some CO_2 gas. The exact nature and structural properties of the ice-to-liquid water transition boundary is not well constrained by observations. It has not been observed by radars, and is dependent on the water inventory of the planet (see Lasue et al. this issue). Also, the crustal lithologic structure, thermal gradient and the salinity of potential deep aquifers are not constrained. The question of salinity is particularly important since it affects the cryosphere by depressing the freezing point of both water ice and clathrates. Salts are well known inhibitors of clathrate formation (e.g. Sloan 1998). Early studies of the chemical and thermodynamic stability of various salts on Mars suggested that NaCl brines (which have a freezing point of 252 K at their eutectic) were the most likely to be found within the crust (Clark and van Hart 1981; Burt and Knauth 2003), although other chloride-rich, multicomponent brines (such as $CaCl_2$, $MgCl_2$ and LiCl) might also exist with freezing points as low as 210 K (Brass 1980; Clark and van Hart 1981; Burt and Knauth 2003). The Wet Chemistry Lab (WCL) and Thermal and Evolved Gas Analyzer (TEGA) analyses from the Phoenix Lander showed that Martian soils contain up to 1 wt% of perchlorate (Hecht et al. 2009). The presence of perchlorate is important because of its high deliquescence, its relative chemical inertia with possible concentration buildups over time and its very low eutectic temperature of 203 K when considering its $Mg(ClO_4)_2$ form (Hecht et al. 2009). Depending on the types of salts that are present in the aquifers of the crust, the depth of the base of the cryosphere reached may be reduced by a factor 3 (Clifford et al. 2010), modifying significantly the clathrate formation and transition as shown in Fig. 3 with a solid water ice to liquid brine phase transition line shifted towards the left of the diagram. This inhibition of clathrate formation by the presence of salts will translate into an upward shift of the clathrate stability zone in Fig. 3 (see Elwood Madden et al. 2007 for detailed discussion of this process). This shift will have a minimal effect on the trajectory of the clathrate presented in the figure as changes in temperature and pressure induced by the presence of salt will be modified by a factor of 2 or 3 (e.g. Prieto-Ballesteros et al. 2005). It should, however, have a significant effect for the near-surface stability of clathrates and could also change the structure of the clathrates formed by modifying the way the cages get occupied by the gas molecules (Sloan 1998). In all cases, the kinetics of formation of clathrates at the base of the cryosphere is expected to be mainly constrained by the range of pressure and temperature relevant to the depth where the cryosphere ends.

Experimental studies under conditions relevant to the Martian case (about 273 K and from 1 to 7 MPa) as summarized in Table 1 give rates between 5.5×10^{-7} and 7×10^{-5} $\mathrm{mol\,m^{-2}\,s^{-1}}$ for the formation of methane clathrates (Kuhs et al. 2006; Gainey and Elwood Madden 2012) and from 0.7×10^{-5} to 1×10^{-4} $\mathrm{mol\,m^{-2}\,s^{-1}}$ for the formation of carbon dioxide clathrates (Chun and Lee 1996; Leeman and Elwood Madden 2010). Such constants of formation would lead to formation timescales between about 10^5 and 5×10^7 years for forming a 10 km deep clathrate cryosphere, taking into account the molar content of a cubic meter of clathrate (Max and Clifford 2000). These timescales are short relative to geological times. The formation time of such a cryosphere will thus be limited by the water accretion rate of the cryosphere, and the rates of diffusion and dissociation of clathrates as discussed in the next section.

An interesting analogy can be made with clathrate reservoirs forming within sediments on the sea floor (Haeckel et al. 2004), fed by a CH_4 flux from below. Whatever process emplaces CH_4 clathrates in the permafrost layer, the consequence of repeated episodes of hydrothermal production of CH_4 is to progressively enrich the cryosphere with CH_4, similar to what is observed in deep oceanic sediments. While CH_4 may ultimately sublimate to the atmosphere following the slow rise of the permafrost heated from below, the formation of clathrates in the Martian cryosphere may also be continuous, resulting in constant replenishment of the clathrates located near the surface which release methane to the atmosphere. Over geological times, atmospheric carbon could be recycled to crustal carbonates, with further decomposition and reduction of CO_2 by H_2 produced by serpentinization, and further release of the produced CH_4 to the atmosphere (Chassefière and Leblanc 2011). In this model, the cryosphere would play the role of a buffer, storing methane produced from below and releasing it to the atmosphere, playing an active role in the carbon cycle on Mars.

2.2.2 Cryosphere Clathrate Lifetime

As previously mentioned, the time required to form a 10 km deep clathrate cryosphere is short compared to geological timescales. The dissociation of clathrates under non equilibrium conditions occurs at similar rates. Recent experiments at near-Martian conditions have shown that the dissociation rates of methane clathrates under high pressure and low temperature are in the 3×10^{-6}–10^{-4} $\mathrm{mol\,m^{-2}\,s^{-1}}$ range (Gainey and Elwood Madden 2012). In the case of CO_2 clathrates, values are in the 4–9 $\times 10^{-5}$ range under high pressure conditions (Giavarini et al. 2007; Leeman et al. 2010) and can increase up to 3×10^{-3} under low pressure conditions (Leeman et al. 2010). If dissociation at low pressure is the main mechanism of clathrate destruction, then typical lifetimes of the cryosphere would range from 10^3 to 10^6 years. However these rates will quickly slow over time as pure water ice replaces the clathrates and diffusion becomes the dominant transport mechanism of the clathrate species. The lifetime of a clathrate-rich cryosphere will therefore be limited by the diffusion rates

Table 1 Clathrates formation and dissociation rates reported in the literature

Guest molecule	Process	Rate (mol m^{-2} s^{-1})	E_a (kJ/mol)	T (K)	P (MPa)	Source
CH_4	formation	NA-no surface area	61.4	253–273	6.9	Wang et al. (2002)
CH_4	formation	5.5×10^{-7}–3.5×10^{-5}	92.8	245–270	3–6	Kuhs et al. (2006)
CH_4	formation	4.3×10^{-6}–7.0×10^{-5}	32.3	234–262	1.4–3.5	Gainey and Elwood Madden (2012)
CO_2	formation	1.47×10^{-4}		250	0.745	Leeman and Elwood Madden (2010)
CO_2	formation	0.69–0.71 $\times 10^{-5}$		275.2–279.2	2–3.5	Chun and Lee (1996)
CO_2	formation	NA-no surface area	73.3	271–275	3.25	Ota et al. (2005)
CO_2	formation	NA-no surface area	27.2	230–263	6.2	Henning et al. (2000)
CH_4	decomposition	NA-no surface area	14.5	271–275	3.25	Ota et al. (2005)
CH_4	dissociation	3.0×10^{-6}–9.9×10^{-5}	32.7	222–265	0.1–2	Gainey and Elwood Madden (2012)
CO_2	dissociation	4–9 $\times 10^{-5}$		270–273	0.1–0.3	Giavarini et al. (2007)
CO_2	dissociation	3×10^{-3}		250	3×10^{-4}	Leeman et al. (2010)

which are smaller than either the formation or dissociation rates. Furthermore, the values for dissociation and formation rates are approximately of the same magnitude, which suggests the cryosphere clathrate content will be in steady state with timescales limited by diffusion as the clathrate species are replenished from underground sources.

As shown in Table 2, experimental and theoretical studies have shown that the diffusion coefficient of gases through water ice, or through clathrates, is very slow, though values given in the literature show orders of magnitude uncertainties. Moreover, these values strongly depend on the temperature at which the diffusion coefficients have been measured or obtained by numerical simulations. Note that in the latter case, the diffusion coefficients also depend on the interaction-potential model used to calculate the water-water and the water-guest interactions. For the case of methane diffusion through clathrates, values range from 1.0×10^{-15}–3.5×10^{-14} $m^2\,s^{-1}$ (Kuhs et al. 2006; Peters et al. 2008) to 3×10^{-12}–3.8×10^{-10} $m^2\,s^{-1}$ (Brodskaya and Sizov 2009), while the CO_2 diffusion rate through clathrates is about 10^{-12} $m^2\,s^{-1}$ (Demurov et al. 2002; Uchida et al. 2003).

For the case of diffusion through water ice, values of 10^{-14}–10^{-11} $m^2\,s^{-1}$ are given for methane (Hori and Hondoh 2003; Komai et al. 2004; Ikeda-Fukazawa et al. 2004) but can be as high as 3×10^{-7}–4×10^{-6} $m^2\,s^{-1}$ (Sun and Chen 2006) and they have been reported to be around 10^{-14}–10^{-15} $m^2\,s^{-1}$ for CO_2 (Takeya et al. 2000; Rohde and Price 2007). However, values of about 10^{-11} $m^2\,s^{-1}$ have also been reported around 270 K using molecular dynamics simulations (Ikeda-Fukazawa et al. 2004). If we take into account the values presented here and mostly obtained at relatively high temperatures (i.e., between 250 and 270 K), the removal timescale of clathrates from a 10 km deep cryosphere could be anywhere from 2×10^4 to 10^{12} years. Moreover, taking into account that diffusion coefficients decrease exponentially with temperature, these timescales are certainly lower limits. The discrepancy shown in the diffusion values indicates the difficulty of such studies and more experimental work would be required to constrain these rates. While the values for the lifetime of a stable clathrate cryosphere are poorly constrained, high values of millions of years or more are conceivable and could lead to the trapping of ancient methane and CO_2 in the subsurface of the planet for geologic timescales.

Another important phenomenon to take into consideration is the diffusion of water from the Martian regolith, which would lead to the destabilization of the clathrate matrix and consequently to the release of the trapped gases. Water diffusivity rates through porous regolith under Martian surface conditions were found experimentally to be between 4×10^{-5} and 1.4×10^{-3} $m^2\,s^{-1}$ (Hudson et al. 2007). These values are orders of magnitude higher than the ones measured for the diffusion of clathrate host gas and certainly dominate the sublimation effects of a clathrate cryosphere. Simulations of cryosphere stability at depth have shown that water recondensation on the pores could significantly slow diffusion through the Martian regolith to geological timescales (Mellon et al. 1997) while sophisticated two-dimensional simulations of water transport in the Martian subsurface seem to indicate a potential depletion of the cryosphere around equatorial latitudes in about 10^8–10^9 years (Grimm and Painter 2009). Combining all these considerations, the cryosphere timescales are constrained by the subsurface dissociation of clathrates under non-equilibrium conditions, keeping in mind that water desiccation processes ultimately play an important role when they take place, due to their probably faster kinetics.

Table 2 Diffusion coefficients of molecules through clathrate and ice reported in the literature

Guest molecule	Host phase	Diffusion rate (m^2/s)	E_a (kJ/mol)	T (K)	P (MPa)	Source
CH_4	clathrate	2.5×10^{-15}–3.5×10^{-14}	52.1	245–270	3–6	Kuhs et al. (2006)
CH_4	clathrate	7×10^{-15}		250	4	Peters et al. (2008)
CH_4	clathrate	1×10^{-15}		200–210	4	Peters et al. (2008)
CH_4	clathrate	3×10^{-12}–3.8×10^{-10}		200–220		Brodskaya and Sizov (2009)
Kr	clathrate	3×10^{-12}–2.3×10^{-9}		200–230		Brodskaya and Sizov (2009)
H_2	clathrate	1–4×10^{-12}	3	250–265	1–20	Okuchi et al. (2007)
CO_2	clathrate	10^{-12}		273		Demurov et al. (2002)
CO_2	clathrate	10^{-12}		268		Uchida et al. (2003)
CO_2	ice	8.3×10^{-16}		263		Takeya et al. (2000)
CO_2	ice	10^{-14}		242	6.2	Rohde and Price (2007)
CO_2	ice	10^{-11}		270		Ikeda-Fukazawa et al. (2004)
CH_4	ice	10^{-13}		268		Komai et al. (2004)
CH_4	ice	10^{-14}				Hori and Hondoh (2003)
CH_4	ice	10^{-11}				Ikeda-Fukazawa et al. (2004)
CH_4	ice	3×10^{-7}–4×10^{-6}		262–270		Sun and Chen (2006)
H_2	ice	8×10^{-8}		242	6.2	Rohde and Price (2007)
N_2	ice	2×10^{-10}	5.1	248		Ikeda-Fukazawa et al. (2005)
N_2	ice	2×10^{-13}		263		Ikeda et al. (1999)
O_2	ice	4.3×10^{-13}		263		Ikeda et al. (1999)
O_2	ice	4.3×10^{-9}	9.7	248		Ikeda-Fukazawa (2004)
O_2	ice	10^{-11}				Hori and Hondoh (2003)
N_2	ice	10^{-12}				Hori and Hondoh (2003)
He	ice	10^{-9}	11.4	258–268		Satoh et al. (1996)
Ne	ice	10^{-10}		258–268		Satoh et al. (1996)
Ar	ice	10^{-11}		258–268		Satoh et al. (1996)

2.3 Mechanisms of Methane Release into the Atmosphere and Links with the Seasonal Cycle

2.3.1 Destabilization of Clathrate and Methane Release

There is a major temporal and geographical paradox with the release models of methane. According to climate models, methane must be released in small quantities and at certain periods of the year to be destroyed rapidly (Lefèvre and Forget 2009). This would be apparently compatible with seasonal destabilization of the clathrate layer and subsequent release of methane from the subsurface to the atmosphere. However, some methane plumes have been observed in the equatorial regions, around Nili Fossae, where thermodynamic models and various observations indicate a lack of water ice, although it remains possible that this ice is out of reach of the instruments (Mumma et al. 2009). Despite thermodynamically unfavorable conditions, shallow ice has been observed poleward of 30° latitude (Bandfield 2007; Byrne et al. 2009). Therefore, following the ice cemented soil model, clathrates could be present in a metastable state at lower latitudes than thermodynamic models suggest. Nevertheless, the main problem remains the lack of methane release at high latitudes, except over the north polar cap during northern summer (Geminale et al. 2011), suggesting that the corresponding clathrates are either out of reach (too deep) or non-existent. If the methane is released from clathrates in the equatorial regions, then the destabilization mechanism is not controlled by seasonal cycles, because they would be below the depth affected by seasonal temperature changes (typically a depth around a meter, e.g. Ulrich et al. (2010)).

Some detailed studies combining thermodynamics with kinetics of clathrate destabilization and subsequent diffusion through the soil have concluded that pure methane clathrate is not compatible with the methane release fluxes observed on the surface (Elwood Madden et al. 2011). This study suggests that contamination by other gases, including H_2S, would stabilize clathrates in the subsurface and explain the observed fluxes. It would be interesting to combine such new kinetic models with complex thermodynamics to test if the most realistic compositions are compatible with the present-day fluxes.

Another issue is the release process of methane into the atmosphere, with a release flux observed to be variable from place to place and from time to time (Mumma et al. 2009; Fonti and Marzo 2010; Geminale et al. 2011). The geothermal gradient, which controls the stability of clathrates, depends on the structure of the subsurface and is not expected to change over short time periods of a few weeks or months. The mechanisms of spatial and temporal variability of the methane release flux are not understood. Some form of seismic or tectonic activity, as well as impacts, could open cracks connecting the surface (diffusive region) to the subsurface where the clathrates are stocked, allowing a sudden release of methane to the atmosphere. Alternatively, if a clathrated cryosphere is in contact with the atmosphere at several locations, and if the release rate of methane is variable in time due to local heterogeneities of clathrates, sporadic releases at various places may occur. Over much longer timescales, changes due to variations of obliquity, local heatflow or salinity of the water involved in clathrate formation could significantly affect the release rate of methane to the atmosphere.

But the most striking feature of the methane atmospheric field is its seasonal variability, described in the next subsection, with significantly more gaseous methane in the atmosphere during northern spring and summer (Fig. 2). If the sources of methane were located at high latitudes, such a seasonal variability could result from the surface obstruction of the porous network by seasonal glacial deposits preventing methane from being released to the atmosphere. However, the efficiency of such a mechanism is not well established. A strong increase of the methane atmospheric methane content has been observed from Martian orbit by the PFS instrument on Mars Express over the north polar cap during local summer, which has been interpreted as the sign of a methane reservoir associated with the polar cap (Geminale et al. 2011). On the other hand, methane plumes observed from Earth are localized in northern summer at low and mid-latitudes (Mumma et al. 2009). Nevertheless, if the hypothesis of an atmospheric source for gaseous methane described in the next subsection were true, the locations of detected gaseous methane plumes could be different from the locations of the delivery of methane (under solid form) from the subsurface to the atmosphere, and a polar origin for methane cannot be ruled out. But, even if the source of methane were polar, the hypothesis that the seasonal variability of methane is controlled by the sealing effect of seasonal glacial deposits cannot in any event explain the spatially patchy structure of the methane field, and we consider such an explanation as not satisfactory.

Another possibility is a seasonal variation of CH_4 generated by alternate formation of CO_2-CH_4 clathrates on the seasonal polar caps, which is thermodynamically possible. From a thermodynamic study of the stability and composition of such clathrates, and an estimate of the amount of water seasonally exchanged between the two polar caps, Herri and Chassefière (2012) showed that at most a few percent (up to 3 %), of the atmospheric content of methane can be seasonally exchanged between the two hemispheres. The true value is probably much smaller and the effect of the seasonally alternate formation of clathrates on the CH_4 atmospheric content is therefore small, if not negligible. In a similar way, the adsorption of CH_4 in the regolith cannot result in a significant seasonal variation of the atmospheric CH_4 (Meslin et al. 2011).

A seasonal fluctuation of typically $\sim$10 ppb of the atmospheric methane mixing ratio is observed on Earth (see e.g. Khalil and Rasmussen 1990; Chen and Prinn 2006). Emissions from wetlands, with a significant contribution of high-latitude wetlands, is a major source of atmospheric methane, about 30 % of the global emission (Chen and Prinn 2006). The methane content presents a minimum during spring and summer, contrary to the case of Mars, and the relative amplitude of variation is only $\pm$0.5 % on Earth, whereas it is $\sim$50 % on Mars (see next subsection). Interestingly, soil freezing in Earth's high-latitude region results in a sharp emission of methane during one month at the beginning of fall, in October, probably due to the exsolution of methane during water freezing process (Mastepanov et al. 2008). On Mars, no liquid water is present at the surface of the planet and no such event is expected. The case of Mars is thus quite different.

In the next subsection, we suggest that both seasonal variability and temporal and spatial heterogeneities of atmospheric methane may be due to control by meteorology of the release of gaseous methane from metastable atmospheric particles of methane clathrates.

2.3.2 Delivery of Metastable Particles of Methane Clathrate to the Atmosphere

Alternatively to the clathrate dissociation mechanism, it has been suggested that CH_4 could be delivered to the atmosphere in the form of small (submicron) metastable clathrate particles, rather than gaseous CH_4 (Chassefière 2009). There are two major reasons for postulating that methane could be, totally or partially, released to the atmosphere under solid form.

First, the global correlation between CH_4 mixing ratio and water vapor abundance as well as water ice content, as observed by PFS on Mars-Express (Geminale et al. 2008), is difficult to explain if the sources of gaseous CH_4 are fixed vents at the surface. Because the distribution of water vapor in the atmosphere of Mars is mainly controlled by atmospheric dynamics, such a correlation acts in favour of an atmospheric source of gaseous

CH_4. Suspended atmospheric clathrate particles may decompose to the gas phase due to condensation/sublimation processes related to cloud activity (Chassefière 2009). Complex processing of such particles serving as condensation nuclei by condensation, particle settling, further sublimation and recycling through the clouds could result in their progressive erosion. Further analysis of PFS data (Geminale et al. 2011) confirms that there is a seasonal variation of the atmospheric methane content, with more CH_4 during northern spring and summer, and that peak mixing ratios (up to 45 ppb) of CH_4 are reached above the north polar cap during its sublimation to the atmosphere, where the peak water vapor abundance is observed. Although a systematic correlation between CH_4 and water vapor is not found in the most recently published PFS data (Geminale et al. 2011), the global coincidence between CH_4 and water vapor maxima during northern spring and summer, and above north polar cap during springtime sublimation, is confirmed and seems confidently established. A link between water vapor condensation, therefore cloud activity, and decomposition of clathrate to gas phase in the atmosphere, as proposed above, could explain this coincidence, and the observed seasonal behaviour of the atmospheric CH_4 content. This is supported by recent experiments suggesting that methane clathrate may directly sublimate into the atmosphere instead of forming gas and ice, under martian conditions (Blackburn et al. 2009). The hypothesis of submicron metastable CH_4 clathrate particles decomposing in the atmosphere under the effect of condensing water vapor is a possible explanation of the observed CH_4 seasonal behaviour. These clathrate particles could form the submicron particle population evidenced from an analysis of a Viking limb image taken at LS = 176° (Montmessin et al. 2002).

Second, the phenomenon of "anomalous preservation" (also named "incomplete dissociation" in its partial form) of clathrates below the ice point has been experimentally observed for more than 20 years for a great variety of clathrate systems, including CH_4 clathrate (Stern et al. 2001), and has potential application for successful retrieval of natural gas hydrates. It has been experimentally shown that CH_4 clathrate particles of a few hundred microns radius preserve most of their structure during rapid release of the pressure of the confining system where they have been synthesized (Stern et al. 2001). Upon heating through the ice point (273 K), the particles rapidly decompose and release all their residual CH_4, which is a property common to all anomalously preserved clathrates studied up to now. It has been experimentally shown that the water ice produced at the surface of clathrate particles by decomposition could play a role by sealing the crystals and preventing further dissociation. A recent study, using a different clathrate synthesis experimental system (Zhang and Rogers 2008), yielding very pure aggregates of micron-size particles showed that CH_4 clathrate crystals, provided they are compact and with minimal structural defects, don't significantly decompose when the pressure decreases. In this case, fractures and void spaces through the clathrate mass are so small that the surrounding clathrate layers, whose mechanical strength much exceeds water ice strength, are postulated to be able to sustain the high internal equilibrium pressure within defect spaces. For example, from theoretical considerations, a 1 μm diameter sphere of clathrate would require only a very thin shell of water ice of 0.03 μm to contain its internal equilibrium pressure. This outer shell could be formed as well of clathrate, with no outer water ice shell needed to prevent the particle from dissociating (Zhang and Rogers 2008). The metastability of isolated micron-size CH_4 clathrate particles, rather than polycrystalline assemblages of such small particles, still remains to be characterized in Martian atmospheric conditions by appropriate laboratory measurements. After remaining at 1 bar and 268 K during 10 days, less than 1 % dissociation is observed, almost all CH_4 remaining within the particle in clathrate form (Zhang and Rogers 2008). During the sublimation of the upper layers of the CH_4 clathrate-rich cryosphere to the atmosphere, clathrate inclusions could preserve their clathrate structure, and small clathrate particles could be released to the atmosphere and survive days or weeks suspended in the atmosphere before being eroded and decomposed to the gas phase. Possibly, decomposition could result in the formation of an outer ice layer preventing CH_4 from escaping to the atmosphere. If so, CH_4 could remain trapped within the outer icy shell under gaseous and/or clathrate form in the central region of the particle until water ice sublimates. Cloud processes would therefore play a major role in eroding and destabilizing particles, finally releasing the CH_4 contained in the particles to the gas phase (Chassefière 2009). It could explain the patchy morphology of the atmospheric CH_4 field (Geminale et al. 2011). Local cloud processes, occurring through a uniform ambient haze of small suspended clathrate crystals, would result in local pulses of gaseous CH_4 produced in the middle atmosphere.

The decomposition process remains to be understood in detail (Chassefière 2009). Generally speaking, the process of methane clathrate decomposition involves two steps: first, the destruction of the clathrate host lattice at the surface of the particle, and second, the desorption of the methane molecule from the surface (Kim et al. 1987). As the decomposition progresses, the particle shrinks and methane gas is generated at the solid surface, then released to the gas phase. It has been suggested (Chassefière 2009) that the crystalline structure of the clathrate host lattice at the surface of the particle is progressively damaged by condensation of atmospheric water vapor, resulting in an increasing fragility of the external layers of the particles and their final decomposition. Due to the deposition of a pure water ice condensation shell at the surface of the clathrate particles during condensation, and to the fact that clathrate destruction occurs at the surface of the particle, the release of methane to the atmosphere could be inhibited at an early stage of condensation, or at least slowed down, by the rapidly growing pure water ice shell. Provided this shell becomes thick enough before the clathrate particle inside has decomposed, the clathrate structure could be preserved in the central part of the particle. The methane molecules remaining in the particles would be released, totally or partially, to the atmosphere below the condensation level, during the sublimation phase of settling ice particles. In this way, some methane may be potentially released to the atmosphere during both condensation and sublimation phases. Such processes would have to be characterized by further laboratory work. This hypothesis is not exclusive of the release of some gaseous CH_4 directly to the atmosphere if decomposition occurs as soon as the water ice matrix sublimates. If so, both gas and particles could be released to the atmosphere.

2.3.3 Related Water and Carbon Cycles

In this section, we briefly describe how a potential carbon cycle linked to the presence of clathrates in the cryosphere could be occurring on Mars. In the previous sections, we have seen that the rates of clathrate formation at the base of the cryosphere and the rates of clathrate dissociation in the near subsurface have similar values and could allow a stable clathrate reservoir within the cryosphere over geological timescales. Simulations of the orbital history of Mars show that the obliquity and eccentricity cycles can lead to long periods of high obliquity ($>45°$) implying insolation conditions very different from those observed today (Laskar 2004). Under such conditions, alternate cycles of CO_2 and CH_4 clathrate dissociation could occur at the poles and near the equator (Root and Elwood Madden 2012). It has been suggested that the methane observed in the atmosphere of Mars (Formisano et al. 2004; Krasnopolsky et al. 2004; Mumma et al. 2009) could originate from methane clathrate dissociation in the equatorial regions (see Sect. 2.3.1). At the same time, the conditions near the poles of Mars are suitable for clathrate formation (Longhi 2006; Chassefière and Leblanc 2011). Polar basal melting has been suggested as a process to replenish deep aquifers on Mars (Clifford 1987). Should the ice deposited near the poles contain a significant fraction of CO_2 and CH_4 clathrates formed during cycles of the planet, then the basal melting of the polar caps would release liquid CO_2 and CH_4 that could be trapped within the cryosphere (Longhi 2006) until a new cycle restarts after slow diffusion through the cryosphere. Such processes can lead to the formation of deep carbon reservoirs stable over geological timescales and replenishment of the deep aquifers in gas that would replenish in turn the cryosphere from below via convective or unsaturated zone circulation processes.

The inventory of carbon on Mars and its evolution through time are poorly known. According to upper atmosphere models for early Mars (Tian et al. 2009), the timescale for 1 bar of CO_2 to be removed from the early Martian atmosphere through thermal (hydrodynamic) escape would have been only a few million years until the end of carbon hydrodynamic escape, 4.1 Gyr ago. If Mars has been endowed with amounts of CO_2 similar to those found on Venus and the Earth ($\sim$100 bars), most of the initial CO_2 inventory of Mars would have been lost within the first few 100 Myr after its formation. But some carbon may have been stored in carbonates through weathering of basalts over timescales of a few million years (Pollack et al. 1987), similar to thermal escape timescales (Lammer et al. this issue). Despite the importance of escape, substantial amounts of carbon could therefore have been trapped in surface carbonate reservoirs during the Noachian. Later volcanism would have released to the atmosphere, not only mantle carbon, but also a significant fraction of carbon produced by thermal decomposition of carbonates, as occurred on Earth. If so, the cumulative amount of CO_2 released to the atmosphere by volcanism over the whole history of Mars could be significantly larger than the 0.5–1 bar value derived from mantle thermal evolution models (Grott et al. 2011), typically a few bars. The cumulative amount of CO_2 lost to space by non-thermal escape since the end of hydrodynamic escape should not exceed 10 mbar (Chassefière and Leblanc 2011). The quantity of CO_2 contained in the atmosphere and in the south polar cap (Phillips et al. 2011) is of the same order of magnitude, that is 10–20 mbar. Up to 140 mbar of CO_2 could be adsorbed in the regolith (Fanale and Cannon 1979). Therefore, most of the few bars of CO_2 present on Mars at the end of the Noachian must be trapped in the subsurface, either in the form of carbonates, or as CO_2 and/or CH_4 clathrates. It has been suggested that the total amount of CH_4 cycled through the subsurface by hydrothermal processes may have been up to, due to gravity, an equivalent CO_2 pressure of 0.2 to 2 bars (Chassefière and Leblanc 2011). The carbon content of possible subsurface carbonate reservoirs is totally unknown, but could reach one to several bars according to geochemical modelling (Griffith and Shock 1995; Niles et al. 2012).

3 Role of Clathrates in Fractionating Atmospheric Noble Gases

Musselwhite and Lunine (1995) and Swindle et al. (2009) pointed out that if clathrates form on Mars, this may result in strong noble gas fractionation. This may be quantitatively illustrated with the use of a thermodynamic statistical model utilized in the recent literature to determine the composition of clathrates which may form in various planetary environments (Lunine and Stevenson 1985; Thomas et al. 2007, 2008, 2009; Swindle et al. 2009; Mousis et al. 2009, 2010, 2011). This approach allows us to calculate the relative abundances of Ar, Kr and Xe incorporated in clathrates formed at the Martian surface/atmosphere interface under given temperature and pressure conditions. If clathrate formation is strongly related to the Martian seasonal cycle, then we should expect significant variations of the noble gas abundance ratios.

3.1 Thermodynamic Statistical Model

Our model follows a classical statistical mechanics method to relate the macroscopic thermodynamic properties of clathrates to the molecular structure and interaction energies. It is based on the original ideas of van der Waals and Platteeuw (1959) for clathrate formation, which assume that trapping of guest molecules into cages corresponds to the three-dimensional generalization of ideal localized adsorption. This approach is based on four key assumptions:

1. The host molecule's contribution to the free energy is independent of the clathrate occupancy. This assumption implies that the guest species do not distort the cages.
2. (a) The cages are singly occupied (i.e. one molecule/cage). (b) Guest molecules rotate freely within the cage.
3. Guest molecules do not interact with each other.
4. Classical statistics is valid, i.e., quantum effects are negligible.

In this formalism, the fractional occupancy of a guest molecule K for a given type t ($t =$ small or large) of cage can be written as

$$y_{K,t} = \frac{C_{K,t} P_K}{1 + \sum_J C_{J,t} P_J}, \tag{1}$$

where the sum in the denominator includes all the species which are present in the initial gas phase. $C_{K,t}$ is the Langmuir constant of species K in the cage of type t, and P_K is the partial pressure of species K. This partial pressure is given by $P_K = x_K \times P$ (we assume that the sample behaves as an ideal gas), with x_K the mole fraction of species K in the initial gas phase, and P the total gas pressure, which is dominated by CO_2.

The Langmuir constant depends on the strength of the interaction between each guest species and each type of cage, and can be determined by integrating the molecular potential energy within the cavity as

$$C_{K,t} = \frac{4\pi}{k_B T} \int_0^{R_c} \exp\left(-\frac{w_{K,t}(r)}{k_B T}\right) r^2 dr, \tag{2}$$

where R_c represents the radius of the cavity (assumed to be spherical), k_B the Boltzmann constant, T the temperature and $w_{K,t}(r)$ is the spherically averaged potential (here the Kihara potential) representing the interactions between the guest molecules K and the H_2O molecules forming the surrounding cage t. This potential $w(r)$ can be written for a spherical guest molecule, as (McKoy and Sinanoğlu 1963)

$$w(r) = 2z\epsilon\left[\frac{\sigma^{12}}{R_c^{11} r}\left(\delta^{10}(r) + \frac{a}{R_c}\delta^{11}(r)\right) - \frac{\sigma^6}{R_c^5 r}\left(\delta^4(r) + \frac{a}{R_c}\delta^5(r)\right)\right], \tag{3}$$

with

$$\delta^N(r) = \frac{1}{N}\left[\left(1 - \frac{r}{R_c} - \frac{a}{R_c}\right)^{-N} - \left(1 + \frac{r}{R_c} - \frac{a}{R_c}\right)^{-N}\right]. \tag{4}$$

In Eq. (3), z is the coordination number of the cell. Parameters z and R_c, which depend on the structure of the clathrate (I or II) and on the type of the cage (small or large), are given in Table 3. The intermolecular parameters a, σ and ϵ describe the guest molecule-water interactions in the form of a Kihara potential. Table 4 summarizes the interaction parameters of the main molecules in the Martian atmosphere which are used in the present study. This compilation of parameters supersedes the one used by Swindle et al. (2009) since

Table 3 Parameters for the cavities. R_c is the radius of the cavity (values taken from Parrish and Prausnitz 1972). b represents the number of small (b_s) or large (b_ℓ) cages per unit cell for a given structure of clathrate (I or II), z is the coordination number in a cavity

Clathrate structure	I		II	
Cavity type	small	large	small	large
R_c (Å)	3.975	4.300	3.910	4.730
b	2	6	16	8
z	20	24	20	28

Table 4 Parameters for Kihara potentials used in this work. σ_{K-W} is the Lennard-Jones diameter, ϵ_{K-W} is the depth of the potential well, and a_{K-W} is the radius of the impenetrable core, for the guest-water pairs

Molecule	σ_{K-W} (Å)	ϵ_{K-W}/k_B (K)	a_{K-W} (Å)	Reference
CO_2	2.97638	175.405	0.6805	Sloan and Koh (2008)
N_2	3.13512	127.426	0.3526	Sloan and Koh (2008)
O_2	2.7673	166.37	0.3600	Parrish and Prausnitz (1972)
CH_4	3.14393	155.593	0.3834	Sloan and Koh (2008)
CO	3.1515	133.61	0.3976	Mohammadi et al. (2005)
Ar	2.9434	170.50	0.184	Parrish and Prausnitz (1972)
Kr	2.9739	198.34	0.230	Parrish and Prausnitz (1972)
Xe	3.32968	193.708	0.2357	Sloan and Koh (2008)

it contains interaction parameters for CO_2, N_2, CH_4, CO and Xe that have been updated by recent experiments (Mohammadi et al. 2005; Sloan and Koh 2008).

Finally, the mole fraction f_K of a guest molecule K in a clathrate can be calculated with respect to the whole set of species considered in the system as

$$f_K = \frac{b_s y_{K,s} + b_\ell y_{K,\ell}}{b_s \sum_J y_{J,s} + b_\ell \sum_J y_{J,\ell}}, \tag{5}$$

where b_s and b_l are the number of small and large cages per unit cell respectively, for the clathrate structure under consideration, with the sum of the mole fractions of enclathrated species normalized to 1.

The calculations are performed using temperature and pressure conditions along the stability boundary at which the multiple guest clathrates are formed. The corresponding temperature and pressure values ($T = T_{mix}^{diss}$ and $P = P_{mix}^{diss}$) can be read from the dissociation curve of the multiple guest clathrates. Here, the dissociation pressure P_{mix}^{diss} of a multiple guest clathrate is calculated from the dissociation pressure P_K^{diss} of a pure clathrate of guest species K as (Lipenkov and Istomin 2001):

$$P_{mix}^{diss} = \left[\sum_K \frac{x_K}{P_K^{diss}} \right]^{-1}, \tag{6}$$

where x_K is the molar fraction of species K in the gas phase. The dissociation pressure P_K^{diss} is derived from laboratory measurements and follows an Arrhenius law (Miller 1961):

$$\log(P_K^{diss}) = A + \frac{B}{T}, \tag{7}$$

Table 5 Parameters A and B needed in Eq. (7)

Molecule	A	B (K)
CO_2	10.100	−1112.0
N_2	9.860	−728.5
O_2	10.500	−888.1
CH_4	9.886	−951.2
CO	9.770	−732.02
Ar	9.344	−648.7
Kr	9.028	−793.7
Xe	9.547	−1208.0

Table 6 Relative abundances of CH_4, CO_2, O_2, CO, Kr, Xe, Ar and N_2 in the initial gas phase (x_K) and in structure I CO_2-dominated clathrate (f_K). These ratios are calculated at $P = 7$ mbar, and at the corresponding temperature on the multiple guest dissociation curve. Except for CH_4, whose considered abundance is relatively high, all others gas phase mixing ratios are representative of the present-day Martian atmosphere (see text)

Species	x_K	f_K	Abundance ratio
CH_4	1×10^{-3}	8.48×10^{-5}	8.48×10^{-2}
CO_2	0.955	9.99×10^{-1}	1.046
N_2	2.71×10^{-2}	7.33×10^{-5}	2.70×10^{-3}
Ar	1.61×10^{-2}	1.41×10^{-4}	8.76×10^{-3}
Kr	2.01×10^{-7}	5.97×10^{-8}	0.297
Xe	8.03×10^{-8}	1.22×10^{-6}	15.19
O_2	1.31×10^{-3}	4.02×10^{-6}	3.07×10^{-3}
CO	7.03×10^{-4}	4.52×10^{-6}	6.43×10^{-3}

where P_K^{diss} and T are expressed in Pa and K, respectively. The constants A and B used in the present study (see Table 5) have been fitted to the experimental data used by Swindle et al. (2009).

3.2 Seasonal Noble Gas Variability in the Atmosphere

Table 6 gives the ratios between the relative abundances f_K of the different species in structure I clathrate and their initial gas phase abundances x_K. Except for CH_4, the abundance of which is set relatively high, all other gas phase mixing ratios x_K are representative of the present-day Martian atmosphere and derive from Viking in situ measurements (Swindle et al. 2009). Note that each abundance ratio has been determined at a given point on the dissociation curve of the multiple guest clathrate, which corresponds to the present average atmospheric pressure on Mars, i.e., $P = 7$ mbar. Table 6 shows that the trapping efficiency of CO_2 remains high in the multiple guest clathrate, with an abundance ratio still larger than 1. On the other hand, the abundance ratios of CH_4, N_2, O_2 and CO are much lower than 1, meaning that the trapping efficiency of these molecules is quite low in the multiple guest clathrate, regardless of initial gas phase abundances. Concerning the noble gases, Table 6 shows that Xe would be strongly partitioned into the clathrate, while virtually all of the Ar would remain in the atmosphere, and the behavior of Kr would be intermediate. Although Table 6 is calculated for a methane abundance of 0.1 %, variations in methane abundance in

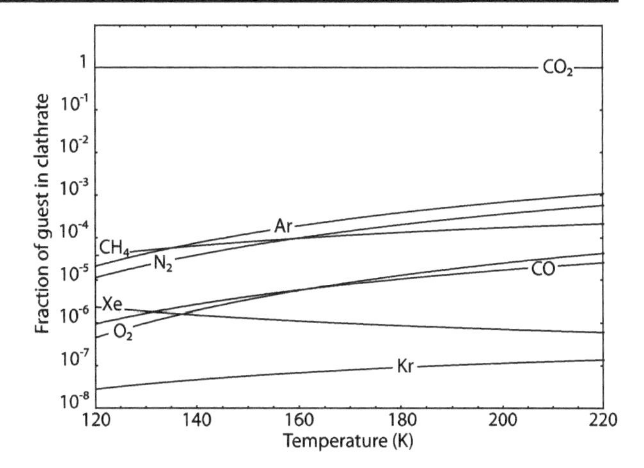

Fig. 5 Relative abundances of CO_2, CO, CH_4, N_2, O_2, Ar, Kr and Xe as a function of temperature in structure I clathrates formed from the gas phase detailed in Table 6

the 0–1 % range have little influence on the relative fractions of other trapped species, due to the low propensity of CH_4 to be trapped in this multiple guest clathrate. Figure 5 represents the evolution with temperature of the relative abundances of CO_2, CO, CH_4, N_2, O_2, Ar, Kr and Xe trapped in structure I clathrate formed from the initial gas phase composition given in Table 6. This figure shows that the trapping behavior is the same for each situation, that is the relative abundances of Ar, N_2, CH_4, O_2, CO and Kr slightly increase in clathrate when the formation temperature increases, whereas those of CO_2 and Xe slightly decrease, irrespective of the initial gas phase abundances. Moreover, due to their small size, Ar and N_2 are always poorly trapped in clathrate irrespective of the temperature.

The fractionation of noble gases due to trapping in Martian clathrates may have at least two implications: first, Martian meteorites have unexplained noble gas signatures which could be the result of sequestration of noble gases in clathrates; second, if clathrates form seasonally and incorporate noble gases, the effect should be detectable by a spacecraft lander anywhere on the surface of Mars by monitoring the noble gas abundances over the course of a Martian year. Of the various types of Martian meteorites, shergottites appear to have trapped Martian atmosphere with a Kr/Xe ratio similar to that of the terrestrial atmosphere, although it is not clear whether or not the similarity has genetic significance (e.g. Pepin and Porcelli 2002; Dauphas 2003). The nakhlites and Allan Hills 84001, on the other hand, appear to have trapped Martian atmosphere with a Kr/Xe ratio several times lower than Earth's. Several different explanations for the distinction have been put forward (Swindle 2002), including the preferential incorporation of Xe into clathrates that could play a part. Musselwhite and Swindle (2001) suggested that since the different types of meteorites were ejected from Mars at different times (the shergottites in the last 5 Myr, the others 10 Myr or more ago), the abundance of clathrates could have changed as the martian climate changed. Later, Swindle et al. (2009) considered the possibility of seasonal variability in clathrate formation, and suggested that the difference between the shergottites and the other martian meteorites could be related to the season of impact ejection. In this case, the Kr/Xe ratio in meteorites would be representative of the value fixed by CO_2-dominated clathrate at the time of its formation on Mars. Regardless of the effects on meteorites, Swindle et al. (2009) pointed out that if clathrates can form seasonally, the ratios of the noble gases within the atmosphere could change significantly. The amount of CO_2 in the atmosphere varies seasonally, with the maximum approximately 30 % higher than the minimum (Tillman 1988). If even 10 % of the CO_2 variation is a result of clathrate formation, the Kr/Xe ratio could change by a factor of two as a result of the higher propensity of Xe (compared to Kr) to be incorporated in clathrate (Swindle et al. 2009). Assuming the lower atmosphere mixes on a reasonably rapid timescale, this would easily be measurable by a mass spectrometer on a spacecraft lander, such as the Sample Analysis at Mars package on the Mars Science Laboratory. Although the Viking lander measured the noble gases in the Martian atmosphere, and did not detect any seasonal variation, the reported uncertainties on the abundances of Kr and Xe are each roughly one order of magnitude, so substantial variations could have gone undetected (Swindle et al. 2009).

The relative concentrations of minor gases also vary seasonally due to the cyclic formation/disappearance of CO_2 frost in both hemispheres. The GRS instrument onboard Mars Odyssey (Sprague et al. 2004, 2007; Forget et al. 2008) observed that Ar mixing ratios change by a factor of 6 enhancement over south polar latitudes near the onset of southern winter. No comparable feature has been reported for the northern polar winter. This enhancement in argon is likely due to CO_2 condensing and leaving behind non-condensible (or more volatile) species in the atmosphere (Sprague et al. 2004). This process, akin to a local distillation, generates a relative increase of the non-condensible species in places and at seasons when CO_2 forms as frost. The capability of enhanced argon to spread towards lower latitudes heavily depends on the horizontal mixing strength of the atmosphere. Such seasonal signatures have been detected near the equator by the Spirit and Opportunity rovers using the APXS instrument (Economou 2008; Arvidson et al. 2011), indicating that non-condensible enrichment subsequent to CO_2 condensation has a global influence. However, such an enrichment process affects all non-condensible species equally and should therefore be easily separated from the differential clathrate formation process.

4 Implications for the Geomorphology of Mars

Clathrate dissociation has been proposed to be involved in the process of formation of several geomorphic features on Mars. Here we review some of these features. It is important to note that any particular single landform may be the result of different sets of processes, i.e. equifinality, so that our goal is to show the geomorphological clues that are found consistent with clathrate formation/destabilization.

4.1 Chaotic Terrains and Catastrophic Floods

Chaotic terrains are irregularly shaped terrains commonly present on Mars in late Hesperian and Amazonian period but rarely on Earth. They consist of sets of angular blocks of various size (kilometers to tens of kilometers) sometimes separated by linear features, each block preserving the upland surface (Sharp 1973). Chaotic terrain is attributed primarily to localized collapse of the cratered upland due to removal of subsurface material, either ground ice, clathrate, or magma. Chaotic terrains are connected to large troughs and outflow channels, generally near the Martian dichotomy boundary. The analogy between the outflow channels on Mars and catastrophic floods in the Channeled Scablands of eastern Washington (USA) is based on the size, regional anastomosis, low sinuosity, streamlined hills and craters, and scour upstream of flow obstacle (Baker and Milton 1974; Baker 2009). Based on this analogy, the water discharge can be estimated at up to 10^9 m^3/s (Carr 1979; Kleinhans 2005) on a time scale of 10^3 to 10^6 years (Kleinhans 2005). Some analyses indicate multiple stages of chaos formation, such as in Xanthe Terra (Rodriguez et al. 2005).

The catastrophic dissociation of carbon dioxide clathrate has been proposed as the origin of catastrophic floods on Mars, explaining the large water discharge and the missing volume of the chaotic terrains (Milton 1974; Hoffman 2000). The dissociation of the base of the clathrate layer by an increase of geothermal flux may produce gas and liquid that could escape from the subsurface (Max and Clifford 2001). The thermal decomposition of 1 m^3 of gas hydrate yields $\sim$164 m^3 of gas (at STP) and 0.87 m^3 of liquid water (Clifford and Parker 2001). This process will lead to gentle venting, removing a part of the clathrate layer, to catastrophic decompression if the resulting pressure drop destabilizes the clathrate in a positive retroaction. Dillon et al. (1998) have identified some chaotic features on in the Earth's sea-floor that may be due to this catastrophic disruption.

Alternatively, the water released by the disruption of an aquifer confined below the water ice permafrost could also produce such surface features (Carr 1979). Ganges Chaos formation has been interpreted as a brittle frozen ice/clathrate layer above liquid CO_2/H_2O that has been destabilized creating rapid exsolution (Rodriguez et al. 2006). In this scenario, the melt water is present in the subsurface due to the heat flow (Carr 1979; Clifford and Parker 2001) at a depth depending on the amount of salt and perchlorate (Clifford et al. 2010). For the equatorial region, the cryosphere depth is at 0 km for perchlorate eutectic, at 2 km for sodium chloride eutectic and at 4 km for pure water (see Fig. 4 in Clifford et al. 2010). When an insulating lens of hydrate (such as methane clathrate hydrate or hydrate salt) is present in the shallow subsurface, melt water can be stable below the anomaly (Kargel et al. 2007). Furthermore, the complex shape of the chaos may be explained by a fluidization of the material when the gas release was intense enough to create a stress state liquifying the sediment (Komatsu et al. 2000). The detection of small volcanoes in Hydrates Chaos indicates that the destabilization could been produced by a magmatic intrusion (Meresse et al. 2008) but many plausible causes of destabilization have been proposed including climatic change (Clifford and Parker 2001), internal heat flux (Clifford and Parker 2001), fracture propagation (Rodriguez et al. 2006), and seismic activity (Nummedal and Prior 1981; Tanaka 1999). Another possibility is the melting of the ice permafrost by dikes (McKenzie and Nimmo 1999). However, the total volume of ground ice in this early period of Mars seems to be insufficient to account for the magnitude of the outflows (Clifford and Parker 2001). Max and Clifford (2001) argued that only a confined aquifer scenario is realistic, with possible clathrate in the permafrost.

The presence of a clathrate layer is controversial since the sublimation/melting of ground ice could result in similar surface features. Rare areas of chaos are not connected to any outflow channel and thus the presence of liquid phase is not required. For instance Galaxia Chaos has been proposed to have formed by sublimation of ice under the lava (Pedersen and Head 2011). In addition, the decomposition of clathrate (with a latent heat of fusion of $\sim$73 kJ/mole) is more difficult than melting ground ice ($\sim$6 kJ/mole; Clifford and Parker 2001).

4.2 Mud Volcanoes

Occurring on land, sea and lake floor, mud volcanoes are meter to kilometer scale conical features with intermittent mud flowing from the top (Mazzini 2009). They are formed by ejection of gas, water and sediments from the subsurface due to overpressured conditions.

On Earth, mud volcanoes may be formed due to rapid sedimentation, magmatic/geothermal destabilization, tectonic/subsidence stress, seismic shaking, or geochemical reaction such as clay dehydration or organic metamorphism (Milkov 2000; Dimitrov 2002; Skinner and Mazzini 2009). The gas composition is mainly CH_4 ($\sim$60 %), CO_2 and N (Dimitrov 2002). Interestingly, Earth sea-floor mud volcanoes are associated with methane clathrate formation by both hydrothermal (using only exhumed fluids) or by metasomatic (using a mixture of local and exhumed fluids) processes (Milkov 2000) because at the sea floor the stable condition of gas and water is in the clathrate form. The analogy with Martian features has been recently reviewed, including related features of pockmarks and spring mounds (Skinner and Mazzini 2009; Komatsu et al. 2011). On Mars, the origin of fluid may be due to the clathrate destabilization in the subsurface. Mud volcanism has been proposed in several regions such Acidalia Planitia (Farrand et al. 2005; Oehler and Allen 2010), Chryse (Tanaka 1997; Rodríguez et al. 2007), and Cydonia Mensae (Farrand et al. 2005) but no activity has been observed at present time. Since the scale of the feature is relatively small (up to 2 km), it is not feasible to date its formation by crater counting methods.

5 Work in Progress and Future Work

Below we detail the future measurements at Mars that could enable the detection of clathrates. We also discuss the further theoretical modeling and laboratory experiments that are required given the lack of constraints on clathrate formation and dissociation conditions (equilibrium curves, kinetics), and their composition when they result from a gas containing multiple compounds.

5.1 Future Measurements at Mars

The atmosphere of Mars is the most accessible place to search for clathrates, based on the hypothesis that CH_4 may be released to the atmosphere in the form of metastable submicron-size clathrate particles (Chassefière 2009). In the framework of this hypothesis, correlations are expected between the CH_4 mixing ratio, the H_2O vapor mixing ratio, the H_2O ice particle abundance (as already evidenced by PFS measurements on Mars-Express, see e.g. Geminale et al. 2008), and the CH_4 clathrate particle abundance. A combination of IR and UV spectroscopy is required for measuring both the gas phase and particles from a spacecraft. Because clathrate particles are expected to be very small, behaving as Rayleigh scatterers, their detection and identification are not straightforward. The infrared spectrum of CH_4 clathrates has been measured in the laboratory for very low temperatures prevailing in interstellar clouds (Dartois et al. 2010), but remains to be characterized at Martian temperatures. Alternatively, the use of UV spectroscopy (in solar occultation geometry) and/or spectropolarimetry from orbit may permit detection of a specific population of small particles, which could be identified with CH_4 clathrate particles provided the expected correlations with water vapor, water ice particles and gaseous CH_4 are in evidence. Suspended small metastable CO_2 clathrate particles could similarly be searched for. Such particles can be possibly detected also from landers by GCMS-type instruments (e.g. on MSL).

The second accessible place to search for clathrates is the icy surface of the polar caps, where CO_2-dominated clathrates, including methane and noble gases, may be stable during some periods of the seasonal cycle (see e.g. Chastain and Chevrier 2007; Thomas et al. 2009; Swindle et al. 2009; Herri and Chassefière 2012). A change in the structure and composition of the ice resulting from clathrate formation could possibly induce changes in the spectrum of the reflected solar light. Such measurements would require high-resolution IR spectral imaging from orbit. Searching for inhomogeneities of the CH_4 mixing ratio in possible local atmospheric plumes forming during the sublimation of polar clathrate pockets would be similarly useful. These measurements may be difficult to achieve, in particular if the amount of gas involved in clathrate formation and/or the spatial scale of clathrate deposits are small.

Subsurface clathrates can be directly reached only from landers by drilling and compositional analysis. Interestingly, an indirect way to detect the presence of subsurface clathrates, released to the atmosphere in the course of the progressive sublimation of the cryosphere, is to search for plumes of trace gases, which may occur over ancient volcanic areas where volcanic and/or hydrothermally produced gases could have been stored in clathrates. Characterizing outgassing, and more generally gas exchange between the subsurface and the atmosphere, is a major challenge for future Mars exploration. It may be achieved by the combination of:

1. An orbiter equipped with high resolution/ high sensitivity IR/UV spectrometers able to detect in addition to H_2O vapor a broad suite of atmospheric trace gases, including not only CH_4 but also heavier hydrocarbons (C_2H_2, C_2H_4, C_2H_6), sulfur compounds (SO_2, H_2S), chlorine compounds and their isotopes, similar to the planned Exomars 2016 Trace Gas Orbiter.
2. Small landers devoted to the detection and monitoring of noble gases, in particular Xe and Kr, which may present a seasonal variation due to the formation of clathrates and/or adsorption in the regolith, and ^{222}Rn (to be measured by an alpha spectrometer) which has been indirectly detected on Mars (Meslin et al. 2011) and may be used as a tracer of outgassing. Such landers, equipped with a high sensitivity mass spectrometer devoted to the molecular, elemental and isotopic composition of the atmosphere (H_2O, CH_4, trace gases, $^{12}C/^{13}C$, D/H,...), would provide a complete picture of trace gases, their isotopic ratios, and correlations between each other at local scale, and a temporally highly-resolved monitoring of these abundances and correlations along one or several seasonal cycles. Such measurements could give information on the origin of CH_4 (serpentinization, biotic production,...), and possible other trace gases. The use of radon-222, which has a short radioactive half-life (3.8 days), could allow determination of the location of outgassing sources, and therefore the position of potential clathrate reservoirs if positive correlations with trace gases were found. Such measurements have been planned in the frame of the Mars-Next mission, consisting of a network of four landers, studied by ESA a few years ago in the framework of the now cancelled ESA/NASA Mars exploration program.

5.2 Future Laboratory Experiments

While considerable work has been done to investigate the thermodynamic stability of CO_2 and CH_4 clathrates under terrestrial conditions ($T > 270$ K, $P > 1$ kPa), little experimental work has been conducted under low temperature, low pressure conditions to determine equilibrium curves for clathrate, water ice, and guest phase (CO_2, CH_4, H_2S, CO, O_2, noble gases, etc.) relationships and test theoretical models of mixed clathrate stability, indicative of near surface conditions on Mars. Preliminary experiments conducted by Blackburn et al. (2009) suggest that methane clathrate may sublimate into the atmosphere, rather than dissociating to form ice and gas under Mars surface pressure conditions. If near-surface clathrates are sublimating rather than dissociating, water vapour is released together with methane to the atmosphere. Although it could result in an enhancement of water vapor in the regions of the atmosphere where methane is observed, this mechanism hardly explains the global correlation between methane and water vapor discussed in Sect. 2.3.2. Indeed, the typical amplitude of water vapor fluctuations in the atmosphere is a few hundred ppm, namely three orders of magnitude above the possible enhancement due to clathrate sublimation.

Further experiments investigating the rate and nature of clathrate decomposition under low pressure conditions analogous to Mars are also required to further constrain the potential role of anomalous preservation in the near subsurface and atmosphere. Laboratory pressure release experiments conducted at temperatures from 240–270 K have documented metastable preservation of CO_2 and CH_4 clathrates at 1 bar (Stern et al. 2001). Subsequent neutron diffraction and scattering experiments have demonstrated that the anomalous preservation observed under these temperature and pressure conditions is a result of limited diffusion rates through the outer layer of ice that forms on the samples as the clathrate undergoes initial rapid dissociation (Takeya et al. 2005; Ogienko et al. 2006; Falenty and Kuhs 2009; Takeya and Ripmeester 2010). However, under near-surface or atmospheric conditions on Mars, water ice may be actively sublimating, limiting the efficacy of the ice surface to maintain internal guest concentrations sufficient for clathrate to remain stable. On the other hand, some kinds of metastable particles of CH_4 clathrates synthetized in the laboratory don't show the formation of outer layers of water ice (Zhang and Rogers 2008), and such particles could be stable everywhere in Mars' atmosphere. Therefore, additional experiments are needed to characterize clathrate decomposition mechanisms at low temperatures.

At the base of clathrate reservoirs (see details concerning the formation of this reservoir in Sect. 2.2), the aqueous chemistry of the porewaters also affects the stability of clathrates (Elwood Madden et al. 2007). As salts are added to the system through water-rock interactions, evaporation, or eutectic freezing, the activity of water decreases, resulting in gas hydrate destabilization. While the thermodynamics of salt addition have been investigated, the rate of clathrate dissociation in aqueous systems with decreasing activity of water remains unexplored.

Formation/dissociation rates of pure CO_2 and CH_4 clathrates and diffusion rates of various molecules useful in the planetary science context have been reported over a wide range of pressures and temperatures in the literature (see Tables 1 and 2). While many of these experiments were conducted under terrestrial analog conditions, their results can be used to constrain the rate of clathrate formation and dissociation at the base of potential clathrate reservoirs, where temperatures and pressures are similar to seafloor or permafrost conditions observed on Earth. It is generally accepted that the formation of clathrates starts with nucleation. A definite molecular picture for clathrate nucleation is not yet available despite recent progress in the water–gas system, but for nucleation from ice the situation is even less clear (Walsh et al. 2009; Jacobson et al. 2010). This point is important because the rate of growth depends on the geometry of the system and the physical state of reacting phases from which ice/gas are investigated. For example, a recent study of the kinetics of CO_2 clathrate formation from frostlike powders composed of micrometer-sized ice particles at low temperatures shows that the formation of CO_2 clathrates on Mars' polar cap, in the temperature range from 150 to 170 K, is very slow (Falenty et al. 2011). At 150 K, the timescale for the formation of CO_2 clathrates is in the range from a few hundred years to a few ten thousand years depending on the radius of clathrate particles. At such a slow rate, clathrates cannot be formed on a seasonal time scale. Moreover, in natural systems, the composition of the gas phase is likely not pure, resulting in mixed gas clathrate compounds. The kinetics of mixed gas clathrate systems has recently received some attention from the carbon sequestration community (Ota et al. 2005). However, little experimental work has been done to determine the effects of other gases on clathrate kinetics in mixed gas systems. Indeed, mechanisms of mixed gas clathrate formation and dissociation have been mainly explored from a theoretical point of view and these investigations remain to be assessed experimentally. In particular, differences in formation, dissociation, and diffusion rates between different gas compositions may result in significant fractionation of gases as they pass through a clathrate reservoir, acting as a cold, semi-permeable sieve for gases produced in the deep subsurface (Ikeda et al. 1999).

5.3 Improvement of the Treatment of Interaction Energies in Thermodynamic Statistical Models

The model used here and in numerous previous studies to predict the clathrate cage occupancies requires calculation of the Langmuir constants (Eq. (2)) based on a very simple expression of the interaction energy between the guest and the surrounding water molecules (Eqs. (3) and (4)). Its accuracy thus strongly depends on the accuracy of the parameters used to calculate this interaction and also on the type of molecules considered. Indeed, the potential expressions given in Eqs. (3) and (4) assume that the guest–water interactions can be represented by dispersion-repulsion interactions between spherical molecules. Whereas this assumption is reasonable for noble gases or spherical molecules such as methane trapped in clathrates, it appears of course questionable when considering non-spherical molecules like CO_2 or polar molecules such as SO_2 or CO. Moreover, these interactions are spherically averaged over the clathrate cages, thus disregarding the real crystallographic positions of the water molecules forming the cages. Because it has been recently shown that these assumptions can strongly affect the calculations of cage occupancies (Thomas et al. 2009, 2010), it is interesting to briefly comment here on the improvements that could be made on the treatment of interaction energies between guest and water molecules in clathrates. We envision several types of improvements.

First, it is important to note that even the simplest approach (spherically averaged dispersion-repulsion potentials) can give acceptable results if it is based on accurate effective parameters. Indeed, it is often possible to accurately represent complex systems by means of simple potentials, provided that their parameters have been correctly fitted either to experimental results or to very accurate calculations like high level *ab initio* calculations. These parameters are called effective parameters because they are able to accurately represent the complexity of the system under investigation by using simple potential expressions. However, problems often come from using effective parameters fitted to experimental results obtained at temperature and pressure conditions far from those corresponding to the astrophysical conditions of interest, or from using parameters fitted on *ab initio* results obtained on very simple systems (e.g., taking into account a very small number of molecules). This raises the question of the transferability of these parameters to real systems. As a consequence, the first way to improve the credibility of the calculations based on the simplest approach is to get more reliable effective parameters (Bazant and Trout 2001; Papadimitriou et al. 2007).

Second, it is possible to improve the potential model used in the determination of the Langmuir constants (Eq. (1)). Several modifications have been suggested in the literature, listed according to increasing complexity.

1. Guest-water interactions. An atom-atom description of the guest-water interactions can be more accurate than using the simple averaged potentials of Eqs. (2) and (3) (Klauda and Sandler 2002; Anderson et al. 2004; Sun and Duan 2005; Thomas et al. 2010). In this case, the potential expression given in Eq. (2) should be replaced by

$$w_{K,t} = \sum_{k=1}^{N_t} \sum_{j=1}^{N_W} \sum_{i=1}^{N_K} \left\{ 4\epsilon_{ij} \left[\left(\frac{\sigma_{ij}}{r_{ij_k}} \right)^{12} - \left(\frac{\sigma_{ij}}{r_{ij_k}} \right)^{6} \right] + \frac{1}{4\pi\epsilon_o} \frac{q_i q_j}{r_{ij_k}} \right\}, \tag{8}$$

   where r_{ij_k} measures the distance between the *i*th site of the guest molecule and the *j*th site of the *k*th water molecule in the clathrate lattice. N_t represents the total number of water molecules taken into account in the calculations, whereas N_W and N_K represent the number of sites occupied by a water or a guest molecule, respectively. ϵ, σ and q are

the parameters of the dispersion-repulsion (here, written in the Lennard-Jones form) and Coulomb (limited to charge-charge interactions) potentials. This approach allows taking into account an angle-dependent part of the molecular interactions. However, the convergence of these atom-atom interactions should be carefully addressed in the calculations, and it is thus recommended to account for as many water molecules as possible when calculating the guest-water interaction energy (Klauda and Sandler 2002; Thomas et al. 2010).
2. Thermal effects. The thermal motions of the guest and water molecules should be taken into account in the calculations of the Langmuir constant (Pimpalgaonkar et al. 2011). This can be achieved by performing temperature-dependent numerical simulations and several studies on clathrate structure and thermodynamics have thus been reported in the literature following the pioneering works of Tester and co-workers, based on the Monte Carlo technique (Tester et al. 1972) and of Tse et al., by means of the molecular dynamics method (Tse et al. 1983). Although most of the simulations addressed simple guest clathrate properties (see for instance the recent work of Sarupria and Debenedetti 2011), it might be also possible to simulate multiple guest clathrates occupancy using, for instance, the Grand Canonical Monte Carlo method (Tanaka 1998);
3. Flexibility and multiple occupancies. More subtle refinements can also be introduced in these numerical simulations such as the internal flexibility of the guest and water molecules (Castillo-Borja et al. 2008) and the possibility for multiple occupancy of the cages (Alavi et al. 2006a, 2006b). Although these refinements could be, in principle, used in the calculations of the integral of Eq. (1), it is important to note that they are not taken into account in the main assumptions of the van der Waals and Platteeuw theory. As a consequence, using Langmuir constants calculated in this way could be questionable for the determination of multiple guest clathrate occupancy.

Finally, another way to increase the relevance of the calculations is to improve the thermodynamic statistical model itself, going beyond its main assumptions (i.e., beyond the Langmuir model). For instance, recent modifications of the original van der Waals and Platteeuw theory have included accounting for multiple occupancy of the cages (Tanaka et al. 2004), inclusion of guest-guest interactions (Klauda and Sandler 2003; Belosludov et al. 2009) and of the influence of guest molecules on the host lattice (Belosludov et al. 2009), as well as volume increase of clathrate due to occupancy of the cages (Ballard and Sloan 2002). A recent paper has also investigated the possible influence of nuclear quantum effects in simulations of clathrates, showing that these effects should be included for simulations performed at temperatures below 150 K (Conde et al. 2010). It is important to note that some of these improvements depend on the availability, accuracy, and transferability of the required parameters. As a consequence, the relevance of the various approaches presented in this paragraph must actually be tested, case by case, by comparison with experimental data or with in situ measurements when available.

6 Conclusions

In this paper, we have reviewed the different scenarios involving the role of clathrates in the release of CH_4 in the Martian atmosphere. We have discussed the trapping mechanism that may have enabled the formation and long term stability of CH_4 clathrate at the surface and in the shallow subsurface of Mars. In this scenario, methane formed clathrates close to the surface in the past, at a time when its atmospheric partial pressure was sufficiently high. These clathrates, possibly buried under thin layers of regolith, would have survived until now. An alternative mechanism is formation of clathrates much deeper in the crust—at the base of the cryosphere—supplied by production of gaseous methane in deep hydrothermal systems. Such deep clathrate formation might even be happening today. Alternatively, CH_4 could be delivered to the atmosphere in the form of small (submicron) metastable clathrate particles, rather than as gaseous CH_4. We have also reviewed the role of clathrates in fractionating Ar, Kr and Xe present in the atmosphere. If clathrate formation is strongly related to the Martian seasonal cycle, then we should expect significant seasonal variations of the noble gas abundance ratios.

Despite the fact that clathrates are invoked to play a major role in all these situations, there is no direct evidence for their presence. Important clathrate reservoirs might exist in the Martian cryosphere but their dielectric constants are too close to that of water ice to allow to them to be easily distinguished at radar wavelengths. The detection of a liquid layer predicted to be overlain by a clathrate layer is also very difficult to perform. The radar sounders' capabilities are limited by horizontal resolution or orbital coverage (complete for MARSIS, about 50 % so far for SHARAD), making most of the sub-kilometer objects difficult to detect. Also, the optimal conditions for subpermafrost groundwater detection, such as a sharp transitional boundary, a well saturated crustal porosity (>10 %) and a low surface roughness, characterize about 20 % of the planet's surface (Picardi et al. 2003). From the observation that effective penetration depth seems to be lower than that predicted over most of the planet, possibly down to 100 m in dry volcanic terrains (Heggy et al. 2006), Farrell et al. (2009) show that crustal material with a plausibly high electrical conductivity would explain the lack of detection of the presumed subpermafrost water-table so far.

Acknowledgements O.M. acknowledges support from CNES. E.C. and A.L. acknowledge support from CNRS EPOV interdisciplinary program. T.S. acknowledges support from NASA Fundamental Research, and J.I.L. from JPL's Distinguished Visiting Scientist Program. We wish to thank the organizers of the ISSI workshop for having been able to gather scientists from various fields around the Martian geochemistry. The authors are indebted to P.-Y. Meslin, C. Sotin, M. Toplis and M. Trainer whose comments and suggestions greatly improved this manuscript. M.E.E.M. acknowledges support from the NASA Planetary Geology and Geophysics Program.

References

S. Alavi, J.A. Ripmeester, D.D. Klug, Molecular dynamics simulations of binary structure H hydrogen and methyl-tert-butylether clathrate hydrates. J. Chem. Phys. **124**, 204707 (2006a)

S. Alavi, J.A. Ripmeester, D.D. Klug, Molecular dynamics simulations of binary structure II hydrogen and tetrahydrofurane clathrates. J. Chem. Phys. **124**, 014704 (2006b)

B.J. Anderson, J.W. Tester, B.L. Trout, Accurate potentials for argon-water and methane-water interactions via ab initio methods and their application to clathrate hydrates. J. Phys. Chem. B **108**, 18705–18715 (2004)

R.E. Arvidson et al., Opportunity Mars rover mission: overview and selected results from Purgatory ripple to traverses to Endeavour crater. J. Geophys. Res. **116**, E00F15 (2011)

S.K. Atreya, P.R. Mahaffy, A.-S. Wong, Methane and related trace species on Mars: origin, loss, implications for life, and habitability. Planet. Space Sci. **55**, 358–369 (2007)

V.R. Baker, The channeled scabland: a retrospective. Annu. Rev. Earth Planet. Sci. **37**, 393–411 (2009)

V.R. Baker, D.J. Milton, Erosion by catastrophic floods on Mars and Earth. Icarus **23**, 27–41 (1974)

A.L. Ballard, E.D. Sloan, The next generation of hydrate prediction: I. Hydrate standard states and incorporation of spectroscopy. Fluid Phase Equilib. **194**, 371–383 (2002)

J.L. Bandfield, High-resolution subsurface water-ice distributions on Mars. Nature **447**, 64–67 (2007)

S. Barabash, A. Fedorov, R. Lundin, J.-A. Sauvaud, Martian atmospheric erosion rates. Science **315**, 501 (2007)

M.Z. Bazant, B.L. Trout, A method to extract potentials from the temperature dependence of Langmui constants for clathrate-hydrates. Physica A **300**, 139–173 (2001)

R. Belosludov, O.S. Subbotin, H. Mizuseki, Y. Kawazoe, V.R. Belosludov, Accurate description of phase diagram of clathrate hydrates at the molecular level. J. Chem. Phys. **131**, 244510 (2009)
D.G. Blackburn, R. Ulrich, M.E. Elwood Madden, J.R. Leeman, V.F. Chevrier, Experimental study of the kinetics of CO_2 hydrate dissociation under simulated martian conditions, in *Lunar and Planetary Institute Science Conference Abstracts*, vol. 40 (2009), p. 1341
G.W. Brass, Stability of brines on Mars. Icarus **42**, 20–28 (1980)
D. Breuer, S. Labrosse, T. Spohn, Thermal evolution and magnetic field generation in terrestrial planets and satellites. Space Sci. Rev. **152**, 449–500 (2010)
E.N. Brodskaya, V.V. Sizov, Molecular simulation of gas hydrate nanoclusters in water shell: structure and phase transitions. Colloid J. **71**(5), 589–595 (2009)
D.M. Burt, L.P. Knauth, Electrically conducting, Ca-rich brines, rather than water, expected in the Martian subsurface. J. Geophys. Res. **108**, 8026 (2003)
S. Byrne et al., Distribution of mid-latitude ground ice on Mars from new impact craters. Science **325**, 1674 (2009)
M.H. Carr, Formation of Martian flood features by release of water from confined aquifers. J. Geophys. Res. **84**, 2995–3007 (1979)
M.H. Carr, Channels and valleys on Mars: cold climate features formed as a result of a thickening cryosphere. Planet. Space Sci. **44**, 1411–1423 (1996)
F. Castillo-Borja, R. Vásquez-Román, U. Bravo-Sánchez, The effect of flexibility on thermodynamic and structural properties in methane hydrates. Mol. Simul. **34**, 661–670 (2008)
E. Chassefière, Metastable methane clathrate particles as a source of methane to the Martian atmosphere. Icarus **204**, 137–144 (2009)
E. Chassefière, F. Leblanc, Methane release and the carbon cycle on Mars. Planet. Space Sci. **59**, 207–217 (2011)
B.K. Chastain, V. Chevrier, Methane clathrate hydrates as a potential source for Martian atmospheric methane. Planet. Space Sci. **55**, 1246–1256 (2007)
Y.-H. Chen, R.G. Prinn, Estimation of atmospheric methane emissions between 1996 and 2001 using a three-dimensional global chemical transport model. J. Geophys. Res. **111**, D10307 (2006)
V. Chevrier, F. Poulet, J.-P. Bibring, Early geochemical environment of Mars as determined from thermodynamics of phyllosilicates. Nature **448**, 60–63 (2007)
M.-K. Chun, H. Lee, Kinetics of formation of carbon dioxide clathrate hydrates. Korean J. Chem. Eng. **13**(6), 620–626 (1996)
B.C. Clark, D.C. van Hart, The salts of Mars. Icarus **45**, 370–378 (1981)
S.M. Clifford, Polar basal melting on Mars. J. Geophys. Res. **92**, 9135–9152 (1987)
S.M. Clifford, A model for the hydrologic and climatic behavior of water on Mars. J. Geophys. Res. **981**, 10973–11016 (1993)
S.M. Clifford, T.J. Parker, The evolution of the Martian hydrosphere: implications for the fate of a primordial ocean and the current state of the northern plains. Icarus **154**, 40–79 (2001)
S.M. Clifford, J. Lasue, E. Heggy, J. Boisson, P. McGovern, M.D. Max, Depth of the Martian cryosphere: revised estimates and implications for the existence and detection of subpermafrost groundwater. J. Geophys. Res. **115**, 7001 (2010)
M.M. Conde, C. Vega, C. McBride, E.G. Noya, R. Ramirez, L.M. Sesé, Can gas hydrate structures be described using classical simulations? J. Chem. Phys. **132**, 114503 (2010)
E. Dartois, D. Deboffle, M. Bouzit, Methane clathrate hydrate infrared spectrum. II. Near-infrared overtones, combination modes and cages assignments. Astron. Astrophys. **514**, A49 (2010)
N. Dauphas, The dual origin of the terrestrial atmosphere. Icarus **165**, 326–339 (2003)
A. Demurov, R. Radhakrishnan, B.L. Trout, Computations of diffusivities in ice and CO_2 clathrate hydrates via molecular dynamics and Monte Carlo simulations. J. Chem. Phys. **116**, 702–709 (2002)
W.P. Dillon, W.W. Danforth, D.R. Hutchinson, R.M. Drury, M.H. Taylor, J.S. Booth, Evidence for faulting related to dissociation of gas hydrate and release of methane off the southeastern united states. J. Geol. Soc. **137**, 293–302 (1998)
L.I. Dimitrov, Mud volcanoes-the most important pathway for degassing deeply buried sediments. Earth-Sci. Rev. **59**, 49–76 (2002)
T.E. Economou, Mars atmosphere argon density measurement on MER mission. LPI Contrib. **1447**, 9102 (2008)
M.E. Elwood Madden, S.M. Ulrich, T.C. Onstott, T.J. Phelps, Salinity-induced hydrate dissociation: a mechanism for recent CH_4 release on Mars. Geophys. Res. Lett. **341**, 11202 (2007)
M.E. Elwood Madden, P. Szymcek, S.M. Ulrich, S. McCallum, T.J. Phelps, Experimental formation of massive hydrate deposits from accumulation of CH_4 gas bubbles within synthetic and natural sediments. Mar. Pet. Geol. **26**, 369–378 (2009)

M.E. Elwood Madden, J.R. Leeman, M.J. Root, S. Gainey, Reduced sulfur-carbon-water systems on Mars may yield shallow methane hydrate reservoirs. Planet. Space Sci. **59**, 203–206 (2011)

S. Emmanuel, J.J. Ague, Implications of present-day abiogenic methane fluxes for the early Archean atmosphere. Geophys. Res. Lett. **341**, 15810 (2007)

A.G. Fairén, A.F. Davila, L. Gago-Duport, R. Amils, C.P. McKay, Stability against freezing of aqueous solutions on early Mars. Nature **459**, 401–404 (2009)

A. Falenty, W.F. Kuhs, Self-preservation of CO_2 gas hydrates-surface microstructure and ice perfection. J. Phys. Chem. B **113**(49), 15975–15988 (2009)

F. Falenty, G. Genov, T.C. Hansen, W.F. Kuhs, A.N. Salamentin, Kinetics of CO_2 hydrate formation from water frost at low temperatures: experimental results and theoretical model. J. Phys. Chem. C **115**, 4022–4032 (2011)

F.P. Fanale, W.A. Cannon, Mars—CO_2 adsorption and capillary condensation on clays: significance for volatile storage and atmospheric history. J. Geophys. Res. **84**, 8404–8414 (1979)

W.H. Farrand, L.R. Gaddis, L. Keszthelyi, Pitted cones and domes on Mars: observations in Acidalia Planitia and Cydonia Mensae using MOC, THEMIS, and TES data. J. Geophys. Res. **110**, 5005 (2005)

W.M. Farrell, J.J. Plaut, S.A. Cummer, D.A. Gurnett, G. Picardi, T.R. Watters, A. Safaeinili, Is the Martian water table hidden from radar view? Geophys. Res. Lett. **36**, 15206 (2009)

S. Fonti, G.A. Marzo, Mapping the methane on Mars. Astron. Astrophys. **512**, A51 (2010)

F. Forget, R.M. Haberle, F. Montmessin, B. Levrard, J.W. Head, Formation of glaciers on Mars by atmospheric precipitation at high obliquity. Science **311**, 368–371 (2006)

F. Forget, E. Millour, L. Montabone, F. Lefevre, Non condensable gas enrichment and depletion in the Martian polar regions. LPI Contrib. **1447**, 9106 (2008)

V. Formisano, S. Atreya, T. Encrenaz, N. Ignatiev, M. Giuranna, Detection of methane in the atmosphere of Mars. Science **306**, 1758–1761 (2004)

S.R. Gainey, M.E. Elwood Madden, Kinetics of methane clathrate formation and dissociation under Mars relevant conditions. Icarus **218**, 513–524 (2012)

A. Geminale, V. Formisano, M. Giuranna, Methane in Martian atmosphere: average spatial, diurnal, and seasonal behaviour. Planet. Space Sci. **56**, 1194–1203 (2008)

A. Geminale, V. Formisano, G. Sindoni, Mapping methane in Martian atmosphere with PFS-MEX data. Planet. Space Sci. **59**, 137–148 (2011)

C. Giavarini, F. Maccioni, M. Politi, M. Santarelli, CO_2 hydrate: formation and dissociation compared to methane hydrate. Energy Fuels **21**(6), 3284–3291 (2007)

R.V. Gough, J.J. Turley, G.R. Ferrell, K.E. Cordova, S.E. Wood, D.O. Dehaan, C.P. McKay, O.B. Toon, M.A. Tolbert, Can rapid loss and high variability of Martian methane be explained by surface H_2O_2? Planet. Space Sci. **59**, 238–246 (2011)

L.L. Griffith, E.L. Shock, A geochemical model for the formation of hydrothermal carbonates on Mars. Nature **377**, 406–408 (1995)

R.E. Grimm, S.L. Painter, On the secular evolution of groundwater on Mars. Geophys. Res. Lett. **362**, 24803 (2009)

M. Grott, A. Morschhauser, D. Breuer, E. Hauber, Volcanic outgassing of CO_2 and H_2O on Mars. Earth Planet. Sci. Lett. **308**, 391–400 (2011)

V.C. Gulick, Magmatic intrusions and a hydrothermal origin for fluvial valleys on Mars. J. Geophys. Res. **1031**, 19365–19388 (1998)

M. Haeckel, E. Suess, K. Wallmann, D. Rickert, Rising methane gas bubbles form massive hydrate layers at the seafloor. Geochim. Cosmochim. Acta **68**, 4335–4345 (2004)

B.C. Hahn, S.M. McLennan, E.C. Klein, Martian surface heat production and crustal heat flow from Mars Odyssey Gamma-Ray spectrometry. Geophys. Res. Lett. **381**, 14203 (2011)

I. Halevy, D.P. Schrag, Sulfur dioxide inhibits calcium carbonate precipitation: implications for early Mars and Earth. Geophys. Res. Lett. **36**, 23201 (2009)

M.H. Hecht, S.P. Kounaves, R.C. Quinn, S.J. West, S.M.M. Young, D.W. Ming, D.C. Catling, B.C. Clark, W.V. Boynton, J. Hoffman, L.P. DeFlores, K. Gospodinova, J. Kapit, P.H. Smith, Detection of perchlorate and the soluble chemistry of Martian soil at the Phoenix lander site. Science **325**, 64–67 (2009)

E. Heggy, S.M. Clifford, R.E. Grimm, C.L. Dinwiddie, D.Y. Wyrick, B.E. Hill, Ground-penetrating radar sounding in mafic lava flows: assessing attenuation and scattering losses in Mars-analog volcanic terrains. J. Geophys. Res. **111**, 6 (2006)

R.W. Henning, A.J. Schultz, V. Thieu, Y. Halpern, Neutron diffraction studies of CO_2 clathrate hydrate: formation from deuterated ice. J. Phys. Chem. A **104**(21), 5066–5071 (2000)

J.-M. Herri, E. Chassefière, Carbon dioxide, argon, nitrogen and methane clathrate hydrates: thermodynamic modelling, investigation of their stability in Martian atmospheric conditions and variability of methane trapping. Planet. Space Sci. (2012, in press)

J.-M. Herri, M. Cournil, E. Chassefière, Thermodynamic modelling of clathrate hydrates in the atmosphere of Mars, in *Proceedings of the 7th International Conference on Gas Hydrates (ICGH 2011)*, Edinburgh, Scotland, UK (2011), pp. 17–21

N. Hoffman, White Mars: a new model for Mars' surface and atmosphere based on CO_2. Icarus **146**, 326–342 (2000)

A. Hori, T. Hondoh, Theoretical study on the diffusion of gases in hexagonal ice by the molecular orbital method. Can. J. Phys. **81**, 251–259 (2003)

T.L. Hudson, O. Aharonson, N. Schorghofer, C.B. Farmer, M.H. Hecht, N.T. Bridges, Water vapor diffusion in Mars subsurface environments. J. Geophys. Res. **112**, 5016 (2007)

T. Ikeda, H. Fukazawa, S. Mae, L. Pepin, P. Duval, B. Champagnon, V.Y. Lipenkov, T. Hondoh, Extreme fractionation of gases caused by formation of clathrate hydrates in Vostok Antarctic ice. Geophys. Res. Lett. **26**, 91–94 (1999)

T. Ikeda-Fukazawa, Diffusion of nitrogen gas in ice Ih. Chem. Phys. Lett. **385**, 467–471 (2004)

T. Ikeda-Fukazawa, K. Kawamura, T. Hondoh, Mechanism of molecular diffusion in ice crystals. Mol. Simul. **30**, 973–979 (2004)

T. Ikeda-Fukazawa, K. Fukumizu, K. Kawamura, S. Aoki, T. Nakazawa, T. Hondoh, Effects of molecular diffusion on trapped gas composition in polar ice cores. Earth Planet. Sci. Lett. **229**, 183–192 (2005)

L.C. Jacobson, W. Hujo, V. Molinero, Amorphous precursor in the nucleation of clathrate hydrates. J. Am. Chem. Soc. **132**(33), 11806–11811 (2010)

J.S. Kargel, R. Furfaro, O. Prieto-Ballesteros, J.A.P. Rodriguez, D.R. Montgomery, A.R. Gillespie, G.M. Marion, S.E. Wood, Martian hydrogeology sustained by thermally insulating gas and salt hydrates. Geology **35**, 97 (2007)

J.F. Kasting, Planetary science: update: the early Mars climate question heats up. Science **278**, 1245 (1997)

M.A.K. Khalil, R.A. Rasmussen, Atmospheric methane: recent global trends. Environ. Sci. Technol. **24**, 549–553 (1990)

H.H. Kieffer, B.M. Jakosky, C.W. Snyder, M.S. Matthews, Mars. Mars (1992)

H.H. Kieffer, T.N. Titus, K.F. Mullins, P.R. Christensen, Mars south polar spring and summer behavior observed by TES: seasonal cap evolution controlled by frost grain size. J. Geophys. Res. **105**, 9653–9700 (2000)

H.C. Kim, P.R. Bishnoi, R.A. Heidemann, S.S.H. Rizvi, Kinetics of methane hydrate decomposition. Chem. Eng. Sci. **42**(7), 1645–1653 (1987)

J.B. Klauda, S.I. Sandler, Ab initio intermolecular potentials for gas hydrates and their predictions. J. Phys. Chem. B **106**, 5722–5732 (2002)

J.B. Klauda, S.I. Sandler, Phase behavior of clathrate hydrates: a model for single and multiple gas component hydrates. Chem. Eng. Sci. **58**, 27–41 (2003)

M.G. Kleinhans, Flow discharge and sediment transport models for estimating a minimum timescale of hydrological activity and channel and delta formation on Mars. J. Geophys. Res. **110**, 12003 (2005)

T. Komai, S.-P. Kang, J.-H. Yoon, Y. Yamamoto, T. Kawamura, M. Ohtake, In situ Raman spectroscopy investigation of the dissociation of methane hydrate at temperatures just below the ice point. J. Phys. Chem. B **108**(23), 8062–8068 (2004)

G. Komatsu, J.S. Kargel, V.R. Baker, R.G. Strom, G.G. Ori, C. Mosangini, K.L. Tanaka, A chaotic terrain formation hypothesis: explosive outgas and outlow by dissociation of clathrate on mars, in *Lunar and Planetary Institute Science Conference Abstracts*, vol. 31 (2000), p. 1434

G. Komatsu, G.G. Ori, M. Cardinale, J.M. Dohm, V.R. Baker, D.A. Vaz, R. Ishimaru, N. Namiki, T. Matsui, Roles of methane and carbon dioxide in geological processes on Mars. Planet. Space Sci. **59**, 169–181 (2011)

V.A. Krasnopolsky, J.P. Maillard, T.C. Owen, Detection of methane in the Martian atmosphere: evidence for life? Icarus **172**, 537–547 (2004)

W.F. Kuhs, D.K. Staykova, A.N. Salamatin, Formation of methane hydrate from polydisperse ice powders. J. Phys. Chem. B **110**(26), 13283–13295 (2006)

H. Lammer, E. Chassefière, K. Özgur, A. Morschhauser, P.B. Niles, O. Mousis, P. Odert, U.V. Möstl, D. Breuer, V. Dehant, M. Grott, H. Gröller, E. Hauber, L. Binh San Pham, Outgassing history and escape of the Martian atmosphere and water inventory. Space Science Rev. (2012, in press)

J. Laskar, A comment on "Accurate spin axes and solar system dynamics: climatic variations for the Earth and Mars". Astron. Astrophys. **416**, 799–800 (2004)

J. Lasue, N. Mangold, E. Hauber, S. Clifford, W. Feldman, O. Gasnault, S. Maurice, O. Mousis, Quantifying the Martian hydrosphere: current evidence, time evolution and implications for the habitability. Space Science Rev. (2012, in press)

J.R. Leeman, M.E. Elwood Madden, CO_2 clathrate formation and dissociation rates below 273 K. Geochim. Cosmochim. Acta **74**, A576 (2010)

J.R. Leeman, D.G. Blackburn, M.E. Elwood Madden, R. Ulrich, V. Chevrier, CO_2 clathrate dissociation rates below the freezing point of water, in *Lunar and Planetary Institute Science Conference Abstracts*, vol. 41 (2010), p. 1418
F. Lefèvre, F. Forget, Observed variations of methane on Mars unexplained by known atmospheric chemistry and physics. Nature **460**, 720–723 (2009)
Q. Li, W.S. Kiefer, Mantle convection and magma production on present-day Mars: effects of temperature-dependent rheology. Geophys. Res. Lett. **34**, 16203 (2007)
V.Ya. Lipenkov, V.A. Istomin, On the stability of air clathrate-hydrate crystals in subglacial Lake Vostok, Antarctica. Materialy Glyatsiol. Issled. **91**, 129–133 (2001)
J. Longhi, Phase equilibrium in the system CO_2–H_2O: application to Mars. J. Geophys. Res. **111**, 6011 (2006)
J.I. Lunine, D.J. Stevenson, Thermodynamics of clathrate hydrate at low and high pressures with application to the outer solar system. Astrophys. J. Suppl. Ser. **58**, 493–531 (1985)
J.R. Lyons, C. Manning, F. Nimmo, Formation of methane on Mars by fluid-rock interaction in the crust. Geophys. Res. Lett. **321**, 13201 (2005)
M. Mastepanov, C. Sigsgaard, E.J. Dlugokencky, S. Houweling, L. Ström, M.P. Tamstorf, T.R. Christensen, Large tundra methane burst during onset of freezing. Nature **456**, 628–631 (2008)
M.D. Max, S.M. Clifford, The state, potential distribution, and biological implications of methane in the Martian crust. J. Geophys. Res. **105**, 4165–4172 (2000)
M.D. Max, S.M. Clifford, Initiation of Martian outflow channels: related to the dissociation of gas hydrate? Geophys. Res. Lett. **28**, 1787–1790 (2001)
A. Mazzini, Mud volcanism: processes and implications. Mar. Petroleum Geol. **26**, 1677–1680 (2009)
P.J. McGovern, S.C. Solomon, D.E. Smith, M.T. Zuber, M. Simons, M.A. Wieczorek, R.J. Phillips, G.A. Neumann, O. Aharonson, J.W. Head, Correction to "Localized gravity/topography admittance and correlation spectra on Mars: implications for regional and global evolution". J. Geophys. Res. **109**, 7007 (2004)
D. McKenzie, F. Nimmo, The generation of Martian floods by the melting of ground ice above dykes. Nature **397**, 231–233 (1999)
V. McKoy, O. Sinanoğlu, Theory of dissociation pressures of some gas hydrates. J. Chem. Phys. **38**(12), 2946–2956 (1963)
M.T. Mellon, R.J. Phillips, Recent gullies on Mars and the source of liquid water. J. Geophys. Res. **106**, 23165–23180 (2001)
M.T. Mellon, B.M. Jakosky, S.E. Postawko, The persistence of equatorial ground ice on Mars. J. Geophys. Res. **1021**, 19357–19370 (1997)
S. Meresse, F. Costard, N. Mangold, P. Masson, G. Neukum (The HRSC Co-I Team), Formation and evolution of the chaotic terrains by subsidence and magmatism: hydraotes chaos, Mars. Icarus **194**, 487–500 (2008)
P.-Y. Meslin, R. Gough, F. Lefèvre, F. Forget, Little variability of methane on Mars induced by adsorption in the regolith. Planet. Space Sci. **59**, 247–258 (2011)
A.V. Milkov, Worldwide distribution of submarine mud volcanoes and associated gas hydrates. Mar. Geol. **167**, 29–42 (2000)
S.L. Miller, The occurrence of gas hydrates in the solar system. Proc. Natl. Acad. Sci. USA **47**, 1798–1808 (1961)
D.J. Milton, Carbon dioxide hydrate and floods on Mars. Science **183**, 654–656 (1974)
M.A. Mischna, M.I. Richardson, R.J. Wilson, D.J. McCleese, On the orbital forcing of Martian water and CO_2 cycles: a general circulation model study with simplified volatile schemes. J. Geophys. Res. **108**, 5062 (2003)
A.H. Mohammadi, R. Anderson, B. Tohidi, Carbon monoxide clathrate hydrates: equilibrium data and thermodynamic modeling. AIChE J. **51**, 2825–2833 (2005)
F. Montmessin, The orbital forcing of climate changes on Mars. Space Sci. Rev. **125**, 457–472 (2006)
F. Montmessin, P. Rannou, M. Cabane, New insights into Martian dust distribution and water-ice cloud microphysics. J. Geophys. Res. **107**, 5037 (2002)
O. Mousis, J.I. Lunine, C. Thomas, M. Pasek, U. Marboeuf, Y. Alibert, V. Ballenegger, D. Cordier, Y. Ellinger, F. Pauzat, S. Picaud, Clathration of volatiles in the solar nebula and implications for the origin of Titan's atmosphere. Astrophys. J. **691**, 1780–1786 (2009)
O. Mousis, J.I. Lunine, S. Picaud, D. Cordier, Volatile inventories in clathrate hydrates formed in the primordial nebula. Faraday Discuss. **147**, 509–525 (2010)
O. Mousis, J.I. Lunine, S. Picaud, D. Cordier, J.H. Waite, K.E. Mandt, Removal of Titan's atmospheric noble gases by their sequestration in surface clathrates. Astrophys. J. Lett. **740**, L9 (2011)
O. Mousis, J.I. Lunine, E. Chassefière, F. Montmessin, A. Lakhlifi, S. Picaud, J.-M. Petit, D. Cordier, Mars cryosphere: a potential reservoir for heavy noble gases? Icarus **218**, 80–87 (2012)

M.J. Mumma, G.L. Villanueva, R.E. Novak, T. Hewagama, B.P. Bonev, M.A. DiSanti, A.M. Mandell, M.D. Smith, Strong release of methane on Mars in northern Summer 2003. Science **323**, 1041 (2009)

D. Musselwhite, J.I. Lunine, Alteration of volatile inventories by polar clathrate formation on Mars. J. Geophys. Res. **1002**, 23301–23306 (1995)

D.S. Musselwhite, T.D. Swindle, Is release of Martian atmosphere from polar clathrate the cause of the nakhlite and ALH84001 Ar/Kr/Xe ratios? Icarus **154**, 207–215 (2001)

P.B. Niles, D.C. Catling, G. Berger, E. Chassefière, B.L. Ehlmann, J.R. Michalski, R. Morris, S.W. Ruff, B. Sutter, Geochemistry of carbonates on Mars: implications for climate history and nature of aqueous environments. Space Sci. Rev. (2012). doi:10.1007/s11214-012-9940-y

D. Nummedal, D.B. Prior, Generation of Martian chaos and channels by debris flows. Icarus **45**, 77–86 (1981)

D.Z. Oehler, C.C. Allen, Evidence for pervasive mud volcanism in Acidalia Planitia, Mars. Icarus **208**, 636–657 (2010)

A.G. Ogienko, A.V. Kurnosov, A.Y. Manakov, E.G. Larionov, A.I. Ancharov, M.A. Sheromov, A.N. Nesterov, Gas hydrates of argon and methane synthesized at high pressures: composition, thermal expansion, and self-preservation. J. Phys. Chem. B **110**(6), 2840–2846 (2006)

T. Okuchi, I.L. Moudrakovski, J.A. Ripmeester, Efficient storage of hydrogen fuel into leaky cages of clathrate hydrate. Appl. Phys. Lett. **91**, 171903 (2007)

M. Ota, Y. Abe, M. Watanabe, R.L. Smith, H. Inomata, Methane recovery from methane hydrate using pressurized CO_2. Fluid Phase Equilib. **228**, 553–559 (2005)

C. Oze, M. Sharma, Have olivine, will gas: serpentinization and the abiogenic production of methane on Mars. Geophys. Res. Lett. **321**, 10203 (2005)

N.I. Papadimitriou, I.N. Tsimpanogiannis, A.G. Yiotis, T.A. Steriotis, A.K. Stubos, On the use of the Kihara potential for hydrate equilibrium calculations, in *Physics and Chemistry of Ice*, ed. by W. Kuhs (RSC, London, 2007), p. 476

W.R. Parrish, J.M. Prausnitz, Dissociation pressures of gas hydrates formed by gas mixtures. Ind. Eng. Chem. Process Des. Dev. **11**(1), 26–35 (1972) [Erratum: Ind. Eng. Chem. Process Des. Dev. **11**(3), 462 (1972)]

G.B.M. Pedersen, J.W. Head, Chaos formation by sublimation of volatile-rich substrate: evidence from Galaxias Chaos, Mars. Icarus **211**, 316–329 (2011)

R.O. Pepin, D. Porcelli, Origin of noble gases in terrestrial planets. Rev. Mineral. Geochem. **47**, 191–246 (2002)

B. Peters, N.E.R. Zimmermann, G.T. Beckham, J.W. Tester, B.L. Trout, Path sampling calculation of methane diffusivity in natural gas hydrates from a water-vacancy assisted mechanism. J. Am. Chem. Soc. **130**, 17342–17350 (2008)

R.J. Phillips et al., Massive CO_2 ice deposits sequestered in the south polar layered deposits of Mars. Science **332**, 838 (2011)

G. Picardi, D. Biccari, A. Cicchetti, R. Seu, J. Plaut, W.T.K. Johnson, R.L. Jordan, D.A. Gurnett, R. Orosei, E. Zampolini, Mars advanced radar for subsurface and ionosphere sounding (MARSIS). EGS—AGU—EUG Joint Assembly, 9597 (2003)

H. Pimpalgaonkar, S.K. Veesam, N. Punnathanam, Theory of gas hydrates: effect of the approximation of rigid water lattice. J. Phys. Chem. A **115**, 10018–10026 (2011)

J.J. Plaut, A. Safaeinili, J.W. Holt, R.J. Phillips, J.W. Head, R. Seu, N.E. Putzig, A. Frigeri, Radar evidence for ice in lobate debris aprons in the mid-northern latitudes of Mars. Geophys. Res. Lett. **360**, 2203 (2009)

J.B. Pollack, J.F. Kasting, S.M. Richardson, K. Poliakoff, The case for a wet, warm climate on early Mars. Icarus **71**, 203–224 (1987)

O. Prieto-Ballesteros, J.S. Kargel, M. Fernández-Sampedro, F. Selsis, E.S. Martínez, D.L. Hogenboom, Evaluation of the possible presence of clathrate hydrates in Europa's icy shell or seafloor. Icarus **177**, 491–505 (2005)

J.A.P. Rodriguez, S. Sasaki, R.O. Kuzmin, J.M. Dohm, K.L. Tanaka, H. Miyamoto, K. Kurita, G. Komatsu, A.G. Fairén, J.C. Ferris, Outflow channel sources, reactivation, and chaos formation, Xanthe Terra, Mars. Icarus **175**, 36–57 (2005)

J.A.P. Rodriguez, J. Kargel, D.A. Crown, L.F. Bleamaster, K.L. Tanaka, V. Baker, H. Miyamoto, J.M. Dohm, S. Sasaki, G. Komatsu, Headward growth of chasmata by volatile outbursts, collapse, and drainage: evidence from Ganges chaos, Mars. Geophys. Res. Lett. **331**, 18203 (2006)

J.A.P. Rodríguez, K.L. Tanaka, J.S. Kargel, J.M. Dohm, R. Kuzmin, A.G. Fairén, S. Sasaki, G. Komatsu, D. Schulze-Makuch, Y. Jianguo, Formation and disruption of aquifers in southwestern Chryse Planitia, Mars. Icarus **191**, 545–567 (2007)

R.A. Rohde, P.B. Price, Diffusion-controlled metabolism for long-term survival of single isolated microorganisms trapped within ice crystals. Proc. Natl. Acad. Sci. USA **104**(24), 16592–16597 (2007)

M.J. Root, M.E. Elwood Madden, Potential effects of obliquity change on gas hydrate stability zones on Mars. Icarus **218**, 534–544 (2012)

S. Sarupria, P.G. Debenedetti, Molecular dynamics study of carbon dioxide hydrate dissociation. J. Phys. Chem. A **115**, 6102–6111 (2011)

K. Satoh, T. Uchida, T. Hondon, S. Mae, Diffusion coefficient and solubility measurements of noble gases in ice crystals, in *National Institute of Polar Reasearch-Proc. NIPR Symp. Polar Meteoral. Glaciol*, vol. 10 (1996), pp. 73–81

E.J. Sauter, S.I. Muyakshin, J.-L. Charlou, M. Schlüter, A. Boetius, K. Jerosch, E. Damm, J.-P. Foucher, M. Klages, Methane discharge from a deep-sea submarine mud volcano into the upper water column by gas hydrate-coated methane bubbles. Earth Planet. Sci. Lett. **243**, 354–365 (2006)

R.P. Sharp, Mars: fretted and chaotic terrains. J. Geophys. Res. **78**, 4073–4083 (1973)

J.A. Skinner Jr., A. Mazzini, Martian mud volcanism: terrestrial analogs and implications for formational scenarios. Mar. Petroleum Geol. **26**, 1866–1878 (2009)

E.D. Sloan, *Clathrate Hydrates of Natural Gases* (Marcel Dekker, New York, 1998)

E.D. Sloan, C.A. Koh, *Clathrate Hydrates of Natural Gases*, 3rd edn. (CRC/Taylor & Francis, Boca Raton, 2008)

A.L. Sprague, W.V. Boynton, K.E. Kerry, D.M. Janes, D.M. Hunten, K.J. Kim, R.C. Reedy, A.E. Metzger, 'Mars' south polar ar enhancement: a tracer for south polar seasonal meridional mixing. Science **306**, 1364–1367 (2004)

A.L. Sprague, W.V. Boynton, K.E. Kerry, D.M. Janes, N.J. Kelly, M.K. Crombie, S.M. Melli, J.R. Murphy, R.C. Reedy, A.E. Metzger, 'Mars' atmospheric argon: tracer for understanding Martian atmospheric circulation and dynamics. J. Geophys. Res. **112**, 3 (2007)

L.A. Stern, S. Circone, S.H. Kirby, W.B. Durham, Anomalous preservation of pure methane hydrate at 1 atm. J. Phys. Chem. B **105**, 1756–1762 (2001)

C.-Y. Sun, G.-J. Chen, Methane hydrate dissociation above 0 °C and below 0 °C. Fluid Phase Equilib. **242**, 123–128 (2006)

R. Sun, Z. Duan, Prediction of CH_4 and CO_2 hydrate phase equilibrium and cage occupancy from ab initio intermolecular potentials. Geochim. Cosmochim. Acta **69**, 4411–4424 (2005)

T.D. Swindle, Some puzzles about what noble gas components were mixed into the nakhlites, and how, in *Unmixing the SNCs: Chemical, Isotopic, and Petrologic Components of the Martian Meteorites* (2002), pp. 57–58

T.D. Swindle, C. Thomas, O. Mousis, J.I. Lunine, S. Picaud, Incorporation of argon, krypton and xenon into clathrates on Mars. Icarus **203**, 66–70 (2009)

S.Y. Takeya, J.A. Ripmeester, Anomalous preservation of CH_4 hydrate and its dependence on the morphology of hexagonal ice. ChemPhysChem **11**, 70–73 (2010)

S. Takeya, T. Hondoh, T. Uchida, In situ observation of CO_2 hydrate by X-ray diffraction. Ann. N.Y. Acad. Sci. **912**, 973–982 (2000)

S. Takeya, T. Uchida, J. Nagao, R. Ohmura, W. Shimada, Y. Kamata, T. Ebinuma, H. Narita, Particle size effect of CH_4 hydrate for self-preservation. Chem. Eng. Sci. **60**(5), 1383–1387 (2005)

K.L. Tanaka, Sedimentary history and mass flow structures of Chryse and Acidalia Planitiae, Mars. J. Geophys. Res. **102**, 4131–4150 (1997)

H. Tanaka, A novel approach to the stability of clathrate hydrates: grand canonical Monte Carlo simulation. Fluid Phase Equilib. **144**, 361–368 (1998)

K.L. Tanaka, Debris-flow origin for the Simud/Tiu deposit on Mars. J. Geophys. Res. **104**, 8637–8652 (1999)

H. Tanaka, T. Nakatsuka, K. Koga, On the thermodynamic stability of clathrate hydrates IV: double occupancy of cages. J. Chem. Phys. **121**, 5488–5493 (2004)

J.W. Tester, R.L. Bivins, C. Herrick, Use of Monte-Carlo in calculating thermodynamic properties of water clathrates. AIChE J. **18**, 1220 (1972)

C. Thomas, O. Mousis, V. Ballenegger, S. Picaud, Clathrate hydrates as a sink of noble gases in Titan's atmosphere. Astron. Astrophys. **474**, L17–L20 (2007)

C. Thomas, S. Picaud, O. Mousis, V. Ballenegger, A theoretical investigation into the trapping of noble gases by clathrates on Titan. Planet. Space Sci. **56**, 1607–1617 (2008)

C. Thomas, O. Mousis, S. Picaud, V. Ballenegger, Variability of the methane trapping in Martian subsurface clathrate hydrates. Planet. Space Sci. **57**, 42–47 (2009)

C. Thomas, S. Picaud, V. Ballenegger, O. Mousis, Sensitivitiy of predicted gas hydrate occupancies on treatment of intermolecular interactions. J. Chem. Phys. **132**, 104510 (2010)

F. Tian, J.F. Kasting, S.C. Solomon, Thermal escape of carbon from the early Martian atmosphere. Geophys. Res. Lett. **36**, 2205 (2009)

J.E. Tillman, Mars global atmospheric oscillations: annually synchronized, transient normal-mode oscillations and the triggering of global dust storms. J. Geophys. Res. **93**, 9433–9451 (1988)

B.J. Travis, N.D. Rosenberg, J.N. Cuzzi, On the role of widespread subsurface convection in bringing liquid water close to Mars surface. J. Geophys. Res. **108**, 8040 (2003)
J.S. Tse, M.L. Klein, I.R. McDonald, Dynamical properties of the structure-I clathrate hydrate of xenon. J. Chem. Phys. **87**, 2096–2097 (1983)
T. Uchida, S. Takeya, L.D. Wilson, C.A. Tulk, J.A. Ripmeester, J. Nagao, T. Ebinuma, H. Narita, Measurements of physical properties of gas hydrates and in situ observations of formation and decomposition processes via Raman spectroscopy and X-ray diffraction. Can. J. Phys. **81**, 351–357 (2003)
R. Ulrich, T. Kral, V. Chevrier, R. Pilgrim, L. Roe, Dynamic temperature fields under Mars landing sites and implications for supporting microbial life. Astrobiology **10**, 643–650 (2010)
J.H. van der Waals, J.C. Platteeuw, Clathrate solutions, in *Advances in Chemical Physics*, vol. 2 (Interscience, New York, 1959), pp. 1–57
M.R. Walsh, C.A. Koh, E.D. Sloan, A.K. Sum, D.T. Wu, Microsecond simulations of spontaneous methane hydrate nucleation and growth. Science **326**, 1096–1098 (2009)
X.P. Wang, A.J. Schultz, Y. Halpern, Kinetics of methane hydrate formation from polycrystalline deuterated ice. J. Phys. Chem. A **106**(32), 7304–7309 (2002)
K. Zahnle, R.S. Freedman, D.C. Catling, Is there methane on Mars? Icarus **212**, 493–503 (2011)
G. Zhang, R.E. Rogers, Ultra-stability of gas hydrates at 1 atm and 268.2 K. Chem. Eng. Sci. **63**, 2066–2074 (2008)

Space Sci Rev (2013) 174:251–300
DOI 10.1007/s11214-012-9947-4

Geochemical Reservoirs and Timing of Sulfur Cycling on Mars

Fabrice Gaillard · Joseph Michalski · Gilles Berger · Scott M. McLennan · Bruno Scaillet

Received: 16 February 2012 / Accepted: 20 October 2012 / Published online: 30 November 2012

Abstract Sulfate-dominated sedimentary deposits are widespread on the surface of Mars, which contrasts with the rarity of carbonate deposits, and indicates surface waters with chemical features drastically different from those on Earth. While the Earth's surface chemistry and climate are intimately tied to the carbon cycle, it is the sulfur cycle that most strongly influences the Martian geosystems. The presence of sulfate minerals observed from orbit and in-situ via surface exploration within sedimentary rocks and unconsolidated regolith traces a history of post-Noachian aqueous processes mediated by sulfur. These materials likely formed in water-limited aqueous conditions compared to environments indicated by clay minerals and localized carbonates that formed in surface and subsurface settings on early Mars. Constraining the timing of sulfur delivery to the Martian exosphere, as well as volcanogenic H_2O is therefore central, as it combines with volcanogenic sulfur to produce acidic fluids and ice. Here, we reassess and review the Martian geochemical reservoirs of sulfur from the innermost core, to the mantle, crust, and surficial sediments. The recognized occurrences and the mineralogical features of sedimentary sulfate deposits are synthesized

F. Gaillard (✉) · B. Scaillet
Univ dOrléans, ISTO, UMR 7327, 45071 Orléans, France
e-mail: gaillard@cnrs-orleans.fr

F. Gaillard · B. Scaillet
CNRS/INSU, ISTO, UMR 7327, 45071 Orléans, France

F. Gaillard · B. Scaillet
BRGM, ISTO, UMR 7327, BP 36009, 45060 Orléans, France

J. Michalski
Planetary Science Institute, Tucson, 85719 AZ, USA

G. Berger
CNRS/Université de Toulouse, IRAP, 14, avenue Edouard Belin, 31400 Toulouse, France

S.M. McLennan
Department of Geosciences, State University of New York at Stony Brook, Stony Brook, NY 11794-2100, USA

and summarized. Existing models of formation of sedimentary sulfate are discussed and related to weathering processes and chemical conditions of surface waters. We also review existing models of sulfur content in the Martian mantle and analyze how volcanic activities may have transferred igneous sulfur into the exosphere and evaluate the mass transfers and speciation relationships between volcanic sulfur and sedimentary sulfates. The sedimentary clay-sulfate succession can be reconciled with a continuous volcanic eruption rate throughout the Noachian-Hesperian, but a process occurring around the mid-Noachian must have profoundly changed the composition of volcanic degassing. A hypothetical increase in the oxidation state or in water content of Martian lavas or a decrease in atmospheric pressure is necessary to account for such a change in composition of volcanic gases. This would allow the pre mid-Noachian volcanic gases to be dominated by water and carbon-species but late Noachian and Hesperian volcanic gases to be sulfur-rich and characterized by high SO_2 content. Interruption of early dynamo and impact ejection of the atmosphere may have decreased the atmospheric pressure during the early Noachian whereas it remains unclear how the redox state or water content of lavas could have changed. Nevertheless, volcanic emission of SO_2 rich gases since the late Noachian can explain many features of Martian sulfate-rich regolith, including the mass of sulfate and the particular chemical features (i.e. acidity) of surface waters accompanying these deposits. How SO_2 impacted on Mars's climate, with possible short time scale global warming and long time scale cooling effects, remains controversial. However, the ancient wet and warm era on Mars seems incompatible with elevated atmospheric sulfur dioxide because conditions favorable to volcanic SO_2 degassing were most likely not in place at this time.

Keywords Sulfur · Mars · Basalt · Mantle · Sediment · Redox · Sulfate · Water · Core

1 Introduction

A striking feature of the surface of Mars revealed by in-situ and remote sensing instruments is the overwhelming abundance of sulfur (Clark and Baird 1979; Foley et al. 2008; Brückner et al. 2008; King and McLennan 2010), predominantly in its oxidized mineralogical form—sulfate—that covers terrains dated from the late Noachian and Hesperian epochs (Gendrin et al. 2005). Sulfate minerals are most likely predominantly of sedimentary origin, although hydrothermal occurrences may also exist. The surface waters that produced such deposits were most likely acidic (Fairen et al. 2004; Chevrier et al. 2007) and ultimately related to sporadic events possibly triggered by intermittent volcanic eruptions in a thin atmosphere. The emerging picture for Mars is that of a planet whose surface geochemistry and possibly its climate was dominated by the sulfur chemical cycle (e.g., Settle 1979; Clark and Baird 1979; Wänke and Dreibus 1994; King et al. 2004; Halevy et al. 2007; McLennan and Grotzinger 2008; Johnson et al. 2008; King and McLennan 2010; McLennan 2012). This picture contrasts with Earth for which the carbon cycle is believed to control the dynamics of chemical and climate processes in the near-surface environment (e.g., Berner 1995). Yet, on Mars, the period conducive to widespread sulfate deposits was preceded by a period marked by sedimentary processes depositing clay minerals (Ehlmann et al. 2012, this issue) and possibly carbonates, although the overall importance of carbonate minerals remains to be defined (Ehlmann et al. 2008; Michalski and Niles 2010). Such a sequence of mineralogical eras undoubtedly reflects a sequence of surface geochemical eras (Jakosky and Phillips 2001; Poulet et al. 2007; Bibring et al. 2006), that in turn reflects complex and evolving exchanges between the various Martian reservoirs: mantle-crust-atmosphere-hydrosphere.

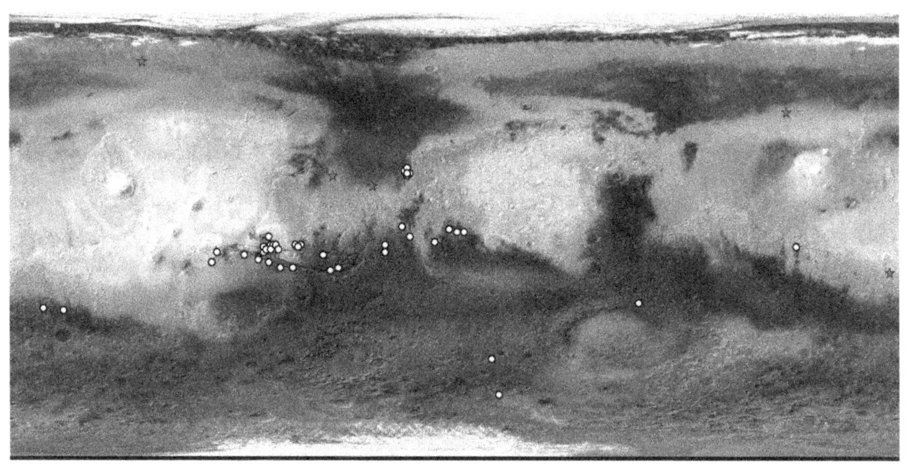

Fig. 1 A survey of sulfate detections to date is shown over a global map of surface albedo measured by TES. The detections are grouped into 3 categories shown symbolically, corresponding to layered Hesperian-Noachian sulfates (*yellow circles*), polar sulfates (*green circles*), and sulfates measured or inferred within soils at landing sites (*red crosses*)

The purpose of this paper is to summarize the most recent advances about Martian sulfur, from source to sink. We also attempt to step beyond a conventional review paper by suggesting links and hypotheses that allow several observations to be connected. Accordingly, we also propose a sequence of events relating igneous Mars to surficial Mars. Finally, we draw comparisons and highlight several differences between the surface chemistry of Mars and that of Earth.

Because we are concerned with the fate of a single element, sulfur, through the various Martian reservoirs, from the innermost core to the outermost sediments, various processes need to be considered, from early accretion, core formation and mantle differentiation, mantle melting and basalt eruption, to volcanic degassing and atmospheric processes, and finally to sedimentary processes. We believe this approach is justified because of the ubiquitous role of sulfur in surficial Martian processes and because sulfur in itself is a very complex chemical element, with many different redox states (−II, 0, +I, +II, +IV, +VI). Depending on its redox state, sulfur would potentially have very different chemical behaviors as a function of redox conditions (Gaillard and Scaillet 2009). Also noteworthy is that the behavior of sulfur is influenced by other chemical elements or chemical parameters, which unavoidably requires us to deviate in places from the sulfur-only perspective.

Accordingly, the paper is organized with an observation to model perspective. We first present and discuss sedimentary Mars, essentially based on numerous recent observations. This part of the paper characterizes the inventory of sedimentary sulfate deposits and evaluates the processes that may have generated these deposits. We then discuss models of the deep martian interior and igneous Mars, including an examination of volcanic degassing of sulfur. The last sections discuss the exchanges of sulfur between the different reservoirs and how these exchanges may have varied through time and how they may have affected martian climate.

Table 1 Detections of sulfate minerals on Mars from orbital data

Description	Lat.	Long.	Elev.	Minerals	References
Noctis Labrynthis	−11	261.7	3500	PHS, MHS, S, P	Weitz et al. 2011
Gale Crater mound	−5	137.5	−4000	PHS, MHS, P	Thomson et al. 2011; Milliken et al. 2010
Ophir Chasma	−4.2	286	−2000	PHS, MHS, FeOx	Wendt et al. 2011
	−4.5	287	−600	PHS, MHS, FeOx	Wendt et al. 2011
	−3.5	287.1	−2500	PHS, MHS, FeOx	Wendt et al. 2011
	−4.4	288.4	−4500	PHS, MHS, FeOx	Wendt et al. 2011
Coprates Chasma	−13	295.1	80	PHS, MHS	Fueten et al. 2011
Columbus Crater	−29.4	194	900	PHS, MHS	Wray et al. 2011
Cross Crater	−30.1	202.4	700	PHS, MHS	Wray et al. 2011
Capris Chasma	−13.3	312.6	−1500	PHS, MHS	Flahaut et al. 2010b; Gendrin et al. 2005
Polar till	82	115	−4500	PHS	Masse et al. 2010
Polar dunes	82	200	−4200	PHS	Langevin et al. 2005; Fishbaugh et al. 2007
Mawrth Vallis	24.2	341.6	−3000	PHS, MHS	Wray et al. 2010
Mawrth Vallis	22.9	341.5	−3200	PHS, MHS	Wray et al. 2010
Mawrth Vallis	25.4	339.7	−3350	J	Farrand et al. 2009
Mawrth Vallis	25.5	340.7	−3585	J	Michalski et al. 2010
Layered plains deposits	−8.16	307.3	1824	FeSO4	Le Deit et al. 2010
Opportunity landing site	−1.95	354.5	−1383	J, PHS	Glotch et al. 2006a
Meridiani Plaunum	1	4	−1300	PHS, MHS	Wiseman et al. 2010; Poulet et al. 2008; Wray et al. 2009
	1	1	−1200	PHS, MHS	Wiseman et al. 2010; Poulet et al. 2008
	2	358.5	−1380	PHS, MHS	Wiseman et al. 2010; Poulet et al. 2009
Aram Chaos	3	339.2	−2700	PHS, MHS	Lichtenberg et al. 2010; Glotch and Christensen 2005; Masse et al. 2008
Phoenix landing site	68.2	234.25	−4115	MgSO4, CaSO4	Kounaves et al. 2010
Noctis Labrynthis	−7.3	263.9	2000	MHS/PHS/?	Mangold et al. 2010
Ius Chasma	−8.5	280.6	−3950	PHS, MHS, S	Roach et al. 2010b
Layered plains deposits	−8.3	274.8	4280	S, FeSO4	Weitz et al. 2010; Le Deit et al. 2010; Milliken et al. 2008
Layered plains deposits	−9.6	280.8	3880	S, FeSO4	Weitz et al. 2010; Le Deit et al. 2010; Milliken et al. 2008
Layered plains deposits	−6.8	283.5	4448	S, FeSO4	Weitz et al. 2010; Le Deit et al. 2010; Milliken et al. 2008

Table 1 (*Continued*)

Description	Lat.	Long.	Elev.	Minerals	References
Layered plains deposits	−4	296.5	2320	S, FeSO4	Weitz et al. 2010; Le Deit et al. 2008
Juventae Chasma	−4.4	297.6	−857	PHS, MHS	Bishop et al. 2009
	−4.6	296.9	−1357	PHS, MHS	Bishop et al. 2009
S. Highlands	−49.2	14.5	500	S-Z	Wray et al. 2009
	−63.2	18.2	2247	PHS, MHS	Wray et al. 2009
Candor Chasma	−5	283.5	550	PHS, MHS	Murchie et al. 2009b; Mangold et al. 2008; Bibring et al. 2007
	−6	283.8	3000	PHS, MHS	Murchie et al. 2009b; Mangold et al. 2008; Bibring et al. 2007
	−6	286	910	PHS, MHS	Murchie et al. 2009b; Mangold et al. 2008; Bibring et al. 2007
Miyamoto Crater	−3.2	352.5	−1954	PHS, MHS	Wiseman et al. 2008
Gusev Crater	−14.57	175.5	−1920	FeSO4	Lane et al. 2008; Johnson et al. 2007
Melas	−10.5	285.2	−100	PHS, MHS	Gendrin et al. 2005
	−12.5	290.3	−2500	PHS, MHS	Gendrin et al. 2005
Ophir Chasma	−4.3	288.3	−4500	PHS, MHS	Gendrin et al. 2005
Candor Chasma	−6.4	288.8	−2300	PHS, MHS	Gendrin et al. 2005
Hebes Chasma	−1.2	284.8	−3052	MHS	Gendrin et al. 2005
Capris Chasma	−13.9	310	−3690	PHS, MHS	Gendrin et al. 2005
Iani Chaos	−1.3	342.3	−2000	PHS, MHS	Gendrin et al. 2005; Glotch and Rogers 2007
Aureum Chaos	−3.5	332.5	−3780	PHS, MHS	Gendrin et al. 2005; Glotch and Rogers 2007
Arisinoes Chaos	−7.3	331.6	−3090	PHS, MHS	Gendrin et al. 2005; Glotch and Rogers 2007

2 Inventory and Nature of Sulfate Deposits on Mars

2.1 Overview

The earliest surface exploration by the Viking spacecraft revealed high sulfur contents that were interpreted as evidence for sulfate minerals in a widely homogenized layer of regolith (Clark et al. 1977, 1993). This interpretation was supported by analyses of surface chemistry by the Mars Pathfinder and Soujourner Rover, which also detected relatively high levels of sulfur in regolith and on rock surfaces (Rieder et al. 1997). Since that time, sulfur and sulfate minerals have been detected in a number of geological settings on Mars using both orbital remote sensing and in-situ analyses (Fig. 1—global map) (Table 1—list of detections)

(Gendrin et al. 2005; Langevin et al. 2005; Clark et al. 2005; Bibring et al. 2006; Murchie et al. 2009a). Deposits can be categorized into one of 5 groups: (a) Hesperian layered sulfates (Squyres and Knoll 2005; Clark et al. 2005; McLennan et al. 2005), (b) Interior Layered Deposits (ILDs), (c) polar deposits, (d) intracrater sediments, and (e) as part of the global dust and regolith (Wang et al. 2006; Lane et al. 2008). Recent results suggest that a sixth type of deposit has been detected: sulfates as secondary vein minerals within silicate bedrock on the rim of Endurance Crater (Squyres et al. 2012). Taken together, these observations clearly illustrate that sulfur and sulfate minerals constitute a significant fraction of Mars' aqueous geologic record, from pole to equator, and have played a major role in Mars' sedimentary rock cycle (King and McLennan 2010; McLennan 2012).

Below, we discuss the various deposits of sulfates in more detail. First, we provide summary of progress in detection and mapping of sulfates on Mars from orbital data. Then, we describe some interesting trends in mineral associations between sulfates and other minerals. Lastly, we discuss the geology of the various types of sulfate deposits, organized by deposit type.

2.2 Orbital Detection of Sulfates

Two instruments have provided unambiguous evidence for sulfate minerals on Mars from orbit. The first discoveries where reported using data from the *Obervatoire pour la Minéralogie, l'Eau, les Glaces, et l'Activité* (OMEGA) (Gendrin et al. 2005; Langevin et al. 2005) onboard the Mars Express spacecraft. OMEGA has now mapped a large fraction of the Martian surface at the scale of 100s of meters/pixel (e.g. Fig. 2—4 panel figure) and revealed numerous sulfate deposits on the surface (Bibring et al. 2006; Poulet et al. 2007; Carter et al. 2011). Targeted observations with the *Compact Reconnaissance Imaging Spectrometer for Mars* (CRISM) on the Mars Reconnaissance Orbiter spacecraft have revealed additional sulfate deposits (Fig. 3—CRISM figure), and shown the detections at higher spectral and spatial resolution (~18 m/pixel) (Murchie et al. 2009a). Both spectrometers operate in the visible-near infrared (~0.5–4 μm) region and are therefore sensitive to spectroscopic absorptions that arise from S-O vibrational overtones and combination bands, vibrations associated with bound or adsorbed water, and electronic transitions associated with Fe, if present in the sulfate structure (Lane and Christensen 1998; Cloutis et al. 2006 and references therein).

From a mineralogical perspective, two major categories of sulfates have been detected from infrared remote sensing: monohydrated and polyhydrated sulfates (MHS or PHS, respectively). Specific sulfates such as kieserite, gypsum, jarosite, alunite, szomolnokite, and ferricopiapite have been detected from orbit with varying levels of confidence, but in many cases, it is not possible to distinguish specific minerals beyond PHS or MHS. Mg- and Fe-bearing sulfates are more common than Ca-bearing sulfates (e.g. Fig. 3).

The abundances of sulfate minerals are difficult to interpret from infrared remote sensing data because the interpretation depends not only on the actual abundance, but also the texture of the surface, grain size of particulates, crystallinity, and hydration state. Furthermore, thin rock coatings could potentially mask minerals present as bulk components of a substrate (e.g. Kraft et al. 2003). A simple baseline for sulfate abundances on Mars comes from observations of the Martian soils and dust. Surface chemical measurements suggest that 6–7 % SO_3 is typically present within the globally homogenized dusty soil, which may translate to ~5–10 % sulfate minerals (McSween et al. 2010). These values are also consistent with global mapping of S concentrations in the upper few tens of centimeters of the Martian surface by the Mars Odyssey gamma-ray spectrometer experiment (King and McLennan 2010).

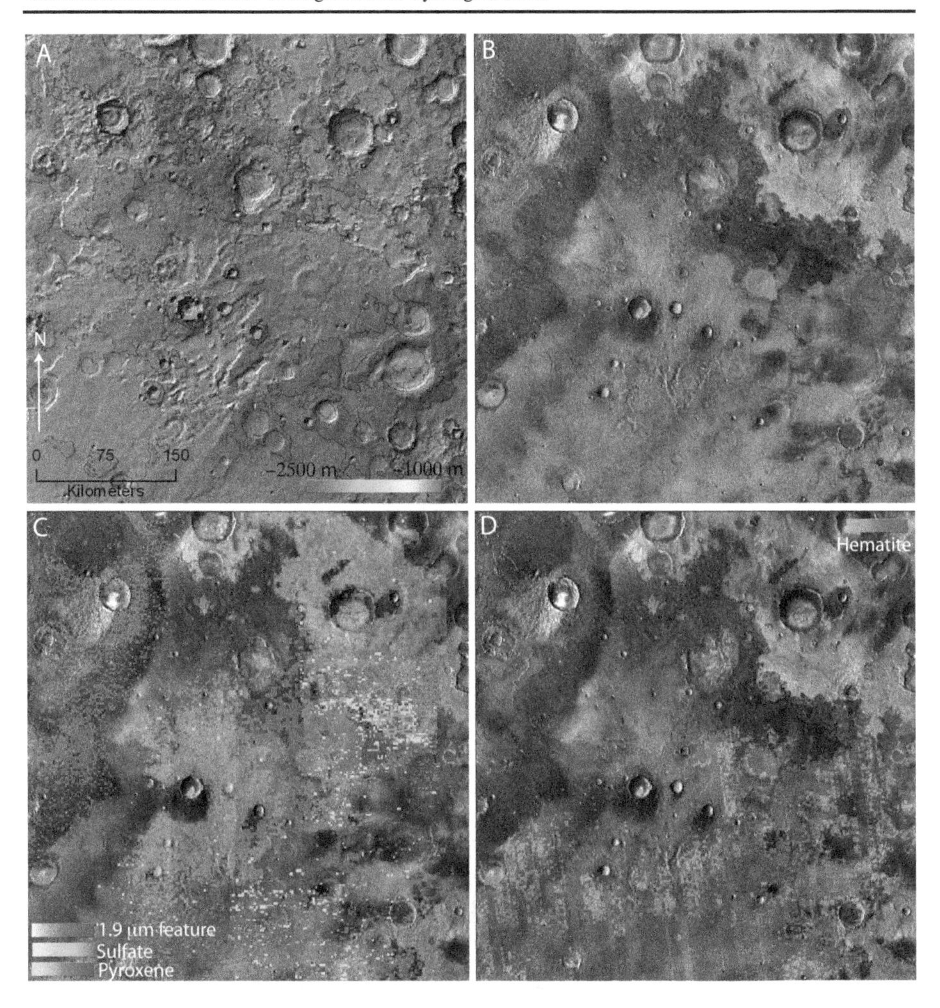

Fig. 2 MOLA topographic data are shown in for the Meridiani Planum area in (**A**). THEMIS nighttime thermal infrared data are draped over daytime infrared data in (**B**). *Warm colors* correspond to surfaces containing coarser grained or more well indurated materials. In (**C**), OMEGA spectral index maps are shown. *Orange colors* correspond to the 1.9-micron index that maps bound water; *blue colors* correspond to pyroxene minerals (after Poulet et al. 2009); and the *green colors* correspond to 2.4-micron index values that correlate with sulfates. In (**D**), TES hematite index data are shown draped onto THEMIS daytime infrared. The scale is from 10 % to 18 % hematite (after Christensen et al. 2001)

OMEGA and CRISM observations of the global dust/soil do not show the presence of sulfates, therefore, this abundance is the minimum detection limit for fine-grained sulfates on Mars from orbital data.

In cases where sulfate is clearly detected, the data can be compared to mathematical mixtures that take into account fundamental spectral properties and estimates of grain size (e.g. Poulet and Erard 2004). For example, dunes in the north polar region of Mars can be modeled with a mixture of ~35 % gypsum and 65 % siliciclastic material (Fishbaugh et al. 2007). In other cases, rocks that are known to be sulfate-bearing from in-situ analyses may contain little or no spectral evidence for sulfates from orbit (Fig. 2). Yet, similar rocks in the same region clearly show evidence for sulfates (Fig. 3 CRISM example). Such differences

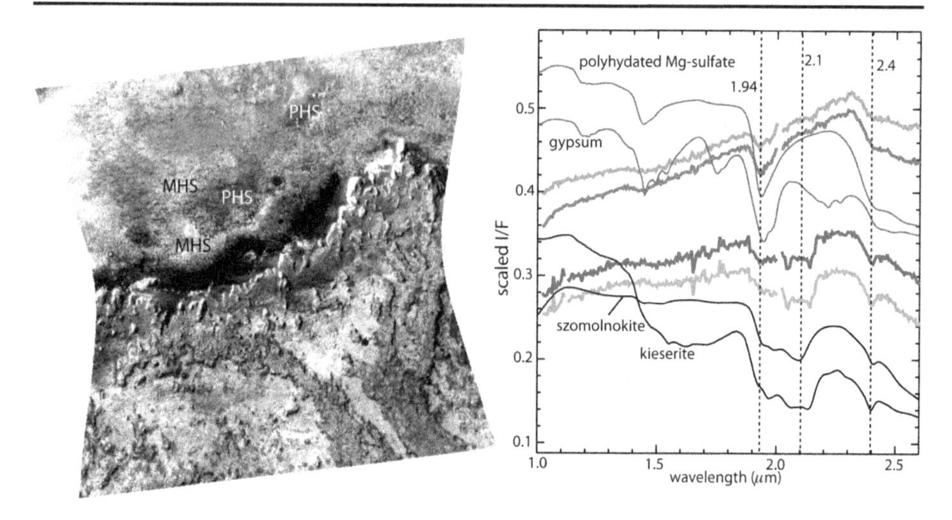

Fig. 3 CRISM cube frt00016f45_07_if165l_trr3 shows sulfates in the Meridiani Planum area. The annotations "PHS" and "MHS" indicate the locates where spectra were extracted to show examples of polyhydrated and monohydrated sulfates, respectively. The spectra were ratioed against a spectrally unremarkable terrain

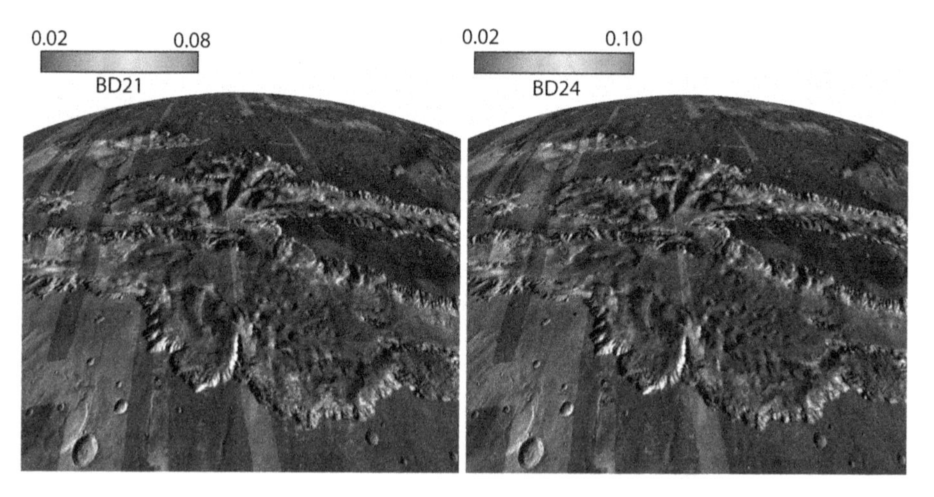

Fig. 4 OMEGA spectral index maps of the central Valles Marineris area are shown. The 2.1 micron index (BD21) corresponds to detections of monohydrated sulfates and the BD24 map corresponds to polyhydrated sulfates

in detectability may be an example of the influence of rock texture and weathering style or surface exposure age on the spectral character of the deposits.

Mineral abundances have also been modeled using orbital thermal infrared remote sensing data from the Thermal Emission Spectrometer (TES) and thermal infrared data from the Mini-TES instruments aboard the Mars Exploration Rovers. These analyses suggest that 20–40 % sulfates are present in layered rocks at Meridiani Planum (Glotch et al. 2006a, 2006b), and ~15 % sulfates in similar rocks in Aram Chaos (Glotch and Christensen 2005). Chemical analyses of rocks at Meridiani Planum also point to abundances of ~20–30 % sulfates (Clark et al. 2005). Taken together, all available data suggest that many "sulfate-rich" rocks on Mars may contain roughly 1/3 sulfate minerals by volume. A range is likely to exist, but it is unlikely that the orbital detections of sulfates could correspond to sulfates that occur only as trace components.

2.3 Mineral Associations

An important consideration for interpreting the geological significance of sulfate minerals on Mars is the associations among the various minerals. In this sense, the most obvious association seems to be the co-occurrence of sulfate minerals with hematite. Coarsely crystalline gray hematite was originally detected in Meridiani Planum with TES data (Christensen et al. 2000), and it is now clear from surface observations that the hematite occurs as "blueberries," or mm-scale spherules (Squyres et al. 2004; Squyres and Knoll 2005), that have eroded out of sulfate-rich bedrock and are now found as a lag deposit on the surface. The thermal infrared observations of these spherules suggest that they contain *c*-axis-oriented hematite, always oriented away from the surface of the spherule (Glotch et al. 2006b). This observation likely supports an interpretation of the hematite spherules as diagenetic concretions within the sulfates (McLennan et al. 2005; Sefton-Nash and Catling 2008).

Further work with TES and OMEGA data has shown that hematite is found in association with sulfates in many locations, including chaos terrains and within the ILDs (Christensen et al. 2001; Bibring et al. 2007; Weitz et al. 2008; Mangold et al. 2008; Le Deit et al. 2008). The nature of this association was investigated experimentally by Tosca et al. (2008), who demonstrated that low temperature oxidation of ferrous and ferric sulfates, in the presence of high ionic strength brines, leads ultimately to the precipitation of ferric oxides. These critical observations suggest that diagenetic hematite formation may be a critical component of sulfate formation on Mars (e.g. Roach et al. 2010a, 2010b; McLennan 2012).

Other mineral associations are observed. In layered terrains on the plains around Valles Marineris, jarosite appears to occur along with hydrated amorphous silica (Milliken et al. 2008). This observation might point to a similar origin as the sulfate within Meridiani Planum deposits, which occurs within siliciclastic rocks that likely contain a component of amorphous silica (Clark et al. 2005; Glotch et al. 2006a). Clay minerals are not commonly associated with sulfates from the orbital remote sensing perspective. However, in a few cases, interlayered sulfates and phyllosilicates are observed. In Columbus Crater, albuminous clays and sulfates are interlayered in crater-floor sediments (Wray et al. 2010). In the Mawrth Vallis channel, Fe/Mg-bearing clays occur stratigraphically above sulfates in certain cases (Wray et al. 2010). Also in the Mawrth Vallis area, aluminous clays occur in deposits that seem to be stratigraphically equivalent with jarosite-bearing deposits (Farrand et al. 2009). The base of the mound within Gale Crater (Mount Sharp) may contain interlayered sulfates and clays, which await exploration by the Curiosity Rover (Milliken et al. 2010).

2.4 Geology

2.4.1 Hesperian Layered Sulfates

Layered sulfates of probable Early Hesperian age are found throughout the western Arabia Terra region, in the greater Sinus Meridiani area (Poulet et al. 2008; Griffes et al. 2007; Wiseman et al. 2010). Also in this category are certain instances of layered crater fill, such as the sulfates found within Gale Crater (Milliken et al. 2010; Thomson et al. 2011). Exploration of such deposits by the Opportunity Rover shows that, in the Meridiani area, the

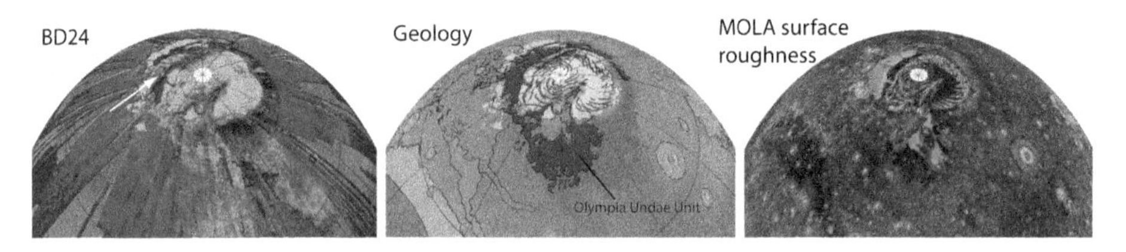

Fig. 5 An OMEGA 2.4 micron spectral index map shows the location of sulfates in the northern polar region of Mars. The locations of the sulfates correspond roughly to locates mapped as the Olymbia Undae unit (Tanaka et al. 2003), and to unique regions in MOLA surface roughness maps

rocks are layered at the decimeter scale, and laminated down to millimeter scale in places (Grotzinger et al. 2005, 2006). Sulfur contents were consistently measured in the range of 20–30 % SO_3, and chemical correlations suggest that Mg-sulfates are the most abundant group of minerals (Clark et al. 2005). However, jarosite was also directly detected from Mossbauer spectroscopy (Morris et al. 2006). The low amount of energy required to grind into the rocks with the Rock Abrasion Tool (RAT) aboard Opportunity indicates that most of the rocks are relatively soft (specific grind energies mostly less than 2 $J\,mm^{-3}$), as might be expected for evaporites (Herkenhoff et al. 2008). Microscale imaging of the rock textures shows that they are composed of fine-grained material. Sand-sized particles are observed, and required for the formation of eolian cross bedding seen in the section (Grotzinger et al. 2005, 2006), however the sand grains themselves may in turn be composed of finer-grained materials, such as recycled sulfate-cemented muds (Grotzinger et al. 2005; McLennan et al. 2005; also see Niles and Michalski 2009). Secondary porosity in the form of crystal molds and other vugs observed in parts of the section point to an extended and complex history involving diagenetic fluids (McLennan et al. 2005; McLennan and Grotzinger 2008).

2.4.2 Interior Layered Deposits (ILDs)

The ILDs are found throughout the Valles Marineris trough system, as well as within other chasmata (Chapman and Tanaka 2001). The deposits consist of massive mounds of layered materials that can reach kilometers in height, rivaling the elevation of the canyon rims in places. They lie unconformably on canyon floor deposits (Quantin et al. 2004), drape onto the canyon walls, and do not exhibit massive extensional faults that would have been formed during tectonic formation of the troughs (Okubo et al. 2008). Therefore, the deposits are interpreted to have formed after the final phase of formation of the Valles Marineris, which likely occurred in the late Noachian. The ILDs are challenging to accurately date due to their irregular geometry, but crater counting suggests a lower bound on their ages of Late Hesperian, and in rare cases, perhaps Early Amazonian (Quantin et al. 2004). They are therefore of similar age to the Hesperian layered deposits and may have a similar origin.

Spectral imaging shows that the ILDs, where they are well exposed, contain evidence for sulfates (Fig. 4), hematite, and locally, silica and phyllosilicates (Gendrin et al. 2005; Bishop et al. 2009; Mangold et al. 2008; Le Deit et al. 2008; Murchie et al. 2009b; Flahaut et al. 2010a; Roach et al. 2010a; Fueten et al. 2011; Weitz et al. 2011). In central Valles Marineris, crude compositional stratigraphy is observed where monohydrated sulfates overlie polyhydrated sulfates (Mangold et al. 2008).

2.4.3 *Polar Sulfates*

Gypsum deposits occur in the massive north polar dune field, Olympia Undae (Langevin et al. 2005; Fishbaugh et al. 2007) (Fig. 5—polar). The dunes are Amazonian in age and are thought to have been derived from erosion of the Basal Unit beneath the north polar cap (Fishbaugh et al. 2007). However, there is no evidence to date for sulfates within the Basal Unit itself. One possibility is that the sulfate formed within the dunes, after the clastic material was derived from the Basal Unit by erosion. Another possibility is that the gypsum is more easily detectable in the sand-sized materials within the dunes than in the competent layers of the basal unit. Yet another possibility is that the sulfates occur within the basal unit, but in lower concentration and that they have been concentrated within the dunes due to their density.

Sulfates have also been detected within sedimentary deposits surrounding other parts of the north polar cap (Masse et al. 2010, 2012). These sulfates have a more subtle spectroscopic signature than those within the Olympia Undae dunefield. Their weaker spectral features could correspond to lower absolute abundances, lower crystallinity, or different physical form (e.g. grain size, surface texture) compared to those in the dune sands. The deposits containing these sulfates are interpreted as sediments that weathered out of the polar cap, or as glacial sediments (Masse et al. 2010, 2012) and have formed in association with ice.

2.4.4 *Intracrater Deposits*

Intracrater sulfate deposits are relatively rare, but have been detected within craters in the Terra Sirenum region of Mars. Here, deposits within the floors of Columbus and Cross Craters contain poly- and monohydrated sulfates that are interbedded with aluminous phyllosilicates and associated with alunite, jarosite, Fe-oxides, and Fe/Mg-phyllosilicates (Wray et al. 2011). Alteration minerals occur within a ring in the interior of Columbus Crater, and within layered deposits exposed by erosion and impact degradation of the floor deposits. Such deposits are interpreted to have formed due to groundwater upwelling that may have fed a deep lake, or a transient, spring-fed environment (Wray et al. 2011).

2.4.5 *Sulfates Within the Soil and Dust*

Sulfur occurs with the Martian soil at all landing sites visited thus far at an average level of ~6.8 % SO_3 (King and McLennan 2010). The mineralogy of this sulfur is not well known, although recent results have shed some light. Results from the Wet Chemistry Laboratory (WCL) aboard the Phoenix Lander have demonstrated the presence of soluble sulfates within high latitude soils (Kounaves et al. 2010). At this site, the most likely salts are epsomite and gypsum. This is in contrast to some soils at the Gusev Crater site, which also contain significant levels of Fe-sulfates (Wang et al. 2006). At the Paso Robles site in Gusev Crater, the Spirit Rover exposed soils from the shallow subsurface with its wheels. Chemical and spectroscopic constraints on the bright soils suggest the presence of hydrated Fe-sulfate similar to ferricopiapite (Lane et al. 2008).

2.5 Summary of Observed Sulfate Deposits

Sulfate deposits are observed or inferred to exist within the global regolith, as well as in discrete geological deposits from pole to equator on Mars. Evidence to date suggests that the deposits can generally be considered siliciclastic materials, in some cases unconsolidated and in other cases, as somewhat competent sedimentary rock. By what geologic processes did these deposits form? Is there a common thread or have various deposits formed from disconnected processes at different times? It is almost certain that the ultimate source for sulfur in these deposits is volcanogenic. But, what were the aqueous conditions under which the minerals precipitated? How much water was involved and from what sources? We address these questions in the following section.

3 Mineralization of Sulfur at Mars Surface

3.1 Chemical Constraints

There is a growing consensus that sulfates formed mostly during the late Noachian to Hesperian, succeeding the era of phyllosilicates in the early-middle Noachian. Since the Hesperian, there is little evidence for formation of water-related minerals, clays or sulfates. Transformation of iron sulfates to iron oxides may be an ongoing, albeit very slow process (Tosca et al. 2008; McLennan 2012). Alteration of mafic minerals on the Earth, Mars or elsewhere is the chemical response of desequilibrated water-rock systems, influenced, sometimes inhibited, by kinetic constraints. Silicate minerals, from the chemical point of view, are oxide mixtures having alkaline properties, mainly because of their alkali and alkaline-earth content. As an example, the pH value buffered by pure CaO is 12.7 at 25 °C, near 12 for MgO and much more for Na_2O and K_2O. So substantial mineralogical transformations are generated by acid agents that can lead to huge deviations from equilibrium, and the amount of alteration phases can be compared, possibly correlated, to the amount of protons added to the system. Because the production of protons, and accompanying acidic fluids, always associates with concomitant anions for the obvious constraint of charge balance, this anion can be implied in the secondary phases formed by the conjugate base of the acid and the conjugate acid of the oxide. The occurrence of sulfate in the Martian soils at Meridiani Planum or elsewhere, argues for the idea that the source of protons that drove the alteration is associated with the sulfur cycle.

3.2 The Burns Formation Example

The presence of sulfur-rich sedimentary rocks at Meridiani Planum, termed the Burns formation, was a major discovery by the Mars Exploration Rover (MER) mission. The sediment contains up to 60 % secondary minerals (amorphous silica, Mg- and Ca-sulfates, jarosite, hematite and possibly chlorides). The presence of jarosite has been cited as prime evidence for low pH conditions since it is known to be stable at pH < 4.

The interpretation of the MER team is that sulfate-bearing sedimentary grains were derived from a weathered basaltic source and cemented by later sulfate-dominated secondary minerals (Squyres et al. 2004; Grotzinger et al. 2005, 2006; McLennan et al. 2005; Metz et al. 2009) with a variety of diagenetic features, including secondary porosity, multiple generations of cements and hematitic concretions. In this model, sulfates form through a variety of evaporative and later diagenetic processes.

A number of alternative scenarios have also been proposed to explain the morphology and chemistry of the Burns formation sediments. Hynek et al. (2002) and McCollom and Hynek (2005) proposed that the Burns formation represented pyroclastic ash flows and air fall whereas Knauth et al. (2005) suggested that the layered deposits formed through debris

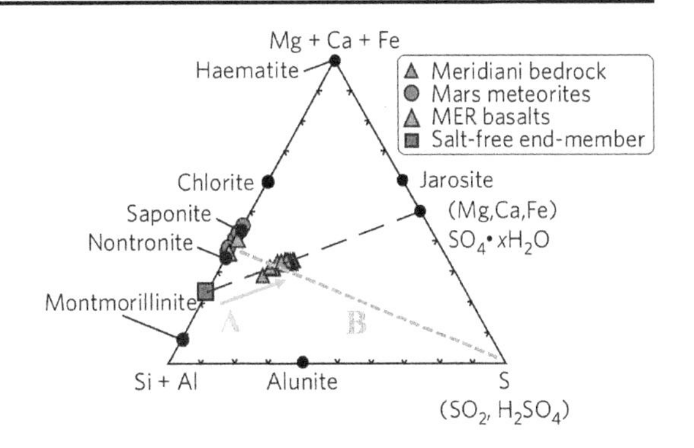

Fig. 6 Evaporite model (MER team, **A**) against by McCollom and Hynek (2005) (**B**) analysis of the Burns formation composition. The evaporite model focuses on the "Meridiani trend" assuming a mixture between an altered basalt (*pink square*) and a sulfate salt. The McCollom and Hynek model assumes a mixture between a pristine basalt and a pure sulfur component

flow following an impact. McCollom and Hynek (2005) interpreted the chemical composition of the Burns sediments as a mixture between a pristine basalt and a pure hydrated sulfur compound (with a composition within the SO_2–O_2–H_2O ternary system), which is in contrast to the eolian-groundwater model that calls for a mixture of altered basalt and sulfate minerals. These two interpretations are presented in a ternary diagram in Fig. 6 where the trend formed by Meridiani sediments suggests a mixture between an aluminosilicate component and a Mg,Fe,Ca sulfate component. The aluminosilicate component does not match the exact composition of a martian basalt but rather appears consistent with an altered basalt. More recently, Niles and Michalski (2009) proposed another scenario in which the deposition at Meridiani Planum of massive ice/dust layered deposits during periods of high obliquity was accompanied by a cryo-concentration of volatile-bearing brines in ice that contained outgassed sulfur-bearing species. In this model, alteration of fine-grained silicates by acidic brines in the ice produces vast quantities of alteration products with limited chemical mobility. In their model, the eolian textures formed from reworking of the sublimation lag during and after removal of ice. The diagenetic textures were generated by water released during dehydration of sulfate minerals that were originally highly hydrated. Such a model has some advantages in that it can explain the formation of layered sulfates in vast mounds that lack obvious provenance, such as the ILDs (Michalski and Niles 2012).

Some of the relevant geochemical issues were discussed by Tosca et al. (2005, 2008) for the chemistry of evaporating brines. On the other hand, Tréguier et al. (2008) and Berger et al. (2009) focused on the source of sulfur and its reaction with a pristine basalt in an in-situ alteration scenario. They argued that SO_3, a strong acid gas resulting from the oxidation of volcanic SO_2 in a dry atmosphere, may have produced a strong and pristine acid solution at the ground through interaction with water produced by ice melting. Based on the statistical analysis (PCA) of the Meridiani chemical compositions and weathering scenario tested by numerical modeling they reproduced the chemical and mineralogical data available for the Burns Formation sediments (Fig. 7), provided that the generated brine leaves the system after a short reaction time and evaporates elsewhere. The oxidation of SO_2 can result from several atmospheric reactions and can be driven back to the surface by acid rains as reported by Schiffman et al. (2006) for the Earth, or can occur at the surface directly assuming a high penetration of the UV radiation through a thin and dry atmosphere. However, as pointed out by Zolotov and Mironenko (2007), the generation of sulfuric acid through oxidation of SO_2 and H_2S is limited by the concentration of photochemically-produced atmospheric oxidants and these authors proposed impact-generated acid rainfalls as an alternative origin of SO_3. But sulfur brought by impactors would nevertheless require an oxidation step to produce SO_3. In both scenarios, volcanic SO_2 or impact-generated sulfur can produce pristine acidic

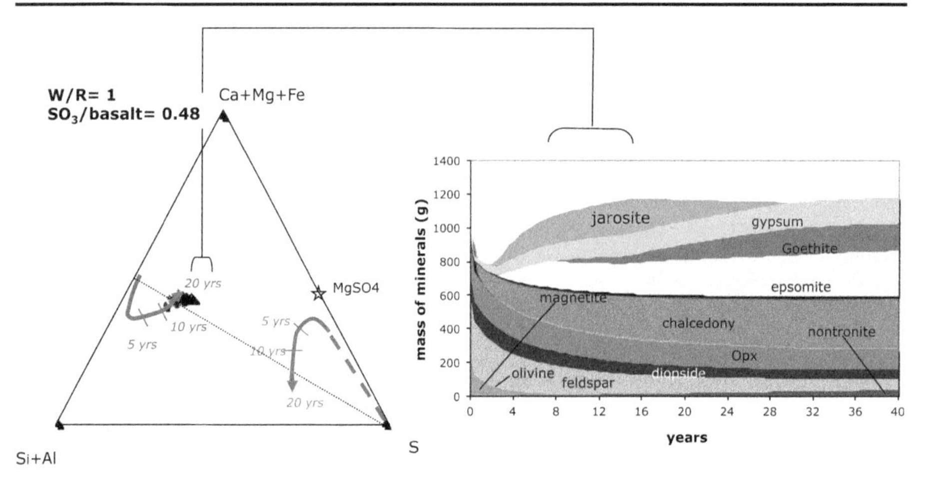

Fig. 7 SO_3-basalt-water kinetic model (Berger et al. 2009) accounting for the "Meridiani trend". Small amount, ephemeral, but strongly acid brine reproduces the chemical and mineralogical features of the Burn Cliff Formation. The variation of mineral composition with time is reported in the right graph. The chemical composition of the solid fraction (*red*) and brine (*blue*) is also showed with time in the *left ternary diagram*, and compared to the Meridiani compositions (*black triangles*). The best fit is obtained when assuming a concentrated sulfuric acid reacting with the rock for a short time, without in-situ evaporation of the resulting brine

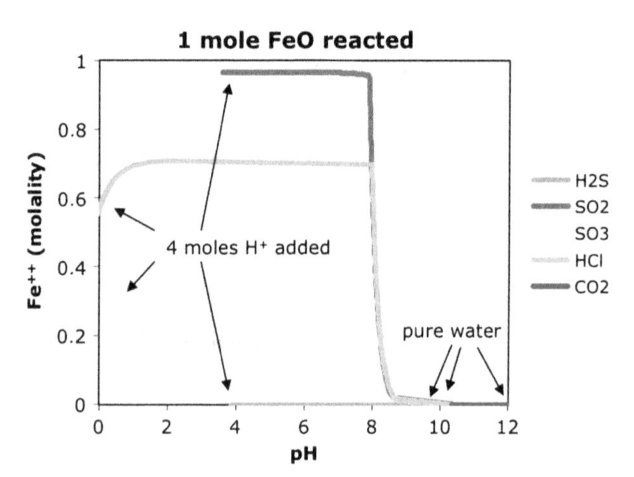

Fig. 8 Aqueous concentration of free Fe^{++} species during dissolution of 1 mole of FeO (the ferrous component in mineral) by fluids of various acidity. The pH of the resulting solution depends on the concentration and also the nature of the acidic reactant. The conjugate anions strongly influence the amount of the aqueous ferric specie following a general equation: $FeO + H_2O + HX \Leftrightarrow aFe^{++}_{aq} + bFeX^{-}_{aq} + cFeX_{2\ aq} + dFeX_{2\ solid}$

water covering the whole surface. In contrast, hydrothermal systems imply acid-base reactions occurring at depth and the resulting fluid reaching the surface is a brine being evolved and partially neutralized. The latter could account for the evaporitic processes suggested by the MER team while the surficial acid weathering scenario can account for the "Meridiani Trend" as described by McCollom and Hynek (2005), i.e. chemical mixing between a pristine basalt and a pure sulfur component, as suggested in Berger et al. (2009) or Niles and Michalski (2009).

The acidic features of the Meridiani sediments were also discussed by Hurowitz et al. (2010). They suggested the oxidation of aqueous ferrous iron as an alternative process to generate acidic solutions at a local scale. However, this process also requires protons to get Fe^{++} in solution in a previous stage. For example, Fig. 8 illustrates the effect of several acid sources on FeO (the ferrous component in mineral). The amount of released ferrous iron depends on the solubility of the conjugate anion salts (chloride, sulfate, carbonate) and the proportion of the complexed forms such as $Fe(OH)^+$, $FeCl_2$, etc. A general equation could be: $FeO + H_2O + HX \Leftrightarrow a\mathrm{Fe}^{++}_{aq} + b\mathrm{FeX}^-_{aq} + c\mathrm{FeX}_{2\ aq} + d\mathrm{FeX}_{2solid}$. The calculations show that a substantial mobilization of Fe^{++} requires substantial addition of protons with a highly soluble conjugate anion. H_2SO_3 appears as the most efficient extracting agent.

Finally, the diversity of the proposed scenarios suggests that, despite the quantitative information collected by Opportunity, the origin of sulfur and the chemical constraints on water-gas-rock interaction remains an important issue for understanding sulfate-bearing sediments at the Martian surface. For example, recently published experimental observations (Dehouck et al. 2012) indicate that alteration of sulfide-bearing basalts may produce mineral assemblies containing sulfate minerals that mimic those identified on Mars. In Sects. 4 and 5, we present several pieces of evidence that volcanic gases can provide most of the sedimentary sulfate and can furthermore trigger the succession of different periods with contrasted surface chemistry.

3.3 SO_2 Versus CO_2 in the Sediments/Soil Records with Time

The formation of sulfate minerals during the late Noachian and Hesperian suggests an alteration process driven by volcanic SO_2. Even in early Noachian terrains where phyllosilicates rather than sulfates are detected, the source of protons for the alteration process, whether SO_3/SO_2 and/or CO_2 is an open question. Carbon-based acids having lower dissociation constants (i.e., weak acids), typically have reaction by-products dominated by clays (Berger et al. 2009). The lack of widespread observations of carbonate minerals in the altered Martian sediments (this point is extensively discussed in the accompanying paper by Niles et al. 2012, this issue) could be explained by the presence of another more acidic compound that precludes the precipitation of Ca,Mg,Fe carbonates. Significant, concentrations of SO_2 in the atmosphere is a reasonable assumption, given the importance of sulfur during the Hesperian era. However, in the context of clay minerals, constraining the in-situ pH (and carbonate precipitation feasibility) is not trivial. Clays, such as smectite, have exchangeable cations in the interlayer positions of their structure, which confer to these minerals a large sorption capacity and make the clays an ion exchanger and pH buffer. Even in the case of low cationic exchange capacities (CEC), i.e. kaolinite, illite, or chlorites, the small particle size confers a huge specific surface area to the material and enhances the consequences of surface chemistry. However, although the acid-base properties of aluminosilicate surfaces and CEC of smectite material are now well known (see for example Tertre et al. 2006), other textural parameters and the accurate estimation of the reacting mineral surface make the prediction of reaction paths and rates difficult.

Another parameter is the differential progression of an SO_3 and CO_2 front within soils and sediments when these two gases are present simultaneously in the lower atmosphere. When SO_3 is not concentrated in the lower atmosphere, it will rapidly be consumed in the superficial layer of the soil, given the high silicate alteration rate at low pH, and will probably not influence the chemistry of the deep sediments. By contrast, CO_2 is a less acidic gas ($pH > 4$) and the solubility of its conjugate salt (carbonate) is highly pH sensitive. CO_2 will not produce carbonate in the SO_3 influenced zone and can subsequently diffuse deeper in the sediments, a process made easier given that rock alteration is slow in the mildly acidic pH-range. This simple analysis, illustrated in Fig. 9, leads to the suggestion that carbonate deposits may exist at depth in the regolith on Mars and, accordingly, could constitute a significant sink for CO_2.

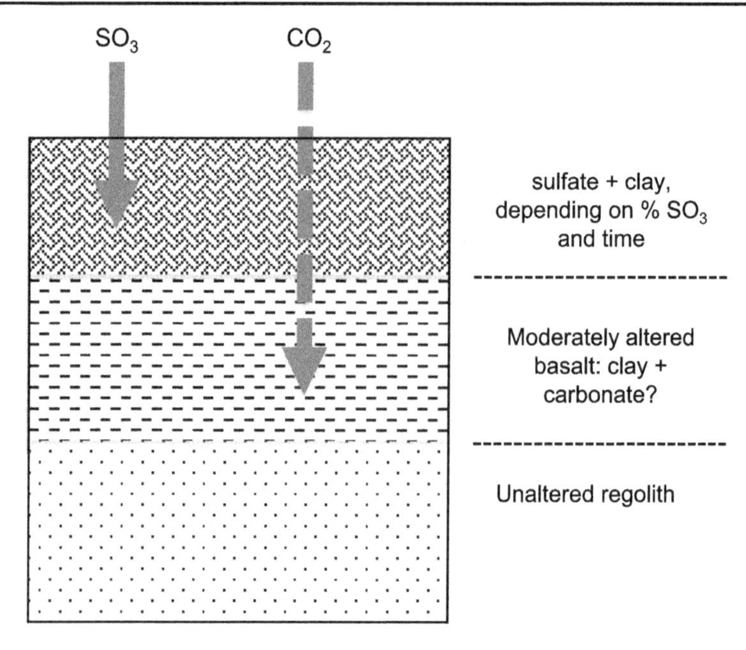

Fig. 9 A possible differential consumption of SO_3 and CO_2 with depth. SO_3 first reacts with the surface of the basaltic regolith, excluding the CO_2 mineralization. CO_2 reacts deeper, below the zone altered by SO_3

The prediction of carbonate precipitation from atmospheric CO_2 should also take into account the water:rock ratio. In the case of a high run-off (low residence time of water within the martian regolith) and/or under near neutral conditions (slow dissolution kinetics) and/or presence of another acid gas (high carbonate solubility at low pH) carbonate precipitation is not expected. An example is shown in Fabre et al. (2011) for the "Snow Ball Earth" aftermath during the Precambrian time where even 10 % atmospheric CO_2 may not have led to continental carbonates.

In conclusion, the mineral evolution of the Martian soil/sediments over geological time can be interpreted as resulting from a global atmospheric change from Noachian to Hesperian, with a decrease of the CO_2/SO_2 ratio.

4 Geochemical Reservoirs of the Interior of Mars

Chondrites, from which terrestrial planets most likely accreted, contain several weight percent of sulfur as sulfide (S content ranging from 2 to 10 wt% for all varieties of chondrites, see Chabot 2004; Gaillard and Scaillet 2009; Ebel 2010). Geochemical observations, however, indicate a general depletion in volatile elements of planetary bodies with respect to chondrites (Righter et al. 2006). This depletion is generally related to the intense early solar activity and/or incomplete condensation during the earliest period of planetary accretion. In some cases, some form of planetary devolatilisation may have accompanied high temperature processes during the accretion process (e.g., giant impacts or volcanic degassing of planetary embryos) and therefore contributed to additional depletion in strongly volatile elements (C-O-H-S), but quantitative constraints on such processes are strongly model dependent. Existing geochemical models point towards sulfur content lower than 5 wt% for bulk Mars (Wänke and Dreibus 1994), whereas similar models for the Earth indicate less than 0.5 wt% sulfur (Dreibus and Palme 1996). The emerging but still poorly constrained conventional wisdom is therefore that Mars must be enriched in sulfur in comparison to the Earth (Dreibus and Wanke 1985; Stewart et al. 2007). Below, we review partitioning of sulfur between Fe-metal, molten silicate, and fluid in order to assess both sulfur reservoirs and fluxes between core, mantle, basalt and evaluate net sulfur transfers by volcanic degassing into the atmosphere.

4.1 Core-Mantle

We discuss here existing studies on the behavior of sulfur during core-mantle equilibration in a magma ocean scenario. The conventional wisdom here is that sulfur contents of the core and mantle of Mars were inherited from a single (or last) equilibration step between metal and silicate at high P–T conditions (Righter and Chabot 2011). Such equilibration is classically addressed by models based on partitioning experiments performed under controlled thermodynamic conditions. Most assessments of partitioning of sulfur between silicate and metal and its application to core-mantle equilibration on planetary bodies are based on the assumption that molten FeS is a good analogue of Fe-metal containing little sulfur. Many influential experiments have indeed studied the partitioning of sulfur between molten silicate and nearly stochiometric FeS (Fei et al. 1995; Li and Agee 1996; Li and Agee 2001; Holzheid and Grove 2002) leading to the implicit analogy between chalcophile and siderophile tendencies. This useful simplification, however, ignores the strongly non-ideal thermodynamic behavior of the Fe-S system, which implies that the energetics of sulfur in molten Fe metal with low S-content cannot simply be extrapolated from that of molten FeS (through a dilution factor, e.g. Holzheid and Grove 2002). The available experimental data bearing on S-partitioning between molten silicate and liquid, S-poor, Fe metal are scarce (Ohtani et al. 1997; Kilburn and Wood 1997; Rose-Weston et al. 2009; Li and Agee 2001). These data are reported in Fig. 10. The simplest thermodynamic treatment that best approximates the partitioning of sulfur between Fe metal and molten silicate can be formulated as (e.g., Gaillard and Scaillet 2009):

$$\mathrm{S}^{\text{metal}} + \mathrm{O}^{2\text{-silicate}} = \mathrm{S}^{2\text{-silicate}} + \frac{1}{2}\mathrm{O}_2 \tag{1}$$

This equilibrium is the sum of the reaction of sulfur equilibrium between silicate melt and gas phases which reads as (O'Neill and Mavrogenes 2002):

$$\mathrm{S}^{2\text{-silicate}} + \frac{1}{2}\mathrm{O}_2^{\text{gas}} = \frac{1}{2}\mathrm{S}_2^{\text{gas}} + \mathrm{O}^{2\text{-melt}} \tag{2}$$

and the dissolution reaction of sulfur gas within molten Fe metal (Wang et al. 1991):

$$\frac{1}{2}\mathrm{S}_2^{\text{gas}} = \mathrm{S}^{\text{metal}} \tag{3}$$

Combining reactions (2) and (3), and using appropriate equilibrium constants, we can write (Gaillard and Scaillet 2009):

$$\ln\frac{X_{S2-}^{\text{silicate}}}{X_S^{\text{metal}}} = -\ln K_3 + \ln\gamma_S^{\text{metal}} + \ln C_S^{\text{silicate}} - \frac{1}{2}\ln f_{O2} \tag{4}$$

Following the notation of O'Neill and Mavrogenes (2002) the fraction of sulfur in the molten silicate $X_{S2-}^{\text{silicate}}$ is expressed in wt ppm whereas the fraction of sulfur in the molten Fe-metal X_S^{metal} is given in atomic fraction. The sulfur capacity (C_s^{silicate}) is a concept introduced

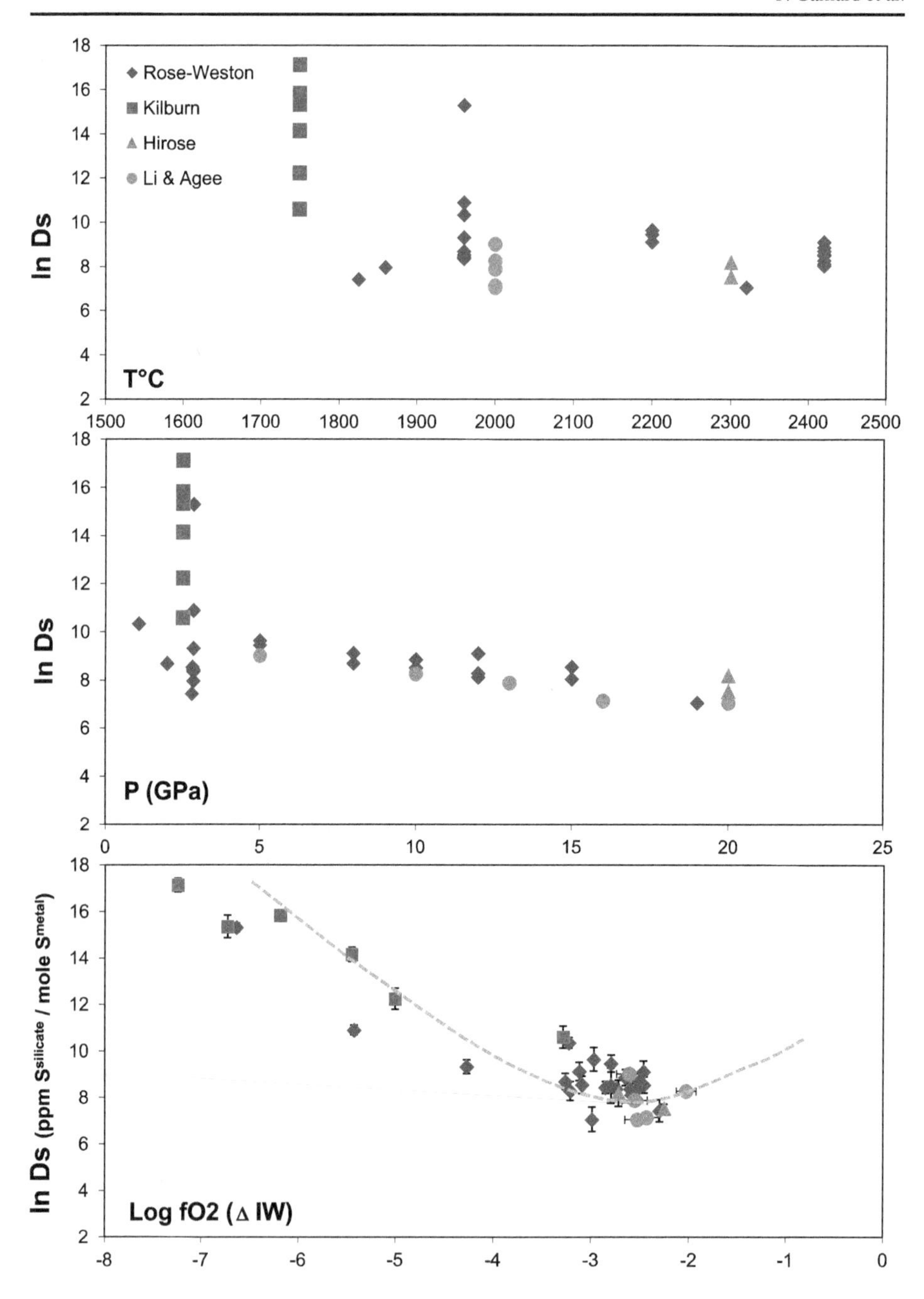

by metallurgists (Fincham and Richardson 1954), defining the ability of silicate melts to dissolve sulfur (by reaction (2)) and its variability with melt composition. C_s^{silicate} has been defined as (Fincham and Richardson 1954):

$$\ln C_S^{\text{silicate}} = \ln X_{S2-}^{\text{silicate}} + \frac{1}{2}\ln f_{O2} - \frac{1}{2}\ln f_{S2} \tag{5}$$

◀ **Fig. 10** Sulfur partitioning between molten silicate and molten Fe-metal (D silicate/metal). The fraction of sulfur in the silicate melt is in ppm-wt S and the fraction of sulfur in the metal is in molar fraction (see (4)). *Top panel* shows sulfur partitioning as a function of temperature and indicates that sulfur becomes slightly more siderophile as temperature increases. *Middle panel* shows sulfur partitioning as a function of pressure and indicates that sulfur become slightly more siderophile as pressure increases. *Bottom panel* shows sulfur partitioning as a function of oxygen fugacity _expressed in log-units relative to oxygen fugacity buffered by the iron-wustite, IW, assemblage_ and indicates that sulfur becomes increasingly lithophile as conditions are increasingly reducing. The combination of the three panels clearly indicates that oxygen fugacity (fO_2) is the prime parameter controlling sulfur partitioning between molten metal and molten silicate. Changes in fO_2 explains the large scatter that partitioning data otherwise shows when plotted as a function of pressure or temperature. In the *bottom panel*, the *dashed line* indicates sulfur partitioning calculated using (4). In agreement with existing experimental data, it predicts that S is increasingly siderophile as fO_2 increases but it also predicts an inversion of the trend as fO_2 is higher than IW-2. Above this fO_2 value, increasing FeO content in the silicate (with increasing fO_2 at metal saturation) implies that sulfur becomes increasingly lithophile with increasing fO_2. There is however, no experimental data existing in this range of redox conditions, which however corresponds to that expected for Mars core-mantle equilibration. Experimental data from Ohtani et al. (1997), Kilburn and Wood (1997) and Rose-Weston et al. (2009); Calculation after Gaillard and Scaillet (2009)

It is well established that the chemical parameter that exerts a prime control on sulfur solubility in silicate melts is their ferrous iron content (FeO). Ferrous iron-rich melts tend to dissolve more sulfur than those poor in FeO. The sulfur capacity, C_S^{silicate} for mafic and ultramafic melts has been extensively studied by O'Neill and Mavrogenes (2002) at 1400 °C. Ln K_3 is the equilibrium constant of reaction (3) whilst γ_S^{metal} is the activity coefficient of sulfur in the molten metal which, according to the interstitial model of Wang et al. (1991), incorporates the effect of temperature. At low S contents (i.e. $X_S^{\text{metal}} < 0.1$), the activity coefficient of S in Fe metal is in the range 1–0.7 decreasing down to 0.3 at $X_S^{\text{metal}} = 0.35$.

Equation (4) appears in the form of a partition coefficient between metal and silicate, a widely used concept in geochemistry to define the pressure and temperature dependence of partitioning properties. However the partition coefficient is here influenced by a large number of additional parameters that are also interdependent: (i) the strongly non-ideal behavior of S in Fe-metal that makes γ_S^{metal} strongly dependent on the bulk S content; (ii) the dependence of (4) on fO_2; (iii) the dependence of C_s on FeO^{melt}; (4) the dependence of FeO^{melt} on fO_2 at Fe-metal saturation. In several studies, the fO_2 dependence of sulfur partitioning between molten silicate and metal has been ignored because of the implicit assumption that liquid FeS and liquids in the S-poor region of the Fe-FeS binary are energetically broadly similar (Holzheid and Grove 2002; Rose-Weston et al. 2009). In fact, metal-silicate S-partitioning data (Fig. 10), display a considerable scatter that is poorly related to variations in pressure and temperature. Figure 10 also shows that increasing temperature and pressure makes sulfur slightly more siderophile, but this effect remains small, even debatable if error bars are considered.

In contrast, when plotted as a function of oxygen fugacity (determined by the equilibrium $\text{Fe} + \frac{1}{2}\text{O}_2 = \text{FeO}$), the data define a single clear trend indicating that sulfur becomes more siderophile as conditions become increasingly oxidizing. The fundamental reason of such a trend is that sulfur dissolves in silicate melts as S^{2-} whereas it is in the S^0 form in the molten Fe-metal. Redox conditions for core-mantle equilibration for Mars may be oxidizing (IW-1.5 for a Martian mantle with 18–20 wt% FeO) relative to those for Earth (IW-2.2 for a Earth's mantle with 8 wt% FeO) (see Righter and Drake 1996). Although no experimental data exist at fO_2 relevant to Mars core-mantle equilibration, equation (4) predicts that sulfur should be less siderophile, as indicated by the dashed lines in Fig. 10.

Figure 11 shows the expected relationship between sulfur in the mantle and sulfur in the core using (4) at 2100 °C–14 GPa. For Mars, these P–T conditions of core-mantle equilibration are after Righter and Chabot (2011) (see also Debaille et al. 2009). The strongly

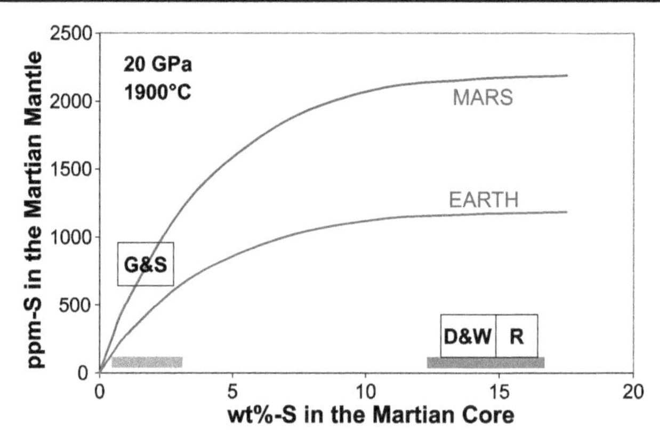

Fig. 11 Likely ranges of sulfur content in the core and mantle of Mars. The *red and blue curves* show sulfur content calculated from the metal–silicate partitioning (4) at 1900 °C, 20 GPa and considering variable bulk sulfur content in the core-mantle system. The *red curve* shows calculation for a Martian core-mantle case (mantle FeO = 18 wt%) and the *blue curve* illustrate an Earth-like core-mantle case (FeO = 8wt%). The *boxes* indicate S-content suggested by previous studies: G&S refers to the estimation of Gaillard and Scaillet (2009) on S content in the mantle and core of Mars; a sulfur content in the Martian core in the range of 14–18 wt% has been suggested by cosmochemical constraints (D&W, Dreibus and Wanke 1985; Wänke and Dreibus 1994: averaging ~14.2 %) or inferred from recent inversions of geodetic data (R, Rivoldini et al. 2011; estimated at 16 ± 2 %). Variations by a factor of 8 on estimations of sulfur content in the core between G&S and R only translate into a factor of 3 in mantle S content because of the strongly non-ideal activity-composition of sulfur in molten Fe-metal

non-linear relationship is due to the non-ideal behavior of sulfur in molten iron. If we adopt a sulfur content in the Martian core to be in the range of 14–18 wt%, as suggested by cosmochemical constraints (Dreibus and Wanke 1985; Wänke and Dreibus 1994: averaging ~14.2 %) or as inferred from inversion of recent geodetic data (Rivoldini et al. 2011; estimated at 16 ± 2 %), the sulfur content in the Martian mantle is calculated to be 2000–2200 ppm. This corresponds to 10–20 times more sulfur than in the Earth's mantle (assuming S-content in the Earth's mantle at 120–250 ppm after Dreibus and Palme 1996; McDonough and Sun 1995; Allegre et al. 2001). It is interesting to note that 15 wt% sulfur in the Martian core corresponds to the eutectic composition at the pressure of the core-mantle boundary (23 GPa) (as defined by Fei et al. 1997; see also Morard et al. 2008). As this eutectic has a very low temperature (1400 K), we could expect that a Martian core with 15 wt% sulfur is presently liquid (Stewart et al. 2007). This would also corroborate gravity constraints that imply an entirely fluid Martian core (Marty et al. 2009).

There are however several reasons to question existing estimates of the sulfur content in the Martian core. Cosmochemical inferences are strongly model-dependent, especially for volatile and siderophile elements, and it is difficult to assess uncertainties. A major factor in suggesting a sulfur rich core is the depletion of chalcophile elements in martian meteorites, suggesting they were stripped into a S-rich core (Righter and Humayun 2012). However, sulfide fractionation from S-rich Martian basalts may also lead to depletion in chalcophile elements and interpretation of element depletions in terms of core-mantle differentiation may be non-unique as increasingly recognized for the Earth (Righter et al. 2006). Geodetic constraints are less model-dependent, but the assessment of the Martian core sulfur content by Rivoldini et al. (2011) assumes ideal mixing for volume properties of sulfur in iron molten core. Recent experimental data collected at 4 GPa (Nishida et al. 2008) on liquid Fe-S mixtures show that the density of molten iron is weakly affected by the addition of up to 20 at% sulfur, which indicates strongly non-ideal volume of mixing. Thermodynamic analyses of the effect of increasing pressure on the liquids of the Fe-rich side of the Fe-FeS system (Buono and Walker 2011) indicate that pressure would tend to make mixing properties more ideal (e.g. Gibbs free energy of mixing between Fe and FeS), but it is actually difficult to retrieve information about the partial molar volume of S from such thermodynamic treatment. If the volume mixing properties of Nishida et al. (2008) still hold at higher pressure (see discussion in Buono and Walker 2011), it is expected that density would be relatively insensitive to sulfur content in the core of Mars, making the Rivoldini et al. (2011) assessment provisional (maximum S-content). In contrast, if non-ideal volume of mixing of the Fe-S system vanishes at pressures higher than 4 GPa (i.e. 10 GPa), the core of Mars is expected to be S-rich and probably still fully molten (Stewart et al. 2007; Rivoldini et al. 2011).

Gaillard and Scaillet (2009) preferred to ignore the high sulfur content of Mars suggested by cosmochemical considerations and instead assumed that the bulk sulfur content of terrestrial bodies are similar. If we consider that Mars has a sulfur content similar to bulk Earth, metal-silicate partitioning calculations shown in Fig. 11 would then indicate likely sulfur contents of 700–900 ppm for the Martian mantle and 1–3 wt% for its core. We can therefore conclude that the sulfur content of the Martian mantle must be in the range of 700–2000 ppm. It is worth noting that, even if large uncertainties for the sulfur content of the core persist, the uncertainties for the sulfur content of the Martian mantle are comparatively much smaller, because of the flattening of the metal-silicate partitioning for high sulfur content. All in all, the S content of the Martian mantle remains well above the sulfur content on the Earth's mantle (Table 2).

4.2 Mantle Melting, Basalts and the Crust

Adopting a sulfur content in the mantle of Mars in the range 700–2000 ppm permits a mass-balance calculation for the maximum content of sulfur in basalts formed upon mantle melting. Assuming 10–20 % of partial melting and considering that sulfur is perfectly incompatible during melting, then between 3,500–18,000 ppm sulfur can be expected in primary mantle Martian basalts (0.35 to 1.8 wt% S). However, the sulfur content in basaltic liquids produced by mantle melting is limited by the saturation in sulfide (FeS). The sulfur content of basalt formed upon mantle melting at sulfide saturation therefore provides the maximum sulfur content that basalts can convey upon ascent to the surface.

The equilibrium between sulfide and basaltic melts can be written as:

$$\mathrm{FeS}^{\mathrm{sulfide}} + \frac{1}{2}\mathrm{O}_2 = \mathrm{FeO}^{\mathrm{basalt}} + \frac{1}{2}\mathrm{S}_2 \qquad (6)$$

O'Neill and Mavrogenes (2002) reformulated the above equilibrium and combined it with (5) so that the sulfur content of basalt at sulfide saturation can be formulated independently of both f_{O2} and f_{S2}:

$$\mathrm{Ln}(X_{S2-})^{\mathrm{ppm\text{-}basalt}} = \Delta G^{\circ}_{(6)}/[\mathrm{RT}] + \ln C_s - \ln a_{\mathrm{FeO}} \qquad (7)$$

The sulfur capacity C_s is the same as in (5), a_{FeO} stands for the activity of ferrous iron in silicate melts (Gaillard et al. 2003a) and $\Delta G^{\circ}_{(6)}$ is the Gibbs free energy of equilibrium for (6).

Such a concept (i.e., that the sulfur content at sulfide saturation does not depend on fO_2) remains valid provided that one considers only ferrous iron in (7) and not total iron as obtained using standard analytical procedures. Accordingly, (7) is strictly fO_2 independent only when conditions are sufficiently reducing (<FMQ) to neglect the presence of ferric

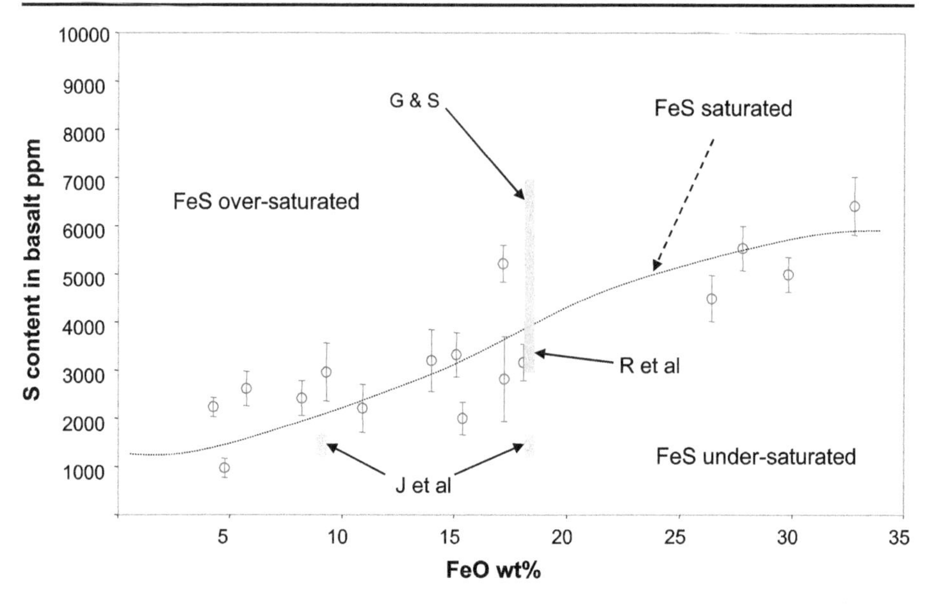

Fig. 12 Sulfur content in basalts saturated in FeS as a function of FeO content in basalts. The *red dots* are experimental data taken from Righter et al. (2009) _ 0.1 MPa to 800 MPa, 1200–1600 °C. The *dashed line* is a line drawn for visual guidance. *Colored boxes* represent estimations of sulfur content in Martian basalts by Gaillard and Scaillet (G&S), Righter et al. (R), and Johnson et al. (J et al.). The likely range of Martian sulfur content defined by the saturation in FeS upon mantle melting is 3500 ppm (this study)

Table 2 Sulfur content in the Martian core, mantle and basalts as deduced from partitioning modeling and comparison with Earth. For Mars, estimations are from this study for the Earth, see text for references

Reservoirs	Mars	Earth
Basalt/basaltic crust	3000–5000 ppm	1000–1500 ppm
Mantle	700–2000 ppm	100–250 ppm
Core	2–16 wt%	0–5 wt%

iron in the melt, which is indeed likely for Martian basalts. Keeping this restriction in mind, the sulfur content of Martian basalts at sulfide saturation is therefore a function of pressure and temperature, but also a function of melt composition, particularly its iron content.

Gaillard and Scaillet (2009) used (7) to show that Martian basalts can contain up to 4000–7000 ppm sulfur upon mantle melting under the P–T conditions inferred from multiple saturation experiments (Monders et al. 2007; Musselwhite et al. 2006). Experiments done by Righter et al. (2009), confirm these calculations, suggesting sulfur contents at the low end of the Gaillard and Scaillet (2009) calculated range (Fig. 12). These studies illustrate how the sulfur content of basaltic compositions increases as a function of total iron content (reported as FeO). Shergottite-like basaltic compositions have FeO in the range 16–20 wt%, which compares favorably to an estimate of the FeO content for the overall Martian crust of 18.2 % (Taylor and McLennan 2009). According to the experiments of Righter et al. (2009) (1–8,000 bar and 1200–1500 °C; that are shown in Fig. 12 by the empty red circles), such iron-rich melts can dissolve up to 3000–5000 ppm sulfur at sulfide saturation. Considering the likely P–T conditions of mantle melting on Mars, as inferred by Monders et al. (2007) and Musselwhite et al. (2006), Righter et al. (2009) concluded that primary martian basalts must contain 3000–4000 ppm dissolved S. Under similar conditions, Earth's mantle-derived basalts can dissolve about 1000–1500 ppm, which was the value adopted by Johnson et al. (2008) as the sulfur content in Martian basalts. The latter value would imply that Martian basalts are undersaturated in FeS during mantle melting, a view which conflicts with simple mass balance arguments. Indeed, given the Martian mantle sulfur concentration inferred above (700–2000 ppm S), and assuming that 10–20 % of mantle partial melting is needed to produce Martian basalts, Martian basalts with 4000 ppm S would not exhaust the mantle in sulfur.

The Martian crust is dominated by basaltic compositions and Taylor and McLennan (2009) concluded that a large fraction of the crust ($\sim$70–90 %) was "primary" and thus formed or was strongly influenced by magma ocean processes (Elkins-Tanton et al. 2005). The sulfur content of the Martian crust can therefore be estimated via the concept of sulfur content at sulfide saturation. We therefore conclude that the Martian crust has a bulk sulfur content of 3500–4000 ppm. This should nevertheless be regarded as a maximum S-content.

5 Volcanic Degassing, Redox State and Water Content of Martian Basalts

Upon magma ascent, the decrease in pressure acts against FeS saturation, whereas cooling should promote FeS stability (Mavrogenes and O'Neill 1999; Holzheid and Grove 2002). For fast rising magmas, such as hot martian basalts, cooling must be limited during ascent. We therefore make the simplifying assumption that all sulfur dissolved in the basalt formed at mantle conditions is entirely conveyed to the surface.

5.1 Degassing Trends in the C-H-S-O System

Based on the above considerations, we adopt the average sulfur content of primary Martian basalts as 3500 ppm, which is admittedly a conservative estimate, and further assume that this amount of sulfur is conveyed by the melt throughout the Martian crust. This simplification may also apply to $\sim$80 % of the primary basaltic Martian crust likely formed as a result of magma ocean processes (e.g., Elkins-Tanton et al. 2005; Taylor and McLennan 2009). The amount of sulfur eventually released into the atmosphere during magma degassing is complex and depends on a variety of parameters that we discuss below. Degassing of a magma ocean has been addressed in Gaillard and Scaillet (2009)

Fig. 13 The efficiency of sulfur degassing from basalts as a function pre-eruptive water content and oxygen fugacity. Pre-eruptive sulfur content in basalts is shown by the *horizontal pink line* at 3500 ppm S. Undegassed sulfur contents in basalts are represented by *horizontal bars* whose top and bottom values respectively correspond to 0.1 and 0.01 bar of degassing conditions. Values are taken from Table 2. *Top panel* shows undegassed sulfur in basalts as a function of pre-eruptive water contents for two pre-eruptive oxygen fugacities (IW; FMQ-1.7). *Bottom panel* shows undegassed sulfur content in basalts as a function of pre-eruptive oxygen fugacity for three pre-eruptive water contents (0.1; 0.2; 0.3 wt% H2O). Pre-eruptive CO_2 content in basalts varies with oxygen fugacity in agreement with experimental constraints on CO_2 content in basalts at graphite saturation (Stanley et al. 2011). *Horizontal boxes* show the range of sulfur contents reported in SNC meteorites (see Meyer (2008) _1300–2600 ppm_ and the sulfur content in basaltic shergottites as evaluated in Righter et al. (2009) _1600 ppm_). These sulfur concentration ranges must be considered with cautions as many of the SNC rocks are cumulates and their sulfur content is poorly representative of that of the parental basaltic melts (Lorand et al. 2005). Nevertheless, degassed basalts with sulfur content at 1600 ppm are obtained if degassing from melts with pre-eruptive water content of 0.4 wt% and pre-eruptive fO_2 at IW and/or if degassing from melts with pre-eruptive water content of 0.2 wt% and pre-eruptive fO_2 at FMQ-1.7. Because FMQ-1.7 represents the upper most fO_2 ranges for shergottites, it implies a minimum water content of 0.2 wt% for their parental melts if we admit that 1600 ppm S is a reasonable estimates for degassed mafic basalts on Mars's surface ▶

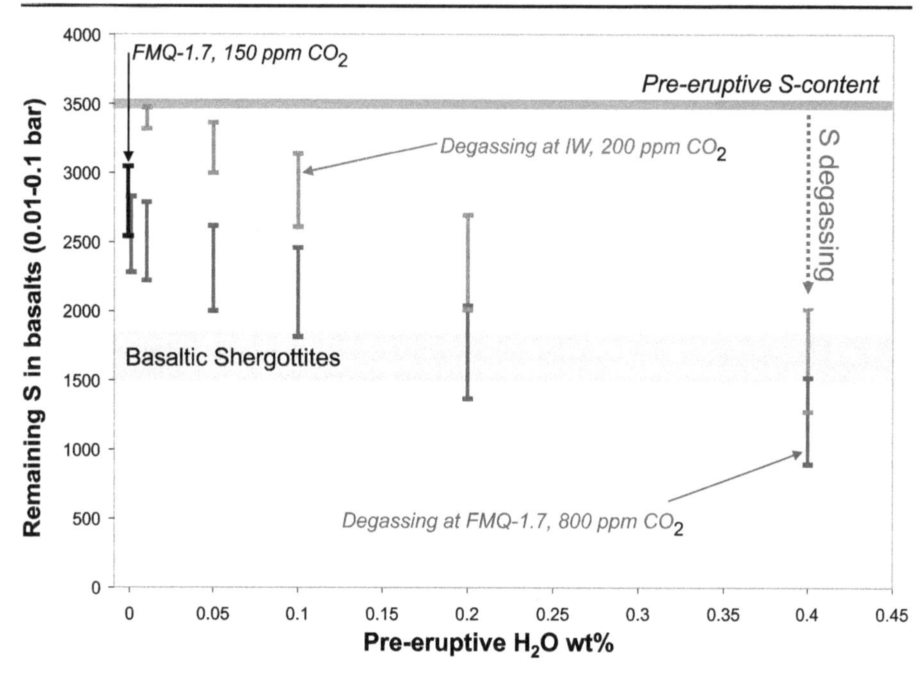

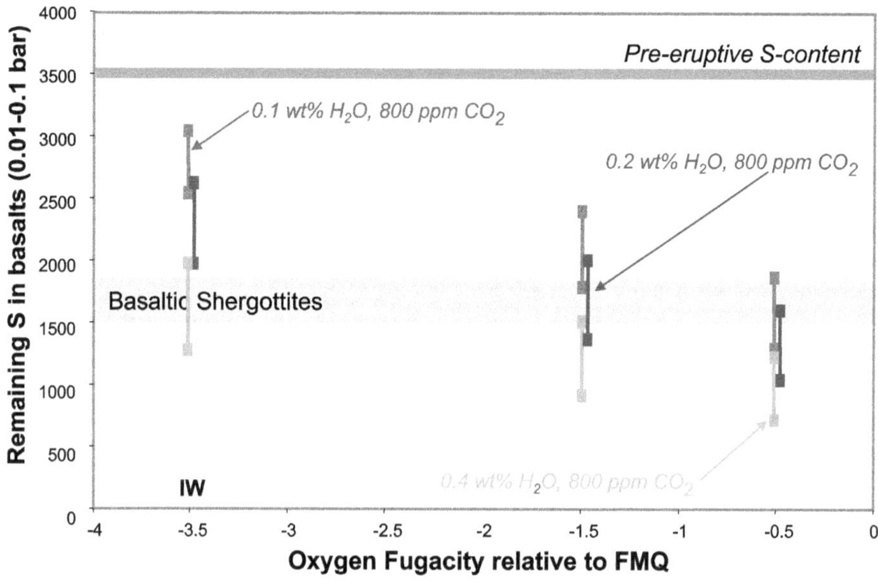

and shown to be unimportant for S-species and instead dominated by CO and H_2 species. We concentrate hereafter on extrusive post-magma ocean magmatism, as defined by Craddock and Greeley (2009), which is dated from the middle Noachian to late Amazonian. Our approach differs from Righter et al. (2009) who considered that most degassed shergottites have between 1500–2000 ppm S and subtracted this number from the sulfur content of primary mantle basalts (3500–4000 ppm) to obtain averaged degassed sulfur. Here, following Gaillard and Scaillet (2009), we simulate equilibrium degassing of primary shergottitic basalts. During degassing, a variable amount of sulfur is lost into the atmosphere, in equilibrium with different gas species, whose relative abundances depend on pressure, oxygen fugacity, water content, as observed on Earth (Gaillard et al. 2011).

Since volatile solubilities are primarily pressure dependent (e.g., Behrens and Gaillard 2006), the pressure at which degassing occurs is of prime importance. We can, for example, distinguish intrusive magmas, which degas at depth, from extrusive ones that release gases directly into the atmosphere. More degassing is expected for the extrusive regime than for the intrusive one, but variations in atmospheric pressure, which possibly occurred throughout Mars' history, must have also greatly influenced the nature and amount of volatiles expelled by sub-aerial volcanic eruptions. We could also consider the influence of the eruptive dynamics as addressed by Wilson and Head (1994). If explosive basaltic eruptions have been facilitated on Mars compared to the Earth due to low atmospheric pressure and lower gravity (Wilson and Head 1994), we may expect that the gas composition of subaerial emissions is inherited from melt-gas equilibria occurring at pressures different (higher) than the atmospheric pressure. Overall, it is probable that Martian eruptive dynamics are not so critical if volatile contents in Martian basalts are small (see below). Plinian or strombolian basaltic eruptions, as occurring in subduction settings on Earth, are associated with pre-eruptive volatile contents generally exceeding 3 wt% H_2O and more than 1 wt% CO_2 (e.g., Aiuppa et al. 2010). Evidence for such elevated concentration levels, typical of arc-volcanoes on Earth, have been so far lacking for Martian basalts.

Table 3, which provides calculations of the amount of volcanic volatiles emitted in ppm-wt of magma erupted (for a sulfur content of the primary basalt of 3500 ppm), illustrates the effect of the pressure of degassing (500 bar for intrusive magma emplacement; 1, 0.1, 0.01 bar for atmospheric degassing under an atmosphere of variable density). Calculations are performed here in the C-O-H-S system and the mass fraction of the following species are shown: CO_2, CO, H_2O, H_2, SO_2, S_2, H_2S. It is noteworthy that, in all cases, calculated CH_4 concentrations are negligible at the pressure-temperature of volcanic degassing even under strongly reduced conditions (Gaillard and Scaillet 2009). It appears that degassing occurring at elevated pressure (500 bar, intrusive magma) emits only C-species. Subaerial degassing occurring at variable pressures, from 1 to 0.01 bar, shows major differences. At 1 bar, little sulfur is emitted in comparison to degassing at 0.01 bar. This means that sulfur mostly remains in the lava in the case of eruption in an atmosphere of 1 bar, whereas if the same lava flows in an atmosphere at 0.01 bar, most of its sulfur will outgas. We emphasize here that such calculations are not aimed at equilibrating the composition of volcanic gases with that of the surrounding atmosphere. The atmosphere is not chemically participating in the process of volcanic degassing but it is physically controlling the final pressure of melt-gas equilibration. Equilibration of the melt with the composition of the atmosphere is impossible as the rate of redox equilibration and diffusion of volatiles into a cooling lava is far too slow (Behrens and Gaillard 2006; Pommier et al. 2010).

Table 3 also shows some chemical effects, which are also important and may interfere with the effect of degassing pressure. The most studied chemical parameter is oxygen fugacity. Under reducing conditions, similar to IW, sulfur is only moderately volatile, and it tends to remain in the basaltic melt even at low pressure. Under oxidizing conditions, similar to FMQ, within the oxygen fugacity range indicated by Martian meteorites (Herd et al. 2002; Herd et al. 2005), sulfur is more volatile and more sulfur can be degassed to the atmosphere (Gaillard and Scaillet 2009). The positive effect of oxygen fugacity on the efficiency of sulfur degassing at atmospheric pressure is shown in Fig. 13, which also illustrates the role of water: for low pre-eruptive water contents, sulfur degasses weakly whereas water-rich melts (by martian standards) efficiently lose much of their sulfur (Gaillard and Scaillet 2009). As

Table 3 Calculated compositions of volcanic gases in ppm wt of Martian magmas. $T = 1300$ °C. Gas compositions are computed following Gaillard et al. (2011), using solubility laws of Iacono-Marziano et al. (2012, in press) for H_2O-CO_2, the work of Gaillard et al. (2003b) for H_2 and the sulfur capacity formulation from O'Neill and Mavrogenes (2002). Speciation in the C-O-H-S gas phase is computed following Shi and Saxena (1992)

Pressure of degassing (bar)	500	1	0.10	0.01
Conditions of magma emplacement	Intrusive magmas	Subaerial ancient Mars	Subaerial recent Mars	Subaerial present-day Mars
IW, 0.01 wt% water, 0.02 wt% carbon dioxide				
H_2O ppm wt	0	3	15	24
H_2	0	0	1	3
H_2S	0	1	3	4
SO_2	0	0	3	20
S_2	0	2	21	165
CO	141	168	167	177
CO_2	75	96	98	82
IW, 0.1 wt% water, 0.02 wt% carbon dioxide				
H_2O	0	274	427	363
H_2	0	14	23	33
H_2S	0	49	70	50
SO_2	0	6	74	156
S_2	0	23	262	770
CO	141	134	138	164
CO_2	74	147	142	100
IW, 0.4 wt% water, 0.02 wt% carbon dioxide				
H_2O	7	2323	2302	1952
H_2	1	58	86	115
H_2S	1	334	244	131
SO_2	0	238	616	1027
S_2	0	339	957	1595
CO	142	93	114	140
CO_2	73	206	172	133
FMQ-1.4, 0.01 wt% water, 0.08 wt% carbon dioxide				
H_2O	0	9	34	39
H_2	0	0	1	2
H_2S	0	2	5	4
SO_2	1	166	332	404
S_2	1	176	548	1078
CO	112	181	262	365
CO_2	559	640	513	352
FMQ-1.4, 0.1 wt% water, 0.08 wt% carbon dioxide				
H_2O	1	389	536	462
H_2	0	7	16	26

Table 3 (*Continued*)

Pressure of degassing (bar)	500	1	0.10	0.01
Conditions of magma emplacement	Intrusive magmas	Subaerial ancient Mars	Subaerial recent Mars	Subaerial present-day Mars
H_2S	0	63	69	44
SO_2	0	233	480	625
S_2	0	251	744	1338
CO	78	132	188	252
CO_2	361	435	346	247
FMQ-1.4, 0.2 wt% water, 0.08 wt% carbon dioxide				
H_2O	4	1076	1173	999
H_2	0	18	35	52
H_2S	1	154	126	72
SO_2	1	431	771	1053
S_2	0	417	966	1544
CO	98	169	237	309
CO_2	479	553	446	334
FMQ-1.4, 0.4 wt% water, 0.08 wt% carbon dioxide				
H_2O	16	2533	2512	2178
H_2	0	42	69	96
H_2S	3	298	200	102
SO_2	1	781	1324	1910
S_2	1	601	1140	1562
CO	113	184	251	320
CO_2	552	619	513	405
FMQ-O.5, 0.1 wt% water, 0.08 wt% carbon dioxide				
H_2O	1	469	605	538
H_2	0	6	14	23
H_2S	0	63	60	36
SO_2	14	783	1095	1319
S_2	4	503	1004	1521
CO	50	160	250	345
CO_2	692	739	599	448
FMQ-O.5, 0.2 wt% water, 0.08 wt% carbon dioxide				
H_2O	6	1164	1271	1112
H_2	0	15	30	46
H_2S	1	140	108	60
SO_2	276	975	1387	1766

Table 3 (*Continued*)

Pressure of degassing (bar)	500	1	0.10	0.01
Conditions of magma emplacement	Intrusive magmas	Subaerial ancient Mars	Subaerial recent Mars	Subaerial present-day Mars
S_2	31	599	1095	1542
CO	22	168	251	338
CO_2	749	725	595	459
FMQ-0.5, 0.4 wt% water, 0.08 wt% carbon dioxide				
H_2O	17	2669	2668	2350
H_2	0	36	61	87
H_2S	3	264	171	85
SO_2	15	1357	1981	2667
S_2	4	704	1137	1404
CO	50	170	243	318
CO_2	692	713	598	480

for the previous set of calculations, the starting S content is fixed at 3500 ppm. The impacts of changes in fO_2 or water content on the amount sulfur outgassed are both positive and comparable in magnitude. Reduced hydrated (IW, 0.4 wt% water) Martian basalts would degas sulfur as efficiently as oxidized dry ones (FMQ-0.5, 0.1 wt% water). Maximum sulfur degassing is obtained for hydrated-oxidized melts (0.4 wt% water, FMQ-0.5) which release into the atmosphere 75 % of their initial S-content. All this discussion is related to Fig. 13 showing degassing at 0.1 to 0.01 bar.

5.2 Defining fO_2-Water Content of Basaltic Shergottites

To date, there are insufficient data in hand to constrain the initial average water content of Martian basalts. Martian meteorites may have contained 140–260 ppm water (McCubbin et al. 2010) and there is evidence that their parental melts may have contained more water (McCubbin et al. 2012). One estimate (McSween et al. 2001) suggests a water content <1.8 wt%, which is likely an upper limit, and which falls within the range of terrestrial magmas: arc-basalts, among the most hydrated magmas on Earth, have about 3 wt% water (e.g., Wallace 2005); MOR-Basalts (Mid Ocean Ridge), which constitute 75 % of the Earth's volcanism, have 0.1 wt% water (e.g., Saal et al. 2002) and hotspot basalts have broadly 0.2–1 wt% (e.g., Dixon et al. 1997).

Since the efficiency of sulfur degassing is being essentially controlled by pre-eruptive water content and fO_2 conditions, it follows that constraints on the sulfur content of degassed Martian basalts can be used to infer the fO_2-water content prevailing during lava emplacement and degassing. To this end, Fig. 13 shows the range of sulfur content analyzed in shergottites (source data compiled Meyers 2008; see in addition Gibson et al. 1985; Zipfel et al. 2000; Lorand et al. 2005; and references in Righter et al. 2009). The figures yield a range from 1300 to 2700 ppm sulfur. A potential limitation of such an approach is that Martian meteorites may not represent the composition of a melt, but instead reflect the effect of cumulate processes. Righter et al. (2009) nevertheless considered that a basaltic shergottite with sulfur content at 1600 ppm, provides a reasonable estimate for degassed Martian basalts. If we assume that 1500–1800 ppm sulfur remains in the shergottite lava after degassing, the fO_2-water content conditions needed to yield such S-content after degassing at final pressure 0.1–0.01 bar are as follows: IW/0.4 wt% H_2O; FMQ–1.5/0.2 wt% H_2O; FMQ–0.5/0.1 wt% H_2O. Given that the upper range of fO_2 estimated for Martian meteorite (FMQ for nakhlites and chassignites; Herd et al. 2002; Herd et al. 2005) does not apply to shergottites, we can eliminate the uppermost fO_2 conditions. We hence conclude that water content of at least 0.2 wt% is required by the sulfur left over in Shergottite magma after degassing. Water content of 0.2 wt% is a minimum because it matches the uppermost fO_2 recorded for Shergottites. Mc Cubbin et al. (2012) also recently provided geochemical indications based on melt/apatite water partitioning that are consistent with water content for the parental melts of shergottites close to 0.2 wt%. If IW is representative of the redox state of shergottite magmas, then Fig. 13 indicates that 0.4 wt% water is needed: more water would be required if more reduced conditions prevailed, which cannot be excluded given that the lowermost fO_2 for shergottite parental melts is IW-0.5 (Herd et al. 2002). Therefore, the range of oxygen fugacity recorded for shergottites, which spans over 4-log units, can be used along with Fig. 13 to infer the corresponding variation in water contents of Martian primitive basalts, from 0.2 to 0.6 wt%, in the fO_2 range FMQ-1.5 to IW-0.5, respectively.

5.3 Changing fO_2 During Degassing of Basaltic Shergottites

There is no consensus on the origin of oxygen fugacity variations for Martian shergottites. Assimilation of crustal material has been suggested (Wadhwa 2001; Herd et al. 2002), but such oxidized crustal material remains to be identified. At the low oxygen fugacity of Martian basalts (relative to Earth's basalts), only a very small amount of ferric iron is present in the melt and the buffering capacity of such melts is low. Furthermore, varying oxygen fugacity from $IW - 0.5$ to $IW + 3$ as recorded in Shergottite rocks implies only moderate changes in ferric-ferrous ratio, with the implications that such fO_2 changes do not require significant redox transfers (or mass transfer of oxygen). In Fig. 14, the fO_2 changes (see also Burgisser and Scaillet 2007; Gaillard et al. 2011) resulting from the degassing of Martian basalts have been calculated for different initial water contents, and two representative initial fO_2 (IW; FMQ-1.5). The initially oxidized melt does not change much in fO_2 for degassing in the range 10–1000 bar, but for pressures lower than 10 bar, SO_2 degassing from S^{2-} in basalts causes a strong fO_2 decrease. The initially reduced melt exhibits a more complex pattern. A severe oxidation in the range 1–1000 bar is calculated and the magnitude of this oxidation correlates with pre-eruptive water content. This is due to the outgassing of water and CO_2 which both decompose into the fluid as H_2 and CO species respectively (which both have very low solubility in the melt, Gaillard et al. 2003b; Morizet et al. 2010). Degassing of H_2S, which is dominant in the pressure range 1–1000 bar (see also Zolotov 2003 for computation of the gas phase) produces no effect on fO_2 (Gaillard et al. 2011). At pressures lower than 10^{-1} bar, sulfur degassing as SO_2 (SO_2 being the dominant S-species in the gas, see also Zolotov 2003) decreases fO_2 as described above. It is noteworthy that reduction trends associated with degassing of sulfur may well explain mineralogical and geochemical observations in nakhlites too (Righter and Humayun 2012; Chevrier et al. 2011). It thus appears that the simple process of volatile degassing upon magma ascent implies fO_2 variations that reproduce the range recorded by shergottites and to a lesser extent in nakhlites. In essence, such results can be used to argue that all shergottites initially derived from parental melts at an fO_2 close to IW and with water and sulfur contents of 0.4–0.6 wt% and 0.35 wt%, respectively. McCubbin et al. (2010) reported a

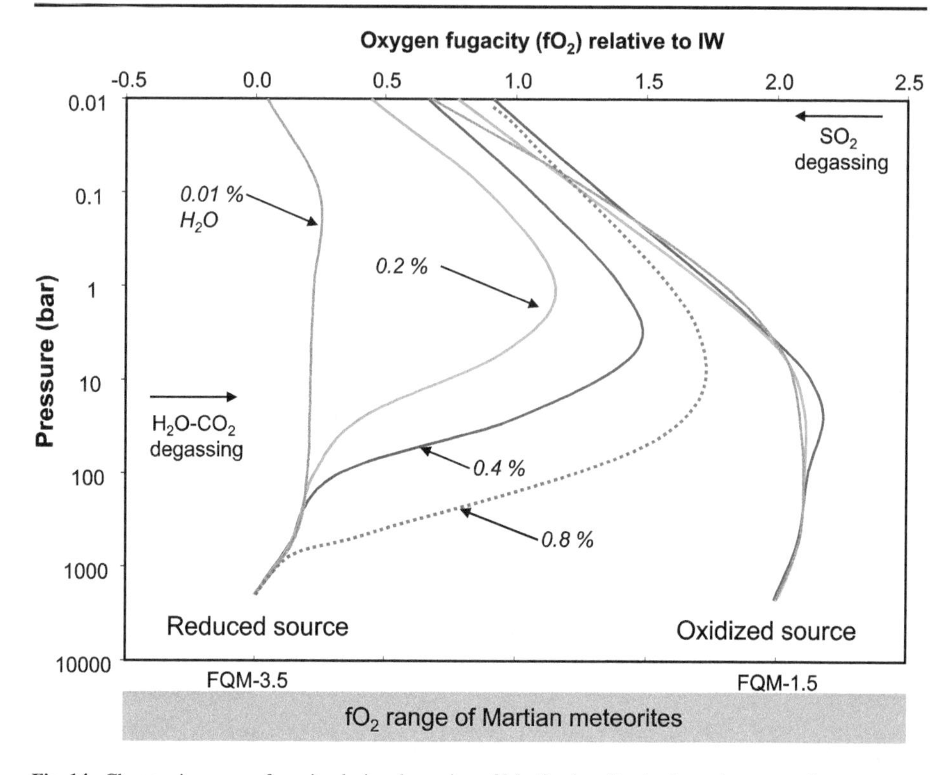

Fig. 14 Changes in oxygen fugacity during degassing of Martian basalts. As degassing occurs in response to decompression, the oxygen fugacity changes are shown as a function of pressure. Two conditions of pre-eruptive oxygen fugacity are considered: IW, the lower range, and FMQ-1.7, the upper range of fO_2 (for Shergottites). From these two initial fO_2 conditions, we also computed several possible pre-eruptive water contents (0.01 to 0.8 wt%). Pre-eruptive CO_2 contents are taken following Stanley et al. (2011). Degassing from oxidized sources produces almost no effect on fO_2 as melts rise through the crust, but near venting conditions, i.e. 2–5 bars, fO_2 strongly decreases as sulfur degases as SO_2. Degassing from reduced sources produces a strong fO_2 increase at crustal depth, which correlates with the pre-eruptive water contents. At venting conditions, SO_2 degassing decreases fO_2. See Burgisser and Scaillet (2007), Gaillard and Scaillet (2009) and Gaillard et al. (2011) for methods

range of water content, 0.4–0.8 wt%, for Chassigny parental magma that are broadly compatible with such estimates. Nakhlites are slightly more oxidized than shergottites and might have experienced degassing with similar impacts on redox state (e.g. SO_2 degassing leading to reduction). It is, however, difficult to provide any estimates of redox state and water content of their source regions or parental melts. Our approach, which only assumes equilibrium degassing, is relatively robust and the resulting inverse approach can account for the available observations. Equilibrium degassing, like equilibrium crystal fractionation or equilibrium melting (Baratoux et al. 2011), provide us with essential constraints as it pinpoints which chemical fractionation trend should have occurred under the sole assumption of thermodynamic equilibrium. It constitutes an essential approach since many of the relevant chemical parameters controlling Martian magmatic processes (e.g., water content, fO_2, source vs. shallow status) are poorly known, largely because of the paucity of direct petrological observations, and also because Martian meteorites may not be straightforwardly reflective of Martian magmatic processes (McSween et al. 2009).

6 Sulfur Emissions from Volcanic Vent to Exosphere

6.1 Sulfur Degassing and Speciation: Ancient vs. Recent Mars

The calculations described above indicate that a melt with 0.35 wt% sulfur (this study), 0.4 wt% water (this study), 200 ppm CO_2 (Stanley et al. 2011) and fO_2 close to IW (this study) may be representative of primary shergottitic basalts. During subaerial volcanism, such melts degas mixtures whose composition depends significantly on venting pressures (Zolotov 2003; Gaillard and Scaillet 2009). Ancient volcanism/magmatism on Mars, such as that associated with crust formation, was voluminous (Taylor and McLennan 2009), intense and likely to have conveyed to the atmosphere an amount of gas sufficient to produce an atmospheric pressure close to 1 bar (Grott et al. 2011). On the other hand, present-day atmosphere on Mars has a low pressure and the shift from a dense to a tenuous atmosphere remains enigmatic (Barabash et al. 2007).

Shergottites (and thus the various parameters inferred above) might not be representative of the Martian crust as a whole (McSween et al. 2009; Taylor and McLennan 2009). Nevertheless, calculated gas compositions (Table 3 and Fig. 15 as a function of pressure), reveal several trends that are independent of the pre-eruptive melt chemical (volatile-free) features. For instance, degassing at a pressure slightly above but of similar magnitude to present day conditions (0.01 bar) produces volcanic gases that are dominated by sulfur species, even if the absolute amount of S emitted per gram of lavas erupted remains small at low fO_2 and low pre-eruptive amount of water (Table 3). Furthermore, sulfur emissions in a tenuous atmosphere (0.01 bar) could have been greatly increased if either basalt water content or its oxidation state, or both, were higher. In detail, an increase in both fO_2 and water content also produces an increase of SO_2 with respect to H_2S and S_2 species.

If degassing occurred at 1 bar, the total sulfur species ($SO_2 + S_2 + H_2S$) amount to $\ll$1000 ppm whatever the conditions of water content and fO_2. In most cases, at 1 bar degassing conditions, sulfur is a minor component of the fluid phase and H_2S is the most abundant sulfur species on a molar basis (Fig. 15, Table 3). Such a high atmospheric pressure scenario applies to the earliest and intense phase of magmatism and degassing that triggered formation of the basaltic crust (Grott et al. 2011). It follows that sulfur emissions during this earliest and abundant volcanic phase on Mars were low with most sulfur being emitted as H_2S (Gaillard and Scaillet 2009; see also Zolotov 2003). We may also consider that a large part of this early magmatism was intrusive and therefore degassed at crustal pressures, even if this is difficult to demonstrate. As shown in Fig. 15 and Table 3, intrusive magmatism is characterized by very low sulfur emissions, all sulfur being expelled as H_2S, whereas the CO and CO_2 emissions are, in contrast, similar irrespective of whether extrusive or intrusive magmatism occurs. Further complications for C-species may arise if magma cooling at depth triggers graphite saturation, but this should only moderately alter the above conclusion on extrusive vs. intrusive degassing of CO and CO_2. It seems likely therefore that the earliest pre-Noachian Martian magmatism associated with the formation of the crust, while voluminous, was associated with only moderate sulfur yields to the atmosphere because conditions of degassing were unfavorable to the release of S-species, and the small amount of S released was emitted as H_2S. In contrast, C-species and water-species, probably dominated early magmatic emissions.

The Tharsis region contains, by mass, a significant fraction of the volcanic material on Mars (Phillips et al. 2001), yet the outgassing of this structure is difficult to quantify because it is unknown what fraction of Tharsis' mass was emplaced as extrusive versus intrusive material (Phillips et al. 2001). Furthermore, while volcanoes in Tharsis may have remained

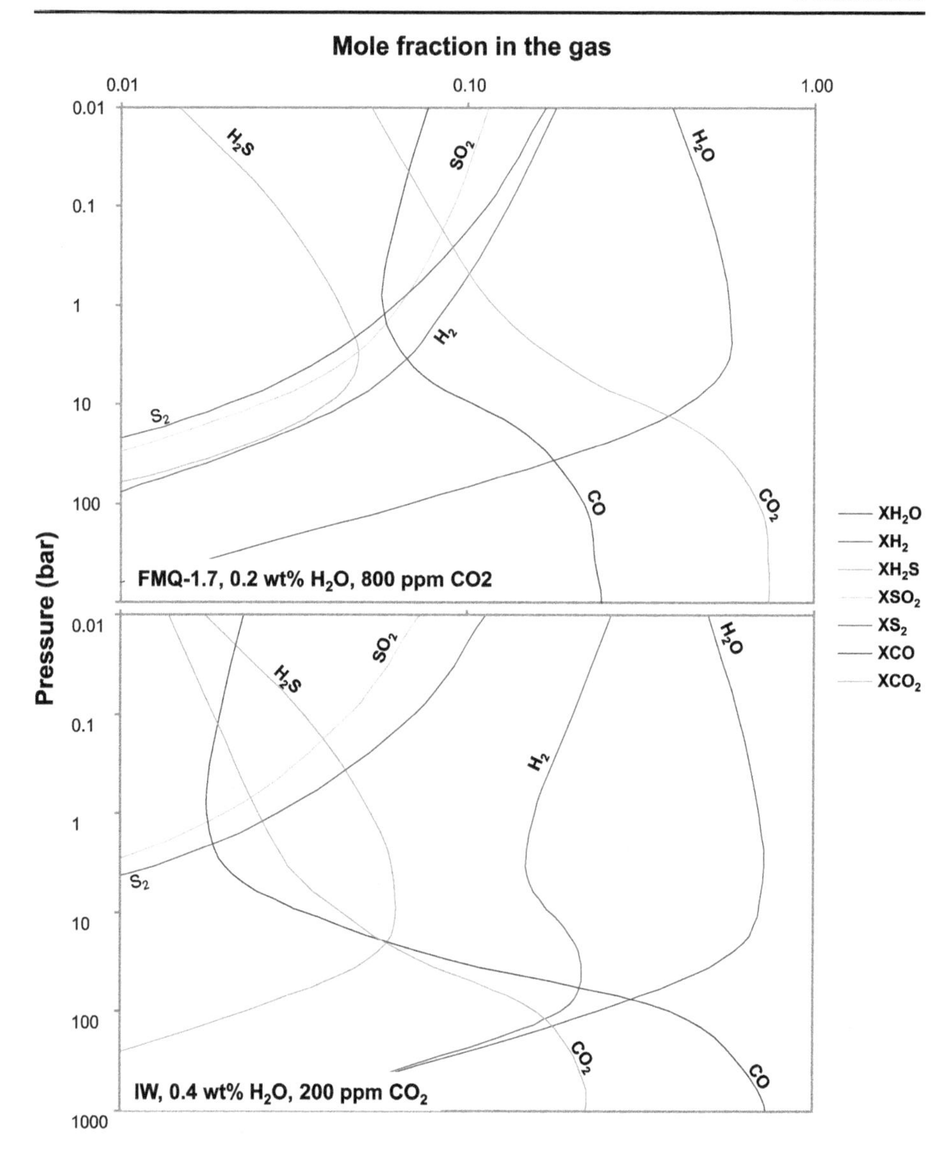

Fig. 15 The composition of Martian volcanic gases as a function of pressure. Two simulations of basalt degassing in the COHS system are shown, representing two pre-eruptive conditions (*Top panel*: FMQ-1.7, 0.2 wt% H_2O, 800 ppm CO_2; *Bottom panel*: IW, 0.4 wt% H_2O, 200 ppm CO_2). These choices are based on observations in Fig. 4 that such pre-eruptive conditions conduct to degassed basalts with 1600 ppm S. Globally, C-species (CO_2 and CO) dominate at pressure higher than a few tens bar. At lower pressure, water-species (H_2O and H_2) dominate. Both simulations show that total sulfur species ($SO_2 + S_2 + H_2S$) concentration in the gas increases as pressure decreases. In detail, however, there is a fundamental change in sulfur speciation at pressure of 0.2–3 bar depending on pre-eruptive conditions. At higher pressure, nearly all sulfur is present as H_2S and at lower pressure, all sulfur is in SO_2 and S_2 forms. This shift in S-speciation is well-known from thermodynamic studies on basalt degassing (see Gaillard et al. 2011 and references therein)

active far beyond the intense early period of magmatism forming >90 % of the crust (Greeley and Schneid 1991; McEwen et al. 1999; Phillips et al. 2001; Fassett and Head 2011;

Craddock and Greeley 2009), it is not well known how much magmatism occurred in this province on early Mars. In fact, due to the low erosion rates that have occurred on Mars since the Hesperian (Golombek et al. 2006), much more is known about younger volcanism than is for the most ancient volcanic crust (Craddock and Greeley 2009). If such subaerial volcanism occurred in an atmosphere similar to that of the present-day, then venting pressures in the range 0.1–0.005 bar are expected. Under such conditions, much more sulfur is introduced into the atmosphere by volcanic degassing than at high atmospheric pressure conditions ($\geq$1 bar). Table 3 indicates that on average 2000–2600 ppm S would be injected by subaerial basaltic eruption at low venting pressure. Furthermore, H_2S constitutes only a minor part of sulfur species (typically <5 % of total S) whereas SO_2 becomes important (typically 30–50 % of total S), if not dominant, for hydrous and oxidized conditions (75 % of total S). Sulfur in the S_2 form is also a major emitted species.

So far we have considered here that the fO_2 of Martian basalts did not significantly vary through time. The suggestion that oxygen fugacity during Martian mantle melting is buffered by graphite saturation (Stanley et al. 2011) calls for limited changes in fO_2 in the mantle source of basalts. At first sight, this contrasts somewhat with the fact that the redox states of Martian meteorites span over 4 orders of magnitude, which suggests that secular changes in Martian mantle redox state cannot be excluded. Secular variations in oxygen fugacity (mantle oxidation state increasing with time) could indeed induce a change in sulfur outgassing and speciation released by volcanic activity. However, Fig. 14 shows that degassing can also account for most of these fO_2 variations. Finally, Zolotov (2003) and Gaillard and Scaillet (2009) have shown that even for reduced basalts, SO_2 is the dominant degassed sulfur species released at low pressure (0.01 bar).

To summarize (Fig. 16), we can distinguish two eras with different amounts and types of volcanic sulfur injected into the atmosphere. (i) Ancient Mars (early to mid-Noachian), with higher fraction of intrusive magmatism and/or with extrusive emissions occurring in a relatively dense atmosphere and/or characterized by a more reduced mantle source, resulting in volcanic gases with low sulfur content, dominated by H_2S; and (ii) recent Mars, Hesperian to Amazonian, dominated by extrusive magmatism and/or degassing within a low density atmosphere and/or more oxidized basaltic eruptions that produced gases dominated by sulfur ($SO_2 + S_2 + H_2S$) with H_2S constituting a negligible fraction. This change in the regime of S delivered by volcanoes assumes the acceptance of several successive geochemical eras on Mars: The early clay period and the more recent sulfate period.

6.2 Estimates of Sulfur Fluxes

An important question is what is the total amount of sulfur that has been degassed over Martian geological history? Among other things, constraining this value constrains in turn the size of the sedimentary reservoir on Mars (McLennan 2012). On Earth, a significant part of near-surface sulfur is recycled back into the mantle by plate tectonic processes leading to a complex S-cycle (Canfield 2004). On Mars, the absence of plate tectonics imposes that degassed S remains at or near the surface, where it has progressively accumulated. Nevertheless, estimating the total S flux is a daunting task. As described above, S degassing speciation and efficiency is highly variable depending on atmospheric (venting) pressure, mantle oxygen fugacity and the overall composition of magmatic gases, all of which are uncertain and likely to have changed over geological time (e.g., Fig. 16). In addition, the S contents of the Martian mantle and mantle-derived magmas (e.g., representativeness of shergottite magmatism for all of Mars), magmatic production rates over geological time, and the relative roles of explosive versus effusive volcanism (Wilson and Head 1994) are still all imperfectly understood.

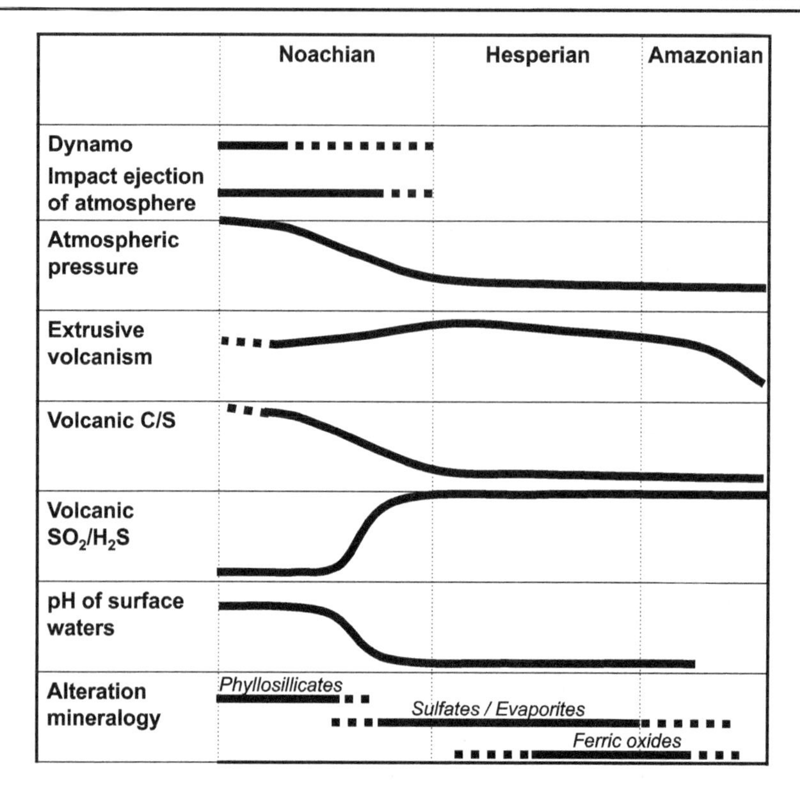

Fig. 16 Schematic flowchart summarizing the timing of volcanic degassing in a general Martian time-framework. Interruption of the core dynamo and impact ejection of the atmosphere are the causes of a pressure decrease of the initial (primordial) atmosphere. Extrusive volcanism is not believed to have significantly decreased from the early Noachian to the late Hesperian. But in response to a decrease of degassing pressure (atmospheric pressure from ca. 1 to ca. 0.05 bar), the C/S ratio of volcanic gases severely decreased and their SO_2 fraction became dominant over H_2S. By the processes described in part 6 and 7, this led to extensive acidification of surface waters, and may have influenced climate in a complex way. The phyllosilicate–sulfate shift may constitute the sedimentary record of such a change in gas compositions

A number of studies have recently estimated global Martian volcanic degassing rates, in each case constrained to be lower limits, and the wide range of derived values clearly illustrate some of the difficulties. Using various constraints on planetary degassing models (see original papers for details), Gaillard and Scaillet (2009) estimated that 5.4×10^{21} g of S had been degassed by subaerial volcanism over geological times whereas Righter et al. (2009) and Craddock and Greeley (2009) estimated S amounts are 1–2 orders of magnitude lower, from 4.5×10^{19} g to 1.7×10^{20} g, respectively.

The estimates of Righter et al. (2009) and Craddock and Greeley (2009) rely on volcanic production rates determined by Greeley and Schneid (1991), which do not consider the early Noachian. McEwen et al. (1999) estimated Noachian volcanic rates and suggested that total Martian volcanism may have been more than a factor of 7 higher than that given by Greeley and Schneid (1991) due to much higher Noachian rates of volcanism. This in turn would lead to comparable increases to the S-degassing estimates of Righter et al. (2009) and Craddock and Greeley (2009). Accordingly, recent estimates, if recalculated to be consistent with the volcanic production rates of McEwen et al. (1999), agree to within about an order of magnitude, falling in the range $\sim 5 \times 10^{20}$ g to $\sim 5 \times 10^{21}$ g. These values are still likely to be minimum estimates because volcanic production estimates of McEwen et al. (1999), corrected for a reasonable intrusive/extrusive ratio (<10), do not account for the entire Martian crust (Taylor and McLennan 2009), the formation of which must have contributed some sulfur to the surface.

Attempting to constrain the size of the Martian sedimentary mass, McLennan (2012) used an entirely different approach by assuming that the proportion of S degassed from the Martian mantle during crust formation and evolution was comparable to the Earth, which Canfield (2004) estimated to be ~ 11 %. The rationale for this approach is that during mantle melting, S is incompatible and Mars is more differentiated (i.e., larger proportion of planet's inventory of incompatible elements reside in the Martian crust compared to terrestrial crust) than Earth (Taylor and McLennan 2009). The overall efficiency of S extraction from crust to exosphere, integrated over geological time was simply assumed to be comparable for the two planets. Correcting for differing masses and likely S concentrations of the terrestrial and Martian primitive mantles, McLennan (2012) arrived at a value of 2.2×10^{22} g of degassed sulfur, about an order of magnitude greater than estimates based on magma degassing (Righter et al. 2009; Gaillard and Scaillet 2009; Craddock and Greeley 2009).

In the discussion above, it was concluded that the "best estimate" for Martian S-degassing rates were $<1{,}000$ ppm for high venting pressures (~ 1 bar), $\sim 2{,}000$–$2{,}600$ ppm for low venting pressures (<0.1 bar) and, at most, a few ppm S for the intrusive components of the magmatism. From the perspective of estimating a global flux, it should be kept in mind that these values were derived in a similar manner to those of Gaillard and Scaillet (2009). The magnitude of S degassing that took place during early magma ocean processes (that led to the primary crust), which could represent as much as 80 % of the total crust (Taylor and McLennan 2009), is shown in Gaillard and Scaillet (2009) to be of minor importance in comparison to the more recent extrusive volcanism.

What do these values imply for total sulfur degassing? A simple model, again likely to be a lower limit, can be constructed in which younger volcanism (Hesperian and Amazonian; 1.97×10^{23} g extrusive magma; Greeley and Schneid 1991) is taken to occur at low atmospheric pressure and thus result in 2,000 ppm S degassing and early volcanism (Noachian; 1.45×10^{24} g extrusive magma; Greeley and Schneid 1991; McEwen et al. 1999) is taken to occur at high atmospheric pressure resulting in 200 ppm S degassing. Since reliable values are unavailable for combined intrusive- and (especially) magma ocean-related magmatism (1.97×10^{25} g using crustal mass estimate of Taylor and McLennan 2009) we adopt another order of magnitude lower value of 20 ppm sulfur (calculated after Gaillard and Scaillet 2009). This leads to a lower limit total S degassing estimate of 1.1×10^{21} g, which is intermediate to the range for previous S-degassing calculations given above and about an order of magnitude less than the estimate of McLennan (2012).

6.3 Global Volcanic C/S Ratio: Mars vs. Earth

An important suggestion connecting the sedimentary records on Mars and its history of volcanic degassing is that ancient volcanic gases must have had high carbon/sulfur ratios possibly similar to those on Earth (see Symonds et al. 1994), whereas recent Martian volcanic gases had C/S ratios ≤ 0.1, which is about 10 times lower than their terrestrial counterparts. This simple mass balance consideration is probably the most likely explanation for the growing body of evidence for Martian surface chemistry dominated by sulfur whereas the Earth's surface is more balanced between hydrogen-carbon-sulfur related chemical processes (Berner 1995, 2005; Halevy et al. 2007; Gaillard and Scaillet 2009). This in turn suggests that the sulfur driven surface chemistry of Mars is a somehow relatively recent evolution, i.e. it operated when the atmospheric pressure dropped below 0.1 bar. If, as often proposed (Melosh and Vickery 1989; Jakosky and Phillips 2001), ancient Mars had a denser atmosphere, then the C/S ratios of magmatic gases were high and more similar to those of the modern Earth. The timing of interruption of the Martian core dynamo may have triggered the loss into space of the early dense atmosphere (Fassett and Head 2011). The Martian dynamo ceased early, but no consensus has been established about the exact timing [Early Noachian to late Noachian, see discussion in Fassett and Head (2011) and also Milbury and Schubert (2010)]. Impact erosion of the atmosphere has occurred throughout the Noachian (Melosh and Vickery 1989) and it may therefore have also contributed to the decreases in atmospheric pressure demanded by numerous observations. Alternatively, carbonate formation in ancient sedimentary or hydrothermalised rocks (Ehlmann et al. 2008; Michalski and Niles 2010), whose importance is not yet clearly identified, might have contributed to a decrease in atmospheric CO_2 pressure down to below 0.1 bar.

The low H/S, low C/S ratios and high SO_2 content of volcanic gases on Mars contrast with extrusive emissions on Earth. It is noteworthy that most extrusive rocks on Earth are emplaced in submarine conditions (average pressure 400 bar). Gases emitted by degassing at 400 bar are sulfur-poor and CO_2 dominated. Only subaerial volcanism significantly contributes to sulfur emissions into the Earth's atmosphere. Gaillard et al. (2011) suggested an increasing amount of subaerial volcanism on Earth is the cause of a major change in composition of volcanic gases that became increasingly sulfur-rich and SO_2-rich, with major impacts on surficial biogeochemistry. The Archean era on Earth has been clearly shown to display limited sulfur cycling whereas through time, sulfur increasingly invaded the exosphere (Lyons and Gill 2010). As summarized in Fig. 16, there are good reasons to believe that ancient Mars (early Noachian) had limited sulfur cycling and was wetter and warmer (Bibring et al. 2006). The large sulfate deposits now widely observed on the Martian surface, and the surface waters from which they were deposited, occurred during the late Noachian-Hesperian times (Fig. 16). To some extent, it is possible that increasing contributions of sulfur to surficial chemical processes is a feature common to both Mars and Earth.

To summarize: (1) most extrusive volcanic degassing on Mars globally occurred at low pressure (<0.1 bar) in contrast to the Earth (where submarine volcanism dominates), (2) Martian basalts must contain more sulfur than terrestrial basalts, and (3) under the reduced conditions prevailing in the Martian mantle, enhanced graphite stability implies low CO_2 content in basalts (Stanley et al. 2011), which is 3–10 times lower than the CO_2 content of Earth's basalts. All this contributes to the low C/S of gases emitted on Mars, which can be lower by a factor of 10 compared to Earth.

The consequences for the chemistry of Martian surface waters are also significant. Low C/S, with sulfur mostly injected as SO_2, would likely create aqueous systems dominated more by some form of a sulfur cycle (i.e., formation of strong S-based acids such as sulfurous and sulfuric acid) than the carbon cycle (weak C-based acids such as carbonic and organic acids). The acidic nature of Martian surface waters and the lack of carbonate minerals in late Noachian-Hesperian sediments in turn may be related to the composition of volcanic gases that for the reasons discussed above were different from those emitted on Earth.

7 Sulfur Cycling: Volcanoes, Atmosphere, Climate, and Prospective

7.1 Sulfur and Climate Models

Farquhar et al. (2000) have shown that some sulfur found in certain SNC meteorites has mass-independent isotopic fractionation, interpreted as evidence of an atmospheric source for sulfur. In detail, atmospheric SO_2 that carries the isotopic mass independent fractionations must be involved in the sulfur transfer from atmosphere to the Martian regolith. Martian atmospheric SO_2, deriving from volcanic activity, may have well influenced the climate on early Martian. Halevy et al. (2007) and Bullock and Moore (2007), following the pioneering study of Wänke and Dreibus (1994), suggested that an SO_2 greenhouse effect may have maintained warm conditions on early Mars that in turn could be reconciled with the clay deposits of the early-mid-Noachian and the formation of valley networks. This warm period goes along with wet conditions and may imply atmospheric pressures greater than about 0.5 bar. Atmospheric simulations performed in several studies corroborated the possibility of warming by SO_2. Johnson et al. (2008) performed calculations showing that atmospheric temperature could increase above +10 °C if a concentration level of ca. 100s ppmv SO_2 is reached in a CO_2 atmosphere of 0.5 bar. But, Johnson et al. (2008) did not consider that SO_2 could react in the atmosphere and produce H_2SO_4 and S_8 aerosols (Settle 1979; Tian et al. 2009). These factors subsequently were considered by Johnson et al. (2009) and imply that the SO_2 lifetime in an ancient, reduced, denser, wetter Martian atmosphere is on the order of 100s years. This may allow transient warm periods to occur after volcanic eruptions, provided that volcanic eruptions supplied enough atmospheric SO_2. The period of warming is however interrupted, and under certain circumstances, compensated by a period of cooling mainly due to H_2SO_4 and S_8 aerosols being formed by photochemical processes from SO_2 and successive reactions with H_2O to form sulfuric acid (Tian et al. 2009). Like Johnson et al. (2009), Tian et al. (2009) used photochemical models but they found SO_2 lifetimes much shorter than predicted by Johnson et al. (2009), typically on the order of several months. It is thus possible that intensive SO_2 emission due to volcanic eruptions produced warm-wet conditions that lasted 1 to 100 years which were followed by cooling and glaciations due to formation of atmospheric S-aerosols. All this may well be consistent with various geomorphological features of Mars (Andrews-Hanna and Lewis 2011).

The overall atmospheric processes are, however, complex and, accordingly, no consensus has yet emerged. The warming effect and the SO_2 lifetime correlate with the amount of atmospheric SO_2 and the total atmospheric pressure (i.e. P_{CO2}). The redox state of the ancient Martian atmosphere as well as its humidity also influences SO_2 lifetime and the extent of H_2SO_4 formation (Johnson et al. 2009; Tian et al. 2009). A possible way to resolve this complexity would be to construct an integrated Martian model in which the initial status of atmospheric models would be imposed by volcanic inputs similar to those calculated here (Table 2).

Johnson et al. (2008) attempted such an integrated approach but their estimates of volcanic sulfur emissions did not consider the high sulfur content in Martian basalts due to their high iron content (see Sect. 3.2). In addition, their inference that SO_2/H_2S ratio equals 1, taken from Halevy et al. (2007), is not consistent with the arguments put forward in this review (see also Gaillard and Scaillet 2009). Accordingly, Johnson et al. (2008) may have underestimated volcanic SO_2 production by a factor 4–6.

Most atmospheric models neglect species other than CO_2–H_2O. All volcanic gas compositions shown in Table 2 indicate that carbon monoxide is present at concentration levels as high as carbon dioxide and that the fraction of H_2 is significant, whereas these species are generally not taken into account. In addition, all atmospheric simulations allow SO_2 concentration and atmospheric pressure to vary independently whereas the above analysis clearly shows that high volcanic SO_2 emissions would occur only if the atmospheric pressure is low (0.1 bar or less). Volcanic degassing in an atmosphere of 1 bar would produce gas dominated by C and H-species with little sulfur (and all sulfur as H_2S). Hence, if ancient warm and wet Mars existed with an atmosphere similar to 1 bar or higher, it follows that SO_2

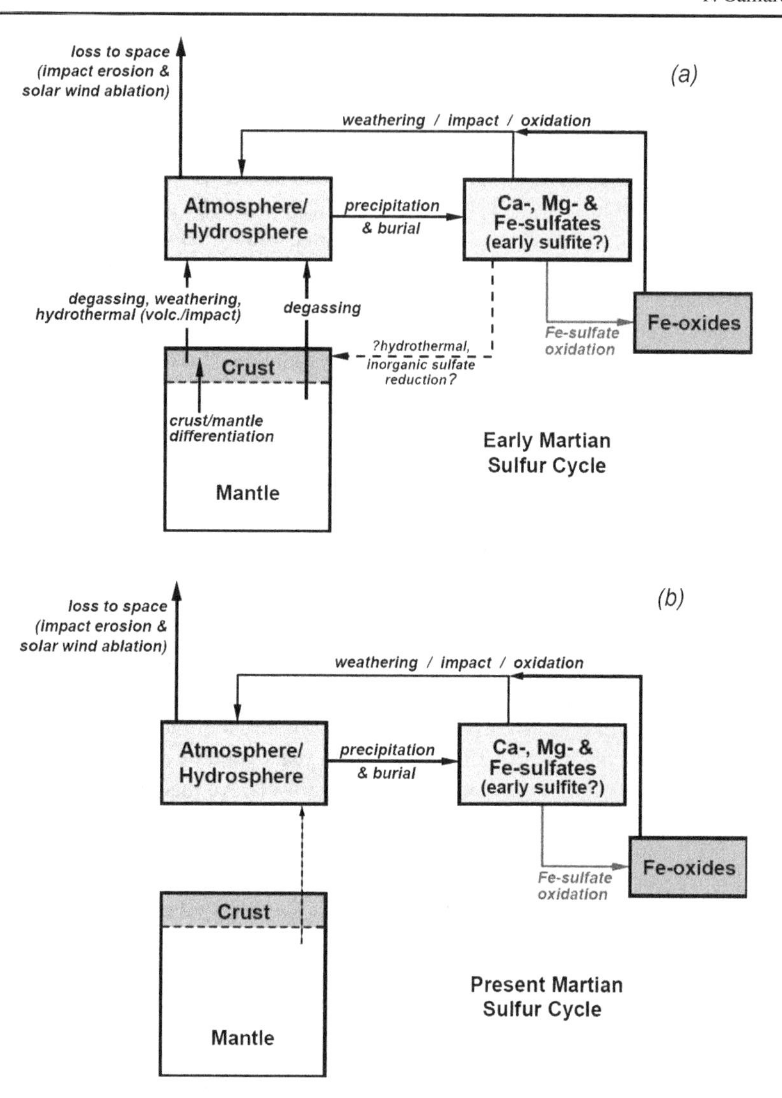

Fig. 17 Simplified model for a possible Martian sulfur cycle. (**a**) Early Mars when abundant sulfur was delivered to the surface through magmatic and related processes. (**b**) Recent Mars after reduction of rated in magmatism and sulfur degassing. During this, sulfur species taking part in surficial processes were likely dominated by sulfur recycling processes with only minor contributions related to magmatic degassing. The possible link between the sulfur and iron cycles at the Martian surface through oxidation of iron sulfates to iron oxides is shown by the *grey arrow*. After McLennan (2012)

from volcanic sources is an unlikely candidate to trigger warm conditions. Furthermore, phyllosilicate deposits that dominated at this time (early-middle Noachian) do not require high activity of sulfur species in the exosphere. Instead, sulfate deposits that appeared later, during the late Noachian and the Hesperian, reveal high activities of oxidized S-species, which further generated acidic surface waters (see parts 1 & 2 and below). There is therefore incompatibility between the period of warm and wet Mars and the conditions conducive to high SO_2 emissions by volcanic eruptions. The early wet-warm Mars must have had atmospheric pressures compatible with CO_2–H_2O+/−H_2S volcanic emissions, whereas the late Noachian, with sulfate deposits, was more likely in a low pressure atmosphere allowing volcanic emissions dominated by SO_2 (Fig. 17). Accordingly, the analysis provided in this paper lends support to the conclusion of Tian et al. (2009) that early Mars must have been kept warm by mechanisms other than volcanic SO_2 greenhouse warming.

It is, however, conceivable that the middle-late Noachian Mars had short episodes of warm periods due to high volcanic SO_2 emissions and its relatively long residence time in a dry atmosphere. If we assume that flood basalts on Mars were comparable to those on Earth, 100 km^3/yr of lava eruption rates is a reasonable estimate (Keszthelyi et al. 2006). Using gas compositions in Table 3, this implies SO_2 degassing of about 10^{17} g per year and total sulfur ($SO_2 + S_2 + H_2S$) emissions of more than twice this value. Tian et al. (2009) suggested that such a value is necessary to elevate the average surface temperature above the freezing point of water. Intermittent warming due to sporadic outbursts of volcanic SO_2 may well be possible during the late Noachian-Hesperian epochs: however, 10^{17} g of SO_2 in an atmosphere with 4 bar CO_2 suggested by Tian et al. (2010) is incompatible with elevated volcanic SO_2 emissions.

To conclude, atmospheric models involving S-species are a fascinating issue for future research on the ancient climate on Mars. The expected efforts will require volcanic emissions and atmospheric chemical physical processes to be related.

7.2 Sulfur Cycling Models

The sulfur-rich character of sedimentary deposits on Mars (see Sect. 2), apparent dominance of sulfate minerals over carbonate minerals (see Sect. 1) and evidence at both Meridiani Planum and Gusev crater for extensive low pH environments in the form of widespread Fe^{3+} mobility under oxidizing conditions have led to suggestions that some form of a sulfur cycle dominated surficial processes over much of Martian geological history. Such a sulfur cycle would likely produce strong S-based acids (e.g., sulfuric acid) and thus surficial processes would be characterized by widespread, relatively low pH conditions (~pH2–pH5). This is in contrast to terrestrial settings, where the carbon cycle, characterized by relatively weak C-based acids (e.g., carbonic acid, organic acids) and modest pH conditions (~pH5–8), dominates. McLennan (2012) recently reviewed the nature of such a potential Martian sulfur cycle and parts of that discussion are briefly summarized below.

Any Martian sulfur cycle would likely be divided into early and late phases with a transition occurring approximately at the time of loss of widespread aqueous conditions and reduction of magmatic rates, and thus sulfur degassing into the atmosphere, sometime >3 Gyr. In addition to being a time of voluminous magmatism (McEwen et al. 1999), the early (>3 Gyr) history of Mars was also characterized by widespread aqueous conditions, possibly facilitated by an early greenhouse effect (but see discussion above). Results from experiments, thermodynamic models, and direct measurements of Martian soils and rocks suggest that a variety of sulfur reservoirs and sulfur cycling processes may have been involved (Fig. 17a).

Acid alteration of basaltic rocks and minerals is now well established for the Martian surface, at least during parts of its early history (McLennan and Grotzinger 2008). Among the major processes that have been documented are low temperature alteration (i.e., weathering) that produced the brines that in turn gave rise to evaporitic minerals such as those in the Burns formation, and higher temperature epithermal to hydrothermal fluids, such as those that have been identified in the Columbia Hills of Gusev crater. Extensive occurrences of sulfate and possible chloride minerals identified from orbit and by in situ techniques on rocks and soils also point to widespread formation of a variety of evaporite minerals across the Martian surface, dominated by Ca-, Mg- and Fe-sulfates of varying hydration states.

Correlations between occurrences of sulfate minerals and iron oxides on the Martian surface identified from orbit and also inferred for the Burns formation sulfate-rich outcrops (i.e., occurrence of hematitic concretions) further suggest a possible genetic link between iron and sulfur cycling (McLennan et al. 2005; Bibring et al. 2007; Tosca et al. 2008). Thus, Tosca et al. (2008) carried out experiments and modeling that evaluated diagenetic oxidation and ageing of iron sulfates to form iron oxides. These oxidation-ageing processes are almost certainly irreversible under Martian surficial conditions and also result in the liberation of sulfur that in turn may be recycled back through the sedimentary system.

As described above, Halevy et al. (2007) have proposed an even earlier version of the sulfur cycle to account for early widespread occurrence of clay minerals, but dearth of carbonates, in the earlier Noachian (e.g., Fig. 7). In this model, relatively reducing atmospheric conditions resulted in inhibition of oxidation of atmospheric SO_2 and thus formation of sulfurous, rather than sulfuric, acids leading to low but more modest pH conditions (~pH4.5–pH5.5). These conditions might allow for the formation of clay minerals while at the same time inhibit precipitation of carbonates. This model predicts formation of widespread sulfite minerals in the early Noachian for which there is no evidence but on the other hand, if formed, would be unlikely to survive later oxidizing conditions.

With the precipitous decline of aqueous activity and volcanic rates sometime before about 3 Gyr, the rate at which sulfur was degassed into the surficial environment likely also diminished. Nevertheless, there is evidence that surficial processes continued to be influenced by some form of a sulfur cycle after this time and through essentially to the present (Fig. 17b). Thus, the chemical compositions of relatively altered present-day rock surfaces and their relatively fresh interiors (exposed by the rock abrasion tool, RAT), analyzed by Spirit in Gusev crater, point to continued low water/rock ratio acid alteration, albeit likely at greatly reduced rates and scales (Hurowitz et al. 2006; Hurowitz and McLennan 2007). The origin of acidity in younger environments is less clear. Although rates of magmatism are minimal during most of the Amazonian, volcanism is generally accepted to be continuing through to the present. As discussed above, under low atmospheric pressures, sulfur degassing is more efficient. Other possible sources include recycling of surface sulfate deposits by impact processes (McLennan et al. 2006; Zolotov and Mironenko 2007) and continued slow recycling of sulfur related to ferrous iron sulfate oxidation processes (Fig. 17).

8 Conclusion

8.1 Secular Changes in Sulfur Outgassing Rates

Numerous deposits of sulfate-bearing, ancient sedimentary rock, as well as the presence of a sulfur-rich global regolith demonstrate that a large cache of S of likely volcanogenic origins is sequestered in the crust. The mineralogy of these deposits generally suggests they formed under water-limited, acidic aqueous conditions. In this review, we provide some constraints on how and when sulfur was delivered to the surface environment.

Our review makes it clear that volcanic sulfur may have been delivered by volcanoes on Mars but only as a consequence of subaerial volcanism that occurred in a low pressure atmosphere (i.e. <0.1 bar). The apparent shift on Mars from an early phase, where clay minerals and carbonates formed more widely to a younger period seemingly dominated by sulfate-bearing sedimentary rocks might be linked to a decrease in atmospheric pressure, consistent with elevated sulfur emissions. We must recognize that interpretation of such secular changes on Mars may be non-unique and degassing simulations show that a change in Martian basalt redox state (increasingly oxidizing) or an increase in their water content could also result in enhanced volcanic sulfur outgassing. Changes in source processes might be expected with cooling of the mantle (Baratoux et al. 2011) but our current understanding of Martian igneous petrology is not sufficient to identify secular changes in water content or redox state with any confidence. Accordingly, we conclude that a decrease in average venting pressure due to a decrease in atmospheric pressure better explains the emission of significant volcanic sulfur in the atmosphere.

The parallel with the Earth is tempting. Like early Mars, early Earth also had limited surficial sulfur activities. Archean oceans on Earth are believed to be nearly sulfur free whereas the Proterozoic is marked by an increase in the sulfur content of surface waters. Sulfate is the second most abundant anion of modern terrestrial seawater (Lyons and Gill 2010). Gaillard et al. (2011) suggested that such secular changes in sulfur delivery on Earth must have been related to modifications of the conditions of volcanic degassing rather than to a change in the volcanic source processes. We may adopt a similar model in the case of Mars. Enhanced volcanic sulfur emissions due to changes in atmospheric pressure (Gaillard and Scaillet 2009) are then likely to have modified the composition of volcanic gases from carbon-dominated to sulfur-dominated. This in turn may have triggered major changes in the surface chemistry and in the nature of sedimentary processes.

8.2 Sulfur on Mars, What's Next?

In spite of the overwhelming abundance of sulfate on Mars's surface, relatively few studies have addressed the fate of this element in the various Martian reservoirs (except for sedimentary deposits, see review and references in this paper and in McLennan 2012). Igneous sulfur, volcanic sulfur, atmospheric sulfur and climatic sulfur have each been addressed in some studies, which contrasts with numerous investigations on the fate of CO_2 or water and its relationships with ancient climate and surface chemistry (see Forget and Pierrehumbert 1997; Phillips et al. 2001; Grott et al. 2011). The examination of our current understanding of sulfur cycling and its impact on climate indicates a need for studies that thoroughly integrate igneous fluxes and atmospheric processes; studies that to date are missing.

Fluids that deposited sedimentary sulfates on Mars also deposited chlorine-bearing minerals (Clark et al. 2005; Squyres and Knoll 2005). The fate of chlorine, like that of sulfur, seems to be connected to acidic brines, abundant and common in the late Noachian. Both sulfur and chlorine most likely originate from volcanic emissions (Gaillard and Scaillet 2009; Filiberto and Treiman 2009). The volcanic degassing of chlorine is however poorly known. Future work will therefore need to address the systematics of multi-component (C-O-H-S-Cl) volcanic degassing from Martian basalts.

The environmental conditions responsible for sulfate deposits on Mars have been addressed recently (King et al. 2004; Tosca et al. 2005; Berger et al. 2009; McLennan 2012) but our review reveals that many of the driving processes remain to be determined. Combined experimentation and thermodynamic/kinetic modeling will allow for the most significant progress. But it remains uncertain the degree to which evaporation or acidic alteration or both are responsible for sedimentary sulfate deposits. The relationship between acidity and redox state of surface waters is also controversial. Finally, the possibility of sulfurous acids rather than sulfuric acids needs to be addressed in greater detail. The occurrence of acidic conditions in water expelled from mines, that are related to elevated discharge of sulfuric acidic (Nordstrom 2011), may be interesting to further investigate as an analogue to Martian brines (King and McSween 2005). The sulfuric acids in such water nevertheless derive from oxidation of pyrite. This is an important difference with the Martian context where most, if not all, of the sulfuric acid may derive from volcanic SO_2 (Berger et al. 2009).

In spite of the many uncertainties, an emerging picture is that martian basalts are poor in CO_2 and poor in water (McCubbin et al. 2010), whereas they are rich in S (Gaillard and Scaillet 2009) and in Cl (Filiberto and Treiman 2009), in comparison to Earth's basaltic rocks. This conclusion needs confirmation but it may indeed explain the S and Cl-rich nature of Mars's surface, which contrasts with the H_2O and CO_2 rich Earth's surface. The implications for the emergence of life and conditions favorable to that emergence need to be discussed in light of these fundamental geochemical differences, whose origin is intimately tied to planet formation and how volatiles (C-O-H-S-Cl) have been accreted to and/or lost during planetary accretion and evolution.

Acknowledgements F. Gaillard is supported by the ERC grant #279790. We acknowledge the editorial handling of Mike Toplis and the helpful reviews by K. Righter, M. Zolotov, and P. King.

References

A. Aiuppa, M. Burton, T. Caltabiano, G. Giudice, S. Guerrieri, M. Liuzzo, F. Mure, G. Salerno, Unusually large magmatic CO_2 gas emissions prior to a basaltic paroxysm. Geophys. Res. Lett. **37**, L17303 (2010). doi:10.1029/2010GL043837

C. Allegre, G. Manhes, E. Lewin, Chemical composition of the Earth and the volatility control on planetary genetics. Earth Planet. Sci. Lett. **185**, 49–69 (2001)

J.C. Andrews-Hanna, K.W. Lewis, Early Mars hydrology: 2. Hydrological evolution in the Noachian and Hesperian epochs. J. Geophys. Res. Planets **116**, E02007 (2011)

S. Barabash, A. Fedorov, R. Lundin, J.A. Sauvaud, Martian atmospheric erosion rates. Science **315**, 501–503 (2007)

D. Baratoux, M.J. Toplis, M. Monnereau, O. Gasnault, Thermal history of Mars inferred from orbital geochemistry of volcanic provinces. Nature **472**, 338–341 (2011). doi:10.1038/nature09903

H. Behrens, F. Gaillard, Geochemical aspects of melts: volatiles and redox behavior. Elements **2**, 275–280 (2006)

G. Berger, M.J. Toplis, E. Treguier, C. d'Uston, P. Pinet, Evidence in favor of ephemeral and transient water during alteration at Meridiani Planum, Mars. Am. Mineral. **94**, 1279–1282 (2009)

R.A. Berner, Chemical weathering and its effects on atmospheric CO2 and climate. Rev. Mineral. **31**, 565–583 (1995)

R.A. Berner, GEOCARBSULF: a combined model for Phanerozoic atmospheric O(2) and CO(2). Geochim. Cosmochim. Acta **70**, 5653–5664 (2005)

J.P. Bibring et al., Global mineralogical and aqueous mars history derived from OMEGA/Mars express data. Science **312**, 400–404 (2006)

J.P. Bibring et al., Coupled ferric oxides and sulfates on the Martian surface. Science **317**(5842), 1206–1210 (2007)

J.L. Bishop et al., Mineralogy of Juventae Chasma: sulfates in the light-toned mounds, mafic minerals in the bedrock, and hydrated silica and hydroxylated ferric sulfate on the plateau. J. Geophys. Res. Planets **114**, E00D09 (2009)

J. Brückner, G. Dreibus, R. Gellert, S.W. Squyres, H. Wänke, A. Yen, J. Zipfel, Mars exploration rovers: chemical compositions by the APXS, in *the Martian Surface: Composition, Mineralogy, and Physical Properties*, ed. by J.F. Bell III (Cambridge University Press, Cambridge, 2008), pp. 58–101

M.A. Bullock, J.M. Moore, Atmospheric conditions on early Mars and the missing layered carbonates. Geophys. Res. Lett. **34**, L19201 (2007). doi:10.1029/2007GL030688

A.S. Buono, D. Walker, The Fe-rich liquidus in the Fe-FeS system from 1 bar to 10 GPa. Geochim. Cosmochim. Acta **75**, 2072–2087 (2011)

A. Burgisser, B. Scaillet, Redox evolution of a degassing magma rising to the surface. Nature **445**, 194–197 (2007)

D.E. Canfield, The evolution of the Earth surface sulfur reservoir. Am. J. Sci. **304**, 839–861 (2004)

J. Carter, F. Poulet, A. Ody, J.P. Bibring, S. Murchie, Global distribution, composition and setting of hydrous minerals on Mars: a reappraisal, in *Lunar and Planetary Science Conference* (2011), p. 2593

N.L. Chabot, Sulfur contents of the parental metallic cores of magmatic iron meteorites. Geochim. Cosmochim. Acta **68**, 3607–3618 (2004)

M.G. Chapman, K.L. Tanaka, Interior trough deposits on Mars: subice volcanoes? J. Geophys. Res. (Planets) **106**(E5), 10087–10100 (2001)

V. Chevrier, J.-P. Lorand, V. Sautter, Sulfide petrology of four nakhlites: Northwest Africa 817, Northwest Africa 998, Nakhla, and Governador Valadares. Meteorit. Planet. Sci. **46**, 769–784 (2011). doi:10.1111/j.1945-5100.2011.01189.x

V. Chevrier, F. Poulet, J.-P. Bibring, Early geochemical environment of Mars as determined from thermodynamics of phyllosilicates. Nature **448**, 60–63 (2007). doi:10.1038/nature05961

P.R. Christensen et al., Detection of crystalline hematite mineralization on Mars by the thermal emission spectrometer: evidence for near-surface water. J. Geophys. Res. (Planets) **105**(E4), 9623–9642 (2000)

P.R. Christensen, R.V. Morris, M.D. Lane, J.L. Bandfield, M.C. Malin, Global mapping of Martian hematite mineral deposits: remnants of water-driven processes on early Mars. J. Geophys. Res. (Planets) **106**(E10), 23873–23885 (2001)

B.C. Clark, A.K. Baird, H.J. Rose Jr., P. Toulmin III, R.P. Christian, W.C. Kelliher, A.J. Castro, C.D. Rowe, K. Keil, G.R. Huss, The viking X ray fluorescence experiment: analytical methods and early results. J. Geophys. Res. **82**(28), 4577–4594 (1977). doi:10.1029/JS082i028p04577

B.C. Clark, A.K. Baird, Is the martian lithosphere sulfur rich. J. Geophys. Res. **84**, 8395–8403 (1979)

B.C. Clark, Geochemical components in martian soil. Geochim. Cosmochim. Acta **57**, 4575–4581 (1993). doi:10.1016/0016-7037(93)90183-W

B.C. Clark, R.V. Morris et al., Chemistry and mineralogy of outcrops at Meridiani Planum. Earth Planet. Sci. Lett. **240**(1), 73–94 (2005)

E.A. Cloutis et al., Detection and discrimination of sulfate minerals using reflectance spectroscopy. Icarus **184**(1), 121–157 (2006)

R.A. Craddock, R. Greeley, Minimum estimates of the amount and timing of gases released into the martian atmosphere from volcanic eruptions. Icarus **204**, 512–526 (2009)

V. Debaille, A.D. Brandon, C. O'Neill, Q.-Z. Yin, B. Jacobsen, Early martian mantle overturn inferred from isotopic composition of nakhlite meteorites. Nat. Geosci. **2**, 548–552 (2009). doi:10.1038/NGEO579

E. Dehouck, V. Chevrier, A. Gaudin, N. Mangold, P.-E. Mathé, P. Rochette, Evaluating the role of sulfide-weathering in the formation of sulfates or carbonates on Mars. Geochim. Cosmochim. Acta **90**, 47–63 (2012)

J.E. Dixon, D.A. Clague, P. Wallace, R. Poreda, Volatiles in alkalic basalts from the North Arch volcanic field, Hawaii: extensive degassing of deep submarine-erupted alkalic series lavas. J. Petrol. **38**, 911–939 (1997)

G. Dreibus, H. Palme, Cosmochemical constraints on the sulfur content in the Earth's core. Geochim. Cosmochim. Acta **60**, 1125–1130 (1996)

G. Dreibus, H. Wanke, Mars, a volatile-rich planet. Meteoritics **20**, 367–381 (1985)

D.S. Ebel, Sulfur in extraterrestrial bodies and the deep earth. In sulfur in magmas and melts: its importance for natural and technical processes. Rev. Mineral. Geochem. **73**, 315–336 (2010)

B.L. Ehlmann, J.F. Mustard, S.L. Murchie, F.F. Poulet, J.L. Bishop, A.J. Brown, W.M. Calvin, R.N. Clark, D.J.D. Marais, R.E. Milliken, L.H. Roach, T.L. Roush, G.A. Swayze, J.J. Wray, Orbital identification of carbonate-bearing rocks on Mars. Science **322**, 1828–1832 (2008)

B.L. Ehlmann et al., Geochemical consequences of widespread clay mineral formation in Mars' ancient crust. Space Sci. Rev. (2012, this issue). doi:10.1007/s11214-012-9930-0

L.T. Elkins-Tanton, P.C. Hess, E.M. Parmentier, Possible formation of ancient crust on Mars through magma ocean processes. J. Geophys. Res. **110**, E12S01 (2005). doi:10.1029/2005JE002480

S. Fabre, G. Berger, A. Nédélec, Continental weathering under high-CO_2 atmospheres during Precambrian times. G-cubed **12** (2011). doi:10.1029/2010GC003444

A.G. Fairen, D. Fernandez-Remolar, J.M. Dohm, V.R. Baker, R. Amils, Inhibition of carbonate synthesis in acidic oceans on early Mars. Nature **431**, 423–426 (2004)

J. Farquhar, J. Savrino, T.L. Jackson, M.H. Thiemnes, Evidence of atmospheric sulfur in the martian regolith from sulfur isotopes in meteorites. Nature **404**, 50–52 (2000). doi:10.1038/35003517

W.H. Farrand, T.D. Glotch, J.W. Rice, J.A. Hurowitz, G.A. Swayze, Discovery of jarosite within the Mawrth Vallis region of Mars: implications for the geologic history of the region. Icarus **204**(2), 478–488 (2009)

C.I. Fassett, J.W. Head, Sequence and timing of conditions on early Mars. Icarus **211**, 1204–1214 (2011)

Y.W. Fei, C.T. Prewitt, H.K. Mao, C.M. Bertka, Structure and density of FeS at high-pressure and high-temperature and the internal structure of Mars. Science **268**, 1892–1894 (1995)

Y. Fei, C.M. Bertka, L.W. Finger, High pressure iron sulfur compound, Fe_3S_2, and melting relations in the Fe–FeS system. Science **275**, 1621–1623 (1997)

J. Filiberto, A.H. Treiman, Martian magmas contained abundant chlorine, but little water. Geology **37**, 1087–1090 (2009). doi:10.1130/G30488A.1

C.J.B. Fincham, F.D. Richardson, The behaviour of sulfur in silicate and aluminate melts. Proc. R. Soc. Lond. **223A**, 40–61 (1954)

K.E. Fishbaugh, F. Poulet, V. Chevrier, Y. Langevin, J.P. Bibring, On the origin of gypsum in the Mars north polar region. J. Geophys. Res. Planets **112**(E7), E07002 (2007)

J. Flahaut, C. Quantin, P. Allemand, P. Thomas, Morphology and geology of the ILD in Capri/Eos Chasma (Mars) from visible and infrared data. Icarus **207**(1), 175–185 (2010a)

J. Flahaut, C. Quantin, P. Allemand, P. Thomas, L. Le Deit, Identification, distribution and possible origins of sulfates in Capri Chasma (Mars), inferred from CRISM data. J. Geophys. Res. Planets **115**, E11007 (2010b)

C.N. Foley, T.E. Economou, R.N. Clayton, J. Brückner, G. Dreibus, R. Rieder, H. Wänke, Martian surface chemistry: APXS results from the Pathfinder landing site, in *The Martian Surface: Composition, Mineralogy, and Physical Properties*, ed. by J.F. Bell III (Cambridge University Press, Cambridge, 2008), pp. 35–57

F. Forget, R.T. Pierrehumbert, Warming early Mars with carbon dioxide clouds that scatter infrared radiation. Science **278**, 1273–1276 (1997)

F. Fueten, J. Flahaut, L. Le Deit, R. Stesky, E. Hauber, K. Gwinner, Interior layered deposits within a perched basin, southern Coprates Chasma, Mars: evidence for their formation, alteration, and erosion. J. Geophys. Res. Planets **116**, E02003 (2011)

F. Gaillard, B. Scaillet, The sulfur content of volcanic gases on Mars. Earth Planet. Sci. Lett. **279**, 34–43 (2009)

F. Gaillard, M. Pichavant, B. Scaillet, Experimental determination of activities of FeO and Fe_2O_3 components in hydrous silicic melts under oxidizing conditions. Geochim. Cosmochim. Acta **67**, 4389–4409 (2003a)

F. Gaillard, B.C. Schmidt, S. Mackwell, C. McCammon, Rate of hydrogen-iron redox exchange in silicate melts and glasses. Geochim. Cosmochim. Acta **67**, 2427–2441 (2003b)

F. Gaillard, B. Scaillet, N.T. Arndt, Atmospheric oxygenation caused by a change in volcanic degassing pressure. Nature **478**, 229–232 (2011)

A. Gendrin, N. Mangold, J.P. Bibring, Y. Langevin, B. Gondet, F. Poulet, G. Bonello, C. Quantin, J. Mustard, R. Arvidson, S. LeMouélic, Sulfates in martian layered terrains: the OMEGA/Mars Express view. Science **307**, 1587–1591 (2005)

E.K. Gibson, C.B. Moore, T.M. Primus, C.F. Lewis, Sulfur in achondritic meteorites. Meteoritics **20**, 503–511 (1985)

T.D. Glotch, A.D. Rogers, Evidence for aqueous deposition of hematite- and sulfate-rich light-toned layered deposits in Aureum and Iani Chaos, Mars. J. Geophys. Res. Planets **112**(E6), E06001 (2007)

T.D. Glotch, P.R. Christensen, Geologic and mineralogic mapping of Aram Chaos: evidence for a water-rich history. J. Geophys. Res. Planets **110**(E9), E09006 (2005)

T.D. Glotch, J.L. Bandfield, P.R. Christensen, W.M. Calvin, S.M. McLennan, B.C. Clark, A.D. Rogers, S.W. Squyres, Mineralogy of the light-toned outcrop at Meridiani Planum as seen by the Miniature Thermal Emission Spectrometer and implications for its formation. J. Geophys. Res. Planets **111**(E12), E12S03 (2006a)

T.D. Glotch, P.R. Christensen, T.G. Sharp, Fresnel modeling of hematite crystal surfaces and application to martian hematite spherules. Icarus **181**(2), 408–418 (2006b)

M.P. Golombek et al., Erosion rates at the Mars Exploration Rover landing sites and long-term climate change on Mars. J. Geophys. Res. **111**(E12), 1–14 (2006)

R. Greeley, B.D. Schneid, Magma generation on Mars: amounts, rates, and comparisons with Earth, Moon, and Venus. Science **254**, 996–998 (1991)

J.L. Griffes, R.E. Arvidson, F. Poulet, A. Gendrin, Geologic and spectral mapping of etched terrain deposits in northern Meridiani Planum. J. Geophys. Res. Planets **112**(E8), E08S09 (2007)

M. Grott, A. Morschhauser, D. Breuer, E. Hauber, Volcanic outgassing of CO2 and H2O on Mars. Earth Planet. Sci. Lett. **308**, 391–400 (2011)

J. Grotzinger et al., Sedimentary textures formed by aqueous processes, Erebus crater, Meridiani Planum, Mars. Geology **34**(12), 1085–1088 (2006)

J.P. Grotzinger, R.E. Arvidson, J.F. Bell, W. Calvin, B.C. Clark, D.A. Fike, M. Golombek, R. Greeley, A. Haldemann, K.E. Herkenhoff, B.L. Jolliff, A.H. Knoll, M. Malin, S.M. McLennan, T. Parker, L. Soderblom, J.N. Sohl-Dickstein, S.W. Squyres, N.J. Tosca, W.A. Watters, Stratigraphy, sedimentology and depositional environment of the Burns formation, Meridiani Planum, Mars. Earth Planet. Sci. Lett. **240**, 11–72 (2005)

I. Halevy, M.T. Zuber, D.P. Schrag, A sulfur dioxide climate feedback on early Mars. Science **318**, 1903 (2007). doi:10.1126/science.1147039

C.D.K. Herd, L.E. Borg, J.H. Jones, J.J. Papike, Oxygen fugacity and geochemical variations in the martian basalts: implications for martian basalt petrogenesis and the oxidation state of the upper mantle of Mars. Geochim. Cosmochim. Acta **66**(11), 2025–2036 (2002)

C.D.K. Herd, A.H. Treiman, G.A. McKay, C.K. Shearer, Light lithophile elements in martian basalts: evaluating the evidence for magmatic water degassing. Geochim. Cosmochim. Acta **69**, 2431–2440 (2005)

K.E. Herkenhoff, M.P. Golombek, E.A. Guinness, J.B. Johnson, A. Kusack, L. Richter, R.J. Sullivan, S. Gorevan, In situ observations of the physical properties of the Martian surface, in *The Martian Surface: Composition, Mineralogy, and Physical Properties*, ed. by J.F. Bell III (Cambridge University Press, Cambridge, 2008), pp. 451–467

A. Holzheid, T.L. Grove, Sulfur saturation limits in silicate melts and their implications for core formation scenarios for terrestrial planets. Am. Mineral. **87**, 227–237 (2002)

J.A. Hurowitz, S.M. McLennan, A $\sim$ 3.5 Ga record of water-limited, acidic conditions on Mars. Earth Planet. Sci. Lett. **260**, 432–443 (2007)

J.A. Hurowitz, W.W. Fischer, N.J. Tosca, R.E. Milliken, Origin of acidic surface waters and the evolution of atmospheric chemistry on early Mars. Nat. Geosci. **3**, 323–326 (2010)

J.A. Hurowitz, S.M. McLennan, N.J. Tosca, R.E. Arvidson, J.R. Michalski, D.W. Ming, C. Schöder, S.W. Squyres, In-situ and experimental evidence for acidic weathering on Mars. J. Geophys. Res. **111**, E02S19 (2006). doi:10.1029/2005JE002515

B.M. Hynek, R.E. Arvidson, R.J. Phillips, Geologic setting and origin of Terra Meridiani hematite deposit on Mars. J. Geophys. Res. **107**(E10), 5088 (2002). doi:10.1029/2002JE001891

G. Iacono-Marziano, Y. Morizet, E. Le-Trong, F. Gaillard, New experimental data and semi-empirical parameterization of H2O-CO2 solubility in mafic melts. Geochim. Cosmochim. Acta **97**, 1–23 (2012). doi:10.1016/j.gca.2012.08.035

B.M. Jakosky, R.J. Phillips, Mars' volatile and climate history. Nature **412**, 237–244 (2001)

S.S. Johnson, A.A. Pavlov, M.A. Mischna, Fate of SO_2 in the ancient Martian atmosphere: implications for transient greenhouse warming. J. Geophys. Res. (Planets) **114**, E11011 (2009). doi:10.1029/2008JE003313

J.R. Johnson, J.F. Bell, E. Cloutis, M. Staid, W.H. Farrand, T. Mccoy, M. Rice, A. Wang, A. Yen, Mineralogic constraints on sulfur-rich soils from Pancam spectra at Gusev crater, Mars. Geophys. Res. Lett. **34**(13), L13202 (2007)

S.S. Johnson, M.A. Mischna, T.L. Grove, M.T. Zuber, Sulfur-induced greenhouse warming on early Mars. J. Geophys. Res. **113**, E08005 (2008). doi:10.1029/2007JE002962

L. Keszthelyi, S. Self, T. Thordarson, Flood lavas on Earth, Io and Mars. J. Geol. Soc. **163**, 253–264 (2006). doi:10.1144/0016-764904-503

M.R. Kilburn, B.J. Wood, Metal-silicate partitioning and the incompatibility of S and Si during core formation. Earth Planet. Sci. Lett. **152**, 139–148 (1997)

P.L. King, H.Y. McSween, Effects of H2O, pH, and oxidation state on the stability of Fe minerals on Mars. J. Geophys. Res. **110**, E12S10 (2005). doi:10.1029/2005JE002482

P.L. King, S.M. McLennan, Sulfur on Mars. Elements **6**(2), 107–112 (2010)

P.L. King, D.T. Lescinsky, H.W. Nesbitt, The composition and evolution of primordial solutions on Mars, with application to other planetary bodies. Geochim. Cosmochim. Acta **68**, 4993–5008 (2004)

L.P. Knauth, D.M. Burt, K.H. Wohletz, Impact origin of sediments at the opportunity landing site on Mars. Nature **438**, 1123–1128 (2005)

S.P. Kounaves et al., Soluble sulfate in the martian soil at the Phoenix landing site. Geophys. Res. Lett. **37**, L09201 (2010)

M.D. Kraft, J.R. Michalski, T.G. Sharp, Effects of pure silica coatings on thermal emission spectra of basaltic rocks: considerations for Martian surface mineralogy. Geophys. Res. Lett. **30**(24), 2288 (2003)

M.D. Lane, P.R. Christensen, Thermal infrared emission spectroscopy of salt minerals predicted for Mars. Icarus **135**(2), 528–536 (1998)

M.D. Lane, J.L. Bishop, M.D. Dyar, P.L. King, M. Parente, B.C. Hyde, Mineralogy of the Paso Robles soils on Mars. Am. Mineral. **93**(5–6), 728–739 (2008)

Y. Langevin, F. Poulet, J.P. Bibring, B. Gondet, Sulfates in the North polar region of mars detected by OMEGA/Mars express. Science **307**(5715), 1584–1586 (2005)

L. Le Deit, S. Le Mouelic, O. Bourgeois, J.P. Combe, D. Mege, C. Sotin, A. Gendrin, E. Hauber, N. Mangold, J.-P. Bibring, Ferric oxides in East Candor Chasma, Valles Marineris (Mars) inferred from analysis of OMEGA/Mars Express data: identification and geological interpretation. J. Geophys. Res. Planets **113**(E7), E07001 (2008)

L. Le Deit, O. Bourgeois, D. Mege, E. Hauber, S. Le Mouelic, M. Masse, R. Jaumann, J.-P. Bibring, Morphology, stratigraphy, and mineralogical composition of a layered formation covering the plateaus around Valles Marineris, Mars: implications for its geological history. Icarus **208**(2), 684–703 (2010)

J. Li, C.B. Agee, Geochemistry of mantle-core differentiation at high pressure. Nature **381**, 686–689 (1996)

J. Li, C.B. Agee, Element partitioning constraints on the light element composition of the Earth's core. Geophys. Res. Lett. **28**, 81–84 (2001). doi:10.1029/2000GL012114
K.A. Lichtenberg et al., Stratigraphy of hydrated sulfates in the sedimentary deposits of Aram Chaos, Mars. J. Geophys. Res. Planets **115**, E00D17 (2010)
J.-P. Lorand, V. Chevrier, V. Sautter, Sulfide mineralogy and redox conditions in some Shergottites. Meteorit. Planet. Sci. Lett. **40**, 1257–1272 (2005)
T.W. Lyons, B.C. Gill, Ancient sulfur cycling and oxygenation of the early biosphere. Elements **6**, 93–99 (2010)
N. Mangold, A. Gendrin, B. Gondet, S. LeMouelic, C. Quantin, V. Ansan, J.P. Bibring, Y. Langevin, P. Masson, G. Neukum, Spectral and geological study of the sulfate-rich region of West Candor Chasma, Mars. Icarus **194**(2), 519–543 (2008)
N. Mangold, L. Roach, R. Milliken, S. Le Mouelic, V. Ansan, J.P. Bibring, P. Masson, J.F. Mustard, S. Murchie, G. Neukum, A late Amazonian alteration layer related to local volcanism on Mars. Icarus **207**(1), 265–276 (2010)
J.C. Marty, G. Balmino, J. Duron, P. Rosenblatt, S. Le Maistre, A. Rivoldini, V. Dehant, T. Van Hoolst, Martian gravity field model and its time variations from MGS and Odyssey data. Planet. Space Sci. **57**, 350–363 (2009)
M. Masse, S. Le Mouelic, O. Bourgeois, J.-P. Combe, L. Le Deit, C. Sotin, J.-P. Bibring, B. Gondet, Y. Langevin, Mineralogical composition, structure, morphology, and geological history of Aram Chaos crater fill on Mars derived from OMEGA Mars Express data. J. Ge ophys. Res. Planets **113**(E12), E12006 (2008)
M. Masse, O. Bourgeois, S. Le Mouelic, C. Verpoorter, L. Le Deit, J.P. Bibring, Martian polar and circum-polar sulfate-bearing deposits: sublimation tills derived from the North Polar Cap. Icarus **209**(2), 434–451 (2010)
M. Masse, O. Bourgeois, S. Le Mouelic, C. Verpoorter, A. Spiga, L. Le Deit, Wide distribution and glacial origin of Polar Gypsum on Mars. Earth Planet. Sci. Lett. **317**, 44–55 (2012)
J. Mavrogenes, H.S.C. O'Neill, The relative effects of pressure, temperature and oxygen fugacity on the solubility of sulfide in magmas. Geochim. Cosmochim. Acta **63**, 1173–1180 (1999)
T. McCollom, B.M. Hynek, A volcanic environment for bedrock diagenesis at Meridiani Planum on Mars. Nature **438** (2005)
F.M. McCubbin, A. Smirnov, H. Nekvasil, J. Wang, E. Hauri, D.H. Lindsley, Hydrous magmatism on Mars: a source of water for the surface and subsurface during the Amazonian. Earth Planet. Sci. Lett. **292**, 132–138 (2010)
F.M. McCubbin, E.H. Hauri, S.M. Elardo, K.E. Vander Kaaden, J.H. Wang, C.K. Shearer, Hydrous melting of the martian mantle produced both depleted and enriched Shergottites. Geology **40**, 683–686 (2012). doi:10.1130/G33242.1
W.F. McDonough, S.S. Sun, The composition of the Earth. Chem. Geol. **120**, 1125–1130 (1995)
A.S. McEwen, M.C. Malin, M.H. Carr, W.K. Hartmann, Voluminous volcanism on early Mars revealed in Valles Marineris. Nature **397**, 584–586 (1999)
S.M. McLennan, J.P. Grotzinger, The sedimentary rock cycle of Mars, in *The Martian Surface: Composition, Mineralogy, and Physical Properties*, ed. by J.F. Bell III (Cambridge University Press, Cambridge, 2008), pp. 541–577
S.M. McLennan, J.P. Grotzinger, J.A. Hurowitz, N.J. Tosca, Sulfate geochemistry and the sedimentary rock record of Mars, in *Workshop on Martian Sulfates as Records of Atmospheric-Fluid-Rock Interactions*. LPI Contribution, vol. 1331 (The Lunar & Planetary Institute, Houston, 2006), p. 54
S.M. McLennan, Geochemistry of sedimentary processes on Mars, in *Mars Sedimentology*, ed. by J.P. Grotzinger, R.E. Milliken. SEPM Special Publication (2012)
S.M. McLennan, J.F. Bell, W.M. Calvin, P.R. Christensen, B.C. Clark, P.A. de Souza, J. Farmer, W.H. Farrand, D.A. Fike, R. Gellert, A. Ghosh, T.D. Glotch, J.P. Grotzinger, B. Hahn, K.E. Herkenhoff, J.A. Hurowitz, J.R. Johnson, S.S. Johnson, B. Jolliff, G. Klingelhöfer, A.H. Knoll, Z. Learner, M.C. Malin, H.Y. McSween, J. Pocock, S.W. Ruff, L.A. Soderblom, S.W. Squyres, N.J. Tosca, W.A. Watters, M.B. Wyatt, A. Yen, Provenance and diagenesis of the evaporite-bearing Burns formation, Meridiani Planum, Mars. Earth Planet. Sci. Lett. **240**, 95–121 (2005)
H.Y. McSween, G.J. Taylor, M.B. Wyatt, Elemental composition of the Martian crust. Science **324**, 736–739 (2009)
H.Y. McSween, I.O. McGlynn, A.D. Rogers, Determining the modal mineralogy of Martian soils. J. Geophys. Res. Planets **115**, E00F12 (2010)
H.Y. McSween, T.L. Grove, R.C. Lentz, J.C. Dann, A.H. Holzheid, L.R. Riciputi, J.G. Ryan, Geochemical evidence for magmatic water within Mars from pyroxenes in the Shergotty meteorite. Nature **409**, 487–490 (2001)

H.J. Melosh, A.M. Vickery, Impact erosion of the primordial atmosphere of Mars. Nature **338**, 487–489 (1989)
J.M. Metz, J.P. Grotzinger, D.M. Rubin, K.W. Lewis, S.W. Squyres, J.F. Bell, Sulfate-rich eolian and wet interdune deposits, Erebus crater, Meridiani Planum, Mars. J. Sediment. Res. **79**, 247–264 (2009)
C. Meyer Jr., Website: http://curator.jsc.nasa.gov/antmet/mmc/index.cfm (2008)
J.R. Michalski, P.B. Niles, Deep crustal carbonate rocks exposed by meteor impact on Mars. Nat. Geosci. **3**, 751–755 (2010)
J. Michalski, P.B. Niles, Atmospheric origin of Martian interior layered deposits: links to climate change and the global sulfur cycle. Geology **40**, 419–422 (2012)
J.R. Michalski, J.P. Bibring, F. Poulet, D. Loizeau, N. Mangold, E.N. Dobrea et al., The Mawrth Vallis region of Mars: a potential landing site for the Mars Science Laboratory (MSL) mission. Astrobiology **10**, 687–703 (2010)
C. Milbury, G. Schubert, Search for the global signature of the Martian Dynamo. J. Geophys. Res. **115**, E10010 (2010)
R.E. Milliken et al., Opaline silica in young deposits on Mars. Geology **36**(11), 847–850 (2008)
R.E. Milliken, J.P. Grotzinger, B.J. Thomson, Paleoclimate of Mars as captured by the stratigraphic record in Gale Crater. Geophys. Res. Lett. **37**, L04201 (2010)
A.G. Monders, E. Médard, T.L. Grove, Phase equilibrium investigations of the Adirondack class basalts from the Gusev plains, Gusev crater, Mars. Meteorit. Planet. Sci. **42**, 131–148 (2007)
G. Morard, D. Andrault, N. Guignot, C. Sanloup, M. Mezouar, S. Petitgirard, G. Fiquet, *In situ* determination of Fe–Fe_3S phase diagram and liquid structural properties up to 65 GPa. Earth Planet. Sci. Lett. **272**, 620–626 (2008)
Y. Morizet, M. Paris, F. Gaillard, B. Scaillet, C-O-H fluid solubility in haplobasalt under reducing conditions: an experimental study. Chem. Geol. **279**, 1–16 (2010)
R.V. Morris, G. Klingelhofer, C. Schroder, D.S. Rodionov, A. Yen, D.W. Ming et al., Mossbauer mineralogy of rock, soil, and dust at Meridiani Planum, Mars: Opportunity's journey across sulfate-rich outcrop, basaltic sand and dust, and hematite lag deposits. J. Geophys. Res. Planets **111**, 27 (2006)
S.L. Murchie, J.F. Mustard, B.L. Ehlmann, R.E. Milliken, J.L. Bishop, N.K. McKeown, E.Z.N. Dobrea, F.P. Seelos, D.L. Buczkowski, S.M. Wiseman, R.E. Arvidson, J.J. Wray, G. Swayze, R.N. Clark, D.J.D. Marais, A.S. McEwen, J.P. Bibring, A synthesis of martian aqueous mineralogy after 1 Mars year of observations from the Mars Reconnaissance Orbiter. J. Geophys. Res. (Planets) **114**, E00D06 (2009a). doi:10.1029/2009JE003342
S.L. Murchie, L. Roach, F. Seelos, R. Milliken, J. Mustard, R. Arvidson, S. Wisema, K. Lichtenberg, J. Andrews-Hanna, J. Bishop, J.P. Bibring, M. Parente Morris R, Evidence for the origin of layered deposits in Candor Chasma, Mars, from mineral composition and hydrologic modeling. J. Geophys. Res. (Planets) **114**, E00D05 (2009b). doi:10.1029/2009JE003343
D.S. Musselwhite, H.A. Dalton, W.S. Kiefer, A.H. Treiman, Experimental petrology of the basaltic shergottite Yamato-980459: implications for the thermal structure of the martian mantle. Meteorit. Planet. Sci. **41**, 1271–1290 (2006)
P.B. Niles et al., Geochemistry of carbonates on Mars: implications for climate history and nature of aqueous environments. Space Sci. Rev. (2012, this issue). doi:10.1007/s11214-012-9940-y
P.B. Niles, J. Michalski, Meridiani Planum sediments on Mars formed through weathering in massive ice deposits. Nat. Geosci. **2**(3), 215–220 (2009)
K. Nishida, H. Terasaki, E. Ohtani, A. Suzuki, The effect of sulfur content on density of the liquid Fe-S at high pressure. Phys. Chem. Miner. **35**, 417–423 (2008)
D.K. Nordstrom, Mine waters: acidic to circumneutral. Elements **7**, 393–398 (2011)
H.S.C. O'Neill, J. Mavrogenes, The sulfide saturation capacity and the sulfur content at sulfide saturation of silicate melts at 1400 °C and 1 bar. J. Petrol. **43**, 1049–1087 (2002)
E. Ohtani, H. Yurimoto, S. Seto, Element partitioning between metallic liquid, silicate liquid, and lower-mantle minerals: implications for core formation of the Earth. Phys. Earth Planet. Inter. **100**, 97–114 (1997)
C.H. Okubo, K.W. Lewis, A.S. McEwen, R.L. Kirk, Relative age of interior layered deposits in southwest Candor Chasma based on high-resolution structural mapping. J. Geophys. Res. Planets **113**(E12), E12002 (2008)
R.J. Phillips, M.T. Zuber, S.C. Solomon, M.P. Golombek, B.M. Jakosky, W.B. Banerdt, D.E. Smith, R.M.E. Williams, B.M. Hynek, O. Aharonson, S.A. Hauck, Ancient geodynamics and global-scale hydrology on Mars. Science **291**, 2587–2591 (2001)
A. Pommier, F. Gaillard, M. Pichavant, Time-dependent changes of the electrical conductivity of basaltic melts with redox state. Geochim. Cosmochim. Acta **74**, 1 (2010)
F. Poulet, S. Erard, Nonlinear spectral mixing: quantitative analysis of laboratory mineral mixtures. J. Geophys. Res. Planets **109**(E2), E02009 (2004)

F. Poulet, C. Gomez, J.P. Bibring, Y. Langevin, B. Gondet, P. Pinet, G. Belluci, J. Mustard, Martian surface mineralogy from Observatoire pour la Mineralogie, l'Eau, les Glaces et l'Activite on board the Mars Express spacecraft (OMEGA/MEx): global mineral maps. J. Geophys. Res. Planets **112**(E8), E08S02 (2007)
F. Poulet, J.P. Bibring, Y. Langevin, J.F. Mustard, N. Mangold, M. Vincendon, B. Gondet, P. Pinet, J.M. Bardintzeff, B. Platevoet, Quantitative compositional analysis of martian mafic regions using the MEx/OMEGA reflectance data. Icarus **201**(1), 69–83 (2009)
F. Poulet, R.E. Arvidson, C. Gomez, R.V. Morris, J.P. Bibring, Y. Langevin, B. Gondet, J. Griffes, Mineralogy of Terra Meridiani and western Arabia Terra from OMEGA/MEx and implications for their formation. Icarus **195**(1), 106–130 (2008)
C. Quantin, P. Allemand, N. Mangold, C. Delacourt, Ages of Valles Marineris (Mars) landslides and implications for canyon history. Icarus **172**(2), 555–572 (2004)
R. Rieder, T. Economou, H. Wanke, A. Turkevich, J. Crisp, J. Bruckner et al., The chemical composition of Martian soil and rocks returned by the mobile alpha proton x-ray spectrometer: preliminary results from the x-ray mode. Science **278**, 1771–1774 (1997)
K. Righter, K. Pando, L.R. Danielson, Experimental evidence for sulfur-rich martian magmas: implications for volcanism and surficial sulfur sources. Earth Planet. Sci. Lett. **288**, 235–243 (2009)
K. Righter, M.J. Drake, Core formation in Earth's Moon, Mars, and Vesta. Icarus **124**, 513–529 (1996)
K. Righter, M.J. Drake, E. Scott, Compositional relationships between meteorites and terrestrial planets, in *Meteorites and the Early Solar System II*, ed. by D.S. Lauretta, H.Y. McSween (University of Arizona Press, Tucson, 2006), pp. 803–828
K. Righter, N.L. Chabot, Moderately and slightly siderophile element constraints on the depth and extent of melting in early Mars. Meteorit. Planet. Sci. **46**, 157–176 (2011). doi:10.1111/j.1945-5100.2010.01140.x
K. Righter, M. Humayun, Volatile Siderophile Elements in Shergottites: Constraints on Core Formation and Magmatic Degassing. 43rd LPSC Program, abstract number 2465 (2012)
A. Rivoldini, T. Van Hoolst, O. Verhoeven, A. Mocquet, V. Dehant, Geodesy constraints on the interior structure and composition of Mars. Icarus **213**(2), 451–472 (2011)
L.H. Roach, J.F. Mustard, M.D. Lane, J.L. Bishop, S.L. Murchie, Diagenetic haematite and sulfate assemblages in Valles Marineris. Icarus **207**(2), 659–674 (2010a)
L.H. Roach, J.F. Mustard, G. Swayze, R.E. Milliken, J.L. Bishop, S.L. Murchie, K. Lichtenberg, Hydrated mineral stratigraphy of Ius Chasma, Valles Marineris. Icarus **206**(1), 253–268 (2010b)
L. Rose-Weston, J.M. Brenan, Y. Fei, R.A. Secco, D.J. Frost, Effect of pressure, temperature, and oxygen fugacity on the metal-silicate partitioning of Te, Se, and S: implications for earth differentiation source. Geochim. Cosmochim. Acta **73**, 4598–4615 (2009)
A.E. Saal, E.H. Hauri, C.H. Langmuir, M.R. Perfit, Vapour undersaturation in primitive mid-ocean-ridge basalt and the volatile content of Earth's upper mantle. Nature **419**, 451–455 (2002)
P. Schiffman, R. Zierenberg, N. Marks, J.L. Bishop, M.D. Dyar, Acid-fog deposition at Kilauea volcano: a possible mechanism for the formation of siliceous-sulfate rock coatings on Mars. Geology **34**, 921–924 (2006)
E. Sefton-Nash, D.C. Catling, Hematitic concretions at Meridiani Planum, Mars: their growth timescale and possible relationship with iron sulfates. Earth Planet. Sci. Lett. **269**, 365–375 (2008)
M. Settle, Formation and deposition of volcanic sulfate aerosols on Mars. J. Geophys. Res. **84**, 8343–8354 (1979)
P.F. Shi, S.K. Saxena, Thermodynamic modelling of the C-H-O-S fluid system. Am. Mineral. **77**, 1038–1049 (1992)
S.W. Squyres, A.H. Knoll, Sedimentary rocks at Meridiani Planum: origin, diagenesis, and implications for life on Mars. Earth Planet. Sci. Lett. **240**(1), 1–10 (2005)
S.W. Squyres et al., The Spirit Rover's Athena science investigation at Gusev crater, Mars. Science **305**, 794–799 (2004)
S.W. Squyres, R.E. Arvidson, J.F. Bell, F. Calef, B.C. Clark, B.A. Cohen, L.A. Crumpler, P.A. de Souza, W.H. Farrand, R. Gellert, J. Grant, K.E. Herkenhoff, J.A. Hurowitz, J.R. Johnson, B.L. Jolliff, A.H. Knoll, R. Li, S.M. McLennan, D.W. Ming, D.W. Mittlefehldt, T.J. Parker, G. Paulsen, M.S. Rice, S.W. Ruff, C. Schröder, A.S. Yen, K. Zacny, Ancient impact and aqueous processes at Endeavour crater, Mars. Science **336**, 570–576 (2012)
B.D. Stanley, M.M. Hirschmann, A.C. Withers, CO2 solubility in Martian basalts and Martian atmospheric evolution. Geochim. Cosmochim. Acta **75**, 5987–6003 (2011)
A.J. Stewart, M.W. Schmidt, W. Van-Westrenen, C. Liebske, Mars: a new core-crystallization regime. Science **316**, 1323–1325 (2007)
R.B. Symonds, W.I. Rose, G.J.S. Bluth, T.M. Gerlach, Volcanic-gas studies: methods, results, and applications, in *Volatiles in Magmas, Reviews in Mineralogy*, vol. 30, ed. by M.R. Carroll, J.R. Holloway, (1994), pp. 1–66

K.L. Tanaka, J.A. Skinner, T.M. Hare, T. Joyal, A. Wenker, Resurfacing history of the northern plains of Mars based on geologic mapping of Mars Global Surveyor data. J. Geophys. Res. Planets **108**(E4), 8043 (2003)

S.R. Taylor, S.M. McLennan, *Planetary Crusts: Their Composition, Origin and Evolution* (Cambridge University Press, Cambridge, 2009), 378 pp.

E. Tertre, S. Castet, G. Berger, M. Loubet, E. Giffaut, Surface chemistry of kaolinite and na-montmorillonite at 25 and 60 °C: experimental study and modelling. Geochim. Cosmochim. Acta **70**, 4579–4599 (2006)

B.J. Thomson, N.T. Bridges, R. Milliken, A. Baldridge, S.J. Hook, J.K. Crowley, G.M. Marion, C.R. de Souza, A.J. Brown, C.M. Weitz, Constraints on the origin and evolution of the layered mound in Gale Crater, Mars using Mars Reconnaissance Orbiter data. Icarus **214**, 413–432 (2011)

F. Tian, M.W. Claire, J.D. Haqq-Misra, M. Smith, D.C. Crisp, D. Catling, K. Zahnle, J.F. Kasting, Photochemical and climate consequences of sulfur outgassing on early Mars. Earth Planet. Sci. Lett. **295**, 412–418 (2009). doi:10.1016/j.epsl.2010.04.016

J.N. Tosca, S.M. McLennan, B.C. Clark, J.P. Grotzinger, J.A. Hurowitz, A.H. Knoll, C. Schröder, S.W. Squyres, Geochemical modeling of evaporation processes on Mars: insight from the sedimentary record at Meridiani Planum. Earth Planet. Sci. Lett. **240**, 122–148 (2005)

N.J. Tosca, S.M. McLennan, M.D. Dyar, E.C. Sklute, F.M. Michel, Fe oxidation processes at Meridiani Planum and implications for secondary Fe mineralogy on Mars. J. Geophys. Res. **113**, E05005 (2008). doi:10.1029/2007JE003019

E. Tréguier et al., Overview of mars surface geochemical diversity through APXS data multidimensional analysis: first attempt at modelling rock alteration. J. Geophys. Res. **113**, E12S34 (2008). doi:10.1029/2007JE003010

M. Wadhwa, Redox states of Mars' upper mantle and crust from Eu anomalies in Shergottite pyroxenes. Science **291**, 1527–1530 (2001). doi:10.1126/science.1057594

P.J. Wallace, Volatiles in subduction zone magmas: concentrations and fluxes based on melt inclusion and volcanic gas data. J. Volcanol. Geotherm. Res. **140**, 217–240 (2005)

A. Wang et al., Sulfate deposition in subsurface regolith in Gusev crater, Mars. J. Geophys. Res. Planets **111**(E2), E02S17 (2006)

C. Wang, J. Hirama, T. Nagasaka, S. Ban-Ya, Phase equilibria of liquid Fe-S-C ternary. ISIJ Int. **11**, 1292–1299 (1991)

H. Wänke, G. Dreibus, Chemistry and accretion history of Mars. Philos. Trans. R. Soc. Lond. A **359**, 285–293 (1994)

C.M. Weitz, M.D. Lane, M. Staid, E.N. Dobrea, Gray hematite distribution and formation in Ophir and Candor chasmata. J. Geophys. Res. Planets **113**, 30 (2008)

C.M. Weitz, R.E. Milliken, J.A. Grant, A.S. McEwen, R.M.E. Williams, J.L. Bishop, B.J. Thomson, Mars Reconnaissance Orbiter observations of light-toned layered deposits and associated fluvial landforms on the plateaus adjacent to Valles Marineris. Icarus **205**(1), 73–102 (2010)

C.M. Weitz, J.L. Bishop, P. Thollot, N. Mangold, L.H. Roach, Diverse mineralogies in two troughs of Noctis Labyrinthus, Mars. Geology **39**, 899–902 (2011)

L. Wendt, C. Gross, T. Kneissl, M. Sowe, J.P. Combe, L. LeDeit, P.C. McGuire, G. Neukum, Sulfates and iron oxides in Ophir Chasma, Mars, based on OMEGA and CRISM observations. Icarus **213**(1), 86–103 (2011)

L. Wilson, J.W. Head, Mars: review and analysis of volcanic eruption theory and relationships to observed landforms. Rev. Geophys. **32**, 221–264 (1994)

S.M. Wiseman, R.E. Arvidson, R.V. Morris, F. Poulet, J.C. Andrews-Hanna, J.L. Bishop, S.L. Murchie, F.P. Seelos, D. Des Marais, J.L. Griffes, Spectral and stratigraphic mapping of hydrated sulfate and phyllosilicate-bearing deposits in northern Sinus Meridiani, Mars. J. Geophys. Res. Planets **115**, E00D18 (2010)

S.M. Wiseman et al., Phyllosilicate and sulfate-hematite deposits within Miyamoto crater in southern Sinus Meridiani, Mars. Geophys. Res. Lett. **35**(19), L19204 (2008)

J.J. Wray, E.Z.N. Dobrea, R.E. Arvidson, S.M. Wiseman, S.W. Squyres, A.S. McEwen, J.F. Mustard, S.L. Murchie, Phyllosilicates and sulfates at Endeavour Crater, Meridiani Planum, Mars. Geophys. Res. Lett. **36**, L21201 (2009)

J.J. Wray, R.E. Milliken, C.M. Dundas, G.A. Swayze, J.C. Andrews-Hanna, A.M. Baldridge, M. Chojnacki, J.L. Bishop, B.L. Ehlmann, S.L. Murchie, R.N. Clark, F.P. Seelos, L.L. Tornabene, S.W. Squyres, Columbus crater and other possible groundwater-fed paleolakes of Terra Sirenum, Mars. J. Geophys. Res. Planets **116**, E01001 (2011)

J.J. Wray, S.W. Squyres, L.H. Roach, J.L. Bishop, J.F. Mustard, E.Z.N. Dobrea, Identification of the Ca-sulfate bassanite in Mawrth Vallis, Mars. Icarus **209**(2), 416–421 (2010)

J. Zipfel, P. Scherer, B. Spettel, G. Dreibus, L. Schultz, Petrology and chemistry of the new shergottite Dar al Gani 476. Meteorit. Planet. Sci. **35**, 95–106 (2000)
M.Y. Zolotov, M.V. Mironenko, Timing of acid weathering on Mars: a kinetic-thermodynamic assessment. J. Geophys. Res. **112**(E7), E07006 (2007). doi:10.1029/2006JE002882
M.Y. Zolotov, Martian Volcanic Gases: are they Terrestrial-like? Lunar and Planetary Science XXXIV, abstract number 1795 (2003)

Space Sci Rev (2013) 174:301–328
DOI 10.1007/s11214-012-9940-y

Geochemistry of Carbonates on Mars: Implications for Climate History and Nature of Aqueous Environments

Paul B. Niles · David C. Catling · Gilles Berger · Eric Chassefière · Bethany L. Ehlmann · Joseph R. Michalski · Richard Morris · Steven W. Ruff · Brad Sutter

Received: 20 January 2012 / Accepted: 29 September 2012 / Published online: 25 October 2012

Abstract Ongoing research on martian meteorites and a new set of observations of carbonate minerals provided by an unprecedented series of robotic missions to Mars in the past 15 years help define new constraints on the history of martian climate with important cross-

P.B. Niles (✉) · R. Morris
Astromaterials Research and Exploration Science, NASA Johnson Space Center, Houston, TX 75080, USA
e-mail: Paul.b.niles@nasa.gov

D.C. Catling
Department of Earth and Space Sciences/Astrobiology Program, University of Washington, Seattle, WA 98195, USA

G. Berger
IRAP, CNRS-Université Toulouse, 31400 Toulouse, France

E. Chassefière
Laboratoire IDES, UMR 8148, Université Paris-Sud, CNRS, 91405 Orsay, France

B.L. Ehlmann
Division of Geological and Planetary Sciences, California Institute of Technology, Pasadena, CA 91125, USA

B.L. Ehlmann
Jet Propulsion Laboratory, California Institute of Technology, Pasadena, CA 91109, USA

J.R. Michalski
Planetary Science Institute, Tucson, AZ 85719, USA

J.R. Michalski
Department of Mineralogy, Natural History Museum, London, UK

S.W. Ruff
School of Earth and Space Exploration, Arizona State University, Tempe, AZ, USA

B. Sutter
Jacobs ESCG, Houston, TX 77258-8447, USA

cutting themes including: the CO_2 budget of Mars, the role of Mg-, Fe-rich fluids on Mars, and the interplay between carbonate formation and acidity.

Carbonate minerals have now been identified in a wide range of localities on Mars as well as in several martian meteorites. The martian meteorites contain carbonates in low abundances (<1 vol.%) and with a wide range of chemistries. Carbonates have also been identified by remote sensing instruments on orbiting spacecraft in several surface locations as well as in low concentrations (2–5 wt.%) in the martian dust. The Spirit rover also identified an outcrop with 16 to 34 wt.% carbonate material in the Columbia Hills of Gusev Crater that strongly resembled the composition of carbonate found in martian meteorite ALH 84001. Finally, the Phoenix lander identified concentrations of 3–6 wt.% carbonate in the soils of the northern plains.

The carbonates discovered to date do not clearly indicate the past presence of a dense Noachian atmosphere, but instead suggest localized hydrothermal aqueous environments with limited water availability that existed primarily in the early to mid-Noachian followed by low levels of carbonate formation from thin films of transient water from the late Noachian to the present. The prevalence of carbonate along with evidence for active carbonate precipitation suggests that a global acidic chemistry is unlikely and a more complex relationship between acidity and carbonate formation is present.

Keywords Mars · Carbonate · Climate · CO_2 · Water · Meteorites · Spectroscopy · Acidity · Atmosphere

1 Introduction

The history and distribution of water and carbon on Mars are closely related to the potential for life outside of the Earth (McKay et al. 1996). Carbonate minerals have long been seen as powerful tools with which to explore these fundamental relationships as they are intimately tied to both the water and the inorganic carbon cycle. Our understanding of the distribution and character of carbonates on the surface of Mars is undergoing rapid changes as several discoveries from recent robotic missions have clarified our view (Boynton et al. 2009; Ehlmann et al. 2008b; Morris et al. 2010). The nature and distribution of carbonates on Mars remains somewhat uncertain and subject to speculation, but some constraints can now be added. It now seems clear that carbonates do not exist as extensive, thick, laterally continuous bedrock units on the surface of the planet similar to those on Earth. Instead local deposits of carbonate have been discovered (Boynton et al. 2009; Ehlmann et al. 2008b; Morris et al. 2010), with the possibility of additional discoveries at smaller spatial scales or in mixed units excavated from the subsurface (Michalski and Niles 2010).

The wide ranging set of observations of carbonate minerals, provided by an unprecedented series of robotic missions to Mars in the past 15 years and ongoing research on martian meteorites, not only defines new constraints on the history of martian climate, but also opens unique windows into primordial martian aqueous environments. While questions about habitability remain unanswered at this time, we are obtaining more and more information about the environments in which water has existed on the martian surface. Here, we review the nature of carbonates detected in meteorites, from orbit, and from landed spacecraft. We then discuss the origin of these carbonates and the potential constraints that they can provide on the history of the martian climate and past environments. Based on this cumulative view of the studies of carbonate on Mars, several cross-cutting themes can be identified and discussed. The number of new discoveries of carbonate across the surface of the planet and the ongoing progress in analysis of carbonates in martian meteorites allow for new constraints to be placed on the martian CO_2 budget. Chemical data from landed missions, orbital remote sensing, and martian meteorites suggest that Mg-, Fe-rich carbonates may be more common on Mars and may have implications for martian aqueous environments. Finally, the work on carbonate formation has implications for planet-wide geochemical processes and especially the importance of acidic aqueous environments.

2 Carbonates in Martian Meteorites/Atmospheric CO_2 Isotopes

There are more than 45 samples of Mars that have now been identified and characterized in our meteorite collections (Meyer 2011). All of these samples are igneous rocks and show some degree of weathering, frequently in the form of carbonate minerals and in low abundance (1 % or less) (Bridges et al. 2001). A strong case has been made that at least some of these samples contain carbonate that formed on Mars and thus preserve the chemical signatures of aqueous environments on that planet. Evidence for a martian origin includes non-terrestrial stable isotopic compositions, petrographical relationships, and atypical chemical compositions that would be extremely difficult to explain as terrestrial contamination (Mittlefehldt 1994; Romanek et al. 1994; Treiman 1995).

2.1 Chemistry and Petrography of Carbonates

The most widely studied and well understood martian carbonates come from meteorites ALH 84001 (4.1 Ga orthopyroxenite), Nakhla (1.3 Ga clinopyroxenite), Governor Valadares (1.3 Ga clinopyroxenite), Lafayette (1.3 Ga clinopyroxenite), and EETA 79001 (173 Ma basalt) (Nyquist et al. 2001) (see also Bouvier et al. 2008, and Bogard and Park 2008 on shergottite age controversy). These rocks contain carbonate phases that cover a diverse chemical range between calcite, magnesite, and siderite (Fig. 1). The carbonates typically occur as small crystals (<100 μm) along cracks and within crushed zones of the rocks. They are sometimes associated with other aqueous alteration phases such as phyllosilicates, iron oxides, and iron sulfides (Bridges et al. 2001).

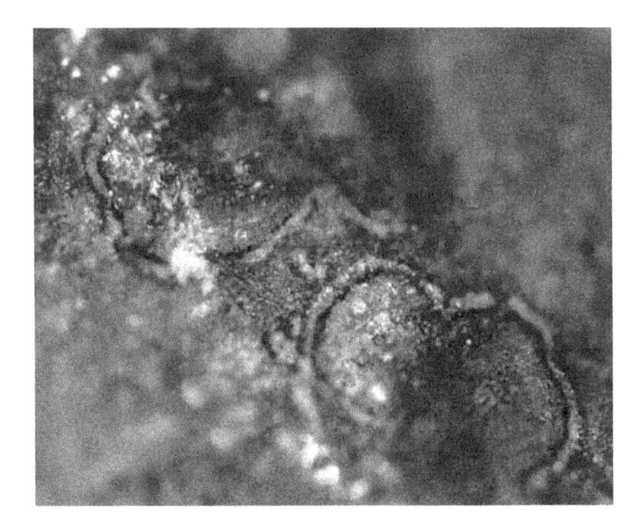

Fig. 1 Photograph of carbonate globules in ALH 84001. Orange central globules correspond to mixed Ca-Fe-Mg-rich carbonate while the outer white rims correspond to Mg-rich carbonate (photo credit Monica Grady)

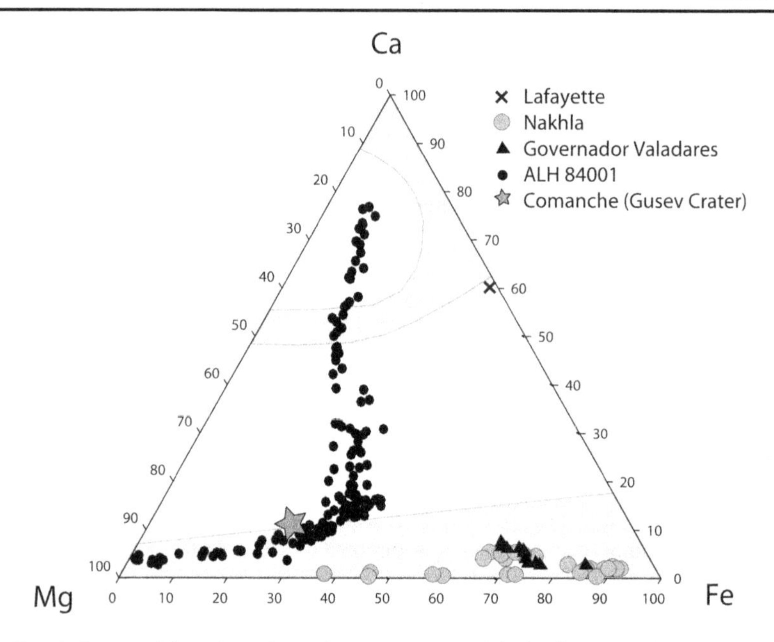

Fig. 2 Chemical composition of martian carbonates as reported in the literature. Figure is adapted from Bridges et al. (2001). The composition of the carbonate in the Comanche outcrop as measured by the Spirit rover (Morris et al. 2010) is also plotted here. The *light orange* fields are calculated stability fields for carbonate minerals at 700 °C (Anovitz and Essene 1987)

The carbonates found in ALH 84001 have unique chemical features which are extremely rare on Earth. ALH84001 is composed of less than 1 % carbonate minerals (Mittlefehldt 1994), which appear in patches and as globules up to ~300 μm in diameter along annealed fractures in the rock as well as within the granular bands (Harvey and McSween 1996; Mittlefehldt 1994; Treiman 1995). The carbonates are orange colored and appear in a variety of different habits. The best-known type of carbonate is the zoned rosettes (Fig. 1), which are pancake shaped and have well-defined Mg-rich rims. These rosettes exhibit strong chemical zoning (Harvey and McSween 1996). Typical concentric chemical zoning consists of an inner core of ankerite (Ca, Fe carbonate) or Ca, Fe, Mg solid solution carbonate, which gradually transitions to white magnesite at the rims (Figs. 1, 2) (Corrigan and Harvey 2004). The carbonates also possess unique bands of micro-scale magnetite crystals that are narrowly concentrated around the outer edges of the carbonate globule.

Among the nakhlites, three pre-terrestrial, chemically distinct populations of carbonate have been measured (Fig. 2). Gooding et al. (1991) reported vein filling carbonate that is very close to pure $CaCO_3$. This has not been reported in any other subsequent studies but may be reflected in isotope analyses (see Sect. 2.2 below). Both Nakhla and Governor Valadares contain Ca-poor and Fe-rich siderite crystals which are typically associated with silicate alteration zones. Lafayette contains Mn-rich siderite which is chemically distinct from the carbonates in Nakhla and Governor Valadares (Bridges and Grady 2000).

The chemical compositions of the ALH 84001 and Lafayette carbonates lie outside of any known equilibrium stability fields (Fig. 2) and have a mixed Ca, Fe, Mg composition that is commonly found in low temperature diagenetic concretions on Earth (Mozley 1989). This is the result of kinetic effects associated with the dehydration of Mg (Lippmann 1973) and rapid crystallization (Valley et al. 1997). The composition is consistent with other evidence

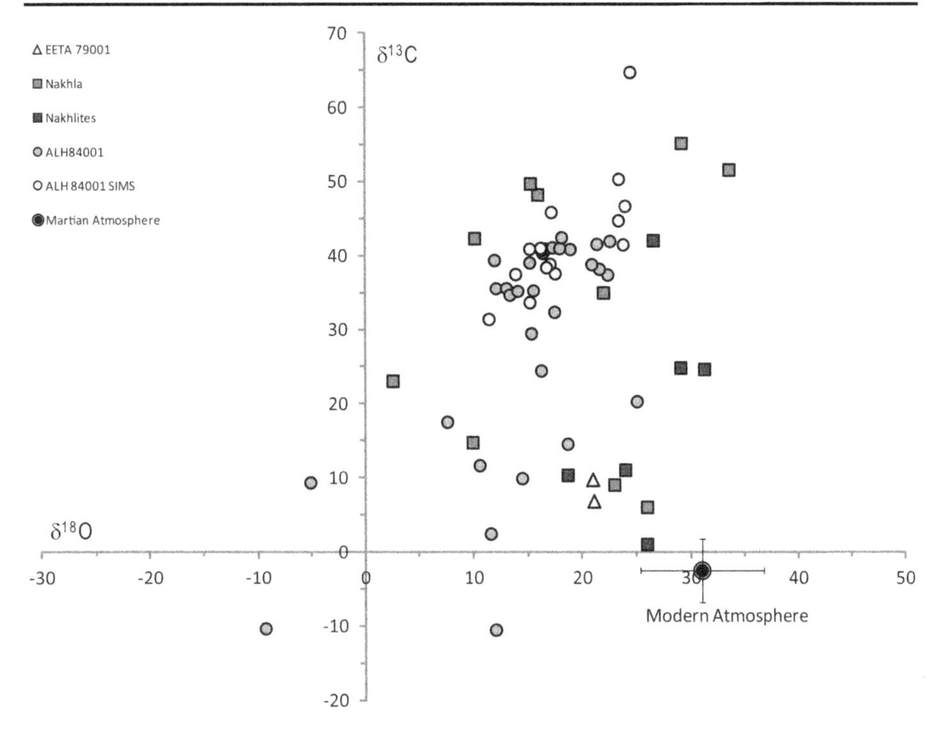

Fig. 3 Carbon and oxygen isotope cross-plot of martian meteorite carbonates after Niles et al. (2010). Data points reflect individual measurements of CO_2 liberated by interaction with phosphoric acid. The scatter in the data likely reflects the variety of different carbonate phases within the martian meteorites. These multiple episodes vary strongly in carbon isotopes and may reflect evolution of the martian atmosphere through time

that suggests that the carbonate-forming aqueous systems were short lived and subject to dynamic environmental changes (Changela and Bridges 2011; Niles et al. 2005; Valley et al. 1997). Experimental work performed to date has succeeded in forming carbonates with a similar chemical composition as the ALH 84001 carbonates (Golden et al. 2000, 2001), but not with the correct variation in isotopic composition. Likewise, analogous carbonates from Spitsbergen show similar chemical compositions but do not show similar isotopic variation (Treiman et al. 2002). Thus far no single laboratory process has been able to re-create both the chemical and isotopic compositions of the ALH 84001 carbonates.

2.2 Carbon and Oxygen Stable Isotopic Compositions of the Carbonates

The carbonates contained in martian meteorites have isotopic compositions that can be highly variable on the micro-scale (Valley et al. 1997) and also contain highly enriched carbon isotope compositions with δ^{13}C values as high as +64 ‰—much more enriched than sedimentary carbonates on Earth, which rarely exceed 10 ‰ (Knauth and Kennedy 2009; Niles et al. 2005; Romanek et al. 1994). The isotopic compositions have been measured using a variety of techniques including acid dissolution, pyrolysis-IRMS, and Secondary Ion Mass Spectrometry (SIMS). So far, only the acid dissolution technique allows for paired δ^{13}C and δ^{18}O analyses of the same sample and the results from those analyses are widely scattered (Fig. 3). Some of this scatter is likely due to the differences in carbonate chemistry (discussed above) and the different measurement methods used in each study. Some is also

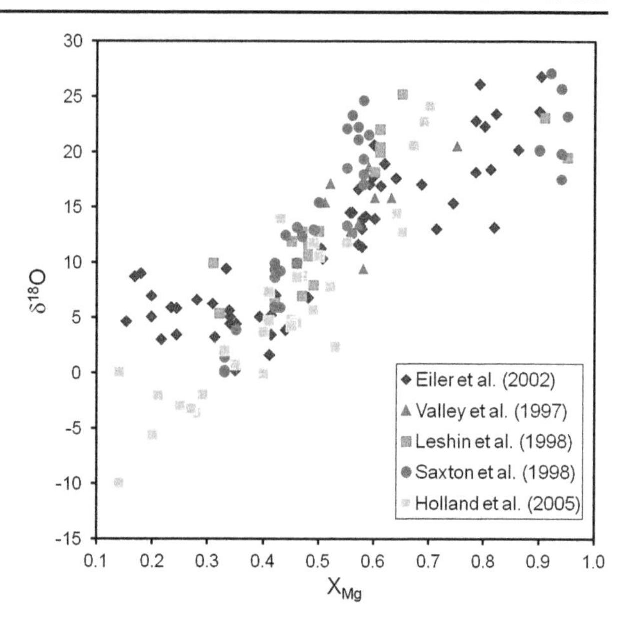

Fig. 4 Ion probe analyses of ALH 84001 carbonates reported in the literature and plotted versus magnesium composition of the analysis spot. Uncertainties on each data point range between 1 to 5 ‰ (2σ) depending on the study.

likely due to terrestrial contamination especially in the meteorites collected in Antarctica. Nevertheless, it is clear from this work that ALH 84001 and the nakhlites have carbonates with very high $\delta^{13}C$ (>40 ‰) (Jull et al. 1995, 2000; Romanek et al. 1994) compared to the composition of the modern martian atmosphere ($\delta^{13}C = -2.5 \pm 4.3$ ‰) (Krasnopolsky et al. 2007; Niles et al. 2010).

SIMS analyses allow for in-situ micro-scale measurement of isotopic composition and can therefore pair isotopic compositions with chemical compositions measured by electron microprobe. The results of these analyses for ALH 84001 show a wide variation in both $\delta^{18}O$ and $\delta^{13}C$ correlated with chemical composition (Fig. 4) (Eiler et al. 2002; Holland et al. 2005; Leshin et al. 1998; Niles et al. 2005; Saxton et al. 1998; Valley et al. 1997). The early-forming, Ca-rich phases were found to have the lowest $\delta^{18}O$ and $\delta^{13}C$ values and the Mg-rich, later-forming phases were found to have the highest $\delta^{18}O$ and $\delta^{13}C$ values (Leshin et al. 1998). The ALH 84001 carbonates have also been recently analyzed for their clumped isotope composition which suggests the carbonates formed between 14 and 22 °C in an evaporative environment (Halevy et al. 2011). The strong dependence of $\delta^{18}O$ and chemical composition is used to calculate $\delta^{18}O$ compositions for the published $\delta^{13}C$ data (Halevy et al. 2011). These values are plotted in Fig. 3 to compare the different analysis techniques which agree well.

Finally, triple oxygen isotope analyses of carbonates in Nakhla, Lafayette, and ALH 84001 have shown that carbonates in these meteorites have a large $\Delta^{17}O$ anomaly indicating that they formed from an atmospheric oxygen reservoir with limited contact with the silicate crust (Farquhar and Thiemens 2000; Farquhar et al. 1998). This not only confirms that these carbonates have a martian origin, but that they formed from water that was more closely associated with the atmosphere than with the silicate rocks. Thus it is unlikely that the carbonates formed from water derived from a subsurface hydrothermal system that was isolated from the atmosphere or long lived groundwater system. Instead, the data favor formation environments that featured ice or rain that had been out of contact with the silicate crust for long periods (Farquhar and Thiemens 2000; Farquhar et al. 1998). Geothermal heating that melted ice and mobilized saline fluids with limited water-rock interaction is a possible way to reconcile the inferences of atmospheric influence from triple oxygen isotope analysis with data suggesting a subsurface environment for the veined carbonates (e.g., Changela and Bridges 2011).

2.3 Formation Environments and Insight into Martian Climate History

A large number of studies have focused on the formation environment of the ALH 84001 carbonates, and two dominant competing views have emerged synthesizing all or most of the available evidence. One hypothesis contends that the ALH 84001 carbonates have a number of characteristics that indicate their low temperature ($< \sim 100$ °C) deposition in a dynamic environment during perhaps several different episodes (Corrigan and Harvey 2004; Halevy et al. 2011; Holland et al. 2005; McSween and Harvey 1998; Niles et al. 2005, 2009; Valley et al. 1997; Warren 1998). Another hypothesis suggests that the carbonates formed in a rapidly cooling environment which was initially at higher temperature (>150 °C) and cooled to ~ 30 °C (Eiler et al. 2002; Leshin et al. 1998; Romanek et al. 1994; Steele et al. 2007), although this is now directly disputed by recently reported clumped isotope results (Halevy et al. 2011).

Both scenarios have features in common, including an aqueous system that is short lived, largely low temperature (<100 °C), dynamically changing, and involves very small amounts of fluid. Neither scenario supports long lived aqueous environments created by warm climatic conditions. Evidence for silicate weathering in ALH 84001 is conspicuously absent for such an old rock (Treiman 1995), supporting the idea that any sustained warm and wet conditions on Mars were likely localized if present at all. Both formation scenarios also suggest that the elevated $\delta^{13}C$ values seen in the ALH 84001 carbonates are derived from the atmosphere at that time, suggesting that the carbon isotopic composition of the atmosphere at ~ 4.0 Ga, which corresponds to the beginning of the Noachian (Frey 2006), was much different from the carbon isotopic composition of the modern atmosphere (Niles et al. 2010).

One possible explanation for the carbon isotope enrichment in the ALH 84001 and nakhlite carbonates is through preferential atmospheric loss of ^{12}C in the Noachian following the loss of the magnetic field >4 Ga (Jakosky et al. 1994). Atmospheric loss would act to increase the $\delta^{13}C$ of the atmosphere early in martian history while the Sun's extreme ultraviolet (EUV) flux remained elevated (Jakosky et al. 1994). However, as the Sun's EUV flux decreased with time, volcanic degassing and carbonate deposition became the dominant processes, bringing the $\delta^{13}C$ of the atmospheric CO_2 back down closer to magmatic values (Grott et al. 2011; Manning et al. 2006; Niles et al. 2010). Another more speculative possibility for the enrichment of ^{13}C in the early atmosphere is that Mars originally had a methane-dominated reduced early atmosphere and minor CO_2 in equilibrium with this atmosphere would have $\delta^{13}C$ values near $+35$ ‰ due to the mass balance between CO_2 and CH_4 (Galimov 2000). A similar mechanism has also been proposed for a terrestrial deposit of $\delta^{13}C$-rich carbonates ($+5.4$ ‰ to $+19.0$ ‰) associated with natural gas-rich shales (Murata et al. 1967). Finally, the possibility exists that the magmatic CO_2 on Mars is substantially enriched in ^{13}C compared to the Earth. However, this is not supported by measurements from martian meteorites which in fact suggest that the carbon isotopic content of magmatic carbon on Mars is depleted in ^{13}C compared to the Earth (Wright et al. 1992). No other mechanisms have been proposed that might explain an enrichment in $\delta^{13}C$ of >50 ‰ for the magmatic CO_2 contained within the planet, and given the available evidence, it is most likely that the $\delta^{13}C$ of the martian atmosphere has changed substantially through martian history as a result of atmospheric loss, carbonate formation, and volcanic degassing.

Nakhla and the Nakhlite meteorite group have been interpreted to derive from different depths of a shallow layered igneous flow or intrusion. This is based on their mineralogy and groundmass textures (Reid and Bunch 1975) as well as variations in their secondary mineralogy (Bridges and Warren 2006). Particularly, Yamato 00593 shows evidence for faster cooling and fewer cumulus phases (Mikouchi et al. 2003) while Lafayette is considered to have cooled the slowest. Furthermore the secondary phases in the nakhlites are not evenly distributed, perhaps also reflecting different source depths in the original cumulate pile (Changela and Bridges 2011).

The different secondary mineral populations in each meteorite have been interpreted to reflect an evolving evaporative environment where Lafayette contained the least soluble phases (most likely derived from initial evaporation) and Nakhla represented the most soluble phases reflecting the highest degree of evaporation. More recently a hydrothermal model has been proposed that suggests that fluids of moderate temperatures (<150 °C) were mobilized from subsurface ice deposits, progressively altering the rocks in the cumulate pile during ascent and evaporation at the surface (Changela and Bridges 2011). Under this model, the elevated $\delta^{13}C$ values in the nakhlite carbonates were likely obtained through a remobilization of more ancient carbonate deposits (Niles et al. 2010) rather than through exposure to the atmosphere during the time of their formation, and the $\Delta^{17}O$ anomaly was inherited from the water source being atmospherically derived.

3 Carbonates Detected on Mars

The identification of carbonates on Mars was for many years limited to those found in the martian meteorites. No carbonates were found in the first global search undertaken with the Thermal Emission Spectrometer (TES) on the Mars Global Surveyor (Christensen et al. 2001). The TES instrument produced near global coverage in the thermal infrared (TIR) region (~6–50 μm, 1670–2000 cm^{-1}) at a spatial resolution of ~3×6 km but did not detect any large-scale (10's of kilometers) occurrences of carbonates at abundances greater than ~10 % (Christensen et al. 2001). This included the long considered candidate on the floor of Pollack crater known as "White Rock" (Ruff et al. 2001). However, TIR spectra show evidence for a low abundance of carbonate minerals in the globally homogenous dust (Sect. 3.1). Subsequent higher resolution observations in the near IR have provided numerous detections of carbonate minerals at the outcrop scale (Sect. 3.2; Fig. 5). In addition, both the Phoenix lander and Spirit rover have detected carbonates on the surface of Mars (Sect. 4). Unfortunately, extensive ground-truthing of remote sensing observations has not occurred, making comparison between remote sensing observations, and landed or sample-based observations difficult to understand.

3.1 Carbonates in Martian Dust

Martian dust is continuously carried into suspension by eolian activity and distributed around the planet through the atmosphere until it settles and accumulates on the surface. TIR measurements of this dust from orbit and by the two Miniature Thermal Emission Spectrometers (Mini-TES) onboard the Mars Exploration Rovers (MER) have demonstrated a common spectral character that suggests uniform mineralogy and particle size wherever the dust is observed (Christensen et al. 2004; Yen et al. 2005). Bandfield et al. (2003) measured TIR spectra of physical mixtures of labradorite as a proxy for martian dust combined with various carbonate minerals and found that a small amount (~2 to 5 weight %) of fine-particulate

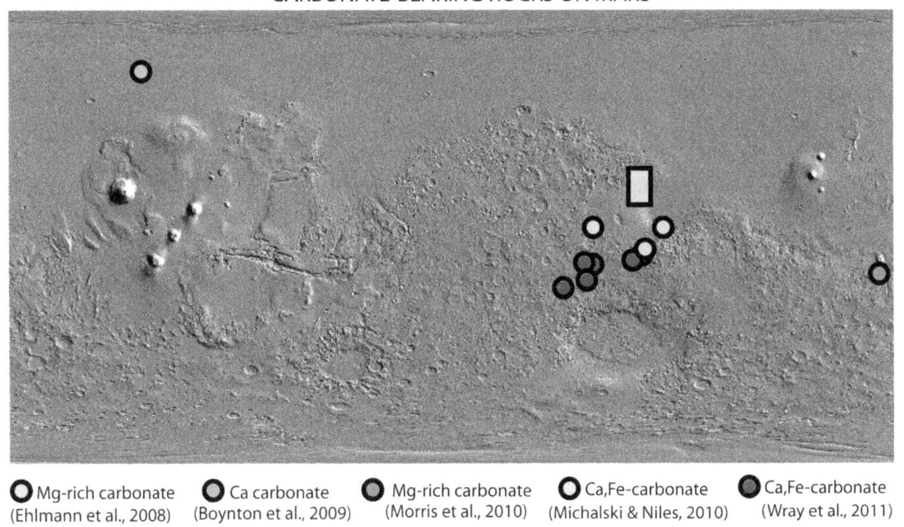

Fig. 5 Map of identified carbonate locations on Mars

magnesite (<10 micron) provided the best fit to distinctive features above 1300 cm^{-1} in TES spectra of dust. Typically such a small abundance of carbonate would not be detectable, but silicates are relatively transparent at high wavenumbers where carbonates exhibit an emissivity peak near 1500 cm^{-1} ($\sim$6.5 μm) that arises due to intense volume reflections at the frequency of C-O stretching vibrations. This is not observed in other mineral groups besides hydrous iron sulfates discussed below; therefore, the emissivity peak in the martian dust is interpreted as evidence for a carbonate admixture in fine-grained silicate material.

The precise wavelength of the emissivity peak is an indication of the composition of the carbonate. Different cations within the carbonates, typically Mg^{2+}, Ca^{2+}, or Fe^{2+}, affect the bond lengths and vibrational frequency of the C-O bonds. The position of the emissivity peak at 6.5 μm in TES spectra of martian dust is most consistent with the presence of Mg-carbonates (Bandfield et al. 2003). Although the abundance of carbonate in martian dust appears low, the ubiquity and uniformity of dust across Mars suggest that it may represent a sink as large as 1–3 bars for atmospheric CO_2 (Bandfield et al. 2003) (see Sect. 6.1 below).

Lane et al. (2004) showed with laboratory data that a similar spectral shape in the 6.5 μm region could be achieved if hydrous iron sulfates (HIS) are present in the dust, however, the spectral feature at 6.5 μm is much weaker in the dust mixtures containing sulfate than it is in mixtures containing carbonate. Therefore the detection of carbonate in the fine grained dust remains the best explanation for the spectral data presented thus far, but it does not rule out some contribution that might be made by HIS. Measurements made by MSL (Mass Science Laboratory) will help clarify this issue.

3.2 Carbonates Detected on the Surface by Remote Sensing

Despite the dominance of volcanic and impact-related landforms, Mars also displays a complex sedimentary record (Malin and Edgett 2000). Spectroscopic data have revealed a diverse suite of silicate and sulfate alteration minerals associated with martian rocks and sediments. In contrast, spectroscopic detection of carbonates has proved challenging, and only recently have data shown the first robust detections. Known occurrences of carbonates detected from orbit can be categorized as: layered-to-massive, irregular deposits of Mg-rich carbonates in olivine-bearing Noachian bedrock units (Ehlmann et al. 2008b), sedimentary deposits that host Fe/Mg smectite clays and carbonates (Ehlmann et al. 2008a, 2009), and layered carbonate-bearing rocks exhumed from several kilometers depth in the crust by impact (Michalski and Niles 2010).

Layered-to-massive deposits of carbonates were detected in Noachian terrains using near-infrared reflectance data from the Compact Reconnaissance Imaging Spectrometer for Mars (CRISM) (Ehlmann et al. 2008b). While 1000s of CRISM images have been analyzed in search for carbonates (Murchie et al. 2009), the detections to date are generally limited to the well-exposed, ancient terrain around the Nili Fossae and slightly further south in Libya Montes and Terra Tyrrhena (Ehlmann et al. 2008b). Throughout this region, the oldest part of the stratigraphy, where visible, consists of Fe-Mg smectite clay and low-calcium pyroxene bearing materials. The Fe/Mg-clay unit is overlain by olivine-rich materials interpreted as impact cumulates from the Isidis melt sheet or fluid, komatiite-like lavas. Moving up in section, an aluminous clay-bearing unit is occasionally present, which is in turn overlaid by a dark, mafic capping unit. The carbonate detections are typically located within the olivine-rich stratigraphic unit. The carbonate is always stratigraphically above the Fe/Mg-clay unit. The relationship between carbonate and aluminous clays is not easily correlated; however, carbonates sometimes underlie the aluminous clays (Ehlmann et al. 2008b, 2009; Mustard et al. 2009). Because of their association with the olivine-rich unit that postdates the Isidis impact and capping by Syrtis Major volcanics that formed in the Early Hesperian (Hiesinger and Head 2004), the carbonates are interpreted as Noachian materials. Sedimentary paleo-lake deposits within Jezero crater also contain carbonates associated with Fe/Mg smectites. These clays and carbonates are likely to have been transported, rather than forming *in situ* in a standing body of water, because they are identical to mineralogic units found in the highlands of the crater's watershed (Ehlmann et al. 2008a).

The evidence for carbonate from CRISM data rests in the co-occurrence of spectral absorptions related to C-O vibrational overtones located at ($\lambda =$) 2.3–2.35, 2.5–2.54, 3.4, and 3.9 μm (Fig. 6) (Calvin et al. 1994; Ehlmann et al. 2008b; Gaffey 1987). As in the thermal infrared, the location of these features is an indication of composition of the carbonates for the same reasons outlined in Sect. 3.1 above. In the case of Nili Fossae carbonates, the band positions at 2.31 μm and 2.51 μm are most consistent with Mg-rich carbonates similar to magnesite, that are hydrated and have only a minor amount of Fe or Ca substitution (Brown et al. 2010). Spectral mixing in the near infrared is nonlinear, and the strengths of spectral absorptions are not a clear indication of mineral abundance.

Three hypotheses have been proposed to explain the origin of the Nili-Fossae-type carbonates. One possibility is that the clay and carbonate units formed simultaneously in a single deep hydrothermal system in zones of different temperature and fluid composition (Brown et al. 2010; van Berk and Fu 2011). A second variant of the hydrothermal model is that the Mg carbonate formed in the shallow subsurface due to diagenetic or low-T hydrothermal alteration of olivine, sometimes in the presence of CO_2; this may be consistent with the occasional detection of serpentine associated with the olivine unit (Ehlmann et al. 2009). The third possibility is that surface weathering occurred, potentially in association with enhanced precipitation and runoff near the end of the Noachian (Ehlmann et al. 2009).

Carbonates have also been detected from orbit within impact craters, in rocks exhumed from deep in the martian crust by impact (Michalski and Niles 2010). The deep crust of Mars is generally poorly exposed because of the lack of a plate tectonic-like mechanism to drive

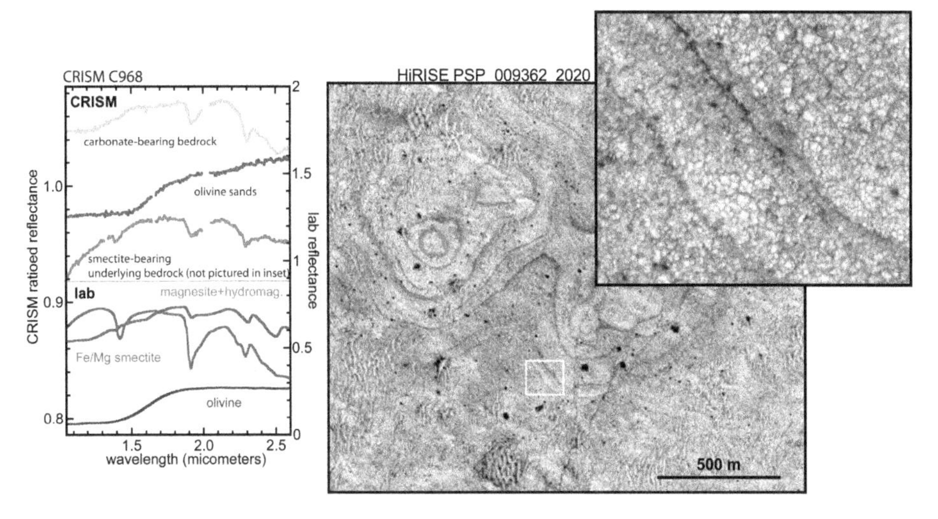

Fig. 6 Carbonate plains in Nili Fossae. A distinctive stratigraphy of olivine and carbonate-bearing rocks overlying Fe/Mg smectite bearing rocks are observed throughout the region, partially obscured by olivine sands. Inset image is 150 m wide and the carbonate bearing bedrock has a banded appearance (Mustard et al. 2009) and is fractured at a few meter scale. [Credits: NASA/UA/HiRise and NASA/JPL/JHUAPL/CRISM]

uplift and resurfacing. One exception is within the central peaks of impact craters, which sample crust at a depth roughly 1/10 the diameter of the crater (Cintala and Grieve 1998; Pilkington and Grieve 1992). Leighton crater, located southwest of the Syrtis Major region of Mars, shows carbonates exhumed from $\sim$6 km depth that are different from those observed in the previously described categories because these deep carbonates appear to be Ca- and/or Fe-rich rather than Mg-rich, as determined by band centers at 2.35 and 2.53 μm (Michalski and Niles 2010). They occur in discrete layers, though these layers were highly deformed and disrupted by the impact event. Carbonates may also be a component of the surrounding phyllosilicate-bearing bedrock, which contains evidence for low-T hydrothermal or metamorphic minerals such as prehnite, pumpellyite or chlorite, and chlorite-smectite mixed layer clays.

Several hypotheses have been proposed to explain the occurrence of these deep carbonates (Michalski and Niles 2010). The most likely scenario is that a stratigraphic section of Fe/Mg-rich clays overlaid by carbonates and then by aluminous clays existed in this region, and that section was buried to great depth by volcanic flows from volcanism in Syrtis Major. Low-T metamorphism occurred due to the higher thermal gradient at that time, and due to heat supplied by the overlying volcanics. A second hypothesis resembles the near-surface hydrothermal alteration proposed for Nili Fossae (Ehlmann et al. 2008b, 2009) and suggests that CO_2-rich fluids descended from the surface and altered the deep (5–10 km) basaltic crustal material upon mixing with deep, reduced crustal fluids. However, the ability to transport sufficient amounts of CO_2 from the surface to depth remains a hurdle for this model (Michalski and Niles 2010).

Recent analyses have suggested that carbonates mixed with clay minerals are exposed in more craters throughout the southern Nili Fossae-Terra Tyrrhena region (Wray et al. 2011). Regardless of the mechanism by which they formed, these materials hint at a possible deep cache of carbonate hidden from view that may account for a significant amount of CO_2 sequestered from the atmosphere.

4 Carbonates Detected *In Situ*

4.1 Comanche Carbonate at Gusev Crater

The Mars Exploration Rover (MER) Spirit identified outcrops rich in Mg-Fe carbonate (16 to 34 wt.%) on the southern slope of Haskin Ridge in the Columbia Hills of Gusev crater (Fig. 7) (Morris et al. 2010). The carbonate identification in the Comanche outcrops is based on the aggregate of mineralogical and chemical data from Spirit's Mössbauer spectrometer (MB), Miniature Thermal Emission Spectrometer (Mini-TES), and Alpha Particle X-Ray Spectrometer (APXS). The excess of light elements (e.g., H_2O, CO_2, or NO_2) from APXS has been interpreted as carbonate on the basis of the mineralogical identifications of the phase by MB and Mini-TES. Assignment of the carbonate to Mg-Fe-rich carbonate was based on the intersection of the combined constraints of MB (Mg-Fe-rich carbonate), Mini-TES (Ca-Mg-rich carbonate), and APXS (low total Ca concentration). The combined MB and APXS data were used to calculate a chemical composition of $Mc_{0.62}Sd_{0.25}Cc_{0.11}Rh_{0.02}$ for the Comanche carbonate, where Mc = magnesite ($MgCO_3$), Sd = siderite ($FeCO_3$), Cc = calcite ($CaCO_3$), and Rh = rhodochrosite ($MnCO_3$). This composition is very similar to the average composition of carbonate found in martian meteorite ALH 84001 of $Mc_{0.58}Sd_{0.29}Cc_{0.12}Rh_{0.01}$ at ~1 wt.% abundance (Mittlefehldt 1994; Treiman 1995) (see also Fig. 2).

Independent calculations using combined MB data, APXS data, and Mini-TES data provide the basis for estimating that carbonate is a major component of the Comanche outcrops at 16 to 34 wt.%. The remote sensing capability of the Mini-TES instrument shows that the carbonate is present throughout the visible Comanche outcrops (~50 m^2) and not just at the two small MB and APXS analysis spots (7 to 30 cm^2).

Other phases associated with the Comanche carbonate are olivine, npOx, and hematite (49 %, 19 %, and 8 % of total Fe, respectively) according to MB and olivine (Fo68) and amorphous silicate (33 % and 33 %, respectively) according to Mini-TES (Morris et al. 2010). Recently, a mixture of Mg-rich carbonate and olivine also was identified in the region of the Comanche outcrops using spectra from the CRISM instrument (Carter and Poulet 2012). The carbonate detections by MER and CRISM at Gusev crater and by CRISM at Nili Fossae (Ehlmann et al. 2008b) thus have in common Mg-Fe-rich chemistry and an association with olivine. Assuming the hematite associated with the Comanche carbonate is a

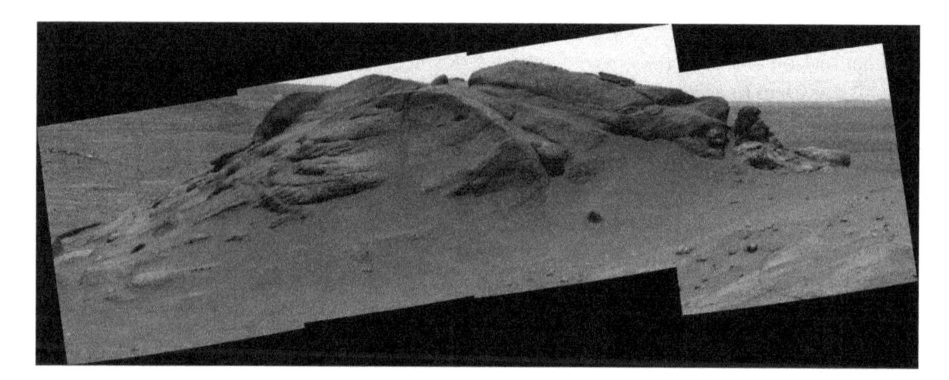

Fig. 7 Pancam false color mosaic of the Comanche outcrop on Gusev crater Mars. Spirit rover sol 695, Pancam sequence P2422 [Credit: NASA/JPL/Pancam]

precipitation product and not a thermal oxidation product of pre-existing magnetite, precipitation of the carbonate and oxides under conditions of low water activity (e.g., hydrothermal and/or cryogenic conditions and/or concentrated brines) are implied. Low water activity of a Mg-rich fluid is the favored condition for Mg-rich carbonates to be formed (see Sect. 6.2, below).

4.2 Phoenix Landing Site Carbonates

The Phoenix Mars lander analyzed several samples of the martian soil in the north polar region near 68°N (Smith et al. 2009b). One of the instruments used in the analysis was the Thermal and Evolved Gas Analyzer (TEGA), which consisted of a high temperature furnace paired with a magnetic sector mass spectrometer. Results from the TEGA analyses were consistent with the detection of carbonate minerals within the soils near the lander, and this was corroborated by analyses from the Wet Chemistry Laboratory (WCL) on the same lander (Boynton et al. 2009; Kounaves et al. 2010a).

TEGA measured an endotherm between 725° and 820° with a corresponding CO_2 release that was consistent with the presence of 3 to 6 % calcite, ankerite, dolomite, and/or another calcium-rich carbonate in the Phoenix landing site soil (Fig. 8) (Boynton et al. 2009; Sutter et al. 2012). In addition, the soil pH (7.7 ± 0.3), [Ca^{2+}], and [Mg^{2+}] measured by the WCL instrument is consistent with the presence of carbonate minerals in the soil (Kounaves et al. 2010b). TEGA also detected a release of CO_2 at a low temperature (between 400° and 680 °C) that has been interpreted to be due to either Fe, Mg

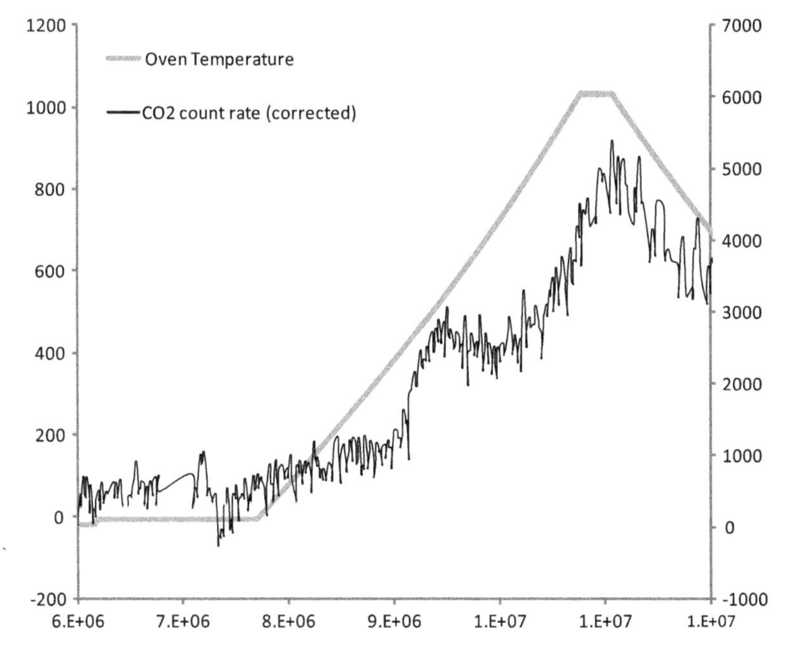

Fig. 8 Measurement of the Wicked Witch soil by TEGA after Boynton et al. (2009). The corrected count rate for CO_2 is the mass 44 count rate measured during the analysis run corrected for background by subtraction of the mass 44 count rate measured during a reanalysis of the same sample on a subsequent day. Two CO_2 peaks are apparent suggesting two separate releases of CO_2. The higher temperature release has been interpreted to be from calcite by Boynton et al. (2009) and Sutter et al. (2012). The lower temperature release is less constrained and could be either an Mg-, Fe-rich carbonate phase or possibly organic matter

carbonate (~1.0 wt.% equivalent carbonate) or an organic material (Boynton et al. 2009; Sutter et al. 2012). One further possibility is that this lower temperature CO_2 release was caused by reaction between carbonate and HCl formed during decomposition of Mg-perchlorate (Cannon et al. 2012).

Several possibilities exist for the formation of the carbonates at the Phoenix landing site. They could have been derived from the Vastitas Borealis and Scandia deposition events, deposited as carbonate-bearing dust, and/or formed by pedogenic processes. The Phoenix landing site is situated on the ejecta from the Heimdall Crater (Heet et al. 2009) that might contain material from both the mid to late Hesperian Vastitas Borealis Formation (VBF) and the early Amazonian Scandia material which are both penetrated by Heimdall Crater (Salvatore et al. 2010; Tanaka et al. 2008). The VBF has been speculated to have formed from sediments generated by outflow channels during the Hesperian (Tanaka et al. 2011) and therefore might contain carbonate minerals eroded during those events or formed during the evaporation or freezing of the outflow channel effluents.

Another possibility is that the carbonates at the Phoenix landing site were originally derived from the Mg-carbonate bearing dust (2–5 wt.%) (Bandfield et al. 2003) on Mars. Subsequent pedogenic processes could result in dissolution and reprecipitation of calcite and/or Ca-Mg carbonate (Sutter et al. 2012). This is a common reaction in soils overlying limestone where the dissolution of carbonate parent rock (e.g., limestone) is followed by reprecipitation of pedogenic carbonate (Rabenhorst et al. 1984).

Finally, the carbonate may have formed in-situ at the Phoenix landing site through interaction of atmospheric CO_2 and ephemeral water films (Boynton et al. 2009). The presence of shallow subsurface ice in this location could provide thin films of water during diurnal and seasonal cycles (Smith et al. 2009a; Zent et al. 2010) which could provide conditions necessary for pedogenic carbonate formation. Obliquity variations may also have created conditions warm enough to melt subsurface ice and provide liquid water for pedogenic carbonate formation. Indeed, perchlorate salts (mainly of Mg form) are known to be present at the Phoenix site, and these allow brines to form at a eutectic temperature of about −68 °C (Hecht et al. 2009).

The various hypotheses for the origin of Phoenix landing site carbonates suggest that such carbonates could be widespread across the northern plains and possibly beyond, although it is unclear whether they are confined to the upper few centimeters that were sampled by the Phoenix lander or if they are present at greater depths. Alkaline pH estimates (7.4–8.7) for the Viking soils based on the Viking Landers' gas exchange experiment (Quinn and Orenberg 1993) could suggest that soil carbonates may in fact be globally present at low enough levels to remain below the detection limits of the orbiting spectrometers.

5 Models of Martian Carbonate Chemistry

Martian carbonates have now been identified from orbit, on the surface, and in martian meteorites; yet the specific environmental conditions under which these carbonates formed remain relatively unconstrained. While specific constraints can be derived from martian meteorites through detailed chemical analysis, it is difficult to determine if these few samples reflect typical conditions elsewhere. Several studies have been conducted examining potential carbonate forming environments that span the range of climatic scenarios from warm/wet to cold and dry. Before the global mapping of mineralogy in the near and thermal infrared, models of martian aqueous chemistry entertained the possibility that carbonates might be globally widespread (Catling 1999), whereas subsequent models have now had to contend with explaining why carbonate outcrops are localized phenomena.

5.1 Warm/Wet Environments

Pollack et al. (1987) predicted that carbonates would be abundant on the surface and would thus indicate the presence of an early dense CO_2 atmosphere. Their model suggested that while CO_2 would rapidly be removed by carbonate formation through weathering of basalt, a dense CO_2 atmosphere (1 to 5 bars) could be maintained by constant recycling of CO_2 back into the atmosphere through rapid burial and thermal decomposition of carbonate minerals. This envisioned a warm, wet Noachian environment with rainfall and standing bodies of water stabilized by a dense CO_2 atmosphere.

This scenario along with others that depend on a dense early CO_2 greenhouse have been challenged by climate models that cannot simulate warm enough conditions using a CO_2 greenhouse as well as the localized nature and relative paucity of carbonates on the surface of Mars. In the early 1970s, Sagan and Mullen (1972) realized that the long-term increase in the luminosity of the Sun predicted by astrophysical theory would have consequences for the evolution of planetary climates and atmospheres. Specifically, over time, the Sun's core contracts and gets denser and hotter as hydrogen is fused into helium. A gradual increase in the temperature-sensitive nuclear fusion reactions has led to a rise in solar luminosity by ~25–30 % since ~4 Ga (Bahcall et al. 2001; Gough 1981). So, if the widespread fluvial activity in the early and middle Noachian had been sustained by a warmer climate on early Mars, the early atmosphere must have been greatly different from today (Pollack et al. 1987).

Modeling a water vapor-CO_2 greenhouse on early Mars has remained problematic. Initial calculations showed a water vapor-CO_2 greenhouse effect in a 5 bar CO_2 atmosphere would be sufficient to sustain liquid water on early Mars with solar luminosity 30 % less than present (Pollack et al. 1987). However, a revision to the calculations indicated that CO_2 ice clouds would start to condense out of the atmosphere at around ~0.4 bar surface pressure so that 5 bar surface pressure would not be attained (Kasting 1991). Subsequently, there was some debate about whether CO_2 ice clouds themselves might have a net warming effect (Forget and Pierrehumbert 1997), but the most comprehensive calculations done to date show that the greenhouse effect from such clouds would, on average, be unable to warm the early martian surface above freezing (Colaprete and Toon 2003). The reasons are that CO_2 cloud particles grow rapidly and precipitate, leading to fast cloud dissipation, while any warming is self-limiting because by heating the air, the clouds cause themselves to disappear. Other greenhouse gas candidates have been proposed along with alternative warming mechanisms (Haberle 1998). A pCO_2 of 1.5–2 bars coupled with ~1 % methane might globally warm early Mars above freezing (Kasting 1997). However, such levels of methane would require a methane source exceeding that of the modern terrestrial biosphere because methane is destroyed relatively rapidly in geologic time by ultraviolet photolysis and oxidation. Others have proposed that volcanic SO_2, which is also a greenhouse gas, could have warmed Mars' early climate (Halevy et al. 2007; Johnson et al. 2008, 2009; Postawko and Kuhn 1986). However, the photochemistry of SO_2 rapidly produces sulfate aerosols (even in reducing atmospheres), and sulfate aerosols reflect sunlight efficiently, producing a net global cooling (Tian et al. 2010). Nonetheless, warming due to SO_2 production by volcanism or impact events may provide brief warming periods (Halevy and Head 2012; Johnson et al. 2009).

5.2 Hydrothermal Environments

Because the hypothesis of thick CO_2 atmospheres on early Mars has various problems, a proposed alternative to a continuous warm, wet climate on early Mars is that much of the aqueous activity on Mars occurred in subsurface hydrothermal systems powered by magmatic or impact activity (Ehlmann et al. 2011; Griffith and Shock 1995; Newsom 1980; Squyres and Kasting 1994). On early Mars, several heat sources for near-surface hydrothermal systems would have existed, including secular planetary cooling (Parmentier and Zuber 2007), volcanism (Phillips et al. 2001; Werner 2009), and large impacts (Frey 2008; Werner 2008). In this scenario, much of the observed alteration to clay minerals (Ehlmann et al., this issue) and formation of thick sulfate deposits would have occurred in groundwater driven systems in a warmer subsurface with only transiently clement surface condition (Ehlmann et al. 2011).

While such systems do not produce the massive sedimentary carbonates once predicted for Mars (Pollack et al. 1987), carbonates may have formed easily in lower temperature subsurface hydrological systems, where they were protected from acidic, oxidizing surface environment. On Earth, hydrothermal systems in mafic rock types result in progressive alkalinization of the hydrothermal fluid; in ultramafic systems, serpentinization occurs, a process in which hydrogen gas and methane gas are produced by the alteration of olivine (Etiope et al. 2011).

A means of communication with the atmosphere or a carbon-rich magmatic fluid is required for carbonate formation at substantial depth in hydrothermal groundwater systems. On Earth, the mixing of the high-pH hydrothermal fluid derived from serpentinization with cold sea water is observed to produce calcite in submarine, ultramafic rock-hosted, carbonate mineral deposits precipitating at hydrothermal vents (Palandri and Reed 2004). Serpentinization of ultramafic rocks could convert crustal carbonate into CH_4 through the reaction of dissolved CO_2 with H_2, similar to what is observed in hydrothermal systems of mid-ocean ridges (Atreya et al. 2007; Oze and Sharma 2005). Hydrothermal alteration of basaltic crust could produce CH_4 and sequester CO_2 in carbonates: the low intrinsic oxygen fugacity (fO_2) of martian basalt may force oxidized carbon supplied by magmatic degassing or crustal carbonates to reduce to CH_4 during water-rock interaction (Lyons et al. 2005). Steele et al. (2007) argued that the carbonates in ALH 84001 formed under similar high temperature, low fO_2 conditions resulting in the simultaneous formation of carbonates and macromolecular carbon.

Calculations using a shergottite-like host rock predict that a great majority of the precipitated carbonate would be in the form of calcite (Griffith and Shock 1995). During the progressive shrinking of the water table in the Hesperian and Amazonian, carbonate deposits may have migrated deeper into the crust, down to the present depth of the water table several kilometers below the surface (Chassefiere and Leblanc 2011a). If 2.5 % carbonate formed and was dispersed throughout the upper 5 km of crust, this would result in the sequestration of 5 bars of CO_2 (Griffith and Shock 1995). This is probably an upper bound for the carbonate content of the crust, but if globally distributed, even lower concentrations over a lesser depth constitute a substantial carbon reservoir, relative to the current 6 mbar atmosphere.

Evidence for this type of hydrothermal circulation may have recently been discovered by the detection of deep crustal carbonates from the spectral analysis of the central peak of the Leighton crater using CRISM data (Michalski and Niles 2010). The carbonates detected in this location alongside hydrated silicates very closely resemble the assemblage predicted by Griffith and Shock (1995). Mg-rich carbonates detected at Nili Fossae are also associated with hydrated silicates and may also have formed from hydrothermal activity (Brown et al. 2010; Ehlmann et al. 2008b; Gaudin et al. 2011).

5.3 Current Martian Conditions

The current atmosphere of Mars is cold enough and at low enough pressure to be well below the triple point of water on average, indicating that liquid water is not stable at the surface (Richardson and Mischna 2005). However, several studies have suggested that transient liquid water may in fact be possible under the current martian climate (Hecht 2002; Ingersoll 1970; Kahn 1985) and there is evidence for the formation of small channels by some sort of liquid in the past few years (McEwen et al. 2011). Thus the current conditions on the martian surface may be sufficient to enable carbonate formation to be an active process on the surface today.

Kahn (1985) suggested that carbonate formation drew down the atmospheric pressure of CO_2, thereby reducing temperatures and inhibiting liquid water and carbonate formation in negative feedback. In that case, the rate of CO_2 sequestration would reach balance with the rate of outgassing. This hypothesis suggests that the current pressure of CO_2 on Mars is regulated by ongoing carbonate formation and requires that carbonate formation be possible under the cold and dry conditions of the modern martian climate.

Several studies suggest that carbonate formation under modern martian conditions is possible and does indeed occur. Carbonates have been formed under simulated martian conditions in the laboratory (Booth and Kieffer 1978; Shaheen et al. 2010). The carbonate in fine dust could result from relatively dry conditions where several hundred monolayers of H_2O around dust grains generate carbonate over geologic time. Laboratory experiments with basaltic fines have shown that carbonate can be produced with 10^2–10^3 monolayers of water around particles in a CO_2 atmosphere, which is equivalent to a moisture content of 0.1 to 0.5 g H_2O per g soil (Stephens 1995; Stephens et al. 1995). For comparison, the moisture content in unvegetated, high-altitude terrestrial soils below the snow line is ~0.05 g/g in the Peruvian Andes and ~0.1 g/g in the Colorado Rockies (King et al. 2008). On Mars, periodic "wet" conditions generated from melted ice over hundreds of millions of years might be responsible for the carbonates in the soil at the Phoenix landing site (Boynton et al. 2009). The existence of recent lava flows (~2 million years old) on Tharsis and Elysium (Neukum et al. 2004; Vaucher et al. 2009) shows that volcanism has been active in the near past (Grott et al. this volume), and Mars may still be internally active, with a potential for carbonate formation in volcanic hydrothermal systems at depth. Finally, many meteorites, both young and old, possess carbonates that have isotopic compositions consistent with formation in equilibrium from modern martian atmospheric CO_2 (Niles et al. 2010) (Fig. 3).

6 Discussion

6.1 CO_2 Budget of Mars

The atmosphere of Mars today is tenuous, consisting mainly of CO_2 at a mean global pressure of about 6 mbar (Haberle et al. 2008). A previously unidentified deposit of solid CO_2 was discovered by the Mars Reconnaissance Orbiter, in the form of bodies of CO_2 ice embedded within the martian south polar layered deposits (Phillips et al. 2011) with an amount equivalent to a global pure CO_2 atmospheric pressure of 4–5 mbar. More substantial amounts of CO_2 could be adsorbed in the regolith, up to $\approx$140 mbar (Fanale and Cannon 1979). CO_2 may also be present in the martian cryosphere in the form of CO_2 clathrate hydrates (Mousis et al., this issue), although in unknown amounts. The amount of CO_2 contained in the identified atmospheric, polar and subsurface reservoirs under its molecular form is probably not in excess of a few tens or one hundred or so millibars.

Like Venus and Earth, Mars should have been endowed with similar amounts of CO_2 during main accretion (Owen and Bar-Nun 1995). Thermal escape of carbon is expected to have removed most of the primordial atmospheric CO_2, with a characteristic time to lose 1 bar of CO_2 in the range from 1 to 10 Myr (Lammer et al., this issue). If most of Mars' atmospheric CO_2 was emplaced during the main accretion phase, the entire inventory could have been lost within 40 Myr, ensuring that any primordial atmosphere was likely to have been removed (Tian et al. 2009). Following this loss to space of CO_2, a secondary atmosphere progressively formed through volcanic outgassing of CO_2 (Lammer et al., this issue). According to a morphological analysis of volcanic landforms on the surface, 0.3 bar of CO_2 would have been outgassed by volcanic eruptions during the last 4 Ga (Craddock and Greeley 2009). Crustal production modeling suggests that a total 0.5–1 bar of CO_2 have been outgassed, mostly during the Noachian (Grott et al. 2011). Finally, modeling using CO_2 contents of martian magmas based on the martian meteorites also predicts between 0 and 1.2 bars of CO_2 outgassing in the past 4.5 Ga (Stanley et al. 2011). Thus, up to $\sim$1.2 bar of magmatic CO_2 has been released to the atmosphere throughout martian history. Recycling of CO_2 from ancient carbonate deposits could allow for higher year to year average outgassing rates, but would not impact the overall inventory available at the surface (Pollack et al. 1987).

Using up-to-date values of Mars' C non-thermal escape rates, no more than 200 mbar of CO_2 may have been removed by escape during the last 4 Ga (Lammer et al. this issue). The recent measurement of the isotopes of the martian atmospheric CO_2 by the Phoenix mission (Niles et al. 2010) clearly shows that escape has not played an important role in fractionating CO_2 during the last 4 Ga, in agreement with most recent model calculations (Chassefiere and Leblanc 2011b). A primitive ^{13}C enrichment possibly due to early sputtering and hydrodynamic escape, would have been later erased through both long-term carbonate formation and dilution from mantle outgassed CO_2 (Gillmann et al. 2011; Niles et al. 2010). The SAM instrument on the MSL mission is poised to potentially make isotopic measurements of Hesperian carbonates (if present) at Gale crater which could further our understanding of the atmospheric evolution of Mars. Furthermore, measurements made by the MAVEN spacecraft will provide stronger constraints on the non-thermal escape of CO_2 from Mars as well as the isotopic fractionation associated with that process.

Removal of the atmosphere through blow off during large impacts is also a possibility and could be important during the early Noachian (Brain and Jakosky 1998; Manning et al. 2006; Melosh and Vickery 1989; Newman et al. 1999). Impact erosion would have removed CO_2 and other gases en masse and would not create isotopic fractionation unlike non-thermal escape processes (Melosh and Vickery 1989). Impact erosion may have removed much of the early volcanically-released CO_2 during the Noachian, but its importance relative to EUV-powered escape processes may be small if EUV-powered processes removed the proto-atmosphere in 10 Myr (Lammer et al. this issue). Thus any carbonates are likely the product of a secondary atmosphere.

Estimates of the total amount of CO_2 deposited as carbonate on the martian surface remain relatively unconstrained. A useful benchmark for making this estimation is that 1 bar of CO_2 can create the equivalent of a planet-wide cover of $\sim$20 m of pure calcite on the surface of Mars. This is calculated using depth $= (P_s M_{CaCO3})/(g M_{CO2} \rho_{CaCO3})$, where a surface pressure of 1 bar is $P_s = 10^5$ Pa, the molar mass of carbon dioxide is $M_{CO2} = 0.044\ \mathrm{kg\,mol^{-1}}$, the molar mass of calcite $M_{CaCO3} = 0.1\ \mathrm{kg\,mol^{-1}}$, the density of calcite is $\rho_{CaCO3} = 2710\ \mathrm{kg\,m^{-3}}$, and g is the gravitational acceleration on Mars. Ultimately this type of calculation needs to be constrained by accurate estimates of the carbonate content of the martian crust which currently don't exist.

One possible method for estimating the size of the global carbonate reservoir is by using the carbonate in the global dust (Bandfield et al. 2003). It is possible that carbonates exposed at the surface and deep carbonates exposed by impact could be the source of carbonates observed in the martian dust (Bandfield et al. 2003; Ehlmann et al. 2008b) especially if the dust on Mars formed primarily due to pulverization of the upper crust of Mars by impact processes. Thus, if the (2–5 %) carbonate detected in the global dust represented the average composition of the upper 1–3 km of the crust of Mars, it is possible that 1–3 bars of CO_2 are sequestered on the surface (Bandfield et al. 2003). The presence of carbonates in the Phoenix soils might support this claim somewhat by showing that the carbonates are not restricted to the fine grained dust and may be globally present in the martian soils as well (see Sect. 4.2 above). However, the depth in the crust at which carbonates are present remains a major uncertainty in this kind of estimate. The extrapolation of the weight abundance of carbonate in the dust downwards through the subsurface bedrock has no justification considering the low carbonate abundances measured in meteorites which are our only samples of the near surface crust of Mars. Assuming ~0.5 wt.% carbonate (i.e., an assumption weighted towards the higher carbonate abundance in ALH84001), an extrapolation through 1 km of the subsurface would give an upper bound of no more than 0.25 bar of sequestered CO_2. Furthermore recent studies (see Sect. 5.3 above) have shown that carbonate may be forming in the modern environment of Mars, making it likely that the proportion of carbonate in the dust and soil has been substantially elevated by Amazonian weathering and is therefore not representative of carbonate deposits at depth.

Another method for assessing the inventory of carbonates on Mars is through observations from orbital data which show that carbonates are not as abundant and widespread as was once anticipated and are not as abundant and widespread as other alteration minerals on Mars such as phyllosilicates, chlorides, and sulfates, although lack of exposure may lead to underestimations of the abundance of all of these phases within ancient rocks as described in Sect. 5.1 above. Carbonate rocks on Earth are prominent on the surface due to the dominance of shelf carbonates in the sedimentary record, which display regional stratigraphic continuity, can be very thick, and contain abundant carbonate minerals, usually calcium-rich. Perhaps a similar situation was once expected for Mars (Fanale and Cannon 1974; Fanale et al. 1982; Pollack et al. 1987), but orbital data do not support this hypothesis. Instead, remote sensing data have revealed that Mg-carbonates are more typical, and that these deposits are relatively rare and possibly related to local weathering and alteration processes.

If Mars hosted a dense CO_2 atmosphere (>1 bar) in the Noachian, large (>20 m global equivalent layer of calcite) carbonate deposits should have been formed during or after the Noachian/Hesperian transition. Based on the current observations summarized above (Sects. 3 and 4), Hesperian carbonate outcrops of any size have not been detected thus far. Of course it is possible that these Hesperian carbonates are simply hidden from view since much of the martian surface is obscured by dust (Ruff and Christensen 2002) that can't be penetrated by the NIR or TIR spectrometers although that has not posed a problem for detections of other alteration minerals such as sulfates, phyllosilicates and clays.

We are limited to observations in areas where the geology is well-exposed and detectable using the available instruments, thus several possibilities exist for "hidden" carbonates on Mars (Craddock and Howard 2002). First of all, the terrain around Nili Fossae is perhaps the best example of well-exposed, ancient terrain on Mars and it might imply that carbonates could also be present in additional buried Noachian units. Hesperian and Amazonian units do not exhibit evidence for carbonates and might indicate that the atmosphere was sparse as far back as the late Noachian. A possibility is that later acidity might have decomposed or prevented Hesperian and Amazonian carbonates from forming (Bibring et al. 2006;

Bullock and Moore 2007; Fairen et al. 2004) but this dissolution process would simply return the CO_2 to the atmosphere where it would have to be reprecipitated as carbonate somewhere else at a later time (see Sect. 6.3 below). Another possibility is that volcanism in the Amazonian might end up producing a sulfate rind on carbonate outcrops through heterogeneous (gas-solid) reactions that turn carbonates into sulfates (Clark and Baird 1979) thus obscuring the carbonate from detection. It may also be possible that the carbonates are widely dispersed and in low enough abundance so that they cannot be detected using the available instrumentation (Blaney and McCord 1989). Finally, micro-roughness common to surfaces of caliche and travertine on Earth reduce spectral contrast of carbonate minerals, potentially masking carbonate detection in the thermal infrared (Kirkland et al. 2002), although this effect has not been shown to be applicable to VNIR spectra.

Because the hypothesis of a thick CO_2 atmosphere on early Mars has various problems, it is possible that the CO_2 budget of Mars has been relatively small. Presuming that the proto-atmosphere and earliest pre-Noachian atmospheres were rapidly lost due to the high solar EUV flux and impact erosion (Lammer et al. this issue), carbonate formation may have been restricted to the earliest periods in martian history. If the subsequent volcanic outgassing and growth of a secondary atmosphere remained small, then the overall amount of CO_2 available for carbonate formation during the Noachian, Hesperian, and Amazonian may have been quite limited (Grott et al. 2011; Manning et al. 2006). Thus the climate of Mars could have been similar to the modern martian climate throughout history (Gaidos and Marion 2003). Another possibility is that there have been a series of intermittent warm, wet episodes caused by heat dumped by very large impacts (Segura and Navarro-Gonzalez 2005; Segura et al. 2002; Toon et al. 2010). In this hypothesis, a temporary greenhouse effect does not rely upon large quantities of CO_2, which has the virtue of being consistent with a general lack of carbonate outcrops. This would result in a series of transient aqueous episodes that temporarily mobilized liquid water on the surface followed by evaporation or sublimation. Other models for intermittent warm periods are those supported by massive volcanic eruptions of CO_2 e.g. (Phillips et al. 2001) and would also have to account for the lack of Hesperian and Amazonian carbonate outcrops on the martian surface.

6.2 Chemistry—Mg-, Fe-rich vs. Ca-, Fe-rich

The wide ranging chemistry of the martian carbonates is in contrast to carbonate chemistry on Earth, which is dominated by Ca-rich phases: calcite, aragonite and dolomite. This could be due to differences in fluid chemistry, but may also be due to differences in carbonate mineral formation environments. Ca-rich phases dominate in terrestrial environments despite the fact that the oceans are Mg-rich, thus the differences in carbonate chemistry between Mars and Earth may be due to the fundamental differences in the aqueous environments on the two planets.

The prevalence of magnesite-rich carbonates on Mars suggests that conditions for magnesite formation may give clues to early martian environments. Primary, anhydrous magnesite is difficult to form as a direct precipitate from a low temperature standing body of water, in contrast to calcite. The reason is that the Mg^{2+} ion in solution is strongly hydrated, so that for the Mg^{2+} ion to enter into the magnesite crystal without bound water requires energy, which presents a "dehydration barrier" (Lippmann 1973). Consequently, hydrates of magnesium carbonate and hydroxyhydrates are the common precipitates from aqueous solution, whereas primary magnesite is absent from marine evaporites. Specifically, the forms of magnesite that are favored with increasing temperature are lansfordite ($MgCO_3{\cdot}5H_2O$), nesquehonite ($MgCO_3{\cdot}3H_2O$) and hydromagnesite ($Mg_5(CO_3)_4(OH)_2{\cdot}4H_2O$ or $Mg_4(CO_3)_3(OH)_2{\cdot}3H_2O$). Such hydrous magnesites often form as the result of the hydrothermal alteration of Mg-rich basic igneous rocks or of serpentinites that were, in turn, originally formed by hydrothermal activity (Pohl 1989; Russell 1996). Ignoring the hydration state for simplicity, carbonation of Mg-olivine (forsterite) proceeds schematically as follows:

$$4Mg_2SiO_4 + 4H_2O + 2CO_2 = 2Mg_3Si_2O_5(OH)_{4(\mathrm{serpentine})} + 2MgCO_{3(\mathrm{magnesite})}$$

Hydrothermal alteration of serpentine into magnesite is seen in the vein swarms of the Piedmont magnesite deposit in the Great Serpentine Belt of New South Wales, Australia (Brownlow and Ashley 1991). Hydromagnesite occurrences where groundwater is in contact with basic igneous rocks and flows into lakes include the carbonate lakes of British Columbia, Canada, the Coorongs of S. Australia, and the East African Rift Valley. A further example is in the highly Mg-rich, alkaline ($pH > 9$) waters of Salda Lake, Turkey, where lake drainage is rimmed on three sides by serpentinites (Russell et al. 1999).

Given the difficulties in forming anhydrous magnesite, it is puzzling as to why it occurs instead of more Ca-rich varieties in ALH 84001 (Mittlefehldt 1994). On Earth and in laboratory experiments magnesite typically indicates elevated temperatures (Lippmann 1973), however the accumulated observations of the ALH 84001 carbonates suggests a low temperature origin. Anhydrous magnesite can also form under conditions of reduced activity of water such as evaporative environments (Canaveras et al. 1998; Pohl 1989). The ALH84001 carbonates have been proposed to form as evaporative precipitates of rapidly receding floodwaters or in a playa lake environment analogous to some low temperature terrestrial environments that host Mg-rich carbonates (McSween and Harvey 1998; Warren 1998).

Unlike the Mg-rich carbonates in nearby Nili Fossae and elsewhere, those excavated from the deep crust in Leighton crater have spectral signatures of siderite ($FeCO_3$) or possibly calcite ($CaCO_3$) (Michalski and Niles 2010). Siderite is typical of a carbonate deposited in anoxic conditions under a moderate pCO_2 (Catling 1999), although the exact pCO_2 required is subject to some uncertainty. One way to set a limit on the pCO_2 is to consider the equilibrium between siderite and hydrous iron silicates, where the latter would form at lower pCO_2 instead of siderite. For example, a precursor to hydrous iron silicates such as greenalite or minnesotaite, which are found after diagenesis and burial, is berthierine ($Fe_2Al_2SiO_5(OH)_4$). One can consider the equilibrium between berthierine, siderite and kaolinite ($Al_2Si_2O_5(OH)_4$), as follows:

$$Fe_2Al_2SiO_5(OH)_4(s) + SiO_2(aq) + 2CO_2(g) = 2FeCO_3(s) + Al_2Si_2O_5(OH)_4(s)$$

Depending on the assumed activity of aqueous silica—perhaps close to saturation—a thermodynamic estimate of the pCO_2 can be calculated (Sheldon 2006). Similar estimates can also be made for the equilibria of siderite with other phyllosilicates (Chevrier et al. 2007). The result is that the pCO_2 boundary between siderite and typical phyllosilicates lies around ~1–10 mbar. Thus, the presence of siderite on Mars only suggests pCO_2 values somewhat higher than today's mean atmospheric pCO_2 or comparable to it, while the formation of certain phyllosilicates would require isolation from a thick carbon dioxide atmosphere, such as in the subsurface. Carbonates excavated from the deep crust may provide a glimpse of a vast record of very ancient, deeply buried sedimentary carbonates from a warmer climate, but a simpler explanation might be local ancient subsurface hydrothermal alteration. Either way, the presence of ferrous iron (Fe^{2+}) in siderite implies a reducing environment, unlike the modern oxidizing atmosphere on Mars.

6.3 Acid vs. Alkaline

The detection of carbonates in Mars rocks and soils has indicated that mildly alkaline geochemical conditions on Mars have existed throughout most of martian history (see Sects. 3.1 and 4.2 above). Acidic mineralogy has been detected on Mars (Farrand et al. 2009; Knoll et al. 2005; Morris et al. 2006; Yen et al. 2008) and has been proposed to indicate that acidic geochemistry has dominated Mars for the past 3.5 Ga, while neutral to alkaline environments were more common earlier during widespread formation of phyllosilicates (Bibring et al. 2006; Hurowitz and McLennan 2007). However, the coexistence of both acidic and alkaline environments on Mars suggests that solution pH has likely been heterogeneous throughout the geologic history of Mars and likely depends on the local environment.

It has been proposed that a global acidic environment dominated by sulfate deposition in the Hesperian (Bibring et al. 2006) could explain the lack of carbonates from that era. Indeed, based on the abundance and character of sulfur on Mars, it is possible that it has played the same role as carbon plays in the weathering process on the Earth (Halevy et al. 2007). Among the possible acidic gases of geological origin, SO_3 is the most soluble and has the lowest acidity constant, its aqueous form being sulfuric acid. It also produces sulfate by reaction with silicate rocks. However, magmatic degassing releases different forms of sulfur such as H_2S, native sulfur or SO_2, but not SO_3, although the latter forms in atmospheric chemistry and hydrates into H_2SO_4 aerosols (Gaillard et al., this issue). Thus one potential way to reconcile the absence of Hesperian aged carbonate outcrops with an early thick CO_2 atmosphere is that sulfuric or sulfurous acids in large, standing bodies of water on early Mars suppressed the formation of carbonates or dissolved existing carbonates (Fairen et al. 2004; Halevy et al. 2007). The dissolution and instability of carbonates in acid is well known and occurs at $pH < 6.2$ at pCO_2 of 1 bar and $pH < 7.7$ at a pCO_2 of 10 mbar (Bullock and Moore 2007). However, while carbonate precipitation may have been suppressed in the Hesperian by widespread acidic conditions this does not provide an explanation for where the atmospheric CO_2 from this time period has gone.

Another problem with the acidic Hesperian hypothesis is that, over time, weathering fluids within a fractured basaltic regolith will tend to neutralize acid, deposit salts, and become alkaline. Mafic or ultramafic materials consume hydrogen ions and end up buffering standing bodies of water at an alkaline pH at moderate temperatures. For example, despite a continuous addition of acid volatiles from volcanism, the terrestrial ocean maintains a pH of ~ 8 because it is buffered by the chemistry of a basaltic seafloor (Macleod et al. 1994). The silicate minerals of basalt are effectively salts of weak acids and strong bases, so that when they dissolve in water the result is a weakly alkaline solution (Stevens and Carron 1948). Pulverized basalt added to water at room temperature produces a pH of roughly 9–10. Finally, in context with the widespread clay minerals on Mars (Ehlmann et al., this issue), widespread acidic solutions are unlikely. Clay minerals, such as smectite, have exchangeable cations in the interlayer position of the structure, conferring swelling properties and large sorption capacity making the clays an ionic exchanger. In contact with an acidic solution, $H+$ exchange with the compensating cations and clays is a well known and efficient alkaline pH buffer. Several studies (Baldridge et al. 2009; Chevrier et al. 2004; Hurowitz et al. 2010; Zolotov and Shock 2005) have proposed potential mechanisms for creating acidic solutions on Mars revolving around weathering of sulfide minerals and oxidation of reduced iron. This is a viable alternative to forming sulfates from atmospheric SO_2 but suggests that sulfate deposits on Mars should be enriched in iron as well as sulfur.

Another possible explanation for the acid-alkaline problem revolves around low-water rock ratios which limit the amount of interaction with the basaltic crust and allow for the coexistence of alkaline and acidic micro-environments in direct proximity (Berger et al. 2009; Niles and Michalski 2009). Low water-rock conditions are also consistent with a cold thin atmosphere which contains minimal amounts of CO_2. This type of environment could allow for local regions of acidic and alkaline alteration and even possibly alternating acidic and alkaline conditions in the same location as may be indicated by the interbedded sulfates and phyllosilicates at Gale crater (Milliken et al. 2010).

7 Summary and Conclusions

The collection of martian carbonates discovered to date does not rule out the possibility of a dense Noachian CO_2 atmosphere but nonetheless does not support it. The Nili Fossae, Leighton crater, and Gusev crater carbonates (Sects. 3.2 and 4.1) are all likely formed during the Noachian time period, and some could potentially have formed earlier. They all represent mixtures of carbonates with igneous minerals, suggesting formation via rock alteration by subsurface fluids rather than sedimentary precipitation. The carbonates in ALH 84001 and the nakhlites (Changela and Bridges 2011; Valley et al. 1997) were also likely formed during brief aqueous events in the subsurface. Taken together, these carbonates are evidence for the presence of subsurface water as opposed to surficial bodies of water, and they likely pre-date the formation of the valley networks and the Noachian-Hesperian transition when the dense CO_2 atmosphere was supposed to have been lost (Carr and Head 2010; Fassett and Head 2008, 2011). The carbonates discovered in the dust (Sect. 3.1) and the martian soils (Sect. 4.2) could represent a much larger crustal reservoir derived from a dense early atmosphere. It is very likely that the carbonate in the dust may be the result of Amazonian weathering processes that have enriched the topmost dust and soils with small amounts of carbonate.

The carbonate record summarized in this review supports a martian aqueous history that is much more limited than what has previously been proposed. Instead of lakes and oceans, we have evidence for carbonate formation from subsurface fluids in the Noachian that declined and transitioned to formation from limited amounts of water in the Hesperian and Amazonian. In many cases the Mg-rich nature of the carbonates suggests low activity of water, which might be due to evaporation or cryo-concentration consistent with an arid environment. While it is clear that acidic conditions were important on Mars, the mineralogic composition of the crust suggests that acidity was limited to local environments or was present under very low water-rock conditions. This more limited view does not require dense CO_2 Noachian atmospheres with the associated abundant Hesperian carbonate deposits but does allow for localized regions of hydrothermal activity occurring in the subsurface volcanic rocks.

While the global view of martian aqueous history presented here may not be ideal with regard to habitability as understood from study of Earth organisms, the longevity of the localized subsurface aqueous activity is poorly constrained and some locales might in fact represent highly desirable astrobiological destinations. The co-occurrence of hydrated phyllosilicates and carbonate minerals in Nili Fossae and Leighton crater in particular, provide strong candidates for the investigation of the habitability of Mars. These sites west of the Isidis basin show strong evidence for the prolonged past presence of liquid water, the presence of carbon, and the presence of heat which are all critical prerequisites for the possible existence of life.

Based on the work presented here, it is possible that the Mars Science Laboratory rover will encounter carbonates at the Gale crater landing site and in the Gale crater sediments.

Carbonate is expected to be present in low concentrations in the dust and soils similar to what was found at the Phoenix landing site and observed from orbit. This could also be true for the sediments in the Gale crater mound. Carbonate minerals may also be present at Gale in association with mafic materials in volcaniclastic materials similar to the carbonates at Gusev crater and what may be present at Nili Fossae. Discoveries of Hesperian-aged carbonate beds or lenses at Gale crater would be unexpected based on the observations described in this review and would provide a compelling counter argument to the interpretation of Mars aqueous history presented here.

Acknowledgements Thoughtful reviews were provided by Ralph Harvey and Jim Bell which greatly improved the manuscript. Thanks to M. Toplis and the ISSI conference organizers for coordinating this review and promoting interesting discussions. E. Chassefière acknowledges support from CNRS EPOV interdisciplinary program.

References

L.M. Anovitz, E.J. Essene, J. Pet. **28**, 389–414 (1987)
S.K. Atreya, P.R. Mahaffy, A.-S. Wong, Planet. Space Sci. **55**, 358–369 (2007)
J.N. Bahcall, M.H. Pinsonneault, B. Sarbani, Astrophys. J. **555**, 990 (2001)
A.M. Baldridge, S.J. Hook, J.K. Crowley, G.M. Marion, J.S. Kargel, J.L. Michalski, B.J. Thomson, C.R. de Souza, N.T. Bridges, A.J. Brown, Geophys. Res. Lett. **36** (2009)
J.L. Bandfield, T.D. Glotch, P.R. Christensen, Science **301**, 1084–1087 (2003)
G. Berger, M.J. Toplis, E. Treguier, C. d'Uston, P. Pinet, Am. Mineral. **94**, 1279–1282 (2009)
J.P. Bibring, Y. Langevin, J.F. Mustard, F. Poulet, R. Arvidson, A. Gendrin, B. Gondet, N. Mangold, P. Pinet, F. Forget, The OMEGA team, M. Berthe, J.-P. Bibring, A. Gendrin, C. Gomez, B. Gondet, D. Jouglet, F. Poulet, A. Soufflot, M. Vincendon, M. Combes, P. Drossart, T. Encrenaz, T. Fouchet, R. Merchiorri, G. Belluci, F. Altieri, V. Formisano, F. Capaccioni, P. Cerroni, A. Coradini, S. Fonti, O. Korablev, V. Kottsov, N. Ignatiev, V. Moroz, D. Titov, L. Zasova, D. Loiseau, N. Mangold, P. Pinet, S. Doute, B. Schmitt, C. Sotin, E. Hauber, H. Hoffmann, R. Jaumann, U. Keller, R. Arvidson, J.F. Mustard, T. Duxbury, F. Forget, G. Neukum, Science **312**, 400–404 (2006)
D.L. Blaney, T.B. McCord, J. Geophys. Res. **94**, 10159–10166 (1989)
D.D. Bogard, J. Park, Meteorit. Planet. Sci. **43**, 1113–1126 (2008)
M.C. Booth, H.H. Kieffer, J. Geophys. Res. **83**, 1809–1815 (1978)
A. Bouvier, J. Blichert-Toft, J.D. Vervoort, P. Gillet, F. Albarède, Earth Planet. Sci. Lett. **266**, 105–124 (2008)
W.V. Boynton, D.W. Ming, S.P. Kounaves, S.M.M. Young, R.E. Arvidson, M.H. Hecht, J. Hoffman, P.B. Niles, D.K. Hamara, R.C. Quinn, P.H. Smith, B. Sutter, D.C. Catling, R.V. Morris, Science **325**, 61–64 (2009)
D.A. Brain, B.M. Jakosky, J. Geophys. Res. **103**, 22689–22694 (1998)
J.C. Bridges, M.M. Grady, Earth Planet. Sci. Lett. **176**, 267–279 (2000)
J.C. Bridges, P.H. Warren, J. Geol. Soc. **163**, 229–251 (2006)
J.C. Bridges, D.C. Catling, J.M. Saxton, T.D. Swindle, I.C. Lyon, M.M. Grady, Space Sci. Rev. **96**, 365–392 (2001)
A.J. Brown, S.J. Hook, A.M. Baldridge, J.K. Crowley, N.T. Bridges, B.J. Thomson, G.M. Marion, C.R. de Souza Filho, J.L. Bishop, Earth Planet. Sci. Lett. **297**, 174–182 (2010)
J.W. Brownlow, P.M. Ashley, Q. Notes, Geol. Surv. New South Wales **82**, 1–20 (1991)
M.A. Bullock, J.M. Moore, Geophys. Res. Lett. **34**, L19201 (2007)
W.M. Calvin, T.V.V. King, R.N. Clark, J. Geophys. Res. **99**, 14659–14675 (1994)
J.C. Canaveras, M.S. Sanchez, R.E. Sanz, M. Hoyos, Sediment. Geol. **119**, 183–194 (1998)
K.M. Cannon, B. Sutter, D.W. Ming, W.V. Boynton, R. Quinn, Geophys. Res. Lett. **39**, L13203 (2012)
M.H. Carr, J.W. Head, Earth Planet. Sci. Lett. **294**, 185–203 (2010)
J. Carter, F. Poulet, Icarus **219**, 250–253 (2012)
D.C. Catling, J. Geophys. Res. **104**, 16453–16469 (1999)
H.G. Changela, J.C. Bridges, Meteorit. Planet. Sci. **45**, 1847–1867 (2011)
E. Chassefiere, F. Leblanc, Earth Planet. Sci. Lett. **310**, 262–271 (2011a)
E. Chassefiere, F. Leblanc, Planet. Space Sci. **59**, 207–217 (2011b)
V. Chevrier, P. Rochette, P.-E. Mathé, O. Grauby, Geology **32**, 1033–1036 (2004)
V. Chevrier, F. Poulet, J.-P. Bibring, Nature **448**, 60–63 (2007)

P.R. Christensen, J.L. Bandfield, V.E. Hamilton, S.W. Ruff, H.H. Kieffer, T.N. Titus, M.C. Malin, R.V. Morris, M.D. Lane, R.L. Clark, B.M. Jakosky, M.T. Mellon, J.C. Pearl, B.J. Conrath, M.D. Smith, R.T. Clancy, R.O. Kuzmin, T. Roush, G.L. Mehall, N. Gorelick, K. Bender, K. Murray, S. Dason, E. Greene, S. Silverman, M. Greenfield, J. Geophys. Res. **106**, 23823–23871 (2001)
P.R. Christensen, M.B. Wyatt, T.D. Glotch, A.D. Rogers, S. Anwar, R.E. Arvidson, J.L. Bandfield, D.L. Blaney, C. Budney, W.M. Calvin, A. Faracaro, R.L. Fergason, N. Gorelick, T.G. Graff, V.E. Hamilton, A.G. Hayes, J.R. Johnson, A.T. Knudson, H.Y. McSween, G.L. Mehall, L.K. Mehall, J.E. Moersch, R.V. Morris, M.D. Smith, S.W. Squyres, S.W. Ruff, M.J. Wolff, Science **306**, 1733–1739 (2004)
M.J. Cintala, R.A.F. Grieve, Meteorit. Planet. Sci. **33**, 889–912 (1998)
B.C. Clark, A.K. Baird, J. Geophys. Res. **84**, 8395–8403 (1979)
A. Colaprete, O.B. Toon, J. Geophys. Res. **108**, 5025 (2003)
C.M. Corrigan, R.P. Harvey, Meteorit. Planet. Sci. **39**, 17–30 (2004)
R.A. Craddock, R. Greeley, Icarus **204**, 512–526 (2009)
R.A. Craddock, A.D. Howard, J. Geophys. Res. **107** (2002)
B.L. Ehlmann, J.F. Mustard, C.I. Fassett, S.C. Schon, J.W. Head, D.J.D. Marais, J.A. Grant, S.L. Murchie, Nat. Geosci. **1**, 355–358 (2008a)
B.L. Ehlmann, J.F. Mustard, S.L. Murchie, F. Poulet, J.L. Bishop, A.J. Brown, W.M. Calvin, R.N. Clark, D.J. Des Marais, R.E. Milliken, L.H. Roach, T.L. Roush, G.A. Swayze, J.J. Wrray, Science **322**, 1828–1832 (2008b)
B.L. Ehlmann, J.F. Mustard, G.A. Swayze, R.N. Clark, J.L. Bishop, F. Poulet, D.J. Des Marais, L.H. Roach, R.E. Milliken, J.J. Wray, O. Barnouin-Jha, S.L. Murchie, J. Geophys. Res. **114** (2009)
B.L. Ehlmann, J.F. Mustard, S.L. Murchie, J.-P. Bibring, A. Meunier, A.A. Fraeman, Y. Langevin, Nature **479**, 53–60 (2011)
J.M. Eiler, J.W. Valley, C.M. Graham, J. Fournelle, Geochim. Cosmochim. Acta **66**, 1285–1303 (2002)
G. Etiope, M. Schoell, H. Hosgorrmez, Earth Planet. Sci. Lett. **310**, 96–104 (2011)
A.G. Fairen, D. Fernandez-Remolar, J.M. Dohm, V.R. Baker, R. Amils, Nature **431**, 423–426 (2004)
F.P. Fanale, W.A. Cannon, J. Geophys. Res. **79**, 3397–3402 (1974)
F.P. Fanale, W.A. Cannon, J. Geophys. Res. **84**, 8404–8414 (1979)
F.P. Fanale, J.R. Salvail, W. Bruce Banerdt, R.S. Saunders, Icarus **50**, 381–407 (1982)
J. Farquhar, M.H. Thiemens, J. Geophys. Res. **105**, 11991–11997 (2000)
J. Farquhar, M.H. Thiemens, T. Jackson, Science **280**, 1580–1582 (1998)
W.H. Farrand, T.D. Glotch, J.W. Rice Jr, J.A. Hurowitz, G.A. Swayze, Icarus **204**, 478–488 (2009)
C.I. Fassett, J.W. Head, Icarus **195**, 61–89 (2008)
C.I. Fassett, J.W. Head, Icarus **211**, 1204–1214 (2011)
F. Forget, R.T. Pierrehumbert, Science **278**, 1273–1276 (1997)
H.V. Frey, J. Geophys. Res. **111** (2006)
H. Frey, Geophys. Res. Lett. **35**, L13203 (2008)
S.J. Gaffey, J. Geophys. Res. **92**, 1429–1440 (1987)
E. Gaidos, G. Marion, J. Geophys. Res. **108** (2003)
E.M. Galimov, Icarus **147**, 472–476 (2000)
A. Gaudin, E. Dehouck, N. Mangold, Icarus **216**, 257–268 (2011)
C. Gillmann, P. Lognonné, M. Moreira, Earth Planet. Sci. Lett. **303**, 299–309 (2011)
D.C. Golden, D.W. Ming, C.S. Schwandt, R.V. Morris, S.V. Yang, G.E. Lofgren, Meteorit. Planet. Sci. **35**, 457–465 (2000)
D.C. Golden, D.W. Ming, C.S. Schwandt, H.V. Lauer, R.A. Socki, R.V. Morris, G.E. Lofgren, G.A. McKay, Am. Mineral. **86**, 370–375 (2001)
J.L. Gooding, M.E. Zolensky, S.J. Wentworth, Meteoritics **26**, 135–143 (1991)
D.O. Gough, Sol. Phys. **74**, 21–34 (1981)
L.L. Griffith, E.L. Shock, Nature **377**, 406–408 (1995)
M. Grott, A. Morschhauser, D. Breuer, E. Hauber, Earth Planet. Sci. Lett. **308**, 391–400 (2011)
R.M. Haberle, J. Geophys. Res. **103**, 28467–28479 (1998)
R.M. Haberle, F. Forget, A. Colaprete, J. Schaeffer, W.V. Boynton, N.J. Kelly, M.A. Chamberlain, Planet. Space Sci. **56**, 251–255 (2008)
I. Halevy, J.W. Head, Punctuated volcanism, transient warming and global change in the late Noachian-Early Hesperian, in *Lunar and Planetary Institute Science Conference Abstracts*, vol. 43 (2012), p. 1908
I. Halevy, M.T. Zuber, D.P. Schrag, Science **318**, 1903 (2007)
I. Halevy, W.W. Fischer, J.M. Eiler, Proc. Natl. Acad. Sci. USA **108**, 16895–16899 (2011)
R.P. Harvey, H.Y. McSween, Nature **382**, 49–51 (1996)
M.H. Hecht, Icarus **156**, 373–386 (2002)
M.H. Hecht, S.P. Kounaves, R.C. Quinn, S.J. West, S.M.M. Young, D.W. Ming, D.C. Catling, B.C. Clark, W.V. Boynton, J. Hoffman, L.P. DeFlores, K. Gospodinova, J. Kapit, P.H. Smith, Science **325**, 64–67 (2009)

T.L. Heet, R.E. Arvidson, S.C. Cull, M.T. Mellon, K.D. Seelos, J. Geophys. Res. **114**, E00E04 (2009)
H. Hiesinger, J.W. Head III, J. Geophys. Res. **109**, E01004 (2004)
G. Holland, J.M. Saxton, I.C. Lyon, G. Turner, Geochim. Cosmochim. Acta **69**, 1359–1370 (2005)
J.A. Hurowitz, S.M. McLennan, Earth Planet. Sci. Lett. **260**, 432–443 (2007)
J.A. Hurowitz, W.W. Fischer, N.J. Tosca, R.E. Milliken, Nat. Geosci. **3**, 323–326 (2010)
A.P. Ingersoll, Science **168**, 972–973 (1970)
B.M. Jakosky, R.O. Pepin, R.E. Johnson, J.L. Fox, Icarus **111**, 271–288 (1994)
S.S. Johnson, M.A. Mischna, T.L. Grove, M.T. Zuber, J. Geophys. Res. **113** (2008)
S.S. Johnson, A.A. Pavlov, M.A. Mischna, J. Geophys. Res. **114** (2009)
A.J.T. Jull, C.J. Eastoe, S. Xue, G.F. Herzog, Meteoritics **30**, 311–318 (1995)
A.J.T. Jull, J.W. Beck, G.S. Burr, Geochim. Cosmochim. Acta **64**, 3763–3772 (2000)
R. Kahn, Icarus **62**, 175–190 (1985)
J.F. Kasting, Icarus **94**, 1–13 (1991)
J.F. Kasting, Science **276**, 1213 (1997)
A.J. King, A.F. Meyer, S.K. Schmidt, Soil Biol. Biochem. **40**, 2605–2610 (2008)
L. Kirkland, K. Herr, E. Keim, P. Adams, J. Salisbury, J. Hackwell, A. Treiman, Remote Sens. Environ. **80**, 447–459 (2002)
L.P. Knauth, M.J. Kennedy, Nature **460**, 728–732 (2009)
A.H. Knoll, M. Carr, B. Clark, D.J. Des Marais, J.D. Farmer, W.W. Fischer, J.P. Grotzinger, S.M. McLennan, M. Malin, C. Schroder, Earth Planet. Sci. Lett. **240**, 179–189 (2005)
S.P. Kounaves, M.H. Hecht, J. Kapit, K. Gospodinova, L. DeFlores, R.C. Quinn, W.V. Boynton, B.C. Clark, D.C. Catling, P. Hredzak, D.W. Ming, Q. Moore, J. Shusterman, S. Stroble, S.J. West, S.M.M. Young, J. Geophys. Res. **115**, E00E10 (2010a)
S.P. Kounaves, M.H. Hecht, J. Kapit, R.C. Quinn, D.C. Catling, B.C. Clark, D.W. Ming, K. Gospodinova, P. Hredzak, K. McElhoney, J. Shusterman, Geophys. Res. Lett. **37**, 09201 (2010b)
V.A. Krasnopolsky, J.P. Maillard, T.C. Owen, R.A. Toth, M.D. Smith, Icarus **192**, 396–403 (2007)
M.D. Lane, M.D. Dyar, J.L. Bishop, Geophys. Res. Lett. **31**, L19702 (2004)
L.A. Leshin, K.D. McKeegan, P.K. Carpenter, R.P. Harvey, Geochim. Cosmochim. Acta **62**, 3–13 (1998)
F. Lippmann, *Sedimentary Carbonate Minerals* (Springer, Berlin, 1973), p. 228
J.R. Lyons, C. Manning, F. Nimmo, Geophys. Res. Lett. **32** (2005)
G. Macleod, C. McKeown, A.J. Hall, M.J. Russell, in *Origins of Life and Evolution of Biospheres*, vol. 24 (1994), pp. 19–41
M.C. Malin, K.S. Edgett, Science **290**, 1927–1937 (2000)
C.V. Manning, C.P. McKay, K.J. Zahnle, Icarus **180**, 38–59 (2006)
A.S. McEwen, L. Ojha, C.M. Dundas, S.S. Mattson, S. Byrne, J.J. Wray, S.C. Cull, S.L. Murchie, N. Thomas, V.C. Gulick, Science **333**, 740–743 (2011)
D.S. McKay, E.K. Gibson, K.L. ThomasKeprta, H. Vali, C.S. Romanek, S.J. Clemett, X.D.F. Chillier, C.R. Maechling, R.N. Zare, Science **273**, 924–930 (1996)
H.Y. McSween, R.P. Harvey, Int. Geol. Rev. **40**, 774–783 (1998)
H.J. Melosh, A.M. Vickery, Nature **338**, 487–489 (1989)
C. Meyer, *The Mars Meteorite Compendium* (NASA Johnson Space Center, Houston, 2011)
J.R. Michalski, P.B. Niles, Nat. Geosci. **3**, 751–755 (2010)
T. Mikouchi, E. Koizumi, A. Monkawa, Y. Ueda, M. Miyamoto, Antarct. Meteor. Res. **16**, 34–57 (2003)
R.E. Milliken, J.P. Grotzinger, B.J. Thomson, Geophys. Res. Lett. **37**, 04201 (2010)
D.W. Mittlefehldt, Meteoritics **29**, 214–221 (1994)
R.V. Morris, G. Klingelhofer, C. Schroder, D.S. Rodionov, A. Yen, D.W. Ming, P.A. de Souza, T. Wdowiak, I. Fleischer, R. Gellert, B. Bernhardt, U. Bonnes, B.A. Cohen, E.N. Evlanov, J. Foh, P. Gutlich, E. Kankeleit, T. McCoy, D.W. Mittlefehldt, F. Renz, M.E. Schmidt, B. Zubkov, S.W. Squyres, R.E. Arvidson, J. Geophys. Res. **111**, E12S15 (2006)
R.V. Morris, S.W. Ruff, R. Gellert, D.W. Ming, R.E. Arvidson, B.C. Clark, D.C. Golden, K. Siebach, G. Klingelhofer, C. Schroder, I. Fleischer, A.S. Yen, S.W. Squyres, Science **329**, 421–424 (2010)
P.S. Mozley, Geology **17**, 704–706 (1989)
K.J. Murata, I.I. Friedman, B.M. Madsen, Science **156**, 1484 (1967)
S.L. Murchie, J.F. Mustard, B.L. Ehlmann, R.E. Milliken, J.L. Bishop, N.K. McKeown, E.Z. Noe Dobrea, F.P. Seelos, D.L. Buczkowski, S.M. Wiseman, R.E. Arvidson, J.J. Wray, G. Swayze, R.N. Clark, D.J. Des Marais, A.S. McEwen, J.-P. Bibring, J. Geophys. Res. **114** (2009)
J.F. Mustard, B.L. Ehlmann, S.L. Murchie, F. Poulet, N. Mangold, J.W. Head, J.P. Bibring, L.H. Roach, J. Geophys. Res. **114**, E00D12 (2009)
G. Neukum, R. Jaumann, H. Hoffmann, E. Hauber, J.W. Head, A.T. Basilevsky, B.A. Ivanov, S.C. Werner, S. van Gasselt, J.B. Murray, T. McCord, The HRSC Team, Nature **432**, 971–979 (2004)
W.I. Newman, E.M.D. Symbalisty, T.J. Ahrens, E.M. Jones, Icarus **138**, 224–240 (1999)

H.E. Newsom, Icarus **44**, 207–216 (1980)
P.B. Niles, J. Michalski, Nat. Geosci. **2**, 215–220 (2009)
P.B. Niles, L.A. Leshin, Y. Guan, Geochim. Cosmochim. Acta **69**, 2931–2944 (2005)
P.B. Niles, M.Y. Zolotov, L.A. Leshin, Earth Planet. Sci. Lett. **286**, 122–130 (2009)
P.B. Niles, W.V. Boynton, J.H. Hoffman, D.W. Ming, D. Hamara, Science **329**, 1334–1337 (2010)
L.E. Nyquist, D.D. Bogard, C.Y. Shih, A. Greshake, D. Stoffler, O. Eugster, Space Sci. Rev. **96**, 105–164 (2001)
T. Owen, A. Bar-Nun, Icarus **116**, 215–226 (1995)
C. Oze, M. Sharma, Geophys. Res. Lett. **32** (2005)
J.L. Palandri, M.H. Reed, Geochim. Cosmochim. Acta **68**, 1115–1133 (2004)
E.M. Parmentier, M.T. Zuber, J. Geophys. Res. **112**, E02007 (2007)
R.J. Phillips, M.T. Zuber, S.C. Solomon, M.P. Golombek, B.M. Jakosky, W.B. Banerdt, D.E. Smith, R.M.E. Williams, B.M. Hynek, O. Aharonson, S.A. Hauck Ii, Science **291**, 2587–2591 (2001)
R.J. Phillips, B.J. Davis, K.L. Tanaka, S. Byrne, M.T. Mellon, N.E. Putzig, R.M. Haberle, M.A. Kahre, B.A. Campbell, L.M. Carter, I.B. Smith, J.W. Holt, S.E. Smrekar, D.C. Nunes, J.J. Plaut, A.F. Egan, T.N. Titus, R. Seu, Science (2011)
M. Pilkington, R.A.F. Grieve, Rev. Geophys. **30**, 161–181 (1992)
W. Pohl, Monogr. Ser. Miner. Depos. **28**, 1–13 (1989)
J.B. Pollack, J.F. Kasting, S.M. Richardson, K. Poliakoff, Icarus **71**, 203–224 (1987)
S.E. Postawko, W.R. Kuhn, J. Geophys. Res. **91**, D431 (1986)
R. Quinn, J. Orenberg, Geochim. Cosmochim. Acta **57**, 4611–4618 (1993)
M.C. Rabenhorst, L.P. Wilding, L.T. West, Soil Sci. Soc. Am. J. **48**, 125–132 (1984)
A.M. Reid, T.E. Bunch, Meteoritics **10**, 317 (1975)
M.I. Richardson, M.A. Mischna, J. Geophys. Res. **110**, E03003 (2005)
C.S. Romanek, M.M. Grady, I.P. Wright, D.W. Mittlefehldt, R.A. Socki, C.T. Pillinger, E.K. Gibson, Nature **372**, 655–657 (1994)
S.W. Ruff, P.R. Christensen, J. Geophys. Res. **107**, 5127 (2002)
S.W. Ruff, P.R. Christensen, R.N. Clark, H.H. Kieffer, M.C. Malin, J.L. Bandfield, B.M. Jakosky, M.D. Lane, M.T. Mellon, M.A. Presley, J. Geophys. Res. **106**, 23921–23927 (2001)
M.J. Russell, Ore Geol. Rev. **10**, 199–214 (1996)
M.J. Russell, J.K. Ingham, V. Zedef, D. Maktav, F. Sunar, A.J. Hall, A.E. Fallick, J. Geol. Soc. **156**, 869–888 (1999)
C. Sagan, G. Mullen, Science **177**, 52–56 (1972)
M.R. Salvatore, J.F. Mustard, M.B. Wyatt, S.L. Murchie, J. Geophys. Res. **115**, E07005 (2010)
J.M. Saxton, I.C. Lyon, G. Turner, Earth Planet. Sci. Lett. **160**, 811–822 (1998)
A. Segura, R. Navarro-Gonzalez, Orig. Life Evol. Biosph. **35**, 477–487 (2005)
T.L. Segura, O.B. Toon, A. Colaprete, K. Zahnle, Science **298**, 1977–1980 (2002)
R. Shaheen, A. Abramian, J. Horn, G. Dominguez, R. Sullivan, M.H. Thiemens, Proc. Natl. Acad. Sci. (2010)
N.D. Sheldon, Precambrian Res. **147**, 148–155 (2006)
P.H. Smith, L.K. Tamppari, R.E. Arvidson, D. Bass, D. Blaney, W.V. Boynton, A. Carswell, D.C. Catling, B.C. Clark, T. Duck, E. DeJong, D. Fisher, W. Goetz, H.P. Gunnlaugsson, M.H. Hecht, V. Hipkin, J. Hoffman, S.F. Hviid, H.U. Keller, S.P. Kounaves, C.F. Lange, M.T. Lemmon, M.B. Madsen, W.J. Markiewicz, J. Marshall, C.P. McKay, M.T. Mellon, D.W. Ming, R.V. Morris, W.T. Pike, N. Renno, U. Staufer, C. Stoker, P. Taylor, J.A. Whiteway, A.P. Zent, Science **325**, 58–61 (2009b)
D.E. Smith, M.T. Zuber, M.H. Torrence, P.J. Dunn, G.A. Neumann, F.G. Lemoine, S.K. Fricke, J. Geophys. Res. **114**, 05002 (2009a)
S.W. Squyres, J.F. Kasting, Science **265**, 744–749 (1994)
B.D. Stanley, M.M. Hirschmann, A.C. Withers, Geochim. Cosmochim. Acta **75**, 5987–6003 (2011)
A. Steele, M.D. Fries, H.E.F. Amundsen, B.O. Mysen, M.L. Fogel, M. Schweizer, N.Z. Boctor, Meteorit. Planet. Sci. **42**, 1549–1566 (2007)
S.K. Stephens, *Carbonate Formation on Mars: Experiments and Models* (California Institute of Technology, Pasadena, 1995), p. 276
S.K. Stephens, D.J. Stevenson, G.R. Rossman, Carbonates on Mars: experimental results, in *Lunar and Planetary Institute Science Conference Abstracts* (1995)
R.E. Stevens, M.K. Carron, Am. Mineral. **33**, 31–49 (1948)
B. Sutter, W.V. Boynton, D.W. Ming, P.B. Niles, R.V. Morris, D.C. Golden, H.V. Lauer Jr., C. Fellows, D.K. Hamara, S.A. Mertzman, Icarus **218**, 290–296 (2012)
K.L. Tanaka, J.A.P. Rodriguez, J.A. Skinner Jr., M.C. Bourke, C.M. Fortezzo, K.E. Herkenhoff, E.J. Kolb, C.H. Okubo, Icarus **196**, 318–358 (2008)
K.L. Tanaka, C.M. Fortezzo, R.K. Hayward, J.A.P. Rodriguez, J.A. Skinner Jr., Planet. Space Sci. **59**, 1128–1142 (2011)

F. Tian, J.F. Kasting, S.C. Solomon, Geophys. Res. Lett. **36** (2009)
F. Tian, M.W. Claire, J.D. Haqq-Misra, M. Smith, D.C. Crisp, D. Catling, K. Zahnle, J.F. Kasting, Earth Planet. Sci. Lett. **295**, 412–418 (2010)
O.B. Toon, T. Segura, K. Zahnle, Annu. Rev. Earth Planet. Sci. **38**, 303–322 (2010)
A.H. Treiman, Meteoritics **30**, 294–302 (1995)
A.H. Treiman, H.E.F. Amundsen, D.F. Blake, T. Bunch, Earth Planet. Sci. Lett. **204**, 323–332 (2002)
J.W. Valley, J.M. Eiler, C.M. Graham, E.K. Gibson, C.S. Romanek, E.M. Stolper, Science **275**, 1633–1638 (1997)
W. van Berk, Y. Fu, J. Geophys. Res. **116**, E10006 (2011)
J. Vaucher, D. Baratoux, M.J. Toplis, P. Pinet, N. Mangold, K. Kurita, Icarus **200**, 39–51 (2009)
P.H. Warren, J. Geophys. Res. **103**, 16759–16773 (1998)
S.C. Werner, Icarus **195**, 45–60 (2008)
S.C. Werner, Icarus **201**, 44–68 (2009)
J.J. Wray, S.L. Murchie, B.L. Ehlmann, R.E. Milliken, K.D. Seelos, E.Z. Noe Dobrea, J.F. Mustard, S.W. Squyres, Evidence for regional deeply buried carbonate-bearing rocks on Mars, in *Lunar and Planetary Science Conference Abstracts*, vol. 42 (2011), p. 2635
I.P. Wright, M.M. Grady, C.T. Pillinger, Geochim. Cosmochim. Acta **56**, 817–826 (1992)
A.S. Yen, R. Gellert, C. Schroder, R.V. Morris, J.F. Bell, A.T. Knudson, B.C. Clark, D.W. Ming, J.A. Crisp, R.E. Arvidson, D. Blaney, J. Bruckner, P.R. Christensen, D.J. DesMarais, P.A. de Souza, T.E. Economou, A. Ghosh, B.C. Hahn, K.E. Herkenhoff, L.A. Haskin, J.A. Hurowitz, B.L. Joliff, J.R. Johnson, G. Klingelhofer, M.B. Madsen, S.M. McLennan, H.Y. McSween, L. Richter, R. Rieder, D. Rodionov, L. Soderblom, S.W. Squyres, N.J. Tosca, A. Wang, M. Wyatt, J. Zipfel, Nature **436**, 49–54 (2005)
A.S. Yen, R.V. Morris, B.C. Clark, R. Gellert, A.T. Knudson, S. Squyres, D.W. Mittlefehldt, D.W. Ming, R. Arvidson, T. McCoy, M. Schmidt, J. Hurowitz, R. Li, J.R. Johnson, J. Geophys. Res. **113** (2008)
A.P. Zent, M.H. Hecht, D.R. Cobos, S.E. Wood, T.L. Hudson, S.M. Milkovich, L.P. DeFlores, M.T. Mellon, J. Geophys. Res. **115**, E00E14 (2010)
M.Y. Zolotov, E.L. Shock, Geophys. Res. Lett. **32** (2005)

Space Sci Rev (2013) 174:329–364
DOI 10.1007/s11214-012-9930-0

Geochemical Consequences of Widespread Clay Mineral Formation in Mars' Ancient Crust

Bethany L. Ehlmann · Gilles Berger · Nicolas Mangold · Joseph R. Michalski · David C. Catling · Steven W. Ruff · Eric Chassefière · Paul B. Niles · Vincent Chevrier · Francois Poulet

Received: 16 December 2011 / Accepted: 10 August 2012 / Published online: 12 September 2012

Abstract Clays form on Earth by near-surface weathering, precipitation in water bodies within basins, hydrothermal alteration (volcanic- or impact-induced), diagenesis, metamorphism, and magmatic precipitation. Diverse clay minerals have been detected from orbital

B.L. Ehlmann (✉)
Division of Geological and Planetary Sciences, California Institute of Technology, Pasadena, CA 91125, USA
e-mail: ehlmann@caltech.edu

B.L. Ehlmann
Jet Propulsion Laboratory, California Institute of Technology, Pasadena, CA 91109, USA

G. Berger
IRAP, CNRS-Université Toulouse, 31400 Toulouse, France

N. Mangold
Laboratoire Planétologie et Géodynamique de Nantes, CNRS/Université de Nantes, Nantes, France

J.R. Michalski
Planetary Science Institute, Tucson, AZ 85719, USA

J.R. Michalski
Mineralogy, Natural History Museum, London, UK

D.C. Catling
Department of Earth and Space Sciences/Astrobiology Program, University of Washington, Seattle, WA 98195, USA

S.W. Ruff
School of Earth and Space Exploration, Arizona State University, Tempe, AZ 85287, USA

E. Chassefière
Laboratoire IDES, UMR 8148, Université Paris-Sud, CNRS, 91405 Orsay, France

P.B. Niles
Astromaterials Research and Exploration Science, NASA Johnson Space Center, Houston, TX 77058, USA

investigation of terrains on Mars and are globally distributed, indicating geographically widespread aqueous alteration. Clay assemblages within deep stratigraphic units in the Martian crust include Fe/Mg smectites, chlorites and higher temperature hydrated silicates. Sedimentary clay mineral assemblages include Fe/Mg smectites, kaolinite, and sulfate, carbonate, and chloride salts. Stratigraphic sequences with multiple clay-bearing units have an upper unit with Al-clays and a lower unit with Fe/Mg-clays. The typical restriction of clay minerals to the oldest, Noachian terrains indicates a distinctive set of processes involving water-rock interaction that was prevalent early in Mars history and may have profoundly influenced the evolution of Martian geochemical systems. Current analyses of orbital data have led to the proposition of multiple clay-formation mechanisms, varying in space and time in their relative importance. These include near-surface weathering, formation in ice-dominated near-surface groundwaters, and formation by subsurface hydrothermal fluids. Near-surface, open system formation of clays would lead to fractionation of Mars' crustal reservoir into an altered crustal reservoir and a sedimentary reservoir, potentially involving changes in the composition of Mars' atmosphere. In contrast, formation of clays in the subsurface by either aqueous alteration or magmatic cooling would result in comparatively little geochemical fractionation or interaction of Mars' atmospheric, crustal, and magmatic reservoirs, with the exception of long-term sequestration of water. Formation of clays within ice would have geochemical consequences intermediate between these endmembers. We outline the future analyses of orbital data, *in situ* measurements acquired within clay-bearing terrains, and analyses of Mars samples that are needed to more fully elucidate the mechanisms of martian clay formation and to determine the consequences for the geochemical evolution of the planet.

Keywords Mars · Clay minerals · Phyllosilicates · Weathering · Alteration · Geochemistry · Mineralogy · Noachian

1 Introduction

The first telescopic observations of Mars, indicated widespread ferric oxides and demonstrated that the surface has been altered from its primary mineralogic composition (Singer et al. 1979; Bell et al. 1990), while the Viking landers provided some circumstantial evidence for salts and a component of clay minerals in the soil (Clark 1978; Banin et al. 1992). More recently, the detection of crystalline hydrous minerals demonstrates that interaction with liquid water has left a mineralogic record from the first billion years of Mars history in the form of widespread clay-bearing terrains and large-scale sulfate deposits (Poulet et al. 2005; Gendrin et al. 2005; Bibring et al. 2006). In particular, clay minerals have been identified in units dating from Mars' Noachian period, indicating a global process or set of processes altering the mostly basaltic bedrock at this time (Poulet et al. 2005; Bibring et al. 2006; Mustard et al. 2008).

As detailed below, most information on clay mineral formation conditions on Mars derives from constraints from orbital datasets with additional constraints derived from study

V. Chevrier
W.M. Keck Laboratory for Space and Planetary Simulation, Arkansas Center for Space and Planetary Science, University of Arkansas, Fayetteville, AR 72701, USA

F. Poulet
Institut d'Astrophysique Spatiale, Université Paris-Sud, Orsay 91405, France

of the martian meteorites and from landing sites visited to date. Our objective herein is to understand the consequences of various geochemical scenarios for clay formation on Mars. Processes leading to clay formation on Earth are important mechanisms in terrestrial geochemical cycles. Ocean and river water chemistry is partly controlled by continental silicate weathering to form clays and clay-forming reactions of seawater with the basaltic seafloor (Spencer and Hardie 1990). Clay formation by weathering also serves as an important feedback on atmospheric chemistry, climate, and sequestration of volatiles in the crust (Kump et al. 2000). To understand the broader geochemical consequences of clay formation on Mars, we focus on two sets of questions. First, what style of alteration/formation process(es) drove clay formation? Second, to what degree did/could clay formation result in geochemical fractionation? What were the implications of clay formation for subsequent evolution of martian geochemical reservoirs? Were new reservoirs created, e.g. sediments/salts? Was the crustal or atmospheric reservoir altered, e.g. by sequestration of water or other volatile species?

We first review the various possible mechanisms of clay formation with reference to the literature from terrestrial studies. We then review observations to date from remote sensing of clay-bearing terrains on Mars, including their mineralogy and stratigraphy, which provide constraints on alteration scenarios and their timing. Next, we examine the likelihood and geochemical consequences of potential scenarios for clay formation on Mars, in light of the data from Mars, lessons from terrestrial analogues, geochemical modeling, and experimental data. Our primary goal is to understand the implications of different scenarios for the evolution and/or formation of martian geochemical reservoirs. Finally, we discuss what further data are needed to distinguish between models for clay formation, including measurements obtainable from future landed, orbital, and sample return missions.

2 Clay Formation on Earth

Clay minerals or phyllosilicates are comprised of sheets of tetrahedrally-coordinated silica and octahedrally-coordinated metal cations in regular, repeating layers (e.g. Meunier 2005). The term "clay" can refer either to the smallest particle size fraction in sediments ($<4\ \mu m$) or phyllosilicate mineralogy. Herein, we use the term "clay" in the latter sense. Clays are hydrated or hydroxylated and sometimes have swelling properties associated with exposure to water. We discuss clay minerals and associated phases (other hydrated silicates, salts, and mineraloids) formed during aqueous alteration.

On Earth, clay minerals form by near-surface weathering, in hydrothermal systems located mostly at sea floor spreading centers and less commonly in continental settings, by diagenesis and metamorphism, or more rarely by direct precipitation in lake basins or from magmatic fluids within cooling lavas. Clay minerals are then eroded, transported, buried, and metamorphosed in processes driven by climate and tectonics (for review see Merriman 2005; Meunier 2005). There are seven major mechanisms by which clay minerals are formed or transformed, described below. Table 1 details expected deposit characteristics for each mechanism with particular attention to traits that may be discernible from orbit, based on examples from terrestrial sites with basaltic precursor materials. The clay formation mechanism—i.e., the composition and quantity of reacting fluids, the chemistry of the precursor rock, temperature of the reaction, distance of transport of fluids, and the degree to which atmospheric volatiles participate in the reaction—determines the composition and petrology of the clays. The formation mechanism and extent of alteration together dictate the geochemical consequences of clay formation, depicted schematically for global geochemical reservoirs for each of the mechanisms detailed below (Fig. 1).

Table 1 Mineralogy and facies of hydrated silicates produced by seven formation mechanisms from a basaltic precursor (references in text)

Setting	Clay mineralogy	Facies and abundance (% vol.)	Setting	Accompanying minerals
Near-surface pedogenic	Fe/Al smectites, kaolinite	Bulk soil component (up to 95 %)	Horizons of leaching and deposition, with alteration lessening with depth	Fe/Al oxides, carbonates, silica, allophone
Near-surface basin	Fe/Mg/Al smectites, kaolinite, illite, chlorite	Layered deposits with sedimentary textures; clay minerals (up to 95 %) mostly detrital, *in situ* formation at very high Si, Al activities	Deposits within a river, lake, or ocean, later exposed by erosion	Other minerals eroded from rocks in basin; evaporites, e.g. chlorides, carbonates, hematite, silica, sulfates (if acidic), potassium feldspar and zeolites (if alkaline)
Hydrothermal (volcanic)	Fe/Mg/Al smectites, kaolinite	Bulk component (variable)	Zoned alteration surrounding fumaroles, vents, and cones	Sulfates (alunite, jarosite), ferric oxides, amorphous silica, allophane, anatase
Hydrothermal (impact)	Saponite, nontronite, celadonite, kaolinite	Fracture fill within breccia, alteration rinds on mineral grains (5–10 % total area)	Beneath the crater floor and rim in fractures and pore spaces of breccia; not exposed unless by erosion	Amorphous silica and altered impact glasses, carbonate, sulfates, sulfides, potassium feldspar quartz, zeolites, native metals
Diagenesis	Illite (from K, Al-rich precursors); Chlorite (from Mg-rich precursors), mixed-layer clays	Bulk rock; veins and pore space (variable)	Bulk rock/sediment altered; sometimes preferential alteration in pore space and veins	silica, original clay minerals
Metamorphism	Fe/Mg smectites, chlorite, zeolites, prehnite, pumpellyite, serpentine, epidote, actinolite	Bulk rock (variable)	100s m to km beneath the surface; not exposed unless by deep erosion or impact excavation	Amorphous silica, zeolite, sulfite, iron oxides, garnet, original rock forming minerals
Magmatic	Fe/Mg smectite, celadonite	Within pores, veins of bulk rock (<15 %)	Formed during final degassing of lavas with substantial volatile content	Silica, primary minerals

2.1 Near-Surface Pedogenesis

Soil formation on Earth involves the soil microbes processing detrital organic matter in addition to mechanical breakdown and abiotic leaching processes. On Mars, "pedogenesis" presumably involves only the latter two processes. The amount and type of alteration observed in terrestrial soil formation varies widely depending on time, starting materials, and degree of leaching. On Earth, pedogenic smectites form in relatively dry conditions from waters with dissolved constituents concentrated by evaporation; kaolinite forms in dilute waters; and gibbsite is an end product of weathering under warm

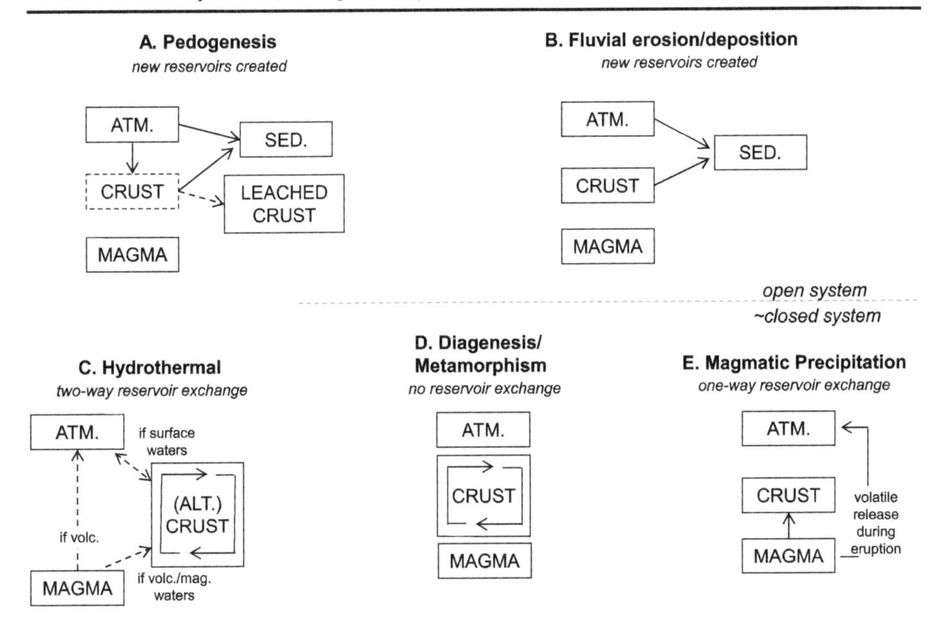

Fig. 1 Schematic of geochemical reservoirs during terrestrial clay formation by scenario. (**a**) Near-surface leaching during pedogenesis fractionates the crustal reservoir, leading to formation of a new sedimentary reservoir. The atmospheric and crustal reservoirs are modified. (**b**) Clay deposition in basins transports materials from the crustal reservoir to basins. A new sedimentary reservoir is formed with input from the crustal and atmospheric reservoirs. The atmospheric reservoir may be modified due to the quantity of volatiles sequestered in salts and hydrated minerals, but in a transport-only scenario, the crustal reservoir is not chemically modified. (**c**) Hydrothermal alteration may occur under open- or closed-system conditions, potentially altering the crust by input from atmospheric or magmatic reservoirs. (**d**) Diagenesis and metamorphism occur in closed system by recirculation of fluids in the crust. (**e**) Clay formation from magmatic precipitation adds to the crustal reservoir, some volatiles are sequestered in clay minerals and others escape to the atmosphere

conditions of very high water/rock ratio. Terrestrial examples of clay-bearing deposits formed by pedogenesis include kaolinite and Al, Fe oxide-rich laterites in tropical soils; smectite-zeolite altered volcanic ashes in the Cascades and New Zealand, and palagonitized recent lava flows in Hawaii (e.g., Singer 1966; Siefferman and Millot 1969; Chamley 1989; Ambers 2001; Ahn et al. 1988; Morris et al. 2000; Schiffman et al. 2000; Schiffman et al. 2002).

For a given protolith, the amount of rainfall and chemistry of the water (pH, Eh, dissolved ions) dictate composition of alteration products. For example, tephra deposits in the altered Keanakoko'i Ash Member at Kilauea volcano in Hawaii (Schiffman et al. 2002; Schiffman et al. 2000) have developed a rind of smectites and other clay minerals. Under acid conditions of pH < 6 owing to rainout of sulfur aerosols, opaline crusts form on outcrop faces of desilicated tephra, whereas under neutral to alkaline conditions of pH 6.5–7.8, the weathering product is dominantly smectites in areas with mean rainfall $<$50 cm/year and kaolinite, allophane and/or imogolite in areas with rainfall $>$250 cm year (Schiffman et al. 2000).

Vertical stratification of heavily leached horizons, depositional horizons, and minimally altered bedrock is expected and diagnostic of near-surface pedogenesis. Such strata in escarpments could be detectable by orbiting instruments. Pedogenesis occurs in an open system environment, usually at high water/rock ratio. The composition of the bedrock is changed as alkali and alkaline cations are dissolved and transported in fluids. Even ultramafic bedrock with a very low proportion of Al can transform into aluminum clay (kaolinite)-rich soils by strong alteration, as observed in Murrin, Australia (Gaudin et al. 2011). Salts including carbonates, chlorides, and sulfates precipitate by reaction of atmospheric volatiles dissolved in the water with dissolved cations.

2.2 Deposition and Formation in Basins

On Earth, alteration products frequently undergo erosion and are deposited in large sedimentary basins in a setting distinct from a non-transported, pedogenic one. Shales, which are a sedimentary end-product of continental weathering and gradation, typically contain 60–70 % clays with lesser amounts of feldspar and quartz. Most clays in these systems are detrital and undergo mineralogical transformation with depth and time (diagenesis), mainly leading the conversion of the detrital clays into illite or less frequently chlorite (see Sect. 2.5). The detrital fraction may also be accompanied by salt precipitation in an evaporitic basin. The dominant salt depends on the fluid chemistry and can be carbonate, sulfate, or chloride. If Al and Si activities are sufficiently high, clays can precipitate. Saline-alkaline waters can lead to zeolite precipitation and formation of chlorite and Mg-smectite from illites (Remy and Ferrell 1989; Hay and Tyser 2001; Bristow and Milliken 2011). Acid lake systems produce sulfates, halite, and hematite but are not generally conducive to formation and preservation of clay minerals. However, clays such as kaolinite, smectite and palygorskite-sepiolite are observed in some acid lakes (Benison et al. 2007; Baldridge et al. 2009; Story et al. 2010). Interbedding of clays and sulfates in some lakes, such as Lake Gilmore in Western Australia might be explained by authigenic sulfates and detrital clay minerals. If fluvial systems were operating on a large scale on early Mars, layered sedimentary strata with salts might be expected to be associated with some clay deposits in large basins.

Another source of sedimentary clay minerals in the deep ocean or inland water bodies originates from altered airborne volcanic ash. Large swaths of the Pacific Ocean floor are covered in aluminous smectites or mixed-layer smectites/illites that derive from volcanic ash (Lisitzin and Rodlfa 1972). Such clays can be distinguished because the devitrification process (i.e., the change from the glassy state) takes time and so the clay minerals are often mixed with ashy shards of volcanic glass. Montmorillonite-bearing altered tuffs forming Wyoming bentonite also form from deposition of ash into standing bodies of water. On Earth, some of the most clay-rich rocks are bentonites, which are derived from pyroclastic materials that are nearly completely altered (90 % clays or more), typically due to deposition into standing water (Chamley 1989).

2.3 Hydrothermal (Volcanic)

Clays that form by hydrothermal, acid-sulfate alteration in volcanic zones are accompanied by ferric oxides and sulfates (Morris et al. 2000; Bishop et al. 2007). Mineral distribution in terrestrial hydrothermal systems commonly exhibits quasi-concentric zoning because temperature and pH vary at spatial scales of tens of square meters around volcanic vents (Swayze et al. 1992, 2002; Guinness et al. 2007). In the basaltic tephras of Mauna Kea, Hawaii, well-crystalline hematite near the center of cinder cones grades to palagonitic material away from the cone. Montmorillonite is the most abundant clay mineral and tends to surround saponite and isolated occurrences of kaolinite and jarosite near cones (Swayze et al. 2002; Guinness et al. 2007). If volcanic hydrothermal systems were active on Mars, zoning and clays accompanied by sulfates and iron oxides would be diagnostic.

Sulfur-poor hydrothermal systems also occur in which intruded magma supplies subsurface heat, but sulfurous gases are not in contact with the water system. In these systems, Fe/Mg smectites and chlorite are the dominant alteration minerals in the bulk rock and celadonite, silica, and zeolites fill rock pore spaces; a smectite-zeolite zone often overlies a chlorite zone (e.g. Wiesenberger and Selbekk 2008; Franzson et al. 2008). Fluids interacting with mafic to ultramafic rock at depth can also generate veins of Fe/Mg clays, as in Uley Mine, South Australia, where nontronite, an iron magnesium smectite clay, infills fractures within amphibolite and gneiss (Keeling et al. 2000). Another type of hydrothermal alteration on Earth occurs under the ocean within the basaltic seafloor. Under these conditions, typical alteration assemblages include serpentines at high temperatures and ultramafic compositions, smectites and chlorite at intermediate temperatures, and nontronite at low temperatures (Alt 1999).

The content of clay in hydrothermal systems is highly variable, ranging from negligible to nearly complete alteration of primary phases to clay minerals. Many intensely hydrothermally altered deposits exhibit striking textural disruption and are penetrated by clays, but the clay minerals may constitute only a small fraction (<5 %) of the rock. Even basalts that have been 100 % converted to secondary minerals by crustal brines may contain only ~25 % clays (Michalski et al. 2007).

2.4 Hydrothermal (Impact)

As in volcanic systems, the occurrence and mineralogy of impact-induced hydrothermal systems depend on water availability, the composition of the target rocks, and the temperature and composition of circulating fluids. Shock transforms some crystalline minerals to other phases (e.g. plagioclase to maskelynite), and heat released by impact generates glasses and, for sufficiently large craters, a coherent impact melt sheet within the crater. In terrestrial craters, a diverse suite of alteration minerals is observed, including clay minerals, carbonates, sulfates, sulfides, zeolites, quartz (amorphous and crystalline), potassium feldspar, iron oxides, and native metals (e.g. Allen et al. 1982b; Ames et al. 2004; Hagerty and Newsom 2003; Larsen et al. 2009). However, a survey of alteration minerals associated with terrestrial impact craters suggests that clays are not widely formed directly by the impact process, and where they occur, they are not abundant phases. For example, studies of altered products of the Manicouagan structure found abundances <15 % smectite in pyroxenes and glasses altered to hematite (Morris et al. 1995). The most common occurrences of alteration minerals formed by impact processes are those that occur in hydrothermally-fed crater lake deposits, where water has flooded the impact structure (Naumov 2005). Hydrothermal activity is restricted to the subsurface beneath craters in the lower parts of the rim, floor, and crater central peak, and if sufficient water is supplied, a hydrothermally circulating crater lake can form (Rathbun and Squyres 2002; Abramov and Kring 2005). Thus, typically, subsequent exhumation of buried alteration minerals would typically be required for detection from Mars orbit of non-lacustrine, impact-formed clays. Heat and shock from impact can also destroy pre-existing clays, from reversible loss of water at low temperatures, to irreversible changes in the crystal structure if temperatures experienced are greater than 600 °C (Milliken and Mustard 2005; Gavin and Chevrier 2010; Che et al. 2011).

2.5 Diagenesis

Water/rock interactions at depth and elevated temperature can generate and modify clay mineral deposits. The early diagenesis of sediments is promoted by bacterial activity leading to sulfate reduction, methane production, and carbonate precipitation. In the zone of intermediate burial (2 to 4 km depth), chemical compaction by pressure-solution of quartz or carbonate grains decreases porosity. The production of CO_2 and aliphatic acids (acids of non-aromatic hydrocarbons) by late organic diagenesis promotes the dissolution of feldspars and the transformation of detrital clays into illite or chlorite (Surdam et al. 1989). Even in the absence of organic material, buried smectites may transform to other phases such as illite and chlorite, during thermal- and pressure-related maturation (Ahn et al. 1988; Ambers 2001; Srodon 1999; Meunier 2005). The degree of transformation depends on time and temperature when sufficient water is available to promote transformations (e.g. Whitney 1990). Provided that sufficient K^+ is available in fluids, dioctahedral smectites such as montmorillonite convert to mixed layer illite/smectite clays, then to end-member illite with more advanced diagenesis at higher temperatures (e.g. Hower et al. 1976; Yau ct al. 1988; Lanson et al. 2002; Fleet and Howie 2006). Provided that sufficient Mg^{2+} is available, trioctahedral smectites such as saponite transform to corrensite (a 1:1 mixed-layer chlorite/smectite) then to chlorite with increasing grade (Merriman and Peacor 1999 and refs. therein). This is a common process on Earth as tectonic activity creates deep sedimentary basins. Clays can also form diagenetically as cements in sandstone (Schneiderhöhn 1965). Clays occur along with other cements (typically silica or calcite) in sandstones, and in this context, constitute 15–30 % of the rock.

2.6 Subsurface Metamorphism

There are poorly-defined boundaries between low-grade metamorphism, hydrothermal activity, and diagenesis with different authors adopting different conventions through time (see Arkai et al. 2003 for a review). In metamorphism, as in diagenesis, small amounts of water interacting with rock lead to the transformation of minerals to phases stable at higher temperatures and pressures. Amphibole minerals (e.g. actinolite), kyanite, and sillimanite occur in high pressure and temperature metamorphism, caused by tectonic activity, e.g. subduction of oceanic plates and thrusting to create mountain ridges (Spear 1995). This type of tectonism has not been observed on Mars. At lower temperatures and pressures ($T <\sim 400\ ^\circ C$ and pressures typically <4 kbar but extending to 8 kbar), sub-greenschist metamorphism occurs and is defined by the thermodynamic stability of prehnite and pumpellyite (Schiffman and Day 1999). The lowest metamorphic temperature and pressure conditions are represented by the zeolite facies, where laumontite and analcime form at lower temperatures in the presence of CO_2-poor or non-CO_2 bearing alkaline fluids (e.g. Hay 1986). These originally subsurface deposits would have to be exposed by erosion or uplift to be detectable from orbit.

2.7 Magmatic

When mafic-ultramafic lavas erupt, they may have volatile contents of a few percent. As these lavas cool, incompatible H_2O becomes concentrated and Fe–Mg clays precipitate directly from the residual liquid that is concentrated in voids remaining in the crystallizing, solidifying lava (Meunier et al. 2010). Nontronite-celadonite and chlorite-saponite have been observed covering solid surfaces, glasses, and grains of pyroxene and apatite in the rock mesostasis. Tens to hundreds of micrometer-wide clay-rich patches exist in lavas at abundances of $\sim$10 % (Meunier et al. 2010).

3 Clay Distribution and Diversity on Mars

Consideration of the seven settings above illustrates that component mineral assemblage, alteration mineral abundance, and stratigraphic setting are constraints that are at least partially ascertainable from orbit and that allow distinguishing among clay mineral formation mechanisms (Table 1) each of which has different consequences for geochemical fractionator (Fig. 1). Below, we consider the nature of martian clay-bearing deposits, reviewing constraints from orbital, landed, and meteorite datasets.

3.1 Clay Detection and Global Distribution

Clay minerals are detectable and mappable in remote sensing data because of distinctive absorptions that occur at infrared wavelengths of light due to vibrations of molecules within the mineral structure. These include fundamental H_2O stretching and bending modes (near 3.0 μm and 6.0 μm) and combinations and overtones of these (near 1.4 μm and 1.9 μm). Metal-OH stretching (near 2.8 μm), bending (16–25 μm), overtone (near 1.4 μm) and combination absorptions (2.2–2.5 μm) are particularly useful in identifying the precise clay mineral because the position and shape of the absorption features depend on the composition of the cations and the configuration of the cation site (e.g. Michalski et al. 2005; Bishop et al. 2008a). For example, in smectite clays, the near-infrared overtones and combination tones occur at 1.41 μm and 2.21 μm for Al-rich montmorillonites, at 1.43 μm and 2.28 μm for Fe-rich nontronites, and at 1.38 μm and 2.32 μm for Mg-rich saponites (Bishop et al. 2008a; Ehlmann et al. 2009).

A global survey for clay minerals at 1 pixel per degree was first undertaken using thermal infrared (TIR) spectra from the Thermal Emission Spectrometer (TES) onboard Mars Global Surveyor, but no definitive identification was made at this scale (Bandfield 2002). Clay minerals were first detected unambiguously in dark deposits and outcrops within Noachian terrains in several locations throughout the southern highlands using visible/near-infrared (VNIR) spectral data from the OMEGA (Observatoire pour la Mineralogie, l'Eau, les Glaces et l'Activité) instrument onboard Mars Express (Poulet et al. 2005, 2007). Subsequent global mapping with OMEGA and the VNIR imaging spectrometer CRISM (Compact Reconnaissance Imaging Spectrometer for Mars) on the Mars Reconnaissance Orbiter (Murchie et al. 2007, 2009) has demonstrated that areas around the Nili Fossae and Mawrth Vallis host the largest ($>$100,000 km^2) regionally extensive clay-bearing outcrops (Loizeau et al. 2007; Mangold et al. 2007; Mustard et al., 2007; Bishop et al. 2008b; Ehlmann et al. 2009; McKeown et al. 2009; Noe Dobrea et al. 2010), while thousands of isolated occurrences of clay minerals in outcrops of smaller size are found throughout the southern highlands (Mustard et al. 2008; Loizeau et al. 2012) as well as associated with large craters of the northern lowlands (Carter et al. 2010) (Fig. 2). The majority of clays detected are associated with impact craters, in central peaks, walls or ejecta. Some large tectonic scarps and heavily eroded terrains expose clay minerals exhumed from depths of hundreds to thousands of meters (Murchie et al. 2009; Mustard et al. 2009; Buczkowski et al. 2010; Ehlmann et al. 2011a). Some sedimentary deposits also host clays (Ehlmann et al. 2008b; Murchie et al. 2009; Milliken and Bish 2010; Dehouck et al. 2010; Ansan et al. 2011; Milliken et al. 2010; Wray et al. 2011). While clays are globally distributed among Noachian terrains (Fig. 2), the major Hesperian volcanic plains within Syrtis Major and the northern plains show no unambiguous spectral evidence for the presence of clay minerals, nor do younger terrains. Instead, sulfates are found within some Hesperian- and Amazonian-aged units around Valles Marineris (Gendrin et al. 2005; Milliken et al. 2008;

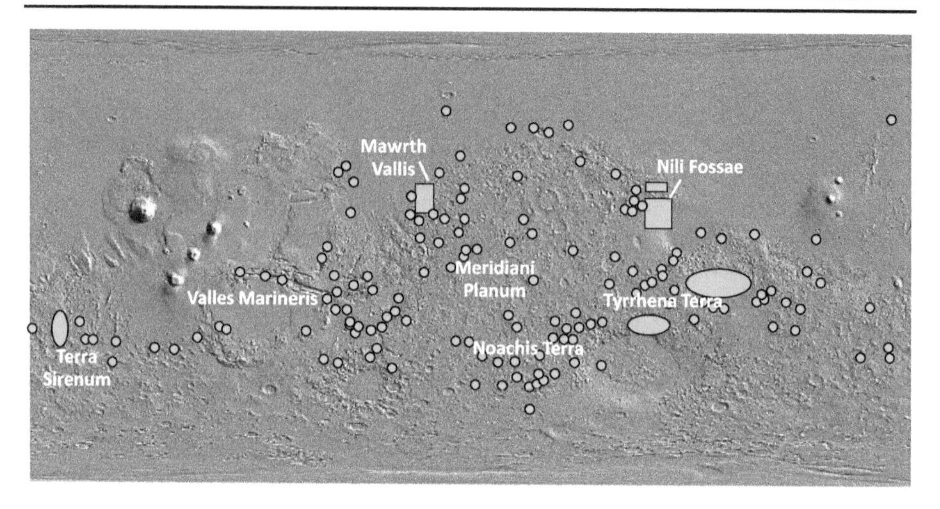

Fig. 2 Schematic distribution of clay mineral detections on Mars

Murchie et al. 2009; Mangold et al. 2010). Sediments in the northern lowlands and Hesperian and Amazonian lava plains show evidence for formation of coatings or rinds of amorphous weathering products (Mustard et al. 2005; Michalski et al. 2006; Skok et al. 2010; Salvatore et al. 2010; Horgan and Bell 2012).

3.2 Estimates of Clay Abundance for Multiple Terrains

Constraining the abundances of clay minerals in clay-bearing rocks on Mars is important for distinguishing between rocks that may be only slightly altered by traces of water versus sedimentary rocks analogous to shales or bentonites (see Sect. 2.2) that may be thoroughly altered and physically processed. Quantitative information is critical for evaluating how geochemical reservoirs are partitioned between clay-bearing rocks and the protoliths from which they were derived. As discussed in Sect. 2, clay abundance is not necessarily an indication of degree of alteration. But, understanding the abundances can help to bracket the types of rocks that contain clays on Mars and their formation mechanisms. The determination of abundances from orbital data is strongly influenced by textural context of the clays within the rocks (as a component of the clastic budget, as cement, as grain coatings, veins and veinlets, etc). Consequently, an alternative indicator of alteration from infrared data can be the presence or absence of primary phases.

Although clay minerals have been identified unambiguously across the planet, their abundance is less clear. For example, at Mawrth Vallis, modeling of VNIR spectra using a nonlinear radiative transfer approach gives estimates of 20–65 % clay for the bulk composition of the iron-magnesium clay bearing unit, the highest for any of the exposures examined to date on Mars (Poulet et al. 2008a). However, clay mineral abundance at Mawrth Vallis does not rise above the approximate detection limit of 15 % by volume based on thermal infrared spectra (TIR), modeled linearly to provide abundance estimates of component phases (Michalski and Fergason 2009). Elsewhere in the southern highlands, including Nili Fossae, Tyrrhena Terra, and Terra Sirenum, clay mineral abundance estimates from VNIR spectra are lower than at Mawrth, between 5–35 % (Poulet et al. 2008a). TIR data are consistent with this, suggesting clay mineral fractions by area of ~30 % or less (e.g., Michalski and Fergason 2009; McDowell and Hamilton 2009; Michalski et al., 2010). Various explanations have been advanced to explain the occasional disparities between TIR and

VNIR results (e.g., Michalski and Fergason 2009; McDowell and Hamilton 2009), but at present the issue remains unresolved. Even so, TIR observations point to clear differences between alteration seen in the Mawrth Vallis region, where spectra are dominated by poorly crystalline alteration phases with little evidence for primary phases, and alteration in the Nili Fossae region, where spectra contain evidence for both clays and primary phases together. Hydrogen abundance in the upper half-meter of the martian surface, acquired by the gamma ray and neutron spectrometers onboard Mars Odyssey at hundreds of kilometer spatial scale, generally shows no correlation with IR-detected exposures of hydrated minerals in mid- and equatorial-latitudes (Feldman et al. 2004; Jouglet et al. 2007; Milliken et al. 2007; Boynton et al. 2008).

3.3 Clay Mineral Diversity and Mineral Assemblages

Over the last decade, the number of alteration minerals identified on Mars and mappable from orbit has expanded significantly, now including hydrated silicates, salts, and oxides/hydroxides (Table 2). In particular, hydrated silicate minerals, i.e. clay minerals and associated phases, have been found to exhibit significant diversity, varying regionally and by geologic setting (Mustard et al. 2008). This diversity in clay minerals and accompanying phases may indicate a diversity of processes and thermodynamic, kinetic, and geochemical parameters responsible for their formation.

Iron-magnesium smectites are the most common clay minerals on Mars, found in over 75 % of locations in which any hydrated silicate is detected (Table 3; Poulet et al. 2005; Mustard et al. 2008; Ehlmann et al. 2011a). Mixed-layer clays, such as chlorite/smectite, are similar spectrally and may also be part of these deposits (Milliken and Bish 2010). Chlorite is the second most commonly detected clay mineral phase. Rarer phases include prehnite, serpentine, illite or muscovite, hydrated silica, and analcime, comprising mineral assemblages that indicate alteration at elevated temperatures. As discussed below, these and chlorite are almost exclusively associated with craters (Ehlmann et al. 2011b).

Although less common than the iron-magnesium clay minerals, aluminum clays have also been detected on Mars. As discussed further below, a few regions exhibit a distinctive stratigraphy of Al clays above Fe/Mg clays. The Al-clays in these settings include a kaolin family mineral, as well as less commonly an Al smectite such as montmorillonite or beidellite (Poulet et al. 2005; Loizeau et al. 2007, 2010; Bishop et al. 2008b; Wray et al. 2008; Ehlmann et al. 2009; McKeown et al. 2009; Bishop et al. 2011). Al-clays also occasionally occur associated with impact craters and within sedimentary basins (Table 2; Wray et al. 2009a).

To the limit of spectral detection capabilities, salts such as carbonates, chlorides, and sulfates are not usually found in the same geologic units as clay minerals (Milliken et al., 2009), an exception being clays and carbonates found within the Leighton crater central peak in southwestern Syrtis Major (Michalski and Niles 2010). However, chlorides, sulfates, and carbonates are found in association with clays in sedimentary deposits (see Sect. 3.5).

3.4 Clays and Craters

The ancient Noachian crust of Mars has been heavily churned by meteorite impacts, and it is mostly within these terrains that clay minerals are found. Most exposures of clay minerals on Mars are in association with craters of diameters <200 km. Craters serve as basins for collecting sediments; we discuss these in Sect. 3.5. Craters can also serve as probes of the subsurface by excavating buried materials from depth and exposing them

Table 2 Alteration minerals discovered from Mars orbit and on the ground

Group/mineral/phase	References[b]
Fe/Mg smectites (e.g., nontronite, saponite)	Poulet et al. 2005; Mustard et al. 2008; Bishop et al. 2008b; Ehlmann et al. 2009; McKeown et al. 2009
Al-smectite (e.g. montmorillonite)	Poulet et al. 2005; Mustard et al. 2008
Kaolin group minerals (e.g. kaolinite, halloysite)	Gondet et al. 2006; Mustard et al. 2008; Bishop et al. 2008b; Ehlmann et al. 2009
Chlorite	Poulet et al. 2005; Mustard et al. 2008; Ehlmann et al. 2009
Serpentine	Ehlmann et al. 2010
High charge Al, K phyllosilicate (e.g. muscovite or illite)	Mustard et al. 2008; Ehlmann et al. 2009
Prehnite	Clark et al. 2008; Ehlmann et al. 2009
Analcime	Ehlmann et al. 2009
Opaline silica	Squyres et al. 2008; Milliken et al. 2008; Ehlmann et al. 2009
Magnesium (iron) carbonate	Bandfield et al. 2003; Ehlmann et al. 2008a; Morris et al. 2010
Calcium carbonate	Boynton et al. 2009; Michalski and Niles 2010[a]
Fe/Mg mono- and poly-hydrated sulfates	Gendrin et al. 2005; Bishop et al. 2009; Murchie et al. 2009
Gypsum	Langevin et al. 2005
Alunite	Swayze et al. 2008
Jarosite	Klingelhofer et al. 2004; Milliken et al. 2008; Farrand et al. 2009
Hydroxylated ferric sulfate (not a named mineral)	Bishop et al. 2009; Lichtenberg et al. 2010
chlorides	Osterloo et al. 2008
perchlorates	Hecht et al. 2009
Goethite	Morris et al. 2006; Farrand et al. 2009
Coarse-grained hematite	Christensen et al. 2000; Klingelhofer et al. 2004; Bibring et al. 2007
Nanocrystalline hematite	Morris et al. 1989; Bell et al. 1990

[a]Orbital near-infrared data are also consistent with an Fe-rich carbonate

[b]Numerous papers been published on these and the references are to the first papers reporting and justifying the detections with particular instruments

at the surface in ejecta, crater rims, and central uplifts (Fig. 3). The cratering process can also serve as an agent for modifying surface composition. Shock and heat transform the properties of impacted materials, generating new minerals or shocked or melted modified phases. Laboratory studies to date have shown that instantaneous shock does not form clays (Allen et al. 1982a), but shocked, glassy materials may be more susceptible to later weathering to form clays. Numerical modeling (Rathbun and Squyres 2002; Abramov and Kring 2005) and studies of terrestrial craters (Newsom 1980) demonstrate that heat from impact can initiate hydrothermal systems at depth, given sufficient supply of water (see Sect. 2.4).

Table 3 For 365 sites with hydrated silicates, percentage of sites with each alteration mineral by geologic setting. Alteration minerals were found in 259 crustal sites (craters, ancient degraded terrain), 45 sedimentary sites (transported, basin-filling materials), and 61 in stratigraphic sections (preserved, coherent stratigraphies with multiple clay-bearing units). Sites tallied are the same as those in Ehlmann et al. (2011a)

	Alteration phase	Crustal (% sites)	Sedimentary (% sites)	In stratigraphy (% sites)
Fe/Mg clays	Fe/Mg smectite	77.6	84.4	95.1
	Chlorite/prehnite	39.0	0	1.6
	Serpentine	3.1	0	6.6
Al clays	Montmorillonite	3.9	4.4	26.2
	Kaolinite	7.7	6.7	36.1
	Al-clays (unspecified)	0	20.0	18.0
Other hydrated silicate	Illite	2.3	0	0
	Silica	5.4	4.4	29.5
	Analcime	1.9	0	0
	Other hydrated (unspecified)	5.4	8.9	1.6
Salts	Carbonate	1.2	4.4	24.6
	Sulfate	0	35.6	3.3
	Chloride	0	20.0	0

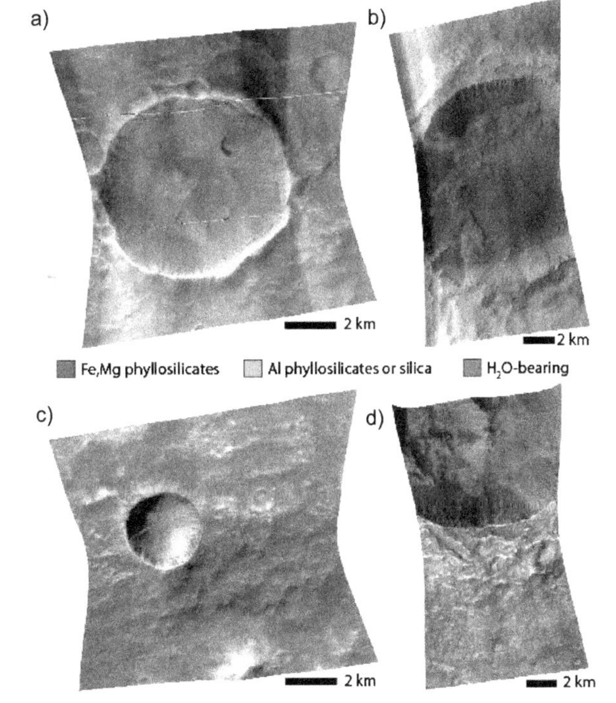

Fig. 3 Clays in craters within Tyrrhena Terra. Crater walls, ejecta, and central peaks have signatures of clay minerals. Parameters (Pelkey et al. 2007) for mapping Fe, Mg clays (D2300), Al clays and silica (BD2200), and hydrated minerals (BD1900) are mapped on infrared albedo for CRISM images (**a**) FRT00013FA4, (**b**) HRL0000D721, (**c**) FRT0000A33C, and (**d**) HRL00013BA7

Thus, clay minerals associated with impact craters could have been formed by either excavation of buried, pre-existing clays within the crust or via local impact-induced hydrothermal circulation. This question has persisted since the first detections of hydrated minerals in crater ejecta (Mangold et al. 2007). In general, observations are consistent with excavation being the primary process responsible for the association of clays with craters (Mangold et al. 2007; Mustard et al. 2008; Fairen et al. 2010; Carter et al. 2010; Ehlmann et al. 2011b). The same kinds of alteration minerals are associated with units in the ejecta and the upper walls, where numerical modeling predicts hydrothermal circulation would be absent or limited, as with the central peak regions, where numerical modeling predicts hydrothermal fluid flow would be most vigorous (Fig. 3; Poulet et al. 2005; Mustard et al. 2008; Ehlmann et al. 2009; Carter et al. 2010; Michalski and Niles 2010, Loizeau et al. 2012). If instead impact-induced hydrothermal activity had generated the clays associated with craters, minerals zoned by temperature and fluid availability would exist mainly within the crater interior (Schwenzer and Kring 2009).

Fe, Mg smectite and/or chlorite are nearly always present in crustal clay units associated with craters (Poulet et al. 2005; Mustard et al. 2008). Distinctive associations of minerals vary from crater to crater, and can include prehnite, hydrated silica, analcime, serpentine, and illite or muscovite, which accompany the smectites and chlorite and indicate excavation of minerals formed at higher temperatures, up to 400 °C (Ehlmann et al. 2009, 2011b). With the exception of silica, these distinctive phases have only been detected in association with craters (Ehlmann et al. 2011a). The materials, associated with craters <200 km in diameter, would originate from maximum depths of ∼5–10 km (Melosh 1989), although multiple impacts, including large basin-forming impacts, may have exposed still deeper materials by repeated churning.

Apart from excavation, craters inside Hesperian plains locally display alteration that could be related to local hydrothermal activity. For example, Toro crater, an impact structure in northern Syrtis Major, displays hydrated minerals including clays and silica in association with possible vents related to hydrothermal circulation (Marzo et al. 2010), although simple impact excavation has also been proposed to explain the distribution of clay minerals in this crater (Ehlmann et al. 2009). Another unnamed 45-km crater that impacted Late Hesperian lava flows displays Fe, Mg smectites on the crater floor and in the lower section of an alluvial fan, possibly indicating aqueous alteration at depth derived from atmospheric snow melted by impact heat (Mangold et al. 2012a). Several craters >40 km in Tyrrhena Terra also have complex assemblages of minerals that could reveal local hydrothermal alteration (Loizeau et al. 2012). Distinguishing hydrothermal alteration from excavation is difficult given the fact that similar minerals can derive from both processes, but careful observations of their setting may lead to additional examples of potential impact-related alteration in the future.

3.5 Clays and Sedimentary/Fluvial Activity

Initial studies found that fluvial valleys and clays were not co-located (Bibring et al. 2006). However, both are found in Noachian-aged terrains almost exclusively, raising the question of whether their formation was coeval. In Nili Fossae, the water–rock interactions that formed clays occurred mostly before the time of the Isidis impact (Mustard et al. 2007; Mangold et al. 2007). Fluvial landforms post-dating Isidis basin in this region are poorly developed, and their formation generated little or no alteration (Mangold et al. 2007; Mangold 2008). In Mawrth Vallis, valley networks incise into clay-bearing terrains without obvious genetic relationships (Fig. 4; Loizeau et al. 2007, 2010; Mangold 2008). Despite the fact that both valley networks and clays are predominantly in Noachian terrains, there is little evidence that valley networks and these alteration products formed at the same time or are characteristic of the same environmental conditions (Fassett and Head 2011).

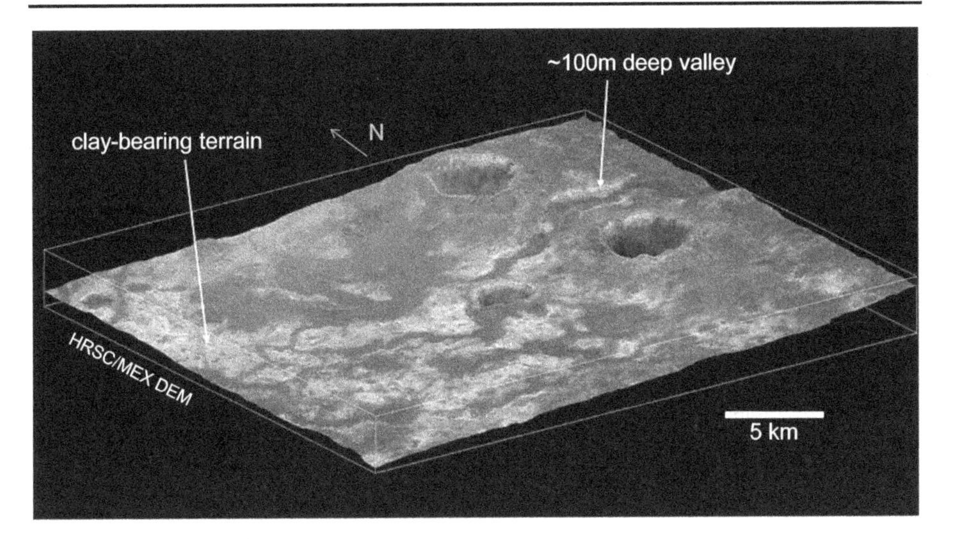

Fig. 4 Valley incising clay-bearing terrain (light-toned outcrops) in the Mawrth Vallis region. A 100-m deep fluvial valley post-dates the clay-bearing unit. The stronger erosion in the western part of the region has degraded the valley to small residual landforms and exhumed clay-bearing rocks. The three-dimensional topography was created by overlaying a single High-Resolution Stereo Camera (HRSC) image on a digital elevation model constructed using HRSC stereo images by V. Ansan (methods in Ansan et al. 2008)

Clays and salts have been detected in several of Mars' many putative paleolakes and fluvial basins. In marked contrast to clays in crustal settings, the hydrated silicates prehnite, analcime, illite/muscovite, and chlorite (except possibly as mixed with smectite) are not detected. Holden, Eberswalde, and other craters contain predominantly Fe, Mg clays (Milliken and Bish 2010). Ismenius Cavus, a depression in Mamers Vallis, displays >300 m thick layered deposits that host clear signatures of Fe, Mg-smectite clays (Dehouck et al. 2010). Some sedimentary basins host clays and salts. Open-basin sediments in the Jezero system, including in its deltas, host Fe, Mg-smectite clays and Mg-carbonate (Ehlmann et al. 2008a, 2009). Terby crater contains layered deposits interpreted as deltaic and composed of Fe, Mg smectites with local zeolites (or sulfates) present in the lowest sections (Ansan et al. 2011). Broad, intercrater depressions in the southern hemisphere with poor drainage connectivity have distinctive units with probable chloride salt units overlying Fe, Mg-clays (Murchie et al. 2009; Glotch et al. 2010; Osterloo et al. 2010). Putative lacustrine sediments within the closed basins of Gale, Columbus, and Cross craters contain sulfate salts, accompanied by Al-clays and/or nontronite (Swayze et al. 2008; Milliken et al. 2010; Wray et al. 2011). At Gale crater, nontronite is interbedded with sulfate and sulfate-iron oxide units at the bottom of a 5-km tall sedimentary mound and will be investigated by the Mars Science Laboratory mission (see Sect. 3.7). Layered deposits in two pits of Noctis Labyrinthus, on the western end of Valles Marineris, also display interlayered sulfates and clays (Weitz et al. 2011; Thollot et al. 2012).

Mineralogic data alone do not permit definitive determination of whether the clay minerals in basins formed *in situ* or were transported. Basins fed by valleys typically host hydrated silicates closely comparable to those in nearby highlands watersheds, consistent with detrital origin (Ehlmann et al. 2009; Milliken and Bish 2010). In contrast, salts found in closed basins are not identifiable in the surrounding terrains, pointing instead to *in situ* precipitation. The timing of deposition may have been coeval to the main fluvial phases on Mars (Late Noachian/Early Hesperian; Fassett and Head 2008) for the oldest lacustrine systems (e.g., Terby, Columbus, or Jezero). But much of the observed alteration may have occurred prior to valley network formation, and clay-bearing sediments in paleolakes could be mainly detrital. In addition, several examples (e.g., Ismenius Cavus, Holden and Eberswalde) show crater retention ages late in the Hesperian (Dehouck et al. 2010; Grant and Wilson 2011; Mangold et al. 2012b) supporting a detrital origin in a climate already too cold to generate *in situ* clay alteration, as shown by the scarcity of alteration in volcanic plains of similar Hesperian ages.

3.6 Key Stratigraphic Sections with Clays, Similarities and Differences

A few geographically extensive stratigraphic sections preserve contacts between different clay-bearing units. These include Mawrth Vallis, Nili Fossae, and Valles Marineris. A distinctive stratigraphy of Al clays above Fe/Mg clays is most typically observed. The Al-clays in these settings include a kaolin family mineral, as well as less commonly an Al smectite (Poulet et al. 2005; Loizeau et al. 2007, 2010; Bishop et al. 2008b; Wray et al. 2008; McKeown et al. 2009; Ehlmann et al. 2009). In Mawrth Vallis and greater Arabia Terra, kaolinite, montmorillonite, and/or silica overlie a nontronite-bearing unit (Loizeau et al. 2007, 2010; Bishop et al. 2008b; Wray et al. 2008; McKeown et al. 2009; Noe Dobrea et al. 2010; Michalski et al. 2010a, 2010b). The exposures are typically in high-standing topography, and the contacts between clay units sometimes follow a pre-existing topographic surface. The nature of these contacts suggests formation by either *in situ* alteration, i.e. leaching, of pre-existing surface materials, or draping by later airfall deposits.

The section of clay-bearing rocks at Mawrth Vallis contains unconformities within the Fe-rich section, indicating that the unit containing the clays was deposited over an extended period of time (Michalski and Noe Dobrea 2007). Around the Nili Fossae, a thin kaolinite-bearing unit overlies both impact-brecciated and sedimentary Fe, Mg-smectite clay units (Fig. 5; Ehlmann et al. 2009; Gaudin et al. 2011). In the eastern portion of this region, Mg-carbonate and serpentine associated with an olivine-rich unit occupy a stratigraphic position similar to that of kaolinite (Ehlmann et al. 2009; Mustard et al. 2009). Al-clays overlying Fe, Mg clays are also exposed high in the walls of Valles Marineris (Murchie et al. 2009; Le Deit et al. 2012) and in other scattered exposures across the southern highlands. In these locations, the clay-bearing units are capped by Hesperian-aged rock units. In other portions of Valles Marineris, near Noctis Labyrinthus, Al clays occur in the lowermost layers of the stratigraphy and are inferred to be a product of volcanically-derived acidic groundwaters later in Mars history (Thollot et al. 2012).

3.7 In Situ Examination of Clay Minerals: Present and Future

The Viking lander missions provided geochemical data (e.g. Toulmin et al. 1977), which indicate that the soils of Mars are mafic in composition but likely with an additional salt component, perhaps added via reaction with volcanic gases (Baird and Clark 1981; Banin et al. 1997). The mafic component could be primary, secondary, or both (e.g. including smectite clays); Viking data are equivocal concerning the mineralogy (Arvidson et al. 1989). Most knowledge of the mineralogy of Mars' rocks and soils is derived from remote telescopic and orbital observations (Table 2) and recent *in situ* measurements from the Mars Exploration Rovers (MERs). The MERs carried both a thermal emission spectrometer (Mini-TES) for mast-mounted remote mineralogic determination and an arm-mounted Mössbauer spectrometer for contact measurement of Fe mineralogy and an alpha particle X-ray spectrometer (APXS) for elemental chemistry. To date, clay minerals have not been identified definitively

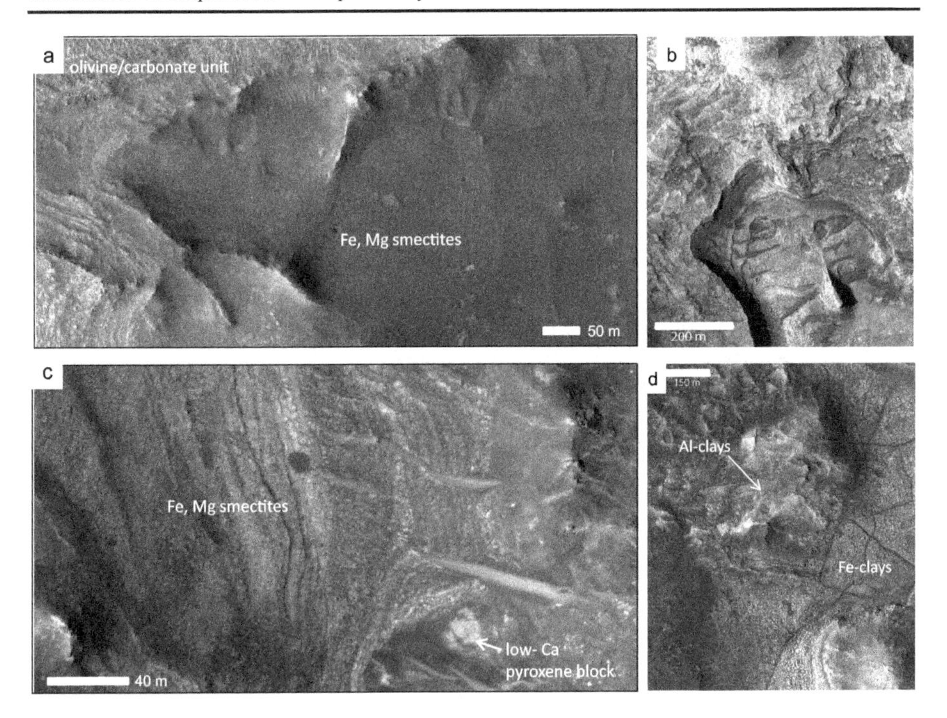

Fig. 5 HiRISE infrared color images of (**a**) crustal clays in Nili Fossae, stratigraphically beneath carbonate, crossed by ridges, and with breccia blocks within the deposit (PSP_002888_2025); (**b**) Mawrth Vallis clays exhibiting meters-scale layering (PSP_001929_2050); (**c**) layered clays within the Nili Fossae trough, overlying a brecciated smectite-pyroxene deposit (ESP_019476_2005); (**d**) layered clays in Mawrth Vallis with Al-rich clays in a butte above fractured, Fe-rich clays (ESP_011383_2030)

in situ at either the Gusev Crater or Meridiani Planum landing sites, despite clear evidence for other aqueously derived phases at both sites (e.g. sulphates, goethite, opaline silica, Mg/Fe carbonate), though have been reported in CRISM VNIR orbital data (Wray et al. 2009b; Carter and Poulet 2012). The Clovis class rocks in the Columbia Hills of Gusev Crater present chemical evidence for the presence of materials with a composition equivalent to that of aluminium clay minerals, specifically montmorillonite (Wang et al. 2006; Clark et al. 2007), although neither mineralogic instrument confirmed this identification. The rocks with montmorillonite-like chemical compositions exhibit TIR spectral character similar to basaltic glass (Ruff et al. 2006), which could either be a primary phase or a non-crystalline altered material. At Meridiani Planum, Mini-TES spectra of the sulfate-rich outcrops were modelled with 10 % smectite (Glotch et al. 2006), but this is not a definitive identification nor was it corroborated via the Mössbauer spectrometer. The 2007 Phoenix lander was not equipped to identify clay minerals. However, Phoenix identified denticles (sawtooth terminations) in atomic force microscopy of soil particles; such denticles are features of pyroxene aqueous alteration that are usually associated with clay mineral formation (Velbel and Losiak 2010).

As of the writing of this manuscript, the Opportunity rover has reached the rim of Endeavour crater, which orbital data indicate host Fe/Mg clays (Wray et al. 2009b), possibly detectable by the rover's Mössbauer spectrometer. The 2011 Mars Science Laboratory, with its ChemMin instrument, a combined XRD/XRF, will visit a location in Gale crater for which orbital identification of clay minerals has been made (Milliken et al. 2010). These rover investigations will allow for the first time the *in situ* investigation of the composition, petrographic texture, and abundance of martian clays.

3.8 Information from Meteorites

Some of the martian meteorites include alteration assemblages with clay minerals. A variety of Mg, Fe-rich alteration phases occur in the nakhlites, including iron oxides, smectites, and amorphous silicates (Gooding 1992). Fe-poor, Al-rich illite clay minerals may also be present in EETA79001 (Gooding 1992), a shergottite meteorite with a crystallization age typically accepted as 173 Myr (Nyquist et al. 2001), though ages at 4.0 Gy have also been proposed for shergottites (Bouvier et al. 2005). In the nakhlites, the secondary mineral assemblages are somewhat younger than the age of the rocks, which crystallized at 1.3 Ga (Swindle et al. 2000; Swindle and Olson 2004). These clay minerals are substantially younger than the widespread Noachian clays and occur in very small abundances; the veins of secondary minerals typically occupy 3–10 % by volume of olivine grains (Changela and Bridges 2011) and <1 % of the bulk rock. Nevertheless, detailed petrologic and compositional studies permit at least this type of later martian clay formation to be better understood.

The nakhlites are believed to come from a single source region that includes meteorites from different excavation depths (Mikouchi et al. 2006). In the nakhlite Lafayette, smectites and Ca-rich siderite ((Fe, Ca)CO_3) occur next to each other in veins. There is a spread in the composition of clay minerals, but the average is similar to saponite and nontronite, and the chemistry of secondary mineral assemblages shows trends with the depth of origin of each meteorite (Changela and Bridges 2011). The abundance of clay minerals and carbonates increases with depth and olivine alteration increases, while the abundance of sulfates decreases.

The chemical compositions of the secondary silicate phases in Nakhla and Lafayette suggest that the clay minerals and other secondary phases did not form via isochemical replacement of igneous minerals but rather from fluids that brought low-CO_2, Fe- and Mg-rich alkaline fluids in from elsewhere (Treiman and Lindstrom 1997; Changela and Bridges 2011). The alteration event recorded in nearly all of the nakhlites was also likely brief and occurred at temperatures <60° based on the $\delta^{18}O$ values of associated carbonates and the modern atmosphere (Treiman and Lindstrom 1997; Saxton et al. 2000; Niles et al. 2010). Textural evidence is preserved from multiple nakhlites of a sequence of rapid cooling and changing water/rock ratios of 1–10 (Changela and Bridges 2011). (For further discussion of alteration in martian meteorites, particularly with regard to carbonate minerals, see Niles et al. (2012)).

4 Geochemical Consequences of Martian Clay Formation

The earliest rock record of Mars, particularly the Noachian, exhibits evidence for clay mineral forming conditions, distinct from later conditions characterized by a mineralogic record characterized by evidence for geochemical cycling of sulfur (see Gaillard et al. 2012). Clay mineral formation on Mars likely occurred by some or all of the same set of mechanisms operating on the Earth to form clays, discussed in Sect. 2. The geochemical consequences for reservoir modification and creation are likely to be similar to those depicted schematically in Fig. 1. However, some aspects of geochemical conditions on ancient Mars preclude direct comparison with the terrestrial rock record. First, the compositions of precursor protoliths and atmosphere differ on Mars. Additionally, the >3.5 Gyr age of the clay-bearing units means that Mars preserves a record of early geologic processes that operated when impact processes were relatively more intense, a period not well-preserved in Earth's rock record. Consequently, evaluating clay formation scenarios from a first principles approach, validated with modeling and laboratory experimentation, is essential. Additional complicating factors in interpreting the martian (and terrestrial) geologic records arise from the fact that clays may have been transported from their original formation environment. Furthermore, altered minerals can suffer post-depositional changes and clay mineral assemblages may reflect changes from later acid weathering (Altheide et al. 2010) or diagenesis (Tosca and Knoll 2009; Ehlmann et al. 2011a, 2011b).

Nevertheless, with improving constraints from orbital analyses as well as new data from *in situ* measurements on Mars, clay formation in multiple, diverse ancient environments has been identified. The critical question is which clay-formation mechanisms were more important and, consequently, how did martian clay formation influence the chemical composition of the crustal and atmospheric reservoirs? Some constraints exist from comparison of the martian clay record with results from laboratory and modeling efforts.

4.1 Constraints on Clay Formation Scenarios

4.1.1 Thermodynamic and Kinetic Considerations

For a chemical system to progress toward a more thermodynamically stable, minimum free-energy state, it must first surmount an energy barrier; the greater the Gibbs free energy of reaction, the greater the activation energy required. The Ostwald Rule derived from this constraint states that if a reaction has multiple pathways and can result in several products, it is the least stable product with free energy closest to the original state that crystallizes first instead of the most energetically stable phase. Hence, with multiple possible pathways, moderate mineralogical transformations with lower activation energies occur more rapidly than more radical changes, even though the more radical changes may be the thermodynamically favored lowest energy state. Because reaction rates are exponentially dependent on temperature, initial transformations are particularly important to understand for cold, martian environments. Carbonate, sulfate, and other salts with ionic bonds have relatively low kinetic barriers for dissolution and precipitation in comparison to minerals comprised of a covalently bonded lattice with Si, Al, and O. Therefore, amorphous aluminosilicate coatings and rinds, rather than clays, are also important alteration products throughout much of Mars history (e.g. Michalski et al. 2006), and the prediction of alteration sequences involving clay minerals requires taking kinetic considerations into account.

On Earth, clay minerals are produced principally during acid-activated pedogenic processes or abiotic temperature-activated alteration. Primary rock-forming minerals are a sink for protons produced by hydrolysis reactions, resulting in production of clay minerals and release of cations, first alkalis, then alkaline earths, and finally iron. On Earth, the proton source (acidity) commonly derives from CO_2, organic acids, or sulfuric acid formed by oxidation of pyrite (Meybeck 1987). On Mars, CO_2 is a potential source of acidity as are oxidation of iron sulfides (Burns and Fisher 1990; Dehouck et al. 2012), oxidation of iron (Baldridge et al. 2009; Hurowitz et al. 2010), and chlorine and sulfur species released by volcanism (Bullock and Moore 2007; Gaillard and Scaillet 2009). Whether or not clays form depends on four parameters: the mass of pristine rock relative to the mass of water (W/R), the amount and source of protons, time, and the supply of the reactants, e.g. from differential dissolution of heterogeneous mineral phases rather than dissolution of a homogeneous glass.

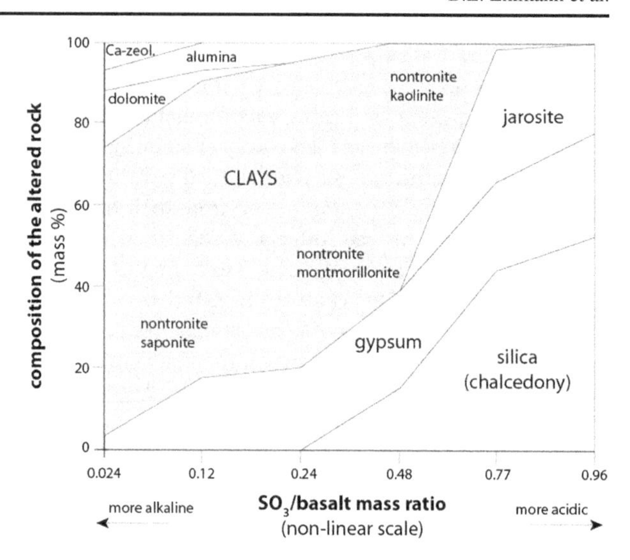

Fig. 6 Thermodynamic prediction for alteration of basalt (Adirondack composition), assuming W/R = 1, 0 °C, congruent dissolution of primary phases, and partial pressures for O_2 and CO_2 observed in the current Mars atmosphere. Figure is adapted from Berger et al. (2009). The parameter SO_3 corresponds to "dry" H_2SO_4, the source of acidity

These kinetic considerations were invoked by Tréguier et al. (2008) and Berger et al. (2009) when considering the nature and amount of secondary phases at Meridiani Planum, a site with sulfates, iron oxides, and primary minerals, detected *in situ* (e.g. Squyres et al. 2004), and clay minerals been detected from orbit in lower stratigraphic units in the region (Poulet et al. 2008b; Wray et al. 2009b). The effect of protons/acid is illustrated in Fig. 6, based on a thermodynamic model reported in Berger et al. (2009), showing the proportions of the secondary phases after the complete reaction of the pristine basalt with an acidic brine obtained by dissolution of SO_3 in water. The scenario has no time-limited kinetic effects and all reactant anions and cations are available. The leftmost part of the plot corresponds to low proton availability where only water is added to the rock, and the alteration assemblage is dominated by formation of clays and zeolite. The right part corresponds to high proton (acidic) conditions supplied by SO_3 dissolved in the fluid.

Thermodynamic calculations predict substantial clay mineral formation up to high concentrations of sulfur relative to basalt (nearly 1:1) (Fig. 6). However, the Fe/Mg clays most commonly observed on Mars are not thermodynamically stable at $pH < 4$ (Chevrier et al. 2007), and laboratory experiments show no formation at $pH < 7$ (Tosca et al. 2008). Modeling including kinetic considerations, i.e. dissolution of primary phases (reactant supply) and precipitation rates of salts (rapid) vs. clays (slow), demonstrates that clays are not a favored product of most short timescale reactions involving sulfur at low water/rock ratio, favoring instead the precipitation of salts (Berger et al. 2009). This results from an assumed high gas/rock ratio relative to water/rock (W/R) ratio, producing a concentrated sulfate brine favorable to the precipitation of sulfate salts.

In general, clay formation indicates long-term interaction of rocks with solution, sufficient for dissolution of primary silicates and crystallization of secondary silicates. The nature of clay minerals formed depends on the W/R and proton source. In basaltic hydrothermal systems, low water/rock ratio reactions lead to propylitic facies (chlorite, Fe/Mg smectites, epidote, prehnite) while higher W/R and/or acidity lead to argillic facies (kaolinite, Fe/Mg smectites, hematite, carbonate) (Berger and Velde 1992). Deep crustal clays excavated by cratering of clay-bearing terrains are more similar to propylitic facies (low W/R, little acidity) as discussed in Sect. 3.4 (Ehlmann et al. 2011a). When instead W/R is low but H^+/rock (acidity) is high, the concentration of the solute will reach the solubility of the secondary salts in which the anion (CO_3, SO_4, Cl) is the proton companion anion. In this scenario, an observed increase in the concentration of a particular anion in an altered rock relative to protolith geochemistry indicates the nature of the acid that drove the alteration, as with S at Meridiani Planum. Weathering at high W/R ratios and short residence time leads to leached horizons with Al clays and Fe oxides, fluid transport, and evaporative salts formed elsewhere. Fe oxides and Al clays, within the clay stratigraphic sections discussed in Sect. 3.6, are more similar to the argillic facies (higher W/R and/or more H^+/R).

4.1.2 Controls on Atmospheric Composition

Knowing how and where martian clays formed originally is vital to any understanding of whether they have implications for the early climate. Surface weathering would entail interaction of the atmospheric and crustal reservoirs (Fig. 1a). An equilibrium pCO_2 of $\sim$5 mbar or greater during clay formation should result in the coexistence of carbonates and Fe/Mg-rich smectites (Chevrier et al. 2007). This coexistence has not been commonly observed, so it has been argued that if we assume that clay minerals formed at the martian surface, the atmosphere of ancient Mars was not CO_2-rich or other greenhouse gases were the dominant atmospheric components necessary to maintain near-surface liquid water (Chevrier et al. 2007). However, the prediction of carbonate precipitation from atmospheric CO_2 also includes other parameters. In the case of high run-off (low residence time of water within the martian regolith), near-neutral pH (slow dissolution kinetics), and/or presence of another acid gas (high carbonate solubility at low pH), carbonate precipitation is not expected (e.g. Fabre et al. 2011).

Plausible candidates for alternative greenhouse gases to provide $\sim$80 K of warming necessary for a continually clement early Mars (Haberle 1998) have not been identified. For comparison, the modern, Holocene Earth has 33 K of greenhouse warming (e.g., Kasting and Catling 2003), primarily from small amounts of H_2O, CO_2, and CH_4. Sufficient quantities of methane on early Mars would require a flux larger than the current biogenic flux on Earth. Similarly, volcanic sulfurous gases, which have been suggested as a possibility for warming the early climate of Mars (Halevy et al. 2007; Johnson et al. 2009), may cool early Mars through the rapid formation of atmospheric aerosols that reflect sunlight to space, similar to the net cooling effect of volcanic sulfur on Earth and Venus (Tian et al. 2010). Existing evolution models of the CO_2 partial pressure similarly suggest that the CO_2 level in the atmosphere of Mars could have been low or moderate at the end of the Noachian, in the range from a few tens of millibars to 1 bar, insufficient for durably maintaining temperatures above freezing (Lammer et al. 2012).

Consequently, environmental conditions of clay formation may have been the result of repeated, transient periods of liquid water-rock interaction at the surface in a relatively cold climate with a sparse atmosphere, with only a slightly thicker atmosphere and warmer surface conditions than Mars today. In this case, most clays may have formed in a subsurface environment from groundwaters isolated from the atmosphere (Ehlmann et al. 2011a). An alternative interpretation of the data is that clay mineral assemblages formed under warm surface conditions may have been modified, i.e. surface-formed, clay-bearing assemblages may have been altered, for example, by removal of accompanying carbonates by acidic weathering and burial and impact churning of Al clay layers.

4.1.3 Geomorphology and Stratigraphy

The geomorphology and stratigraphy of particular martian clay deposits can be examined to help discriminate between the alternative hypotheses for clay formation (Table 1). The majority of clay minerals are exposed by craters and, consequently, the original physical characteristics of the deposit have been disrupted. Sedimentary deposits with clays indicate *in situ* evaporation of fluids leading to the deposition of salts, but as discussed in Sect. 3.5, likely reflect transport of clays from their original formation setting in most cases. A few stratigraphic sections preserve coherent bedding and relationships between geologic units with clay minerals of different composition. Fe/Mg smectite-bearing units range from horizontally layered to brecciated, indicating a range of processes from sedimentary, to extrusive volcanic, to impact disruption. The depth of these layers, both in thickness and in the stratigraphic column suggests that some portion of the alteration to Fe/Mg clays took place in the subsurface (Ehlmann et al. 2011a). In Mawrth Vallis, the uppermost Al-clays and silica bearing units follow the topography and overlie an Fe/Mg smectite-bearing unit (Bishop et al. 2008b; Wray et al. 2008; Loizeau et al. 2010), and similar stratigraphic sections are found in Valles Marineris and Nili Fossae. These Al clays may have formed by processes involving either weathering and pedogenesis or diagenesis by shallow aquifers (<200 m). Draping by volcanic ash, subsequently altered, is one possibility for formation, although whether groundwater or surface waters were the alteration agent cannot be definitely distinguished (McKeown et al. 2009). An alternative possibility may be greater degrees of near-surface leaching, possibly by acidic waters (Bishop et al. 2008b; Ehlmann et al. 2009; McKeown et al. 2009; Noe Dobrea et al. 2010).

4.1.4 Bulk Crust Chemical Composition and Clay Mineral Abundances

At a coarse spatial resolution of a few hundred kilometers, measurements of the upper 0.5 m of the surface by the gamma ray spectrometer onboard the Mars Odyssey orbiter do not show significant fractionation of K and Th, two elements expected to have different mobilities in open system weathering (Taylor et al. 2006, 2010). The observation of clay minerals at abundances <30 % in association with primary minerals pyroxene and plagioclase in most Mars terrains may indicate only partial alteration of the units. Only the Fe-smectite unit at Mawrth Vallis has higher abundances, reaching 65 % clay with little evidence for accompanying primary phases (Poulet et al. 2008a). However, only ~10 of the thousands of exposures of clay minerals have thus far been examined for abundances. Hence, the significance of clay mineral abundance and geochemistry to inferring the geochemistry of alteration may have to await higher resolution geochemical studies and expanded calculations of modal mineralogy.

4.2 Geochemical Consequences

We consider the geochemical consequences of three endmember scenarios that have emerged to explain clay mineral formation on ancient Mars: near-surface (open) systems, ice-dominated systems, and rock-dominated subsurface (closed) systems.

4.2.1 Water in Near-Surface (Open) Systems

Surface or near-surface weathering to form clays has the potential to generate substantial reservoir fractionation, depending on the W/R and hydrological character of the system. If formed from a basaltic substrate, the Al clays found on Mars likely indicate high W/R leaching processes in an open system or from highly acidic waters. Any precipitation on ancient Mars would percolate through the crust, leading to chemical alteration (Fig. 7a). Evidence for such systems exists in Mawrth Vallis (Loizeau et al. 2007, 2010; Bishop et al. 2008b),

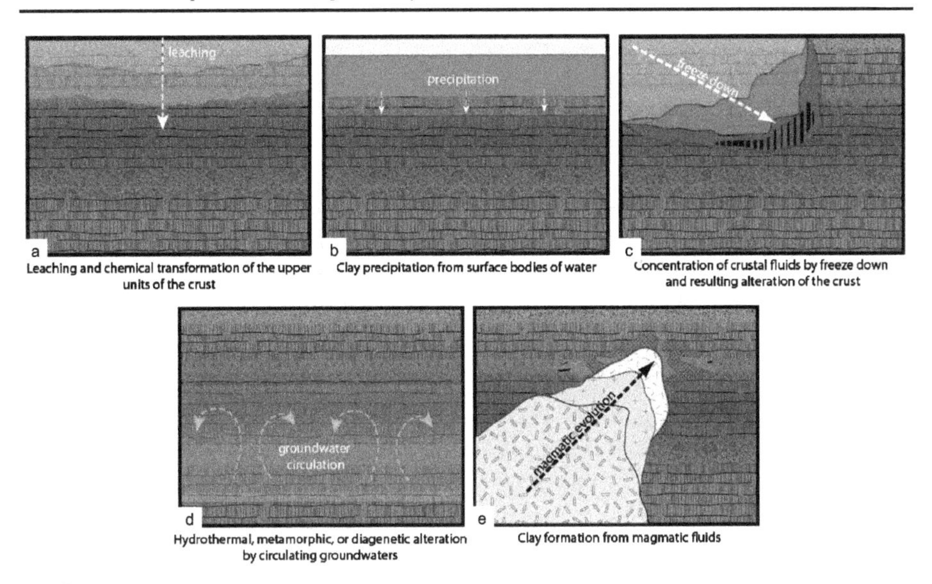

Fig. 7 Scenarios for Martian clay formation

near the Nili Fossae (Ehlmann et al. 2009); and in diverse terrains across the southern highlands (Wray et al. 2009a, 2009b).

As modeled by Milliken et al. (2009), leaching processes result in liberation of cations which then react with anions to form salts, of composition dictated by the composition of the atmosphere. These would include carbonates, sulfates, chlorides, and perchlorates. The result is creation of two reservoirs, a leached crust and a sedimentary reservoir, distinct in composition from the geochemistry of primary bedrock produced by volcanism (Fig. 1). The formation of large salt deposits in basins as water evaporates might be a potential reservoir for atmospheric gases, e.g. CO_2 in carbonates (Fig. 7b), although to date, carbonate-rich sedimentary bedrock units have not been identified (Ehlmann et al. 2008a).

Only a few regions display Al-clay mineralogic assemblages, either because of the lack of extensive global weathering or its poor preservation. Indeed, the difficulty in discriminating near-surface weathering from deeper alteration is a consequence of the subsequent >3.5 billion years of aeolian activity, volcanism, and impact gardening, which have eroded, blanketed and disturbed ancient surfaces. Nevertheless, in this scenario sedimentary units are predicted to have a different composition than altered crust; both would be different from primary crust, and the process of clay formation would have altered the atmospheric reservoir by co-precipitation of salts.

4.2.2 Ice-Dominated Systems

Some models of the evolution of the atmosphere and climate of Mars suggest that Mars lost most of its atmosphere soon after accretion and has not retained a dense enough atmosphere to support extensive, long-lived surface liquid water at any time in its history (Wordsworth et al. 2011; Lammer et al. 2012). In this scenario, an active hydrological cycle driven by warm greenhouse conditions would not supply and replenish the large amounts of water necessary to form the extensive deposits of clay minerals discussed above. Needed water might, however, be supplied by ice.

Obliquity variations change the distribution of ice stability across Mars, allowing for transport of ice and dust from the polar regions to the equator during periods of high obliquity (Jakosky and Carr 1985; Laskar et al. 2002; Head et al. 2003). The resulting deposits of ice and dust might have resembled the modern polar deposits in scale and morphology (Niles and Michalski 2009). In such deposits, cryo-concentration of volatile-bearing brines (e.g. S-rich and Cl-rich) could promote acid weathering of silicate dusts and sand grains inside the deposit (Fig. 7c; Burns 1993; Niles and Michalski 2009). In such a scenario, intra-ice acid weathering would lead to the release of cations and silica into solution leading to the precipitation of salts, the production of leached Si + Al-rich material (either an amorphous residue or crystalline Al-clays), and the possible production of mixed Fe oxide-smectite materials (Burns 1993).

Geochemical consequences of this scenario would include (1) removal of S and Cl volatile species from the atmosphere and sequestration in the crust, specifically within a sedimentary reservoir of different composition from the primary crust; and (2) possible water, Fe, and Mg addition to the primary crust by brines derived from cryo-concentration and acid weathering within the ice.

4.2.3 Rock-Dominated Subsurface Systems

Low-grade metamorphism, hydrothermal activity, and magmatic precipitation are means by which martian clays could form primarily in the subsurface within a crustal reservoir relatively closed with respect to the other reservoirs (Figs. 7d and 7e). Evidence to date indicates that some of the clays, especially the deep crustal units of the southern highlands, may have formed in these rock-dominated systems at temperatures slightly above surface-ambient (Ehlmann et al. 2011a, 2011b).

In metamorphism, the degree of chemical fractionation between parent material and alteration product is small and would result in little geochemical fractionation of Mars' crustal reservoir. If instead clays formed by magmatic precipitation, some volatiles would be released into the atmosphere during eruption of extrusive lavas, but lavas would largely retain their volatiles, precipitated into hydrous minerals during late-stage crystallization. This mechanism can produce new crust up to ~10 % clay by volume (Meunier et al. 2010). Clay formation in hydrothermal systems can occur in chemically open or closed systems. Recirculation of crustal waters can generate clay minerals, as can hydrothermal systems driven by magmatic fluids. If hydrothermal systems are open to meteoric waters, formation of minerals can lead to sequestration of atmospheric volatiles, e.g. CO_2 in minerals precipitated in fractures and H_2O in hydrous silicates.

In general, subsurface, rock-dominated systems are relatively closed, i.e. there is little interaction with other geochemical reservoirs. Subsurface alteration may result in some incorporation of atmospheric species into the uppermost crust, as may be the case for martian meteorites (discussed in Sect. 3.8), but to date, evidence for carbonates and sulfates incorporated in the same deep crustal units as the clays is uncommon (Michalski and Niles 2010). However, a significant effect of water interaction with the crust is that the water itself may be permanently sequestered in hydrous silicates, thus removing it from the atmospheric reservoir and continued cycling.

For example, the hydrothermal alteration of olivine in the serpentinization process is one effective means of sequestering water in the crust, as shown by recent modeling (Chassefière and Leblanc 2011). In the reaction, ferrous iron is oxidized by the water to ferric iron, which typically precipitates as magnetite, while hydrogen from water is reduced to H_2. If fluids vent to the surface, they can release high abundances of CH_4 and other hydrocarbons, provided that CO_2 is dissolved in the fluid and is reduced by H_2, which typically requires a catalyst. On Mars, the generation of H_2 by olivine during serpentinization can be expressed as (Oze and Sharma 2005),

$$(1)\quad \underset{\text{Olivine}}{Mg_{1.5}Fe_{0.5}SiO_4} + 1.17H_2O \rightarrow 0.5\underset{\text{Serpentine}}{Mg_3Si_2O_5(OH)_4} + 0.17\underset{\text{Magnetite}}{Fe_3O_4} + 0.17H_2$$

$$(2)\quad 4H_2 + CO_2 = CH_4 + 2H_2O,$$

assuming reaction of olivine with an Mg # of 0.75. For each released H_2 molecule, 6 H_2O molecules are stored in serpentine, and/or for each released CH_4 molecule, 24 H_2O molecules are stored in serpentine (Chassefière and Leblanc 2011). Over 85 % of the water participating in the serpentinization reaction is ultimately sequestered in the crust. The formation of serpentine and other hydrous silicates in the subsurface is thus a potential crustal sink for H_2O in other martian reservoirs and is a means of sequestering volatiles. High temperatures are not required for this process. Olivine dissolution at low temperatures, generating H_2 and CH_4, has been observed in the laboratory (Neubeck et al. 2011) and in terrestrial ophiolites (Etiope et al. 2011), both cases in which the necessary catalysts are available to promote the reaction.

4.2.4 Subsequent Alteration once Formed

Regardless of the mechanism by which clays formed on Mars, following initial formation, subsequent alteration processes may affect clay-bearing deposits and the geochemical signatures of the original alteration process. There is widespread evidence for physical redistribution of clay minerals by transport in fluvial systems and by excavation during impact cratering (as discussed in Sects. 3.4 and 3.5). Impact cratering can also result in the heat- or shock-induced destruction of clay minerals near the point of impact (Weldon et al. 1982; Hviid et al. 1994; Gavin and Chevrier 2010; Che et al. 2011) or the generation of new and different species of clay minerals by impact-induced or later hydrothermal processes (Schwenzer and Kring 2009).

Chemical alteration may have affected clay-bearing materials following original clay formation, by mechanisms with varying degrees of impact on geochemical reservoirs. Exposure of Fe(II)/Mg smectites formed in an anoxic environment to oxidizing conditions at the present martian surface can lead to transformation to nontronite, an Fe(III) smectite (Burns 1992), although this mechanism would not lead to geochemical fractionation. Experiments with clay weathering in acidic fluids show that clay minerals are fairly resistant to strong acids like H_2SO_4, but do destabilize at low pH (<4) to form salt solutions, which can evaporate to precipitate sulfates, and often amorphous silica residual solid (Fig. 8). Depending on the amount of water, this can be either an open-system or closed system process. The resistance of clay minerals to acidic alteration depends on the nature of the phase. For example, kaolinite is the most resistant and Fe/Mg smectites the least resistant; consequently, acid weathering has been proposed to play a role in the generation of near-surface Al-rich clay units on Mars (Altheide et al. 2010). These scenarios for later clay alteration involve exchange of atmospheric and crustal reservoirs.

Under conditions of deep burial, diagenesis of smectite clays to more stable chlorite or illite phases is an expected transformation if sufficient water and Mg or K, respectively, are available to facilitate the reaction (Tosca and Knoll 2009). For clay diagenesis, the absence of K-rich feldspar—Gusev basalts contain less than 0.1 % K_2O, as reported in McSween et al. (2006)—precludes a significant illitization process, and the conversion of kaolinite to

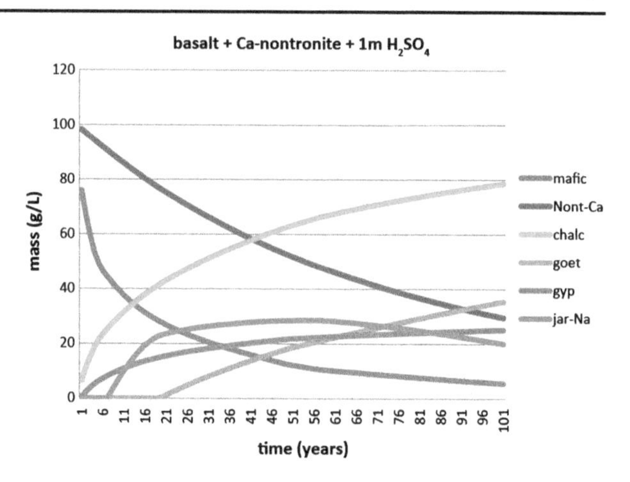

Fig. 8 Low water/rock ratio acid alteration of clay-bearing units. A starting 54 % nontronite, 42 % mafic minerals, 4 % silica, simulating already altered crust. Nontronite is more resilient to later alteration than mafic minerals, but both are eventually replaced by silica, iron oxides, and sulfate salts

dickite, often considered as a spontaneous isochemical reaction, requires the dissolution of feldspars (Lanson et al. 1996). Thus, the more probable clay diagenesis is the chloritization of smectites. There is evidence for such transformations from the widespread presence of chlorite and occasional presence of illite in the southern highlands of Mars (Ehlmann et al. 2009, 2011a, 2011b) as well as the identification of probable mixed layer clays, chlorite/smectite, which is an intermediate product of the transformation (Milliken and Bish 2010). Such transformations occur within the crust in systems not exchanging with atmospheric and magmatic reservoirs.

5 Needed Future Measurements

A suite of possible formation scenarios, each with distinctive geochemical consequences, exists to explain Mars' globally widespread Noachian clays. To resolve the outstanding questions about clay formation environment and concomitant consequences for the evolution of martian geochemical reservoirs requires both detailed studies focused on clay-bearing units as well as broader ranging investigations aimed at understanding the evolution of the coupled Mars geologic-climatic system.

The continued acquisition of high-resolution image and spectroscopic datasets will allow detailed study of relationships between alteration mineral bearing units of different composition. These stratigraphic studies may lead to an understanding of whether clay-bearing units are primarily sedimentary or *in situ* altered volcanics. Further work is needed to develop best practices for modeling modal mineralogy of altered units remotely using the tens of meters scale visible/near-infrared data most conducive to the detection of clay to constrain their abundance. This will permit, not only identification of single minerals, but an understanding of full mineral assemblages most useful for geochemical modeling. Also useful in this regard would be higher resolution studies of the geochemical composition of clay-bearing units in the martian crust.

A combination of modeling and laboratory work is needed to understand the implications of various identified phases. For example, do the commonly detected Fe/Mg smectites require anoxic conditions to form? For how long can different clays persist in the presence of acidic waters at different temperatures? Have clays detected in central peaks of impact craters survived impact, or would they be destroyed by heat and pressure? Furthering these investigations will permit narrower constraints on formation environment.

More detailed *in situ* chemical, mineralogical, and textural data will soon be available for one site with clays, the landing site for the MSL rover mission, Gale crater. The CheMin instrument may identify the presence of smectites and other clay minerals by means of X-ray diffraction as well as provide *in situ* quantitative estimates of complete modal mineralogy, including measurement of accompanying primary phases or salts that might be undetectable with present orbiting instruments (Blake et al. 2012). Coupled with X-ray fluorescence measurements and elemental data from other instruments, the chemistry of the nontronite-bearing unit in Gale crater should be fully explored and evidence for post-formation diagenesis assessed. Trace elements and their mobility in different geochemical scenarios may supplement knowledge gained from major elements. Isotopic data from the gas chromatograph/mass spectrometer on the Sample Analysis at Mars (SAM) instrument suite may constrain alteration mineral formation temperatures or processes and will provide data on the possible presence/abundance of atmospheric CH_4 (Mahaffy et al. 2012). However, data of the last decade have revealed the great diversity of aqueous environments on early Mars, and no single mission can visit the many distinctive mineral/morphologic localities for clay-forming chemical environments on ancient Mars. For example, the question remains open as to whether the Al-over-Fe/Mg clay stratigraphy at several localities on Mars represents a top-down alteration mechanism, a fundamental shift in the depositional setting through time, or both. Definitive resolution of this question will likely require landing at a site with both types of minerals. Furthermore, understanding the nature of possible hydrothermal clays with chlorites and prehnite and their formation pre- or post-impact will likely require landed studies where examination of petrographic textures is possible.

Finally, datasets not focused on clay minerals are needed to constrain the evolution of Mars' atmosphere. Could it have hosted warm, wet surface conditions for an extended period of time or was the surface environment always relatively cold and dry? The 2013 MAVEN (Mars Atmosphere and Volatile EvolutioN) mission will provide isotopic estimates of atmospheric composition that will be useful in constraining loss rates and additional characterization of the atmosphere will be carried out by the mass spectrometer, SAM on MSL. Plausible scenarios can be envisaged where clay mineral assemblages do not tell us about the surface climate but indicate subsurface conditions. Generally, the available evidence and inferences from clay minerals in martian meteorites indicate that these formed in the subsurface rather than in thermodynamic equilibrium with ancient surficial conditions.

Nevertheless, if only a few outcrops of ancient clays were preserved from formation at warmer surface conditions, these outcrops would have a great importance for studies concerning life on Mars, similar to the role played by outcrops in the Barberton and Pilbara for terrestrial geobiology (Buick 2007). *In situ* geologic and geochemical observations at the surface are therefore critical for resolving the issue of the original provenance of clay minerals. The return of samples from Mars to Earth from clay-bearing terrains would provide an alternative means for comprehensive analysis of the origin of clay minerals, permitting use of a suite of laboratory techniques like micron-scale scanning electron microscopy, geochemical and isotopic analyses.

6 Conclusions

The dominant terrestrial clay formation mechanisms are near-surface weathering and hydrothermal processes in seafloor systems, although precipitation of clays in water bodies within basins, diagenesis, metamorphism, and magmatic precipitation also serve to form and modify clays. Like Earth, the geologic record from Noachian Mars also indicates multiple clay-forming processes operating in diverse chemical environments, varying in space and time. Clay formation in subsurface environments is indicated by some stratigraphic units whereas near-surface formation by precipitation-driven weathering or thin films of water derived from ice are favored in other locations. The latter two scenarios would fractionate the martian crustal reservoir into leached and sedimentary components, distinct in composition from primary crust, and would sequester volatiles. In contrast, subsurface formation might not substantially chemically fractionate major elements in the crustal reservoir but could effectively sequester water in the crust. Future modal mineralogy of clay bearing deposits derived from orbital data as well as mineralogic, geochemical, and petrological data acquired *in situ* can be used to discriminate clay formation mechanisms on the basis of their different geochemical consequences. However, no single clay-bearing unit is sufficient to obtain a global understanding of Noachian clay formation processes. Rather, investigations of multiple types of clay-bearing deposits are needed to develop a comprehensive model of the geochemical consequences of clay formation on Mars for its evolution over geologic time.

Acknowledgements Thanks to J. Bell, J. Bishop, and M. Toplis for thorough reviews that improved this manuscript. Thanks also to M. Toplis and the ISSI conference organizers for promoting fruitful interdisciplinary discussion of clays on Mars. E. Chassefière acknowledges support from CNRS EPOV interdisciplinary program.

References

O. Abramov, D.A. Kring, Impact-induced hydrothermal activity on early Mars. J. Geophys. Res. **110**, E12S09 (2005)

J.H. Ahn, D.R. Peacor, D.S. Coombs, Formation mechanisms of illite, chlorite, and mixed-layer illite-chlorite in Triassic volcanogenic sediments from the Southland Syncline. New Zealand. Contrib. Mineral. Petrol. **99**, 82–89 (1988)

C.C. Allen, M.J. Jercinovic, T. See, K. Keil, Experimental shock lithification of water-bearing rock powders. Geophys. Res. Lett. **9**, 1013–1016 (1982a)

C.C. Allen, J.L. Gooding, K. Keil, Hydrothermally altered impact melt rock and breccia: contributions to the soil of Mars. J. Geophys. Res. **87**, 10,083–10,101 (1982b)

J.C. Alt, Very low grade hydrothermal metamorphism of basic igneous rocks, in *Low Grade Metamorphism*, ed. by M. Frey, D. Robinson (Blackwell Scientific, Oxford, 1999), pp. 169–201

T. Altheide, V. Chevrier, E. Noe Dobrea, Mineralogical characterization of acid weathered phyllosilicates with implications for secondary Martian deposits. Geochim. Cosmochim. Acta **74**, 6232–6248 (2010)

R.K.R. Ambers, Relationships between clay mineralogy, hydrothermal metamorphism, and topography in a western cascades watershed, Oregon, USA. Geomorphology **28**, 47–61 (2001)

D. Ames, I.M. Kjarsgaard, K.O. Pope, B. Dressler, M. Pilkington, Secondary alteration of the impactite and mineralization in the basal tertiary sequence, Yaxcopoil-1, Chixulub impact crater, Mexico. Meteorit. Planet. Sci. **39**(7), 1145–1167 (2004)

V. Ansan, N. Mangold, P. Masson, E. Gailhardis, G. Neukum, Topography of valley networks on Mars from the Mars Express High Resolution Stereo Camera digital elevation models. J. Geophys. Res. **113**, E07006 (2008). doi:10.1029/2007JE002986

V. Ansan, D. Loizeau, N. Mangold, S. Le Mouélic, J. Carter, F. Poulet, G. Dromart, A. Lucas, J.-P. Bibring, A. Gendrin, B. Gondet, Y. Langevin, Ph. Masson, S. Murchie, J.F. Mustard, G. Neukum, Stratigraphy, mineralogy, and origin of layered deposits inside Terby crater, Mars. Icarus **211**, 273–304 (2011). doi:10.1016/j.icarus.2010.09.011

P. Arkai, F.P. Sassi, J. Desmons, A systematic nomenclature for metamorphic rocks. 5. Very low- to low-grade metamorphic rocks. Recommendations by the IUGS Subcommission on the systematics of metamorphic rocks. SCMR website (www.bgs.ac.uk/SCMRH) (2003)

R.E. Arvidson, J.L. Gooding, H.J. Moore, The Martian surface as imaged, sampled, and analyzed by the Viking landers. Rev. Geophys. **21**(1), 39–60 (1989)

A.K. Baird, B.C. Clark, On the original igneous source of Martian fines. Icarus **45**, 113–123 (1981)

A.M. Baldridge et al., Contemporaneous deposition of phyllosilicates and sulfates: using Australian acidic saline lake deposits to describe geochemical variability on Mars. Geophys. Res. Lett. **36** (2009). doi:10.1029/2009GL040069

J.L. Bandfield, Global mineral distributions on Mars. J. Geophys. Res. **107**(E6) (2002). doi:10.1029/2001JE001510

J.L. Bandfield, T.D. Glotch, P.R. Christensen, Spectroscopic identification of carbonate minerals in the Martian dust. Science **301**, 1084–1087 (2003)

A. Banin et al., Surface chemistry and mineralogy, in *Mars*, ed. by H.H. Kieffer et al. (Univ. of Arizona Press, Tucson, 1992), pp. 594–625

A. Banin et al., Acidic volatiles and the Mars soil. J. Geophys. Res. **102**, 13341–13356 (1997)

K.J. Benison et al., Sedimentology of acid saline lakes in Southern Western Australia: newly described products and processes of an extreme environment. J. Sediment. Res. **77**, 366–388 (2007)

J.F. Bell, T.B. McCord, P.D. Owensby, Observational evidence of crystalline iron oxides on Mars. J. Geophys. Res. **95**, 14,447–14,461 (1990)

G. Berger, B. Velde, Chemical parameters controlling the propylitic and argillic alteration process. Eur. J. Mineral. **4**, 1439–1454 (1992)

G. Berger, M.J. Toplis, E. Treguier, C. d'Uston, P. Pinet, Evidence in favor of small amounts of ephemeral and transient water during alteration at Meridiani Planum, Mars. Am. Mineral. **94**, 1279–1282 (2009)

J.P. Bibring et al., Global mineralogical and aqueous Mars history derived from OMEGA/Mars Express data. Science **312**, 400–404 (2006)

J.-P. Bibring et al., Coupled ferric oxides and sulfates on the Martian surface. Science **317**, 1206–1210 (2007)

J.L. Bishop, P. Schiffman, E. Murad, M.D. Dyar, A. Drief, M.D. Lane, Characterization of alteration products in Tephra from Haleakala, Maui: A visible-infrared spectroscopy, Mössbauer spectroscopy, XRD, EPMA and TEM study. Clays Clay Miner. **55**(1), 1–17 (2007)

J.L. Bishop, M.D. Lane, M.D. Dyar, A.J. Brown, Reflectance and emission spectroscopy study of four groups of phyllosilicates: smectites, kaolinite-serpentines, chlorites and micas. Clay Miner. **43**, 35–54 (2008a)

J.L. Bishop et al., Phyllosilicate diversity and past aqueous activity revealed at Mawrth Vallis, Mars. Science **321**, 830–833 (2008b)

J.L. Bishop et al., Mineralogy of Juventae Chasma: Sulfates in the light-toned mounds, mafic minerals in the bedrock, and hydrated silica and hydroxylated ferric sulfate on the plateau. J. Geophys. Res. **114**, E00D09 (2009)

J.L. Bishop, W.P. Gates, H.D. Makarewicz, N.K. McKeown, T. Hiroi, Reflectance spectroscopy of beidellites and their importance for Mars. Clays Clay Miner. **59**(4), 376–397 (2011)

D. Blake et al., Characterization and calibration of the CheMin mineralogical instrument on Mars Science Laboratory. Space Sci. Rev. (2012). doi:10.1007/s11214-012-9905-1

A. Bouvier, J. Blichert-Toft, J.D. Vervoort, F. Albarède, The age of SNC meteorites and the antiquity of the Martian surface. Earth Planet. Sci. Lett. **240**, 221–233 (2005)

W.V. Boynton et al., Elemental abundances determined via the Mars Odyssey GRS, in *The Martian Surface: Composition, Mineralogy and Physical Properties*, ed. by J.F. Bell (Cambridge University Press, Cambridge, 2008), pp. 105–124

W.V. Boynton et al., Evidence for calcium carbonate at the Mars Phoenix landing site. Science **325**, 61–64 (2009)

T. Bristow, R.E. Milliken, Terrestrial perspective on authigenic clay mineral production in ancient Martian lakes. Clays Clay Miner. **59**(4), 339–358 (2011)

D.L. Buczkowski et al., Investigation of an Argyre basin ring structure using Mars Reconnaissance Orbiter/Compact Reconnaissance Imaging Spectrometer for Mars. J. Geophys. Res. **115**, E12011 (2010)

R. Buick, The earliest records of life on Earth, in *Planets and Life: The Emerging Science of Astrobiology*, ed. by W.T. Sullivan, J. Baross (Cambridge University Press, Cambridge, 2007), pp. 237–264

M.A. Bullock, J.M. Moore, Atmospheric conditions on early Mars and the missing layered carbonates. Geophys. Res. Lett. **34**, L19201 (2007). doi:10.1029/2007GL030688

R.G. Burns, Dehydroxylated clay silicates on Mars: Riddles about the Martian regolith solved with ferrian saponites. MSATT Workshop on Chemical Weathering on Mars, LPI Tech. Rept., 92-04, 6–7, 1992

R.G. Burns, Rates and mechanisms of chemical-weathering of ferromagnesian silicate minerals on Mars. Geochim. Cosmochim. Acta **57**(19), 4555–4574 (1993)

R.G. Burns, D.S. Fisher, Iron-sulfur mineralogy of Mars: magmatic evolution and chemical weathering products. J. Geophys. Res. **95**(B9), 14,415–14,421 (1990). doi:10.1029/JB095iB09p14415

J. Carter, F. Poulet, J.-P. Bibring, S. Murchie, Detection of hydrated silicates in crustal outcrops in the northern plains of Mars. Science **328**, 1682–1686 (2010)

J. Carter, F. Poulet, Orbital identification of clays and carbonates in Gusev crater. Icarus **219**(1), 250–253 (2012)

H. Chamley, *Clay Sedimentology* (Springer, Berlin, 1989), pp. 391–422

H.G. Changela, J.C. Bridges, Alteration assemblages in the nakhlites: variation with depth on Mars. Meteorit. Planet. Sci. **45**, 1847–1867 (2011)
E. Chassefière, F. Leblanc, Constraining methane release due to serpentinization by the observed D/H ratio on Mars. Earth Planet. Sci. Lett. (2011). doi:10.1016/j.epsl.2011.08.013
C. Che, T.D. Glotch, D.L. Bish, J.R. Michalski, W. Xu, Spectroscopic study of the dehydration and/or dehydroxylation of phyllosilicate and zeolite minerals. J. Geophys. Res. **116** (2011). doi:10.1029/2010JE003740
V. Chevrier, F. Poulet, J.-P. Bibring, Early geochemical environment of Mars as determined from thermodynamics of phyllosilicates. Nature **448**, 60–63 (2007)
P.R. Christensen et al., Detection of crystalline hematite mineralization on Mars by the thermal emission spectrometer: evidence for near-surface water. J. Geophys. Res. **105**, 9623–9642 (2000)
B.C. Clark, Implications of abundant hydroscopic minerals in the Martian regolith. Icarus **34**, 645–665 (1978)
B.C. Clark, R.E. Arvidson, R. Gellert, R.V. Morris, D.W. Ming, L. Richter, S.W. Ruff, J.R. Michalski, W.H. Farrand, A.S. Yen et al., Evidence for montmorillonite or its compositional equivalent in Columbia hills, Mars. J. Geophys. Res. **112**, E06S01 (2007). doi:10.1029/2006JE002756
R.N. Clark et al., Diversity of mineralogy and occurrences of phyllosilicates on Mars. Eos Trans. AGU **89**(53), Fall Meet. Suppl., Abstract P43D-04 (2008)
E. Dehouck, N. Mangold, S. Le Mouélic, V. Ansan, F. Poulet, Ismenius Cavus, Mars: a deep paleolake with phyllosilicate deposits. Planet. Space Sci. **58**(6), 941–946 (2010)
E. Dehouck, V. Chevrier, A. Gaudin, N. Mangold, P.-E. Mathé, P. Rochette, Evaluating the role of sulfide-weathering in the formation of sulfates or carbonates on Mars. Geochim. Cosmochim. Acta **90**, 47–63 (2012)
B.L. Ehlmann et al., Orbital identification of carbonate-bearing rocks on Mars. Science **322**, 1828–1832 (2008a)
B.L. Ehlmann et al., Clay minerals in delta deposits and organic preservation potential on Mars. Nat. Geosci. **1**, 355–358 (2008b). doi:10.1038/ngeo207
B.L. Ehlmann et al., Identification of hydrated silicate minerals on Mars using MRO-CRISM: geologic context near Nili Fossae and implications for aqueous alteration. J. Geophys. Res. **114**, 1–33 (2009)
B.L. Ehlmann, J.F. Mustard, S.L. Murchie, Geologic setting of serpentine deposits on Mars. Geophys. Res. Lett. **37**, L06201 (2010)
B.L. Ehlmann, J.F. Mustard, S.L. Murchie, J.-P. Bibring, A. Meunier, A.A. Fraeman, Y. Langevin, Subsurface water and clay mineral formation during the early history of Mars. Nature **479**, 53–60. doi:10.1038/nature10582 (2011a)
B.L. Ehlmann, J.F. Mustard, R.N. Clark, G.A. Swayze, S.L. Murchie, Evidence for low-grade metamorphism, hydrothermal alteration, and diagnosis on Mars from phyllosilicate mineral assemblages. Clays Clay Miner. **59**(4), 359–377 (2011b)
G. Etiope, M. Schoell, H. Hosgormez, Abiotic methane flux from the Chimaera seep and Tekirova ophiolites (Turkey): understanding gas exhalation from low temperature serpentinization and implications for Mars. Earth Planet. Sci. Lett. **310**, 96–104 (2011)
A.G. Fairen et al., Noachian and more recent phyllosilcates in impact craters on Mars. Proc. Natl. Acad. Sci. USA **107**, 12,095–12,100 (2010)
S. Fabre, G. Berger, A. Nédélec, Continental weathering under high-CO_2 atmospheres during Precambrian times. G-cubed **12** (2011). doi:10.1029/2010GC003444
W.H. Farrand et al., Discovery of jarosite within the Mawrth Vallis region of Mars: implications for the geologic history of the region. Icarus **204**, 478–488 (2009)
C.I. Fassett, J.W. Head, The timing of Martian valley network activity: constraints from buffered crater counting. Icarus **195**, 61–89 (2008)
C.I. Fassett, J.W. Head, Sequence and timing of conditions on early Mars. Icarus **211**, 1204–1214 (2011)
W.C. Feldman et al., Global distribution of near-surface hydrogen on Mars. J. Geophys. Res. **109**, E09006 (2004). doi:10.1029/2003JE002160
M.E. Fleet, R. Howie, *A Rock Forming Minerals, Vol. 3A: Micas* (Geological Society of London, London, 2006). 780pp
H. Franzson, R. Zierenberg, P. Schiffman, Chemical transport in geothermal systems in Iceland: evidence from hydrothermal alteration. J. Volcanol. Geotherm. Res. **173**, 217–229 (2008)
F. Gaillard, B. Scaillet, The sulfur content of volcanic gases on Mars. Earth Planet. Sci. Lett. **279**, 34–43 (2009)
F. Gaillard et al., Geochemical reservoirs and timing of sulfur cycling on Mars. Space Sci. Rev. (2012, this issue)
A. Gaudin, E. Dehouck, N. Mangold, Evidence for weathering on early Mars from a comparison with terrestrial weathering profiles. Icarus **216**(1), 257–268 (2011)

P. Gavin, V. Chevrier, Thermal alteration of nontronite and montmorillonite: implications for the Martian surface. Icarus **208**, 721–734 (2010)
A. Gendrin et al., Sulfates in Martian layered terrains: the OMEGA/Mars Express view. Science **307**, 1587–1591 (2005). doi:10.1126/science.1109087
T.D. Glotch, J.L. Bandfield, P.R. Christensen, W.M. Calvin, S.M. McLennan, B.C. Clark, A.D. Rogers, S.W. Squyres, Mineralogy of the light-toned outcrop at Meridiani Planum as seen by the Miniature Thermal Emission Spectrometer and implications for its formation. J. Geophys. Res. **111**, E12S03 (2006). doi:10.1029/2005JE002672
T.D. Glotch et al., Distribution and formation of chlorides and phyllosilicates in Terra Sirenum, Mars. Geophys. Res. Lett. **37**, L16202 (2010)
B. Gondet et al., First detection of Al-rich phyllosilicate on Mars from OMEGA-MEx. Eur. Geophys. Union Conf., abs. EGU06-A-03691, 2006
J.L. Gooding, Soil mineralogy and chemistry on Mars: possible clues from salts and clays in SNC meteorites. Icarus **99**, 28–41 (1992)
J.A. Grant, S.A. Wilson, Late alluvial fan formation in southern Margaritifer Terra, Mars. Geophys. Res. Lett. **38**, L08201 (2011). doi:10.1029
E.A. Guinness et al., Hyperspectral reflectance mapping of cinder cones at the summit of Mauna Kea and implications for equivalent observations on Mars. J. Geophys. Res. **112**, E08S11 (2007). doi:10.1029/2006JE002822
R.M. Haberle, Early Mars climate models. J. Geophys. Res. **103**, 28467–28479 (1998)
I. Halevy et al., A sulfur dioxide climate feedback on early Mars. Science **318**, 1903–1907 (2007)
J.J. Hagerty, H.E. Newsom, Hydrothermal alteration at the Lonar lake impact structure, India: implications for impact cratering on Mars. Meteorit. Planet. Sci. **38**(3), 365–381 (2003)
R.L. Hay, Geologic occurrence of zeolites and some associated minerals. Pure Appl. Chem. **58**, 1339–1342 (1986)
R.L. Hay, T.K. Tyser, Chemical sedimentology and paleoenvironmental history of Lake Olduvai, a Pliocene lake in northern Tanzania. GSA Bull. **113**(12), 1505–1521 (2001)
J.W. Head, J.F. Mustard, M.A. Kreslavsky, R.E. Milliken, D.R. Marchant, Recent ice ages on Mars. Nature **426**, 797–802 (2003). doi:10.1038/nature02114
M.H. Hecht et al., Detection of perchlorate and the soluble chemistry of Martian soil at the Phoenix lander site. Science **325**, 64–67 (2009)
B.H. Horgan, J.F. Bell III, Widespread weathered glass on the surface of Mars. Geology **40**, 391–394 (2012). doi:10.1130/G32755.1
J. Hower, E.V. Eslinger, M.E. Hower, E.A. Perry, Mechanism of burial metamorphism of argillaceous sediments: 1. Mineralogical and chemical evidence. Geol. Soc. Am. Bull. **87**, 725–737 (1976)
J. Hurowitz et al., Origin of acidic surface waters and the evolution of atmospheric chemistry on early Mars. Nat. Geosci. **3**, 323–326 (2010)
S.F. Hviid, D.P. Agerkvist, M. Olsen, C. Bender Koch, M.B. Madsen, Heated nontronite: possible relations to the magnetic phase in the Martian soil. Hyperfine Interact. **91**, 529–533 (1994)
B.M. Jakosky, M.H. Carr, Possible precipitation of ice at low latitudes of Mars during periods of high obliquity. Nature **315**, 559–561 (1985)
S.S. Johnson, A.A. Pavlov, M.A. Mischna, Fate of SO_2 in the ancient Martian atmosphere: implications for transient greenhouse warming. J. Geophys. Res. **114**, E11011 (2009). doi:10.1029/2008JE003313
D. Jouglet, F. Poulet, J. Mustard, R. Milliken, J.-P. Bibring, Y. Langevin, B. Gondet, Hydration state of the Martian surface as seen by Mars Express OMEGA: 1. Analysis of the 3 mm hydration feature. J. Geophys. Res. **112**, E08S06 (2007). doi:10.1029/2006JE002846
J.F. Kasting, D. Catling, Evolution of a habitable planet. Annu. Rev. Astron. Astrophys. **41**, 429–463 (2003)
J.L. Keeling et al., Geology and characterization of two hydrothermal nontronites from weathered metamorphic rocks at the Uley graphite mine, South Australia. Clays Clay Miner. **48**(5), 537–548 (2000)
G. Klingelhofer et al., Jarosite and hematite at Meridiani Planum from Opportunity's Mössbauer Spectrometer. Science **306**, 1740–1745 (2004)
L.R. Kump, S.L. Brantley, M.A. Arther, Chemical weathering, atmospheric CO_2, and climate. Annu. Rev. Earth Planet. Sci. **28**, 611–667 (2000)
H. Lammer et al., Outgassing history and escape of the Martian atmosphere and water inventory. Space Sci. Rev. (2012, this issue)
Y. Langevin et al., Sulfates in the north polar region of Mars detected by OMEGA/Mars Express. Science **307**, 1584–1586 (2005)
B. Lanson, D. Beaufort, G. Berger, J.C. Lacharpagne, Illitization of diagenetic kaolinite-to-dickite conversion series: the Lower Permian Rotliegend sandstone reservoir, offshore of the Netherlands. J. Sediment. Res. **66**, 501–518 (1996)

B. Lanson et al., Authigenic kaolin and illitic minerals during diagenesis of sandstones: a review. Clay Miner. **37**, 1–22 (2002)
D. Larsen, E.C. Stephens, V.B. Zivkovic, Postimpact alteration of sedimentary breccias in the ICDP-USGS eyreville A and B cores with comparison to the Cape Charles core, Chesapeake Bay impact structure, Virginia, USA. GSA Spec. Pap. **458**, 699–721 (2009)
J. Laskar et al., Orbital forcing of the Martian polar layered deposits. Nature **419**, 375–377 (2002)
K.A. Lichtenberg et al., Stratigraphy of hydrated sulfates in the sedimentary deposits of Aram Chaos, Mars. J. Geophys. Res. **115**, E00D17 (2010)
L. Le Deit et al., Extensive surface pedogenic alteration of the Martian Noachian crust suggested by plateau phyllosilicates around Valles Marineris. J. Geophys. Res. **117**, E00J05 (2012). doi:10.1029/2011JE003983
A.P. Lisitzin, K.S. Rodlfa, Sedimentation in the World Ocean with emphasis on the nature, in *Distribution and Behavior of Marine Suspensions*. Society of Economic Paleontologists and Mineralogists, Special Publication, vol. 17, Tulsa (1972), 218pp
D. Loizeau, N. Mangold, F. Poulet, J.-P. Bibring, A. Gendrin, V. Ansan, C. Gomez, B. Gondet, Y. Langevin, P. Masson, G. Neukum, Phyllosilicates in the Mawrth Vallis region of Mars. J. Geophys. Res. **112**(E8), E08S08 (2007). doi:10.1029/2006JE002877
D. Loizeau, N. Mangold, F. Poulet, V. Ansan, E. Hauber, J.-P. Bibring, B. Gondet, Y. Langevin, P. Masson, G. Neukum, Stratigraphy in the Mawrth Vallis region through OMEGA, HRSC color imagery and DTM. Icarus **205**(2), 396–418 (2010)
D. Loizeau et al., Characterization of hydrated silicate-bearing outcrops in Tyrrhena Terra, Mars. Icarus **219**(1), 476–497 (2012)
P.R. Mahaffy et al., The sample analysis at Mars Investigation and Instrument Suite. Space Sci. Rev. (2012). doi:10.1007/s11214-012-9879-z
N. Mangold et al., Mineralogy of the Nili Fossae region with OMEGA/Mars Express data: 2. Aqueous alteration of the crust. J. Geophys. Res. **112**, E08S04 (2007). doi:10.1029/2006JE002835
N. Mangold, Relationships between clay-rich units and fluvial landforms in the Nili Fossae and Mawrth Vallis, in *Brown-Vernadsky MicroSymposium, Early Climate and Weathering on Mars*, Houston (2008)
N. Mangold, L. Roach, R. Milliken, S. Le Mouélic, V. Ansan, J.P. Bibring, P. Masson, J.F. Mustard, S. Murchie, G. Neukum, A late Amazonian alteration layer related to local volcanism on Mars. Icarus **207**, 265–276 (2010)
N. Mangold et al., Late Hesperian aqueous alteration at Majuro crater, Mars. Planet. Space Sci. (2012a) doi:10.1016/j.pss.2012.03.014
N. Mangold, E.S. Kite, M. Kleinhans, H. Newsom, V. Ansan, E. Hauber, E. Kraal, C. Quantin-Nataf, K. Tanaka, The origin and timing of fluvial activity at Eberswalde crater, Mars. Icarus **220**(2), 530–551 (2012b)
G.A. Marzo et al., Evidence for Hesperian impact-induced hydrothermalism on Mars. Icarus **208**, 667–683 (2010)
M.L. McDowell, V.E. Hamilton, Seeking phyllosilicates in thermal infrared data: a laboratory and Martian data case study. J. Geophys. Res. **114**, E06007 (2009). doi:10.1029/2008JE003317
N. McKeown et al., Characterization of phyllosilicates observed in the central Mawrth Vallis region, Mars, their potential formational processes, and implications for past climate. J. Geophys. Res. **114**, E00D10 (2009)
H.Y. McSween et al., Characterization and petrologic interpretation of olivine-rich basalts at Gusev Crater, Mars. J. Geophys. Res. **111**, E02S10 (2006). doi:10.1029/2005JE002477
J. Melosh, *Impact Cratering: A Geologic Process* (Oxford Univ. Press, Oxford, 1989)
R.J. Merriman, Clay minerals and sedimentary basin history. Eur. J. Mineral. **17**, 7–20 (2005)
R.J. Merriman, D.R. Peacor, Very low-grade metapelite: mineralogy, microfabrics and measuring reaction progress, in *Low-Grade Metamorphism*, ed. by M. Frey, D. Robinson (Blackwell, Oxford, 1999), pp. 10–60
A. Meunier, *Clays* (Springer, Berlin, 2005)
A. Meunier et al., The Fe-Rich clay microsystems in basalt-komatiite lavas: importance of Fe-smectites for pre-biotic molecule catalysis during the Hadean eon. Orig. Life Evol. Biosph. **40**, 253–272 (2010)
M. Meybeck, Global chemical weathering of surficial rocks estimated from river dissolved loads. Am. J. Sci. **287**, 401–428 (1987)
J.R. Michalski, R.L. Fergason, Composition and thermal inertia of the Mawrth Vallis region of Mars from TES and THEMIS data. Icarus **199**, 24–48 (2009). doi:10.1016/j.icarus.2008.08.016
J.R. Michalski, P.B. Niles, Deep crustal carbonate rocks exposed by meteor impact on Mars. Nat. Geosci. **3**, 751–755 (2010)
J.R. Michalski, E.Z. Noe Dobrea, Evidence for a sedimentary origin of clay minerals in the Mawrth Vallis region, Mars. Geology **35**(10), 91–954 (2007)

J.R. Michalski et al., Mineralogical constraints on the high-silica Martian surface component observed by TES. Icarus **174**, 161–177 (2005)
J.R. Michalski et al., Effects of chemical weathering on infrared spectra of Columbia River Basalt and spectral interpretations of Martian alteration. Earth Planet. Sci. Lett. **248**, 822–829 (2006)
J.R. Michalski, S.J. Reynolds, P.B. Niles, T.G. Sharp, P.R. Christensen, Alteration mineralogy in detachment zones; insights from Swansea, Arizona. Geosphere **3**(4), 184–198 (2007)
J. Michalski et al., The Mawrth Vallis region of Mars: A potential landing site for the Mars Science Laboratory (MSL) mission. Astrobiology **10**, 687–703 (2010a)
J. Michalski, F. Poulet, J.-P. Bibring, N. Mangold, Analysis of phyllosilicate deposits in the Nili Fossae region of Mars: Comparison of TES and OMEGA data. Icarus **206**, 269–289 (2010b)
T. Mikouchi et al., Relative burial depths of nakhlites: an update, in *Lunar Planet. Sci. Conf. 37th*, vol. 1865 (2006)
R.E. Milliken, D.L. Bish, Sources and sinks of clay minerals on Mars. Philos. Mag. **90**(17), 2293–2308 (2010)
R.E. Milliken, J.F. Mustard, Quantifying absolute water content of minerals using near-infrared reflectance spectroscopy. J. Geophys. Res. **110**, E12001 (2005). doi:10.1029/2005JE002534
R.E. Milliken, J.F. Mustard, F. Poulet, D. Jouglet, J.-P. Bibring, B. Gondet, Y. Langevin, Hydration state of the Martian surface as seen by Mars Express OMEGA: 2. H_2O content of the surface. J. Geophys. Res. **112**, E08S07 (2007). doi:10.1029/2006JE002853
R.E. Milliken et al., Opaline silica in young deposits on Mars. Geology **36**, 847–850 (2008)
R.E. Milliken et al., Missing salts on early Mars. Geophys. Res. Lett. **36**, L11202 (2009)
R.E. Milliken et al., Paleoclimate of Mars as captured by the stratigraphic record in Gale Crater. Geophys. Res. Lett. **37**, L04201 (2010)
R.V. Morris et al., Evidence for possible pigmentary hematite on Mars based on optical, magnetic, and Mossbauer studies of superparamagnetic (nanocrystalline) hematite. J. Geophys. Res. **94**(B4), 2760–2778 (1989)
R.V. Morris, D.C. Golden, J.F. Bell III, H.V. Lauer Jr, Hematite, pyroxene, and phyllosilicates on Mars: implications from oxidized impact melt rocks from Manicouagan Crater, Quebec, Canada. J. Geophys. Res. **100**, 5319–5328 (1995)
R.V. Morris et al., Mineralogy, composition, and alteration of Mars Pathfinder rocks and soils: Evidence from multispectral, elemental, and magnetic data on terrestrial analogue, SNC meteorite, and Pathfinder samples. J. Geophys. Res. **105**(E1), 1757–1817 (2000). doi:10.1029/1999JE001059
R.V. Morris et al., Mossbauer mineralogy of rock, soil, and dust at Gusev crater, Mars: Spirit's journey through weakly altered olivine basalt on the plains and pervasively altered basalt in the Columbia Hills. J. Geophys. Res. **111**, E02S13 (2006). doi:10.1029/2005JE002584
R.V. Morris et al., Identification of carbonate-rich outcrops on Mars by the Spirit rover. Science **329**, 421–424 (2010)
S. Murchie et al., Compact Reconnaissance Imaging Spectrometer for Mars (CRISM) on Mars Reconnaissance Orbiter (MRO). J. Geophys. Res. **112**, E05S03 (2007). doi:10.1029/2006JE002682
S.L. Murchie et al., A synthesis of Martian aqueous mineralogy after 1 Mars year of observations from the Mars Reconnaissance Orbiter. J. Geophys. Res. **114**, 1–30 (2009)
J.F. Mustard et al., Olivine and pyroxene diversity in the crust of Mars. Science **307**, 1594–1597 (2005)
J.F. Mustard et al., Mineralogy of the Nili Fossae region with OMEGA/Mars Express data: 1. Ancient impact melt in the Isidis Basin and implications for the transition from the Noachian to Hesperian. J. Geophys. Res. **112**, E08S03 (2007). doi:10.1029/2006JE002834
J.F. Mustard et al., Hydrated silicate minerals on Mars observed by the Mars reconnaissance orbiter CRISM instrument. Nature **454**, 305–309 (2008)
J.F. Mustard et al., Composition, morphology, and stratigraphy of Noachian crust around the Isidis basin. J. Geophys. Res. **114**, 1–18 (2009)
M.V. Naumov, Principal features of impact-generated hydrothermal circulation systems: mineralogical and geochemical evidence. Geofluids **5**, 165–184 (2005)
A. Neubeck, N.T. Duc, D. Bastviken, P. Crill, N.G. Holm, Formation of H_2 and CH_4 by weathering of olivine at temperatures between 30 and 70 °C. Geochem. Trans. **12**(6) (2011). doi:10.1186/1467-4866-12-6
H.E. Newsom, Hydrothermal alteration of impact melt sheets with implications for Mars. Icarus **44**, 207–216 (1980)
P.B. Niles, J.R. Michalski, Meridiani Planum sediments on Mars formed through weathering in massive ice deposits. Nat. Geosci. **2**, 215–220 (2009)
P.B. Niles et al., Stable isotope measurements of Martian atmospheric CO_2 at the Phoenix landing site. Science **329**, 1334–1337 (2010)
P.B. Niles, D.C. Catling, G. Berger, E. Chassefière, B.L. Ehlmann, J.R. Michalski, R. Morris, S.W. Ruff, B. Sutter, Geochemistry of carbonates on Mars: implications for climate history and nature of aqueous environments. Space Sci. Rev. (2012, this issue)

E.Z. Noe Dobrea et al., Mineralogy and stratigraphy of phyllosilicate-bearing and dark mantling units in the greater Mawrth Vallis/west Arabia Terra area: Constraints on geological origin. J. Geophys. Res. **115**, E00D19 (2010). doi:10.1029/2009JE003351

L.E. Nyquist et al., Ages and geologic histories of Martian meteorites. Space Sci. Rev. **96**, 105–164 (2001)

M.M. Osterloo et al., Chloride-bearing materials in the southern highlands of Mars. Science **319**, 1651–1654 (2008). doi:10.1126/science.1150690

M.M. Osterloo et al., Geologic context of proposed chloride-bearing materials on Mars. J. Geophys. Res. **115**, E10012 (2010)

C. Oze, M. Sharma, Have olivine, will gas: Serpentinization and the abiogenic production of methane on Mars. Geophys. Res. Lett. **32**, L10203 (2005). doi:10.1029/2005GL022691

S.M. Pelkey et al., CRISM multispectral summary products: Parameterizing mineral diversity on Mars from reflectance. J. Geophys. Res. **112**, E08S14 (2007). doi:10.1029/2006JE002831

F. Poulet et al., Phyllosilicates on Mars and implications for early Martian climate. Nature **481**, 623–627 (2005)

F. Poulet, C. Gomez, J.-P. Bibring, Y. Langevin, B. Gondet, P. Pinet, G. Belluci, J. Mustard, Martian surface mineralogy from Observatoire pour la Minéralogie, l'Eau, les Glaces et l'Activité on board the Mars Express spacecraft (OMEGA/MEx): global mineral maps. J. Geophys. Res. **112**, 08S02 (2007). doi:10.1029/2006JE002840

F. Poulet et al., Abundance of minerals in the phyllosilicate-rich units on Mars. Astron. Astrophys. **487**, L41–L44 (2008a)

F. Poulet, R.E. Arvidson, C. Gomez, R.V. Morris, J.-P. Bibring, Y. Langevin, B. Gondet, J. Griffes, Mineralogy of Terra Meridiani and western Arabia Terra from OMEGA/MEx and implications for their formation. Icarus **195**, 106–130 (2008b). doi:10.1016/j.icarus.2007.11.031

J.A. Rathbun, S.W. Squyres, Hydrothermal systems associated with Martian impact craters. Icarus **157**, 362–372 (2002)

R.R. Remy, R.E. Ferrell, Distributions and origin and analcime in lacustrine mudstones of the Green River Formation, South-Central Uinta Basin, Utah. Clays Clay Miner. **37**(5), 419–432 (1989)

S.W. Ruff et al., The rocks of Gusev Crater as viewed by the Mini-TES instrument. J. Geophys. Res. **111**, E12S18 (2006). doi:10.1029/2006JE002747

M.R. Salvatore, J.F. Mustard, M.B. Wyatt, S.L. Murchie, Definitive evidence of Hesperian basalt in Acidalia and Chryse planitiae. J. Geophys. Res. **115**, E07005 (2010). doi:10.1029/2009JE003519

J.M. Saxton, I.C. Lyon et al., Oxygen isotopic composition of carbonate in the Nakhla meteorite: Implications for the hydrosphere and atmosphere of Mars. Geochim. Cosmochim. Acta **64**(7), 1299–1309 (2000)

P. Schiffman, H.W. Day, Petrological methods for the study of very low grade metabasites, in *Low-Grade Metamorphism*, ed. by M. Frey, D. Robinson (Blackwell, Oxford, 1999), pp. 108–142

P. Schiffman et al., Controls on palagonitization versus pedogenic weathering of basaltic tephra: evidence from the consolidation and geochemistry of the Keanakako'i Ash member, Kilauea Volcano. Geochem. Geophys. Geosyst. **1**, 2000GC000068 (2000)

P. Schiffman et al., Distinguishing palagonitized from pedogenically-altered basaltic Hawaiian tephra: mineralogical and geochemical criteria, in *Volcano-Ice Interactions on Earth and Mars*, ed. by J.L. Smellie, M.G. Chapman (Geological Society, London, 2002), pp. 393–405

P. Schneiderhöhn, Nontronit von Hohen Hagen und Chloropal vom Meenser Steinberg bei Göttingen, in *Tschermaks Mineralogische und Petrograpische Mitteilungen*, vol. 10 (1965), pp. 385–399

S.P. Schwenzer, D.A. Kring, Impact-generated hydrothermal systems capable of forming phyllosilicates on Noachian Mars. Geology **37**, 1091–1094 (2009)

G. Siefferman, G. Millot, Equatorial and tropical weathering of recent basalts from Cameroon: allophones, halloysite, metahalloysite, kaolinite, and gibbsite, in *Proc. of the International Clay Conference*, vol. 1, Tokyo (Israel Universities Press, Jerusalem, 1969)

A. Singer, The mineralogy of the clay fraction from basaltic soils in the Galilee, Israel. J. Soil Sci **17**(1), 136–147 (1966)

R. Singer, T. McCord, R. Clark, J. Adams, R. Huguenin, Mars surface composition from reflectance spectroscopy: a summary. J. Geophys. Res. **84**(B14), 8415–8426 (1979)

J.R. Skok, J.F. Mustard, S.L. Murchie, M.B. Wyatt, B.L. Ehlmann, Spectrally distinct ejecta in Syrtis Major, Mars: evidence for environmental change at the Hesperian-Amazonian boundary. J. Geophys. Res. **115**, E00D14 (2010). doi:10.1029/2009JE003338

F.S. Spear, *Metamorphic Phase Equilibria and Pressure-Temperature-Time Paths* (Mineralogical Society of America, Washington, 1995)

R.J. Spencer, L.A. Hardie, Control of seawater composition by mixing of river waters and mid-ocean ridge hydrothermal brines, in *Fluid-Mineral interactions: A Tribute to H.P. Eugster*, ed. by R.J. Spencer, I.-M. Chou. Geochemical Society Special Publication, vol. 19 (1990), pp. 409–419

S.W. Squyres, R.E. Arvidson, J.F. Bell III, J. Bruckner, N.A. Cabrol, W. Calvin, M.H. Carr, P.R. Christensen, B.C. Clark, L.S. Crumpler et al., The Opportunity Rover's Athena science investigation at Meridiani Planum, Mars. Science **306**(5702), 1698–1703 (2004)

S.W. Squyres et al., Detection of silica-rich deposits on Mars. Science **320**, 1063–1067 (2008)

J. Srodon, Nature of mixed-layer clays and mechanisms of their formation and alteration. Annu. Rev. Earth Planet. Sci. **27**, 19–53 (1999)

S. Story et al., Authigenic phyllosilicates in modern acid saline lake sediments and implications for Mars. J. Geophys. Res. **115**, E12012 (2010). doi:10.1029/2010JE003687

R.C. Surdam, D.B. McGowan, T.L. Dunn, Diagenetic pathways of sandstone and shale sequences. Contrib. Geol. **27**, 21–31 (1989)

G.A. Swayze et al. Ground-truthing AVIRIS mineral mapping at Cuprite, Nevada, in *Summaries of the Third Annual JPL Airborne Geosciences Workshop, Volume 1: AVIRIS Workshop* (JPL Publication 92-14, 1992), pp. 47–49

G.A. Swayze et al., Mineral mapping Mauna Kea and Mauna Loa shield volcanos on Hawaii using AVIRIS data and the USGS tetracorder spectral identification system: Lessons applicable to the search for relict Martian hydrothermal systems, in *Proceedings of the 11th JPL Airborne Earth Science Workshop*, ed. by R.O. Green, (2002), pp. 373–387. JPL Publication 03-4

G.A. Swayze et al., Discovery of the acid-sulfate mineral alunite in Terra Sirenum, Mars, using MRO CRISM: Possible evidence for acid-saline lacustrine deposits? Eos Trans. AGU **89**(53) (2008)

T.D. Swindle, E.K. Olson, 40Ar-39Ar studies of whole rock Naklites: evidence for the timing of formation and aqueous alteration on Mars. Meteorit. Planet. Sci. **39**(5), 755–766 (2004)

T.D. Swindle et al., Noble gases in iddingsite from the Lafayette meteorite: Evidence for liquid water on Mars in the last few hundred million years. Meteorit. Planet. Sci. **35**, 107–115 (2000)

G.J. Taylor et al., Causes of variations in K/Th on Mars. J. Geophys. Res. **111**, E03S06 (2006). doi:10.1029/2006JE002676

G.J. Taylor et al., Mapping Mars geochemically. Geology **38**(2), 183–186 (2010)

P. Thollot et al., Most Mars minerals in a nutshell: Various alteration phases formed in a single environment in Noctis Labyrinthus. J. Geophys. Res. **117**, E00J06 (2012)

F. Tian et al., Photochemical and climate consequences of sulfur outgassing on early Mars. Earth Planet. Sci. Lett. **295**, 412–418 (2010)

N.J. Tosca, A.H. Knoll, Juvenile chemical sediments and the long term persistence of water at the surface of Mars. Earth Planet. Sci. Lett. **286**, 379–386 (2009)

N.J. Tosca et al., Smectite formation on early Mars: experimental constraints, in *Workshop on Martian Phyllosilicates: Recorders of Aqueous Processes?* (2008), abstr. 7030

P. Toulmin III et al., Geochemical and mineralogical interpretation of the Viking inorganic chemical results. J. Geophys. Res. **82**(28), 4625–4634 (1977). doi:10.1029/JS082i028p04625

E. Tréguier et al., Overview of Mars surface geochemical diversity through APXS data multidimensional analysis: First attempt at modelling rock alteration. J. Geophys. Res. **113**, E12S34 (2008). doi:10.1029/2007JE003010

A.H. Treiman, D.J. Lindstrom, Trace element geochemistry of Martian iddingsite in the Lafayette meteorite. J. Geophys. Res. **102**, 1953–1963 (1997)

M.A. Velbel, A.I. Losiak, Denticles on chain silicate grain surfaces and their utility as indicators of weathering conditions on Earth and Mars. J. Sediment. Res. **80**, 771–780 (2010)

A. Wang, R.L. Korotev, B.L. Jolliff, L.A. Haskin, L.S. Crumpler, W.H. Farrand, K.E. Herkenhoff, P.A. de Souza Jr., A.G. Kusack, J.A. Hurowitz et al., Evidence of phyllosilicates in Wooly Patch, an altered rock encountered at West Spur, Columbia Hills, by the Spirit rover in Gusev Crater. J. Geophys. Res. **111**, E02S16 (2006). doi:10.1029/2005JE002516

G. Whitney, Role of water in the smectite-to-illite reaction. Clays Clay Miner. **38**, 343–350 (1990)

R.J. Weldon, W.M. Thomas, M.B. Boslough, T.J. Arhens, Shock-induced color changes in nontronite: implications for the Martian fines. J. Geophys. Res. **97**, 10102–10114 (1982)

T. Wiesenberger, R.S. Selbekk, Multi-stage zeolite facies mineralization in the Hvalfjordur area, Iceland. Int. J. Earth Sci. (2008). doi:10.1007/s00531-007-0296-6

R. Wordsworth, F. Forget, E. Millour, J.-B. Madeleine, Comparison of scenarios for Martian valley network formation using a 3D model of the early climate and water cycle. EPSC abstracts, vol. 6, EPSC-DPS2011-1373, 2011

J.J. Wray, B.L. Ehlmann, S.W. Squyres, J.F. Mustard, R.L. Kirk, Compositional stratigraphy of clay-bearing layered deposits at Mawrth Vallis, Mars. Geophys. Res. Lett. **35**, L12202 (2008). doi:10.1029/2008GL034385

J.J. Wray, S.L. Murchie, S.W. Squyres, F.P. Seelos, L.L. Tornabene, Diverse aqueous environments on ancient Mars revealed in the southern highlands. Geology **37**, 1043–1046 (2009a)

J.J. Wray, E.Z. Noe Dobrea, R.E. Arvidson, S.M. Wiseman, S.W. Squyres, A.S. McEwen, J.F. Mustard, S.L. Murchie, Phyllosilicates and sulfates at Endeavour Crater, Meridiani Planum, Mars. Geophys. Res. Lett. **36**, L21201 (2009b). doi:10.1029/2009GL040734

J.J. Wray et al., Columbus crater and other possible groundwater-fed paleolakes of Terra Sirenum, Mars. J. Geophys. Res. **116**, E01001 (2011)

C.M. Weitz, J.L. Bishop, P. Thollot, N. Mangold, L.H. Roach, Diverse mineralogies in two troughs of Noctis Labyrinthus, Mars. Geology **39**(10), 899–902 (2011)

Y.-C. Yau, D.R. Peacor, R.E. Beane, E.J. Essene, S.D. McDowell, Microstructures, formation mechanisms, and depth-zoning of phyllosilicates in geothermally altered shales, Salton Sea, California. Clays Clay Miner. **36**, 1–10 (1988)

CPSIA information can be obtained
at www.ICGtesting.com
Printed in the USA
BVHW010315081118
532516BV00004B/210/P